MOLECULAR CONTROL MECHANISMS IN STRIATED MUSCLE CONTRACTION

Advances in Muscle Research

Volume 1

Series Editor

G.J.M. Stienen, *Vrije Universiteit, Amsterdam, The Netherlands*

Molecular Control Mechanisms in Striated Muscle Contraction

Edited by

R. John Solaro
University of Illinois at Chicago,
Chicago, Illinois, U.S.A.

and

Richard L. Moss
University of Wisconsin Medical School,
Madison, Wisconsin, U.S.A.

KLUWER ACADEMIC PUBLISHERS
DORDRECHT / BOSTON / LONDON

A C.I.P. Catalogue record for this book is available from the Library of Congress.

ISBN 978-1-4020-0734-7

Published by Kluwer Academic Publishers,
P.O. Box 17, 3300 AA Dordrecht, The Netherlands.

Sold and distributed in North, Central and South America
by Kluwer Academic Publishers,
101 Philip Drive, Norwell, MA 02061, U.S.A.

In all other countries, sold and distributed
by Kluwer Academic Publishers,
P.O. Box 322, 3300 AH Dordrecht, The Netherlands.

Printed on acid-free paper

We dedicate this volume to the memory of our
colleague Dr. Rhea Levine – a superb experimentalist,
an enthusiastic scientist, and a generous friend.

TABLE OF CONTENTS

viii

TABLE OF CONTENTS

PREFACE

Nowhere in biology is the marriage of structure and function as well expressed or understood as in the case of molecules that regulate the contraction and relaxation of striated muscles. In assembling what is known about these molecules, we have focused first on the processes that mediate Ca^{2+} transients in both cardiac and skeletal muscles and then on the processes that translate the Ca^{2+} signal to mechanical activity. The initial chapters by González and Ríos on E-C coupling in skeletal muscle and by Trafford and Eisner on E-C coupling in myocardium together provide a compelling synthesis of the determinants of resting Ca^{2+} levels and mechanisms of Ca^{2+} release and modulation. These chapters describe new and exciting information about the molecular processes that mediate elementary Ca^{2+} release events, i.e., so-called Ca^{2+} sparks. Subsequent chapters address the results of elegant in-depth studies of how molecular processes controlled by Ca^{2+} give rise to mechanical activity. Chapters by Gomez, Harada, and Potter and by Cheung provide detailed insights into the initial trigger for cross-bridge interaction with actin, i.e., Ca^{2+} binding to troponin C. The chapter by Tobacman describes the function and elaborate order of the array of molecules that make up the thin filament, and the chapter by Levine and Kensler does the same for the thick filament. Activation and control of interactions between the thick and thin filaments involve steric, allosteric, and highly cooperative molecular processes, which are described in the chapter by Geeves and Lehrer. The complexity of these processes suggests many potential control points for modulating actin-myosin interactions. Chapters by Fuchs and by Solaro and colleagues address special features of regulation in myocardium in discussions of the modulation of contraction by changes in sarcomere length and by protein phosphorylations. Since muscles ultimately develop force and perform work, additional chapters are dedicated to the regulation of the kinetics of cross-bridge interaction with actin. Moss and colleagues discuss the regulation of the kinetics of force development and Homsher addresses the activation dependence of shortening velocity in muscle and in cell-free model systems. Together these contributions provide comprehensive, up-to-date insights into regulatory mechanisms that control the contractile capabilities of heart and skeletal muscles, as well as lending credence to the idea that there is very little in biology that becomes simpler with experimental investigation. Nonetheless, the insights gained from several decades of investigation have provided a new and evolving framework for understanding muscle function in health and disease.

The editors are grateful to the authors for their long hours of work and for meeting the arbitrary deadlines we set for them! We reserve special thanks to Sue Krey for her persistence, care and outstanding work in editing, formatting and compiling the manuscripts.

R.J.S. Chicago, IL
R.L.M. Madison, WI

INTRODUCTION

Recent advances in Muscle Research have provided novel insight in the molecular mechanism of muscle contraction. The implications and applications of these findings for skeletal, cardiac and smooth muscle are now becoming increasingly clear. In this book series an overview will be presented of the regulation, basic properties and molecular diversity of contractile proteins as well as the structural and biochemical aspects of energy transduction in muscle. These overviews will provide the foundations for contributions linking the basic properties of muscle contraction to contractile function and dysfunction of skeletal, cardiac and smooth muscle in health and disease.

Ger Stienen
Series Editor

ADOM GONZÁLEZ & EDUARDO RÍOS

EXCITATION-CONTRACTION COUPLING IN SKELETAL MUSCLE

1. INTRODUCTION

The chemical reactions underlying the generation of force in muscle are blocked in the resting tissue, and only become possible when free $[Ca^{2+}]$ in the aqueous medium bathing the contractile proteins rises from a resting level of about 150 nM to several μM. This occurs physiologically after every action potential that propagates along the muscle cell membrane. Excitation-contraction (EC) coupling names the processes that link the action potential to the force-producing reactions.

After a summary of early studies, which will serve to introduce the main ideas, we will describe events and mechanisms at the cellular and organellar level. First, the maintenance of a low resting $[Ca^{2+}]$ in the myoplasm, then the causation, through Ca^{2+} release and removal, of the sudden increase in $[Ca^{2+}]$ that precedes the twitch, and the usually rapid reduction to resting levels needed to terminate contraction. We will subsequently proceed in a reductionist way to the description of Ca^{2+} sparks, which are local events but probably involve multiple molecules, and to the examination of function and structure at the molecular level. The chapter will be completed by a section on modulation and modification of function in muscle fatigue.

The cellular studies described in this chapter have been carried out largely on fast-twitch fibers, the only type for which a reasonably complete picture of Ca^{2+} movements can be given.

1.1. Early Studies

The preponderant role of the action potential in the control of the muscle twitch was recognized since the late 19th century, but it was only after world war II that advances accelerated. In 1947, Heilbrunn and Wiercinski showed that intracellular Ca^{2+} injection switches contraction on, while A.V. Hill (1948) argued that diffusion of Ca^{2+} from the outside would not account for the rapid onset of contraction, stressing the need for a mechanism of radial propagation identified a few years later. In 1958 Huxley and Taylor caused local contraction by depolarizing limited regions of a frog fiber's membrane near the I band. This observation located the origin of the

1

R.J. Solaro and R.L. Moss (eds), Molecular Control Mechanisms in Striated Muscle Contraction, 1-48
©2002 Kluwer Academic Publishers, Printed in the Netherlands

hypothetical propagation system, later identified with the transverse tubular (TT) system described by Veratti in 1902 (transl. 1961).

Experiments of Hodgkin and Horowicz (1960) showed shortly thereafter how to produce graded changes in membrane potential by increasing extracellular $[K^+]$ (taking advantage of the selective permeability of the resting muscle membrane). They used such "K contractures" to uncover a monotonic relationship between membrane potential, V_m, and contractile force, thus identifying V_m as the key controlling variable of muscle contraction. These seminal studies also impressed in many of us the notion that activation of Na^+ permeability and of Ca^{2+} release were ultimately similar.

During the sixties, Ebashi and coworkers clarified the role of Ca^{2+} as key to the force generating reactions (reviewed in 1968), while Weber and others established the function of the sarcoplasmic reticulum (SR) as intracellular Ca^{2+} storage and release organelle (Hasselbach, 1964). In the seventies and eighties the focus shifted from TT and SR to the point and functional stage where both come together, the step in which depolarization of the TT membrane causes opening of release channels located in the SR membrane. The discovery process culminated with the identification of the main molecules involved, namely the dihydropyridine receptor (DHPR) of the transverse tubule (rev. by Ríos and Pizarro, 1991) and the ryanodine receptor (RyR) of the sarcoplasmic reticulum (rev. by Meissner, 1994), defining the molecular terms in which the field develops today.

2. EVENTS AT THE CELLULAR AND SUB-CELLULAR LEVELS

The early studies thus established that EC coupling requires low $[Ca^{2+}]_i$ at rest, then a rapid increase, and a return to resting levels in milliseconds. Much effort was therefore devoted to measuring the concentration of Ca^{2+} in the myoplasm, and its changes upon action potential activation or voltage clamp depolarization. Evaluated with optical techniques, considering the interfering effect of intracellular structures that bind the indicator dye, the values of resting $[Ca^{2+}]_i$ range from 100 to 250 nM (Baylor and Hollingworth, 2000).

2.1. The Maintenance of Low Resting $[Ca^{2+}]_i$

The important players in myoplasmic Ca^{2+} movements are illustrated in fig. 1. The voltage-sensitive Ca^{2+} channels of skeletal muscle are of the L-type, implying that their resting Po will be extremely low. Voltage-insensitive "leak" Ca^{2+} channels have been described in cardiac and skeletal muscle (Rosenberg et al., 1988). That they are important in the regulation of resting $[Ca^{2+}]_i$ is indicated by their greater activity in myotubes and adult cells of mice with Duchenne muscular dystrophy, in which an increased proteolysis is a consequence of elevated $[Ca^{2+}]_i$ (Alderton and Steinhardt, 2000). Persistent elevation of $[Ca^{2+}]_i$ triggers the opening of mitochondrial pores and releases cytochrome c, which activates the proteolytic caspases of apoptosis.

The active transport mechanisms that maintain low resting $[Ca^{2+}]_i$ include the $(Ca^{2+}+Mg^{2+})$ - ATPase of the sarcoplasmic reticulum (SERCA), a similar protein in the plasma membrane, and the Na^+/Ca^{2+} exchanger of the plasma membrane. SERCAs are crucial to the rapid restoration of low $[Ca^{2+}]_i$ after a Ca^{2+} transient (and the beat-to-beat diastolic $[Ca^{2+}]_i$ in cardiac muscle). Even though the flow of Ca^{2+} across the plasmalemma is too small to be relevant to EC coupling in the short term, it is essential in Ca^{2+} homeostasis. Ultimately, the total calcium content of a cell can only be controlled by the flow through its boundary, the plasmalemma. It should be clear, therefore, that the SR cannot influence the steady free $[Ca^{2+}]_i$; in the long run, it can only adjust its own concentration to a resting value determined by equality between extrusion from and leak into the myoplasm, across the plasma membrane.

2.1.1. The Plasmalemmal Ca Pump.

A $(Ca^{2+} + Mg^{2+})$-activated ATPase (Schatzmann, 1989) is present in the plasmalemma of most cells. Four isoforms, termed PMCA1 to 4, are known. 1 and 4 are widely distributed, while 2 and 3 are restricted to specific tissues and cell types. PMCA 1, 3 and 4 have been demonstrated in mammalian muscle (Penniston & Enyedi, 1998.) The PMCAs are similar in tertiary structure to the better studied SERCAs, although the sequence homology is only about 40%. The main functional differences are in their stoichiometry (one Ca^{2+} transported per ATP hydrolyzed, while in SERCA it is 2:1), and in the modulation of PMCA activity by calmodulin. In the presence of Ca^{2+}:calmodulin, PMCA's myoplasmic Ca^{2+}-binding K_D is ~ 0.5 µM, while in its absence K_D rises to > 1 µM. This implies that the pumps will have their activity reduced at resting $[Ca^{2+}]_i$, therefore final balance of fluxes at rest will depend to a greater degree on the other mechanism of active transport, Na^+/Ca^{2+} exchange.

2.1.2. The Na^+/Ca^{2+} Exchanger.

In most tissues the exchanger (of which there are isoforms NCX1, 2 and 3) operates at a fixed stoichiometry of 1 Ca^{2+} to 3 Na^+. This exchange is electrogenic, rather than electroneutral, hence it is affected by membrane voltage. Near resting conditions it will extrude Ca^{2+} from the cell, but it becomes an influx pathway at positive voltages. Its equilibrium potential (transmembrane potential at which the net transport by the exchanger vanishes) is termed $E_{Na/Ca}$, and satisfies

$$E_{Na/Ca} = 3E_{Na} - 2E_{Ca} \qquad (1)$$

with

$$E_{Na} = \frac{RT}{F} \ln \frac{[Na^+]_e}{[Na^+]_i} \qquad (2)$$

and

$$E_{Ca} = \frac{RT}{2F} ln \frac{[Ca^{2+}]_e}{[Ca^{2+}]_i} \qquad (3)$$

At $[Ca^{2+}]_e$=2.3 mM and $[Ca^{2+}]_i$ =100 nM, and standard $[Na^+]$ for a mammalian muscle cell, these equations yield 60 mV for E_{Na}, 130 mV for E_{Ca} and -80 mV for $E_{Na/Ca}$. Because resting V_m (-90 mV) is more negative than $E_{Na/Ca}$, there will be a driving force to propel an inward current through the exchanger. In other words, at the resting membrane potential the exchanger will carry out active extrusion of Ca^{2+}, sufficient to balance some passive leaks.

   In addition to this role in setting resting $[Ca^{2+}]$, it may contribute significantly to Ca^{2+} removal after a transient in twitch muscle (Cifuentes et al., 1998). It appears, however, that the Na^+/Ca^{2+} exchangers may play a greater role in contraction-related Ca^{2+} signaling in tonic muscles, as well as in slow-twitch fibers (reviewed by Blaustein and Lederer, 1999).

2.2. Ca^{2+} in the Sarcoplasmic Reticulum

To evaluate the driving forces for Ca^{2+} release it is necessary to monitor $[Ca^{2+}]$ in the SR lumen. Detailed measurements were carried out by Chen et al. (1996), using NMR of Fl. $[Ca^{2+}]_{SR}$ was near 1.5 mM, which is consistent with estimates of Inesi (1994) based on the sharp negative effects on ATP-ase velocity of increasing lumenal $[Ca^{2+}]$ above 2 mM. The concentration gradient driving Ca^{2+} release is therefore of almost exactly four orders of magnitude, from 1.5 mM inside the SR to 150 nM in the myoplasm.

   In frog fast-twitch muscle, where stereologic (Mobley and Eisenberg, 1975) and functional measurements have been carried out in detail, the terminal cisternae (TC) of the SR comprise 4% of cell volume. The SR's total calcium content, estimated at 2.5 mmoles/liter of fiber aqueous volume (Pape et al., 1995) is largely bound to buffers. The concentration of total Ca in the SR can be calculated as 2.5 mM x aqueous fiber volume (0.7 of total volume) divided by TC volume (0.04), or 43.75 mM. The buffer capacity is hence 43.75 mM/1.5 mM, or about 30. This buffer capacity is the role of calsequestrin (MacLennan and Wong, 1971) and minor Ca^{2+}-binding SR proteins.

2.2.1. Calsequestrin
Calsequestrin (*Csq*) is located in the terminal cisternae of the SR (Meissner, 1975), where it forms multimeric chains visible as electron-dense structures near the Ca^{2+} release channels, connected by thin "anchoring filaments" to the Ca^{2+} release channels. Triadin (Caswell et al 1991) and junctin (Jones et al., 1995) are homologous proteins with a lumenal C terminus. It appears that both proteins, separately or together, are part of the anchoring system, constituting both a link that

position Csq and a conduit that directs flow of Ca^{2+} from buffer to channel. (Knudson et al., 1993; Zhang et al., 1997).

Csq contains multiple low affinity binding sites for Ca^{2+}. Its stoichiometry and affinity are somewhat ill-defined (K_D is roughly in the mM range), as the protein molecules tend to form multimers in a Ca^{2+}-dependent process that provides new binding sites of progressively lower affinity as $[Ca^{2+}]$ increases (Wang et al., 1998). In any case, Csq's high capacity, low affinity, rapid equilibration rates, and strategic location, appear to be carefully designed to maintain $[Ca^{2+}]_{SR}$ nearly clamped during release of a substantial portion of the SR Ca^{2+} content. There are indications that Csq may have dynamic interactions, including a change in its Ca^{2+} affinity upon RyR activation (Ikemoto et al., 1991) and reciprocal changes in channel activity induced by Csq (Szegedi et al., 1999). ·

2.2.2. The SR Ca Pump

The gradient of $[Ca^{2+}]$ between the lumen of the SR and the myoplasm is maintained by its Ca pumps. Isoform SERCA1 (in its splice variant a) is the main contributor in adult fast twitch skeletal muscle, while SERCA2a dominates in slow-twitch, cardiac and neonatal fast-twitch muscle (Brandl et al, 1987). A third isoform is also present in skeletal muscle. Inward Ca transport is matched 1:1 by H^+ countertransport, and is hence electrogenic. Because the SR membrane is leaky to many ions, this electric imbalance does not build up potential differences. It will alkalinize the SR lumen, however, which may play a role to balance the counterflow of protons associated with Ca^{2+} release (Kamp et al, 1998)

Each pump is made of a single 110 kDa polypeptide, which removes 2 Ca^{2+} per ATP hydrolyzed. The pump is thermodynamically "tight", it can build up a chemical potential difference for Ca^{2+} matching the free energy released by ATP hydrolysis (~ 10,000-fold $[Ca^{2+}]$ ratio). But, "a motorcycle will not fly even if provided with high energy aviation fuel". Inesi (1994) thus stressed that the pump will generate the full predictable $[Ca^{2+}]$ ratio, provided that its "machine" constraints are satisfied. The constraints, in this case, are the dissociation constants of the two Ca^{2+}-enzyme reactions, at 3 10^{-13} M^2 for Ca^{2+} in the cytoplasm, and 3 10^{-6} M^2 in the SR lumen. Such values dictate, for example, that the enzyme will pump rapidly at $[Ca^{2+}]_{SR} < 10^{-3}$ M and $[Ca^{2+}]_i < 5\ 10^{-7}$ M, but will see its efficiency go down at greater concentrations. These predictions were confirmed by the NMR measurements of $[Ca^{2+}]_{SR}$ mentioned earlier (Chen et al., 1996).

2.3. The Ca^{2+} Transient

As stated, a key event for contractile activation is the increase in myoplasmic free Ca^{2+} concentration. If contraction is to be rapid, the increase in $[Ca^{2+}]_i$ must rapidly reach levels that will drive troponin C close to saturation in a few milliseconds. Equally important is to rapidly terminate the transient. This, and the harmful consequences of a prolonged increase in $[Ca^{2+}]_i$, justifies multiple provisions that contribute to make the "Ca^{2+} transient" large, fast rising, and very brief.

The Ca^{2+} transients vary among different species and different fiber types, and exhibit a steep dependence on temperature. Rome et al. (1996) studied them in some of the fastest vertebrate muscles, including the rattle shaker muscle in rattlesnakes and "sonic" muscle fibers of the toadfish swimbladder, which produce the 200 Hz "boatwhistle" mating call. In these, the Ca^{2+} transient had a peak amplitude of 50 μM and half-magnitude duration of 3.5 ms at 16 degrees. By comparison, Ca^{2+} transients of the toadfish fast-twitch fibers were about 5-fold smaller and longer-lasting. Surprisingly, the EDL muscle of the humble mouse was almost as fast as the sonic muscle (Hollingworth et al., 1996).

Measurements of the Ca^{2+} transient in response to an action potential (eg. Maylie et al., 1987; Hirota et al., 1989) led to the conclusion that some 90% of the regulatory sites in troponin C will be occupied during a twitch, and will remain occupied for tens of milliseconds. The increase in tension that occurs upon successive twitches therefore manifests mechanical phenomena rather than a Ca^{2+} buildup, e.g. stretch of series components or increased cross-bridge binding due to phosphorylation of myosin light chains (see chapter by Moss, et. al.).

2.4. Ca^{2+} Release from the SR

The measurement of Ca^{2+} transients is required to understand quantitatively the control of contraction. Additionally, starting from Ca^{2+} transients it is possible to evaluate rates of Ca^{2+} release, which in turn are needed to determine the state of the release channels, in order to understand their control.

The bottom diagram in fig. 1 illustrates simple quantitative ideas behind the determination of rate of Ca^{2+} release, or release flux (Melzer et al., 1987). The rate $d[Ca^{2+}]_i$ /dt at which $[Ca^{2+}]$ increases in the myoplasmic compartment is equal to the rate at which it enters (*input flux*) minus that at which it is removed from solution (*removal*). In skeletal muscle, by far the main component of input flux is release from the SR. If $[Ca^{2+}]_i$ is measured as a function of time, and the rate of removal from solution is evaluated, then release flux can be calculated as:

$$Release\ flux\ = \frac{d[Ca^{2+}]_i}{dt} + removal\ flux \qquad (4)$$

The evaluation of release flux therefore requires an estimate of removal flux.

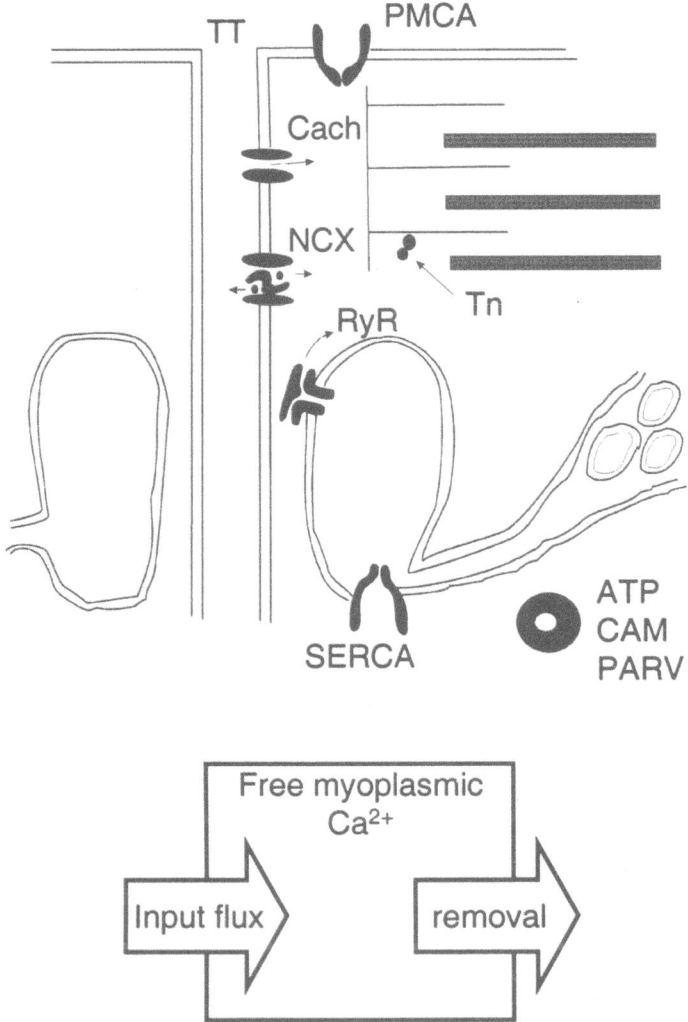

Fig. 1. Contributors to Ca^{2+} flux. In the long term, the cellular Ca^{2+} content is determined by plasmalemmal pathways: Ca^{2+} channels (Cach) including the L-type or DHP receptor, and leak channels open at the resting potential, the $Na^{+}-Ca^{2+}$ exchanger (NCX) and the plasmalemmal Ca^{2+} pump (PMCA). Short term increases are determined by release from the SR through the Ca^{2+} release channel (RyR). In the myoplasm, Ca^{2+} binds to its target sites on troponin C (Tn), to diffusible molecules (ATP, calmodulin, parvalbumin, and extrinsic buffers in experimental conditions) or is transported by the SR pump (SERCA) and the plasmalemmal molecules. The bottom diagram illustrates text equation 4, used to calculate release flux. The rate of change in myoplasmic $[Ca^{2+}]$ is the difference of input and removal fluxes. Release from the SR is the main contributor to input flux, hence eqn. 4.

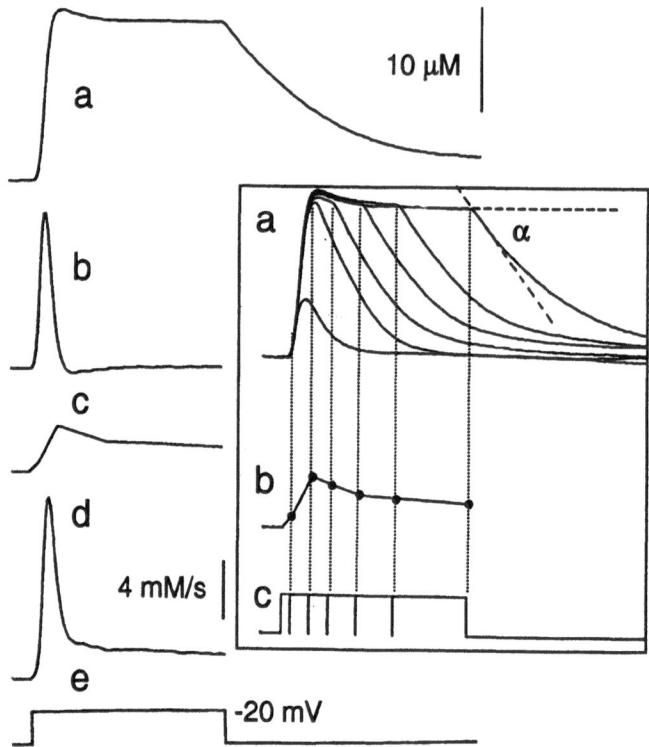

Fig. 2. The empirical calculation of release flux. a) Ca^{2+} transient ($\Delta[Ca^{2+}]_i$ (t)) during a depolarizing pulse of 100 ms, applied to a frog semitendinosus fiber under voltage clamp. $\Delta[Ca^{2+}]_i$ (t) was derived from the change in absorbance of a Ca^{2+}-binding dye. b) the time derivative $d[Ca^{2+}]_i$ /dt. c) removal flux, calculated in the inset. d) release flux, calculated according to eqn. 4, as sum of b and c (multiplied by suitable buffer capacity terms). Note that release flux has an early peak, then decays to a steady level. Inset, the calculation of removal flux at various times during the 100 ms pulse, which is done applying pulses of various shorter durations and measuring the rate of decay of $[Ca^{2+}]_i$ (tan α) after each pulse. Modified from Melzer et al., 1987. (See original for details).

2.4.1. Ca^{2+} Removal from the Myoplasm

The rate of Ca^{2+} removal has been estimated in two ways: an empirical one establishes it by measuring the time course of decay of $[Ca^{2+}]_i$ at the end of a depolarizing pulse (it assumes that release has stopped by then). The removal flux at that time is proportional to the rate of decay of $[Ca^{2+}]_i$ at the beginning of the OFF (slope of angle α in fig. 2 inset; Melzer et al., 1987). By interrupting the depolarization at different durations, the rate of removal can be reconstructed point-by-point. Fig. 2 shows how to then calculate release by addition of $d[Ca^{2+}]_i$ /dt and the removal flux. (Buffer capacity enters this calculation, as described in the original papers).

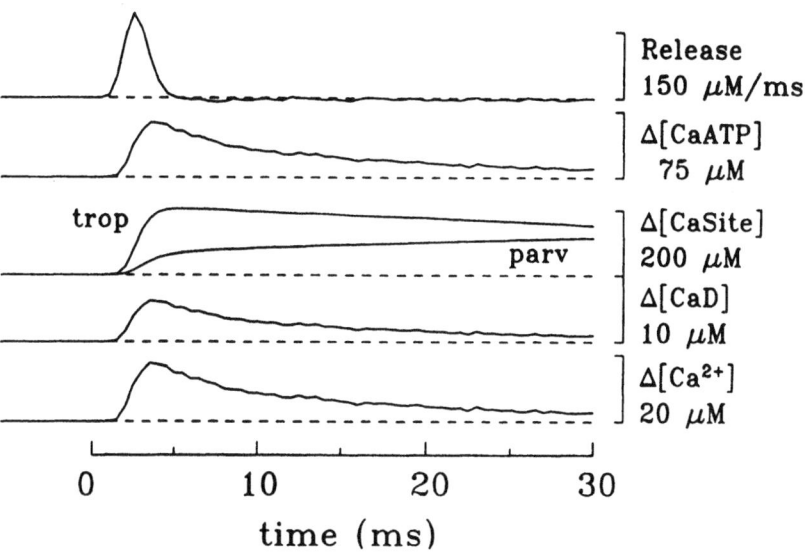

Fig. 3. Synthesis of release from binding fluxes. 1ˢᵗ record from below, Δ[Ca²⁺]ᵢ (t) elicited by an action potential in an intact frog fiber, derived from the change in fluorescence of furaptra. Other records: evolution of the complexes of Ca²⁺ with ATP, troponin C, parvalbumin and the dye. The top record is the sum of the time derivatives of all others, and is approximately equal to a net SR release flux (release minus SR removal). Redrawn from Baylor and Hollingworth, 2000.

A second method synthesizes removal as the time derivative of $[Ca^{2+}]_i$ plus the rate of binding to the known chelating sites on troponin, parvalbumin, calmodulin, ATP, and the extrinsic ligands (including the monitoring dye, and buffers that may have been added). The figure illustrates such calculation in a frog fiber undergoing an action potential (Baylor and Hollingworth, 1996). The second record from below is the optical signal, a change in the fluorescence of the indicator furaptra, which occurs when this rapidly equilibrating ligand binds Ca^{2+}. The signal is proportional, and has been scaled in the figure, to the change $\Delta[CaD]$ in concentration of the CaD(ye) complex. The first record from the bottom is the $\Delta[Ca^{2+}]$ transient, derived from the $\Delta[CaD]$ record using the law of mass action. Because furaptra is a rapidly equilibrating Ca^{2+} chelator, and remains far from saturation, the calculated evolution of $\Delta[Ca^{2+}]_i$ is nearly proportional to that of $\Delta[CaD]$. Then the evolution of the various chelator complexes, driven by the increased $[Ca^{2+}]_i$, can be calculated using reaction rate constants measured in separate experiments. The ones shown are the most relevant. The release flux (top trace) is calculated by differentiating all concentrations with respect to time and adding. As shown, it is brief, and peaks at about 180 mM/s (it would increase total [Ca] in the myoplasm to 180 mM if it continued for 1 second). The release of Ca triggered by an action potential amounts to ~15% of the SR content (Pape et al., 1995). Because the release flux appears to scale linearly with SR content, at least near resting SR load, release will decay during a train of action potentials roughly exponentially.

The brevity of the Ca^{2+} transient is of course a consequence of brevity in the release flux, but also of the efficacy of the removal function. One of the most effective removal devices is parvalbumin, whose Ca^{2+} binding sites (at ~ 1 mM in fast twitch muscle) are largely occupied by Mg^{2+} at rest. An early clue to the role of parvalbumin was its selective distribution in fast muscles. It was especially abundant in fish and amphibia (Gerday, 1982) and this was rationalized as a substitute for Ca^{2+} pumps, which would be more severely affected at the low temperatures of these poikilotherms (Gillis, 1980). Reptiles have little or no parvalbumin because "reptiles are homoiotherms by behavior, they wait to be warmed up before they move" (E.D. Stevens, in Gillis, 1980).

The subtle manner in which parvalbumin works (Robertson et al., 1981) is illustrated by the temporal evolution of its Ca^{2+} complex (fourth record from the top). Unlike the other complexes, [CaParv] increases slowly. This is because the step limiting its formation is the dissociation of Mg^{2+} from parvalbumin sites. Because this rate is independent of $[Ca^{2+}]_i$ the rate of removal by parvalbumin is roughly constant, and proceeds until $[Ca^{2+}]_i$ is very low, due to the high affinity of parvalbumin for Ca^{2+}. Thus the Ca-Mg sites in parvalbumin do not interfere with the initial rise in Ca^{2+} needed to rapidly occupy the target sites in troponin. Then, as Mg^{2+} dissociates, they provide an important sink, operating as a relay station for Ca^{2+} in its way back to the SR.

2.4.2. The Time Course of Ca^{2+} Release

Measured by either method (or by the "triadic gradient", fig.10), the response to a pulse of depolarization goes through an early peak and then relaxes to a lower level (steady phase), which is maintained during the pulse. The rapid decay after the peak reflects an inactivation, that is, a closure of release channels associated with transient loss of their ability to respond (Schneider and Simon, 1988). This process may guard against excessive $[Ca^{2+}]_i$ elevation, as well as unnecessary depletion of the SR. While the time course of this decay depends steeply on temperature, the relative magnitude of peak to steady level does not (Shirokova et al., 1996). The behavior of this ratio is strikingly different in mammals, where it remains at an approximately constant value of 2-4 in the whole range of activating voltages, than in frogs, where it first increases rapidly with voltage, to values as high as 14 at ~ -45 mV, and then decays at higher voltages to near-mammalian values. These differences across taxonomic classes are accompanied, and probably caused, by differences in the molecular makeup of their EC coupling systems (see *RyR3 and sparks*, below).

2.5. The Control of Ca^{2+} Release

Ca^{2+} release occurs at the triad junction, where the TT and SR membrane systems meet, topologically as well as functionally.

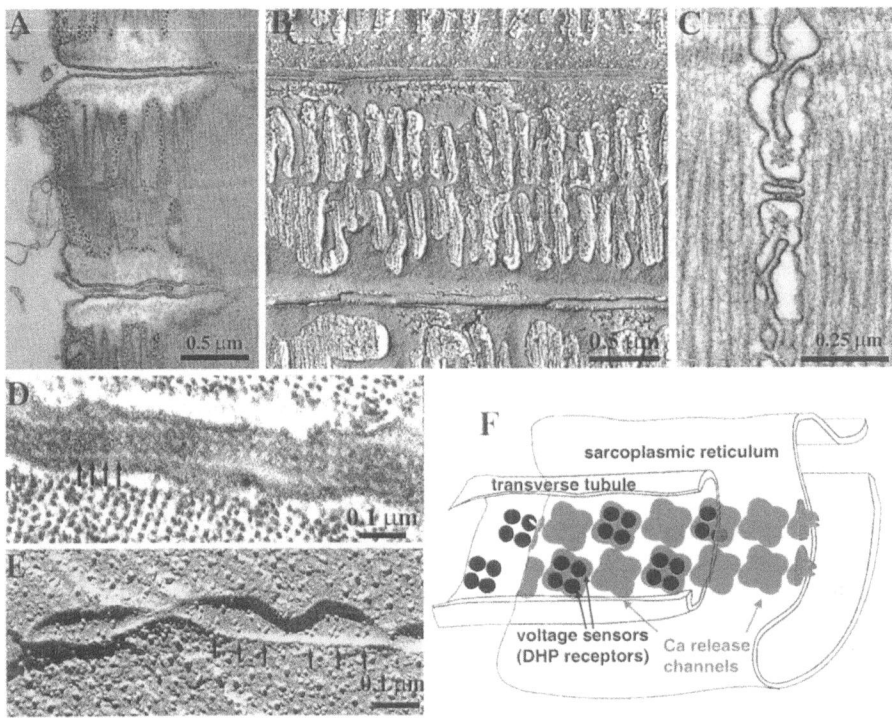

Fig. 4. Membrane systems of EC coupling. A) thin section and B) freeze-fracture, deep etching replica, showing two T tubules and the intervening SR within a sarcomere of a guppy. C) triad of the toadfish swimbladder muscle, showing "feet" in the gaps between T tubule and terminal cisternae, as well as calsequestrin within the cisternae. D) tangential cut of the junctional gap in the guppy, showing feet in orderly double or triple rows. E) cytoplasmic leaflet in freeze-fracture of the T membrane of the swimbladder muscle. Tetrads of particles are separated by double the distance between feet (arrows in D and E). F) interpretation, showing one to one correspondence between T membrane particles (circles, presumably DHP receptors) and protomers of the closely apposed release channel. One channel every other is not in correspondence with DHP receptors. A-E, from Franzini-Armstrong (1999). F, modified from Block et al. (1988). Color version on page 445.

2.5.1. Structure of the Triad
Fig. 4 shows two T tubules in a skeletal muscle cell of a small fish, starting at the plasmalemma (A) and making extensive junctions with terminal cisternae of the SR (B). A cross section in C shows the tubule flanked by two terminal cisternae, constituting a triad. In the cytoplasmic space in-between are dense formations, in pairs, termed feet, known to be part of the Ca^{2+} release channels or ryanodine receptors (RyR). Calsequestrin is visible as dense structures within the cisternae. In D is a tangential view of the T-SR gap, with release channels in double or sometimes triple row. Panel E has the cytoplasmic leaflet of freeze-fractured junctional T membrane, showing particles (L-type Ca^{2+} channels or dihydropyridine receptors --DHPRs) marked by arrows. F depicts how the DHPRs are arranged in tetrads, with sides at an angle of 22 degrees with the TT axis. Individual DHPRs make separate contact with four symmetric sites in the underlying RyRs (which in turn are homotetramers, so there appears to be a one-to-one interaction between DHPRs and RyR protomers). The DHPRs' function is to sense and transmit the change in T membrane potential to the release channels. Remarkably, alternating release channels in the double row do not face voltage sensors, a feature of every skeletal muscle where these two components have been found together (Franzini-Armstrong and Jorgensen, 1994). A crucial difference with cardiac muscle is the lack of tetrad formation in the heart, in spite of the presence of visible DHPR particles.

Some features of the interactions were clarified by microscopy of genetically altered cells. Muscle cells of dyspedic mice (null for the RyR1 gene) develop junctions to which DHPRs are correctly targeted. The same is true of a cell line (1B5) null for RyR 1 and 3 (Takekura et al, 1999). Consistent with the identity of feet and RyRs the junctions do not have feet (hence *dyspedic*). Their DHPRs do not form tetrads (Protasi et al., 1998) indicating a structural role of the DHPR-RyR interaction --and indirectly supporting the idea that the interaction is mechanical, not mediated by a diffusing chemical. It is possible to restore the formation of tetrads in the dyspedic cells by expression of RyR after viral transduction of its cDNA, but this is true only of the skeletal muscle-specific isoform 1 (Protasi et al, 1998, 2000). These observations, which are consistent with other evidences that DHPRs and RyRs do not interact mechanically in heart muscle, mark structural underpinnings of the crucial difference in control mechanisms between skeletal and cardiac muscle.

2.5.2. The Voltage Sensor of the Transverse Ttubular Membrane
The work of Hodgkin and Horowicz alerted researchers of the need for a voltage sensor in EC coupling. As any good voltmeter, such voltage sensor must draw energy from the electric field, by moving charge. The associated current was the first evidence of the sensor's existence. It was first measured by Schneider and Chandler (1973) for skeletal muscle EC coupling, and a few months later by Bezanilla and Armstrong for the Na current of the squid giant axons.

Some properties of a voltage-driven permeability can be used to tell it apart from other signals, such as local Ca^{2+}, or store depletion. Such differences are crucial, for instance, to evaluate the proposed existence of direct control by voltage in cardiac muscle (see following chapter by Eisner, and Howlett et al., 1998.)

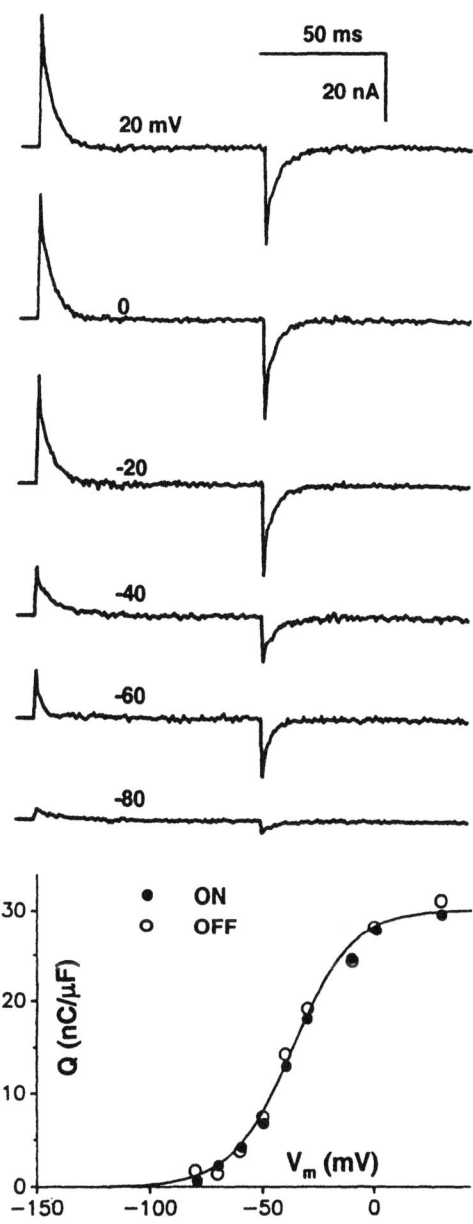

Fig. 5. Intramembranous charge movement. Top, records of charge movement current, in a voltage-clamped frog semitendinousus muscle fiber, upon 100 ms depolarization to the voltages indicated (holding potential, -90 mV). Note that currents are transient, and have no reversal potential. Bottom, charge transfer, integral of the currents, normalized to capacitance of region under voltage control. Continuous line is obtained with eqn. 5. Note equality of transfers in ON and OFF, and saturation at high voltages.

The voltage sensor movement results in an "intramembranous charge movement", which for frog muscle is illustrated in fig. 5. At top are charge movement currents, recorded under strict block of ionic channels and after suitable subtraction of capacitive currents. A defining feature of these currents is that they are transient (there is a finite amount of charge movable in the voltage sensors) and that the charge transferred during the "ON" (proportional to the area under the curve) has to come back during the OFF. This is borne out by the plot of charge transfer (Q) vs applied voltage (V_m), which shows both a rough equality of ON and OFF measures, and the expected saturation at high voltages. A first approximation to the data is the sigmoidal curve, generated with the "Boltzmann equation",

$$Q(V) = Q_{max} \frac{1}{1 + e^{-(V_m - V_t)/K}} \tag{5}$$

This approximation relies on the extremely simplifying assumption that the voltage sensor can only adopt two positions or conformations: "cis", where the mobile charge is closer to the cytoplasmic side and "trans", where the charge has moved outwardly.

The approximation yields three parameters: Q_{max} --the maximum charge that is available to move, some 30 nC/μF in the figure—, V_T -- a voltage of half distribution measuring the spontaneous tendency of the movable charge to be in one position or the other, -42 mV— and a steepness parameter, K, which in a narrow mathematical sense is the voltage increment necessary to increase the number of voltage sensors in the "trans" or activating position by a factor of e (about 10 mV in the example). Physically, the value of K depends on the charge of the elementary particle that responds to voltage (as more heavily charged particles should require smaller changes in voltage to move). Specifically, K is equal to kT / q, where k is Boltzmann's constant, T the temperature and q, the "displacement charge", is the individually moving charge multiplied by the fraction of the transmembrane field traversed in its movement. Initial calculations yielded approximately 2 electron charges per moving unit, which were then combined with Q_{max}, to derive the number of independently moving voltage sensors. That this number turned out roughly similar to the number of "feet" was fortunate --given the coarseness of the approximations-- but it gave plausibility to the postulate of a mechanical link between the two membranes, resident in the feet, and moved by the transmembrane electric field (Chandler et al., 1976, and fig. 8).

Under less restrictive assumptions (Almers, 1978), it can be demonstrated that the open probability of a channel, p_o, becomes in the limit of very negative voltages (and low probabilities) proportional to exp (qV/kT) (this is trivially true for eqn. 5). Using this relationship, q/kT (an "inverse voltage") may be calculated as the limit at low p_o of the derivative of ln p_o with respect to voltage. This "limiting slope" of Ca^{2+} release activation in the frog was $(3.73 \text{ mV})^{-1}$, corresponding to a q of about $6e_0$ (Pape et al., 1995), or 3-fold greater than the initial calculation based on charge movement. Evaluated this way, the gating charge per Shaker K channel was 12 e_0 (Seoh et al., 1996), while that of the L-type Ca channel was ~ 9 e_0 (Noceti et al., 1996). The

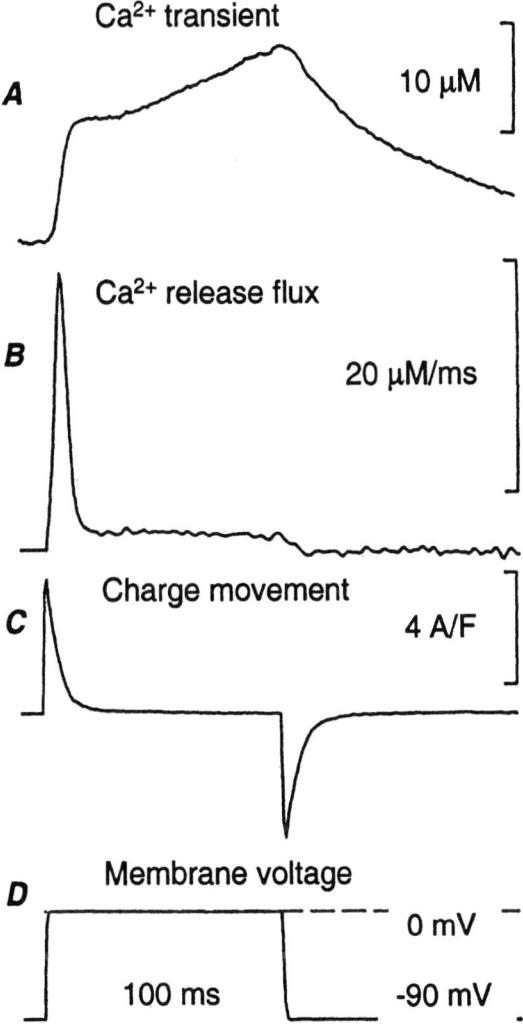

Fig. 6. Causal sequence of events in a Ca^{2+} transient. A) Ca^{2+} transient, time course of $\Delta[Ca^{2+}]_i$ during depolarization of a frog semitendinosus muscle fiber under voltage clamp. The technique (using the low affinity indicator antipyrylazo III) cannot determine actual $[Ca^{2+}]$, but only its change . B) Flux of Ca^{2+} release, derived from record in A as described in text. C) Simultaneously recorded charge movement current, normalized to membrane capacitance. D) measured membrane voltage. (Reproduced from Rios and Pizarro, 1989).

unitary mobile charge is therefore similar for these three voltage sensors. As discussed later, the EC coupling sensor would be expected to have a higher value.

Further insight is given by a comparison of the time course of voltage sensor currents and release flux. The figure represents four records which, starting from the bottom, follow the causal chain of events. D is the applied potential, a 100 ms pulse to 0 mV (from the resting level of -90 mV). This potential change causes movement of the voltage sensor, manifested in the current of charge movement (C). Movement of the voltage sensors is followed rapidly by opening of release channels in the SR, resulting in the fast early increase in Ca release flux (B), and then decay, reflecting release inactivation. The Ca^{2+} transient (A) evolves as a rough time integral of release flux, rising rapidly initially, and then more slowly, as the flux steadies at a lower level.

As seen in the figure, the movement of voltage sensor charge largely precedes the opening of channels. This implies the existence of a subthreshold charge, revealing early stages in the movement of the sensor that are ineffective in channel opening. Upon reaching threshold, release flux then increases beyond proportion to the continued movement of the voltage sensor. The movement of the voltages sensors are therefore not linked one-to-one to the opening of channels gates. The functional advantages of such nonlinearity in the relationship are two-fold: the existence of subthreshold charge stabilizes the resting state (preventing accidental firing), while the marked increase in release beyond threshold contributes to a fast contractile response. An additional nonlinearity, serving the same purpose, is in the "sigmoidal" voltage dependence of the sensor movement, which follows eqn (4).

The term "Calcium Antagonists" names a large group of drugs (including among others the dihydropyridines and phenylalkylamines) that inhibit functions relying on entry of Ca^{2+} to cells. All these drugs bind to the α_1 subunit of L-type Ca^{2+} channels present in many cells, and inhibit their currents by promoting an inactivated state. In skeletal muscle, approximately 300 dihydropyridine receptors reside in every micrometer square of T tubular membrane, where they constitute the particles visible in freeze fracture (fig. 4). One function of these DHPRs is to pass the "slow" membrane Ca^{2+} current (I_{Ca}). The role of the DHPRs as voltage sensors was uncovered studying the effects of nifedipine, which reduced voltage sensor charge movement and simultaneously reduced release flux (Ríos and Brum, 1987). It was then confirmed in a series of experiments using dysgenic mice, homozygothic for the naturally occurring *mdg* mutation, which prevents expression of the pore subunit of the skeletal muscle DHPR (α_{1S}, see "*EC Coupling Molecules*", below). While muscle cells of these mice lack specific DHP binding, L-type Ca^{2+} current, T membrane particles, direct voltage-operated Ca^{2+} release and a good portion of the intramembrane charge movement, all of these characteristics are restored upon expression of α_{1S} (Tanabe et al., 1998; Adams et al., 1990; Takekura et al, 1994).

Dysgenic mice have been used to test several hypotheses relative to the nature of the interaction between voltage sensor and release channel, as well as relationships between structure and function of the voltage sensor, described in a later section. A fundamental result is that expression of the cardiac L-type pore subunit (α_{1C}) restores plasmalemmal Ca^{2+} current and control of Ca^{2+} release in dysgenic muscle cells, but such control is then "cardiac", that is, mediated by entry of Ca^{2+} through the heterologous plasmalemmal channels (Tanabe et al., 1990). That expressio of a cardiac RyR, as described earlier, cannot restore the tetradic arrangement of DHPR particles in dyspedic cells, indicates that the stoichiometric, spatially fixed relationship between DHPR tetrads and RyRs, is characteristic of skeletal muscle molecules, and suggests a mechanical link between DHPRs and RyR.

2.5.3. The Ca^{2+} Release Channel of the Sarcoplasmic Reticulum

Different insights on EC coupling have been obtained from preparations of varying complexity. While release flux is measured in relatively intact cells, substantial advances have come from the work with skinned cells, with the plasmalemma mechanically or chemically removed. Such studies revealed that the SR could be made to release Ca^{2+} by exposing it to somewhat elevated $[Ca^{2+}]$ (Endo et al., 1970). This phenomenon, termed Ca^{2+}-induced Ca^{2+} release (CICR), is now considered a fundamental property of the individual molecules, shared with the other family of intracellular Ca^{2+} channels, the ligand-gated IP_3 receptors.

Still simpler preparations can be derived from muscle homogenates, including suspensions enriched in "triads", which preserve to some extent the SR-TT-SR structure, and purely tubular or *sarco-reticular* fractions. In particular a *heavy SR* fraction derives largely from the terminal cisternae (Meissner, 1975). When its vesicles are fused with planar bilayers, single or multiple cation channels of unusually large conductance and relative low selectivity insert in the membrane (Smith et al., 1985).

As shown in fig. 7 A, the channels' activity depends on Ca^{2+} in a dual way. It increases with $[Ca^{2+}]$ on the cytosolic side, at concentrations between 1 and 100 µM (in good correspondence with the CICR mechanism). Higher concentrations of the ion reduce its open probability. Although it has been proposed that this inhibition by Ca^{2+} could represent at the single channel level the inactivation that follows the peak of whole cell release (fig. 6), the sensitivity of the channels *in vitro* to inhibition by Ca^{2+} cannot account for the rapid kinetics and other properties of inactivation in cells. Panel C in the figure demonstrates a hallmark of these channels, their two-step reaction with the plant alkaloid ryanodine (Sutko and Airey, 1997), which first induces a long-lasting substate of 50% conductance, then, presumably by binding to a site or sites of lower affinity, closes the channel.

Initially isolated by virtue of their high affinity for ryanodine, the channels have been cloned and sequenced. They are homotetramers of polypeptides of about 565,000 molecular mass, encoded by three different genes, and termed the skeletal muscle (RyR1), cardiac (RyR2) and brain (RyR3) isoforms. In lower vertebrates, two isoforms are expressed: α, homologous to RyR1, and β, more similar to mammalian isoform 3. While RyR3 constitutes a good fraction of the chanels in embryonic and neonatal muscle of mammals, it largely disappears in the adult, remaining as a minor fraction in diaphragm, masseter and other muscles. In lower vertebrates, the homologous β appears to have a more important role, as it is present in the adult in about equal proportion as the α isoform, with which it presumably coexists in the same triads (Sutko and Airey, 1996, 1997). RyR3 is more susceptible than RyR1 to activation by Ca^{2+} (Murayama et al., 1999; Fessenden et al., 2000, but see Chen et al., 1997). It is incapable of establishing the "skeletal"-type interaction with the DHPR

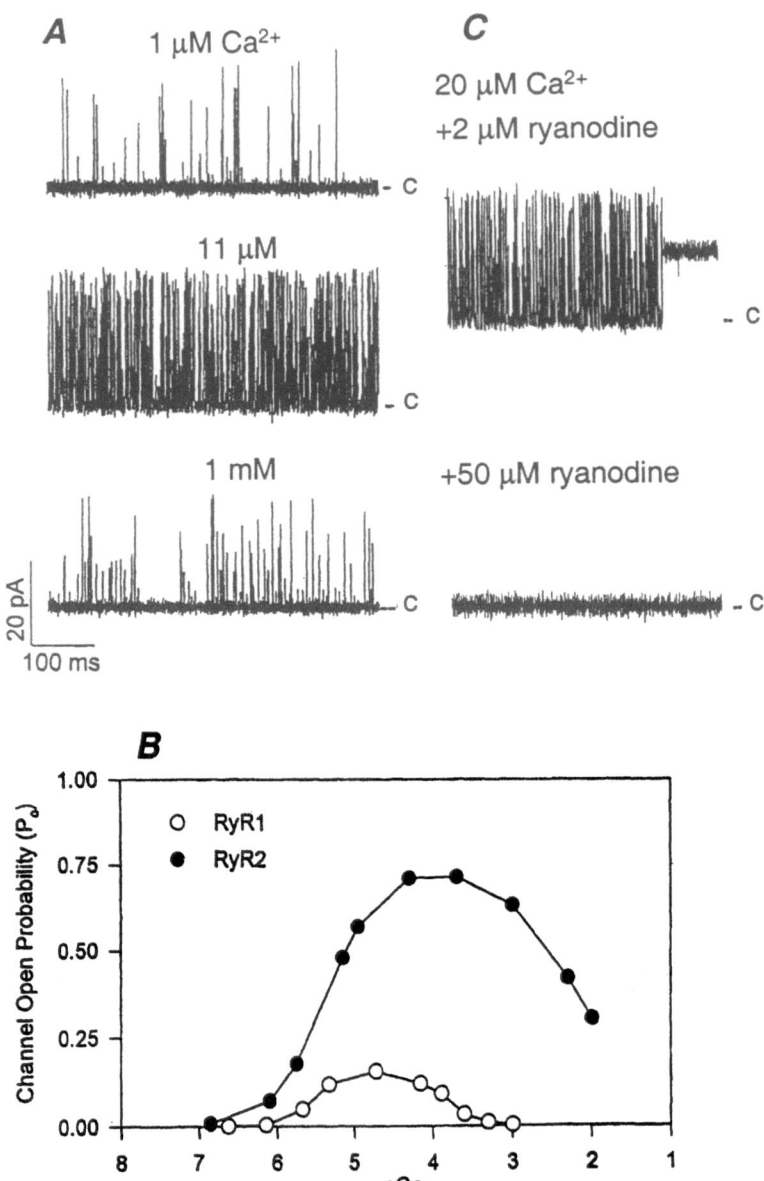

*Fig. 7. Single Ca²⁺ release channel currents. Potassium currents through purified RyR, incorporated in a bilayer held at + 40 mV, in symmetric 250 mM KCl. **c** marks the current level when the channel is closed. A) activation and inhibition by increasing cis (cytoplasmic side) Ca²⁺. B) Open probability of skeletal and cardiac (RyR2) channels vs the log of cis [Ca²⁺]. C) at optimal [Ca²⁺], ryanodine locks the channel in a subconductance state, and closes it at a higher concentration. (From Xu et al., 1998, figs. 2 and 4).*

that underlies direct control by voltage (Fessenden et al., 2000) and does not induce the formation of tetrads of DHPRs (Protasi et al., 1998; see *Structure of the Triad* above). These facts suggest a specialization of the two isoforms. As explained below under *Dual control of Ca^{2+} release,* RyR1 could be directly controlled by voltage sensors, while parajunctional RyR3 channels would be controlled by CICR.

One of the most characteristic features of the RyRs is their ability to interact with many other molecules. The endogenous ligands, many of them claimed to have a role in their physiologic modulation, include mono and divalent cations, anions like Cl^-, sulfhydryl oxidants, nitrosylating agents, adenine nucleotides, and other proteins (including triadin, junctin, calsequestrin, immunophilins, the DHPR, kinases and phosphatases). A long list of exogenous ligands has been shown to modify their state as channels, and used to advantage in physiologic studies, including anions like perchlorate or thiocyanate, components of venoms and toxins such as imperatoxin A (from scorpion venom), bastadins (from sponge extracts) and ryanoids. Xanthines, including caffeine, are powerful promoters of the open state. Sulfhydryl-reducing agents lower the channels open probability while oxidants increase it. 4-chloro-m-cresol, adenosine receptor agonists, digitalis glycosides, and the antiparasitic suramin activate RyRs, while local anesthetics (like tetracaine), polycationic reagents like ruthenium red, and dantrolene (used in the treatment of Malignant Hyperthermia) inhibit channel opening. RyR pharmacology was reviewed by Xu et al. (1998).

2.5.4. Mechanical or Allosteric Control

Many indications that RyRs and DHPRs are mechanically linked were mentioned already. The first proposal of mechanical control was the "plunger" hypothesis of Chandler et al. (1976), according to which the voltage-driven movement of charge in the voltage sensor literally unplugs the release channel. This model, and an alternative "allosteric" one are illustrated in fig. 8. The allosteric model (Ríos et al., 1993) is similar to the classic model of Monod et al. (1965) for the regulation of tetrameric enzymes. In Monod's model, binding of one to four oxygen atoms to hemoglobin increasingly facilitates the transition of the multimeric molecule to a *Relaxed* state. In channel gating, the successive voltage-driven movement of one to four voltage sensors increasingly favors transition of the channel to its open state. The difference between allosteric control and the plunger (or other "isosteric" models) is that these assume a strict correspondence between states of the voltage sensor and states of the release channel. In the allosteric proposal instead, the channel may adopt the open state regardless of the sensors, although with probabilities that depend on the sensors' state. One advantage is that it explains the observed variability in the relationship between movement of intramembrane charge and release activation. Simon and Hill (1992) found that the activation of release, A, was linked to the charge transferred Q by a power relationship $A \propto Q^4$, as if four voltage sensors had to move independently to open a single release channel. Perchlorate, a promoter of Ca release, had a linearizing effect on the relationship, making it similar to a second, even a first power (González and Ríos, 1993). This is accounted for in the model assuming that the anion increases the intrinsic tendency of the release channels to open (rate constant k_L in fig. 8), hence

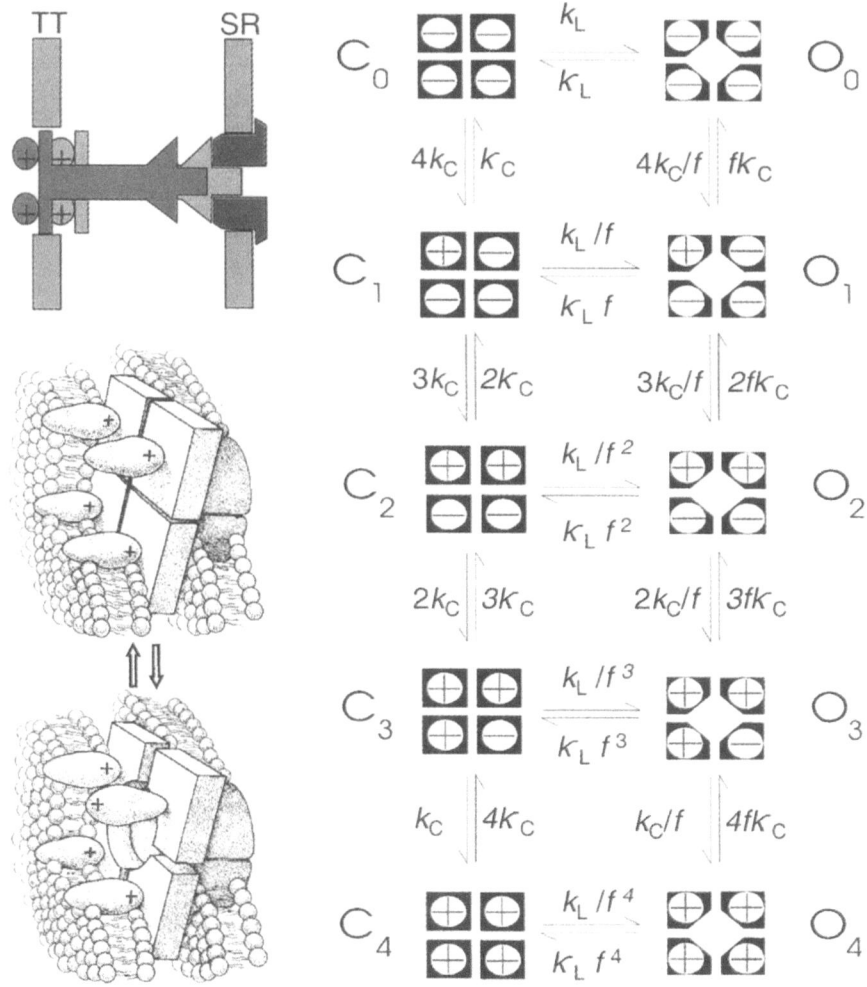

Fig. 8. The evolving view of mechanical coupling. Top left, the "plunger" model of mechanical coupling (Chandler et al., 1976). A molecule that spans the gap, with charges resident in the TT membrane (light green), keeps the SR channel closed, and unplugs the channel when its charges move outwardly upon membrane depolarization (dark green). Bottom and right, an allosteric version of mechanical control (Ríos et al., 1993). Conformational changes in the voltage sensors (depicted for one of them in the drawing) do not deterministically cause, but favor the SR channel opening, depicted as a one-step transition in all four protomers. In the diagram at right C_j and O_j designate, respectively, states of the coupled system with the release channel closed or open. The indicies j correspond to the number of voltage sensors in the "trans" or activating conformation. The rate constants of the close-to-open transition (k_L and k'_L) strongly favor the closed channel, but are multiplied (divided) by the factor 1/f (>1) whenever a voltage sensor assumes the trans conformation. Transitions of the voltage sensors are determined by V_m (through its influence on rate constants k_C and k'_C). The voltage sensor rate constants are made different when the channel is open, to satisfy microscopic reversibility. Modified from Ríos et al., 1993. Color version on page 446.

the activating action of fewer sensors is then sufficient to cause a majority of channels to open.

The observation that perchlorate, a promoter of the movement of voltage sensor's intramembrane charge, is an agonist of the release channel reconstituted in bilayers (Ma et al., 1993) was taken as evidence of a reciprocal effect of the release channels on the voltage sensors.   Indeed, the allosteric model predicts, from rules of thermodynamics, that a promoter of the open state, even if it acts strictly on the release channel, will indirectly favor the movement of the voltage sensor.   In turn, this constituted evidence for a mechanical link in the interaction between voltage sensors and release channels.   A more direct evidence of mechanical interaction was the observation that the dyspedic mouse muscle has little L-type Ca^{2+} current, in spite of a nearly normal endowment of DHPRs, and that the current increases greatly, together with the restoration of voltage-dependent release of Ca^{2+}, upon expression of the skeletal (but not the cardiac) RyR isoform (Nakai et al., 1996).

2.5.5. Control of Inactivation

Ca^{2+} release is subject to two inactivation processes.   One, which occurs in seconds, is brought about by prolonged depolarization.   It appeared first as the process that terminates K contractures, and is associated with an inactivation of the voltage sensor. As such, it is similar in many respects to voltage-dependent inactivation of Na and K channels.

Inactivation of the DHPR induces a characteristic conversion from the normal mode of charge movement (named charge 1) and a second mode, charge 2, characterized by a much more negative V_T and by its inability to cause channel gating (Brum and Ríos, 1987).   This type of inactivation is similar to the fast ("ball and chain") inactivation described for Na (or N-type inactivation in Shaker K) channels, for not involving a substantial movement of charges along the electric field (no intrinsic voltage-dependence), being instead dependent on, and made possible by the sensor's activation.

Unlike fast Na channel inactivation, however, the inactivation of the DHPR does not result in charge immobilization.   The interconversion of charge that occurs instead, is similar to that described during "slow" inactivation of Na channels (Bezanilla et al., 1982) or C-type inactivation of Shaker K channels.   In K channels, this slow process is characteristically dependent on changes in the extracellular solution, and is believed to involve substantial rearrangements of the pore and its external mouth.   Likewise, the inactivation of the voltage sensor of EC coupling is antagonized by extracellular Ca^{2+} or other cations (Brum et al., 1988) with a selectivity sequence strikingly similar to that of permeability through the L-type channel, indicating that sites to which extracellular ions bind in the permeation process stabilize enabled or "primed" conformations of the voltage sensor (Pizarro et al., 1989).

A different inactivation determines the rapid fall of release flux after its peak.   It occurs in milliseconds, and appears not to involve the voltage sensor directly.   Having observed that it continues after conditioning pulses, provided that $[Ca^{2+}]_i$ remains elevated, Schneider and Simon (1987) proposed that it was mediated by Ca^{2+}.   The effector site in the RyR could be the low affinity inhibitory binding locus for Ca^{2+} or Mg^{2+}, revealed in bilayer experiments.

The evidence for Ca^{2+}-mediated inactivation, however, remains weak and indirect. Recent studies have shown inactivation to be "quantal" (a term originated in the field of IP_3 receptors, stressing a number of differences with the "classical" behavior of Na channels). In attempts to adjust a Ca^{2+}-binding model to the process, the apparent K_D of the inactivation site turns out to depend on the conditioning/test pulse pattern (Csernoch et al., 2000). Some of the peculiar aspects of this process may be explained assuming that only the channels that open then inactivate, and that all of those do (in other words, inactivation is strictly and fatally linked to activation; Pizarro et al., 1997). This could be possible if the putative Ca^{2+} site was located at the cytosolic mouth of the channel and its affinity was low, so that only the high $[Ca^{2+}]$ near an open channel could drive the process. On the other hand, a Ca^{2+}-independent inactivation fatally coupled to opening (like a ball-and-chain) has not been ruled out.

2.5.6. Dual Control of Ca^{2+} Release

A mechanism of control of release that uses the Ca^{2+} sensitivities of the RyRs, and their structural duality, manifest in the interaction of voltage sensors with alternate RyRs, was proposed by Ríos and Pizarro (1988). RyRs directly facing voltage sensors would open first, in response to the voltage-driven conformational change in the sensors. Ca^{2+} released from these channels would activate neighboring channels without voltage sensors. The dual control would result in separate kinetic components of flux. Steady release would be carried through directly voltage-operated channels (named "V"), while the peak component would flow through Ca^{2+}-operated ("C") channels.

A mediator role of Ca^{2+} can be tested adding Ca^{2+} buffers at high concentrations. The results of such challenges yielded contradictory results. Our own experiments supported the model, as the presence of fura-2 at 3 mM suppressed the peak component. The buffer also reduced the steady component, indicating that at least part of it was also Ca^{2+}-mediated. Part of the peak could be recovered by simply increasing pulse voltage, suggesting the possibility of activation by Ca^{2+} from multiple V sources. A better challenge of these ideas was possible upon the discovery of Ca^{2+} sparks.

2.6. Ca^{2+} Sparks

Ca^{2+} sparks were discovered by Cheng et al. (1993), imaging in ventricular myocytes the fluorescence of fluo-3, which increases from essentially zero upon binding Ca^{2+}. The use of confocal microscopy, which reduces the thickness of the in-focus region by rejecting out of focus light was crucial. Sparks are brief fluorescence events resulting from Ca^{2+} release, consistent with local increases in $[Ca^{2+}]$ up to several times the resting concentration. They were later found in skeletal (Tsugorka et al., 1995) and in smooth muscle (Nelson et al., 1996).

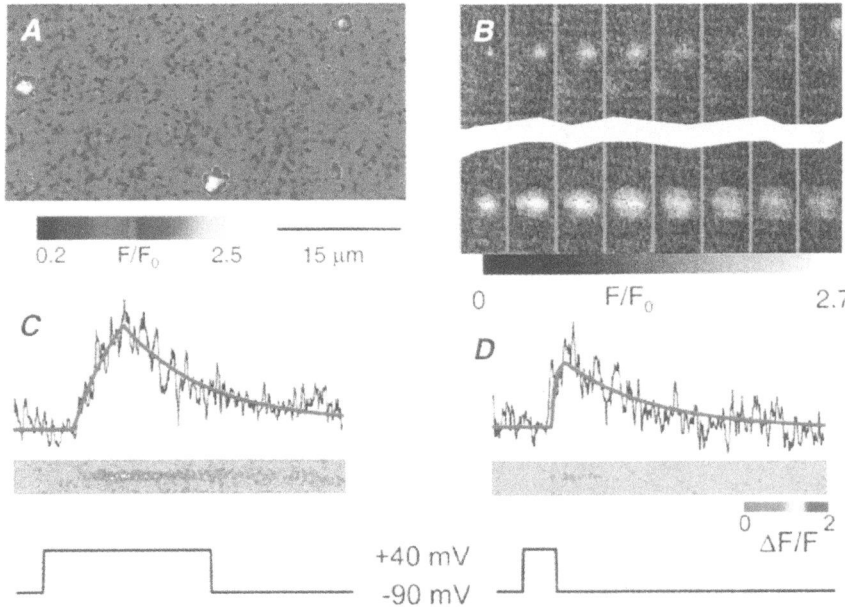

Fig. 9. *Properties of* Ca^{2+} *sparks. A) spontaneous sparks in an xy scan (frog muscle). The acquisition (scan) of the portion of image covering one of the large sparks shown takes < 2 ms, hence these are essentially snapshots of sparks. B) successive images of the same band (~ 5 μm wide), taken at 4 ms intervals. Note the evolution of two sparks, which first increase in intensity and diameter, then decay and disperse, while remaining centered at the same location (Brum et al., 2000). C) the color strip was obtained by scanning repeatedly the same line (~ 6 μm long) in a fiber under voltage clamp, subjected to a 15 ms pulse as indicated. The black trace at top is the time course of fluorescence at the center of the spark. In red are best fit exponentials to the rising and decaying portions of the profile (with boundary times also determined by fitting). D) when elicited by pulses of 3 ms (briefer than the average spark), sparks are briefer, as if terminated by repolarization. A and B, reproduced from Brum et al. (2000); C and D, redrawn from Lacampagne et al. (2000).* **Color version on page 447.**

2.6.1. Spatial and Temporal Aspects of Ca^{2+} Sparks

Fig. 9 shows at left sparks occurring spontaneously in a frog skeletal muscle fiber. The images were obtained by "*xy* scanning", at video rates (~30/s), with a fast confocal system. Most sparks appear circular, consistent with a spherically symmetric signal, which might result either from a single channel, or a channel group too small to be distinguished from a point source.

The evolution of sparks in time can be followed in two ways. Panel B shows repeated images of the same narrow region in the fiber, taken at intervals of 4 ms. Sparks first grow in intensity and radius, then decay while continuing to spread. Most sparks stay in the same location as they evolve (but see *Sparks as unitary events*, below). A greater temporal resolution is obtained by repeatedly scanning the same line, as in the small panels of fig 9 C, D, and in fig. 10. Seen in profile, sparks in such line scan images feature a rising phase, close to 5 ms, followed by a somewhat more prolonged decay. The plots of fig. 9 C, D show that a good fit of this time course is obtained joining two exponential segments, one for the rising and another for the decaying phase (Lacampagne et al., 1998).

The line scan images of figs. 9 C, D and fig.10 were obtained in frog muscle fibers under voltage clamp. In fig. 10, imaging several sarcomeres, the response to a depolarizing pulse consists of an increase in fluorescence that starts at periodic locations (triads) with a clear peak, and then gives way to intermittent sparks, at the same locations. The bottom panel is the "triadic gradient", an average over all triads of the difference in fluorescence between the triad centers and the two flanking regions (Tsugorka et al. 1995). Because fluorescence is roughly proportional to local $[Ca^{2+}]$, this triadic gradient is proportional to the Ca^{2+} concentration gradient between release regions and cytoplasm. This is of course the driving force for diffusion of released Ca^{2+}, hence it is roughly proportional to the time course of release flux, exhibiting the peak and steady components found with whole cell records (i.e. fig. 6 B).

Fig. 10 shows that the time span of the peak of release in every triad is similar to that of sparks. Sharpening this idea, Sham et al. (1998) demonstrated that after a spark, the local release unit is inactivated. Hence, both sparks and the peak of release terminate by inactivation --rather than closing of the channels' activation gate--. The inactivation process may be the same in both, which presents added flanks for understanding this crucial process. The figure also suggests that the peak of Ca^{2+} release is constituted by many superimposed sparks

2.6.2. Role of CICR in Spark Generation

Sparks appear to be mediated by Ca^{2+}. Klein et al. (1996) found a correlation between frequency of "spontaneous" sparks and $[Ca^{2+}]_i$, increased near a membrane lesion. Increased $[Mg^{2+}]$ (González et al., 2000a), and BAPTA (González et al., 2000c), both of which should interfere with CICR, inhibited sparks. They also inhibited the peak of release, in agreement with the idea that sparks and peak depend on CICR. These inhibitors, as well as tetracaine (Shirokova and Ríos, 1997), spared a continuous, event-less mode of release, which carried less flux than the steady release in reference. The results support a dual control by voltage and Ca^{2+}, which would be left operating

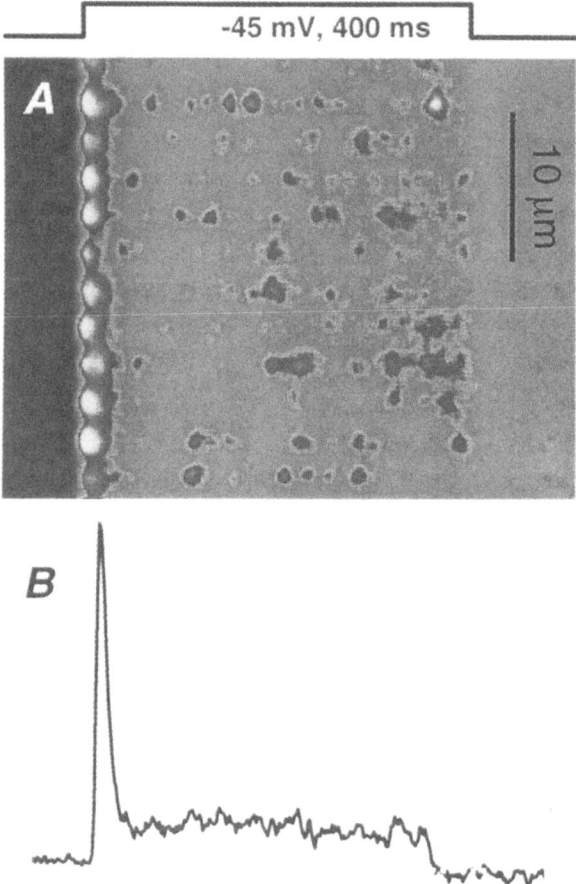

Fig. 10. *Local and global aspects of Ca*$^{2+}$ *release. A, line scan image of a frog fiber stimulated by the pulse at top. Scanning, in the longitudinal direction (parallel to the fiber axis), is much slower than in fig. 9 (it takes 600 ms to acquire the image). Note release centered at periodic positions along the fiber (triads). B, the average difference in fluorescence between the triad regions and the flanking sarcomere centers (triadic gradient), is approximately proportional to release flux (Tsugorka et al., 1995). Unpublished records of N. Shirokova, A. González and E. Ríos. Color version on page 448.*

with just its voltage component in the presence of CICR inhibitors. Though consistent in general with the earlier models (Ríos & Pizarro, 1988), the study of sparks showed clearly that CICR participates in the steady component of Ca^{2+} release as well.

2.6.3. Sparks as Unitary Events of Ca Release

Sparks were thought to be functional units of Ca^{2+} release, much the same as single channel currents are the discrete components of whole cell current. A quantitative test of this proposition was carried out (Klein et al., 1997) recording sparks during pulses of different voltage (after inactivating most voltage sensors to make sparks scarce at all potentials, thus increasing their mutual separation and improving their definition). Under such conditions of low spark frequency, the main effect of increasing voltage was the accumulation of sparks early in the pulse. This procedure yielded a frequency of sparks in time at different voltages, and a spark time course --which turned out to be voltage-independent--. If spark generators did not interact with each other (and the number and density of activated voltage sensors did not affect spark shape or time dependence), then the whole cell Ca^{2+} release should be reproduced by the following recipe: simple superposition of sparks, at the measured time-dependent frequency. Release reconstructed in this way had a small peak at -40 mV, while in fully primed cells the peak was already large at -60 or -55 mV. This discrepancy indicates that multiple voltage sensors may cooperate to elicit sparks, and is consistent with the fact, noted earlier, that increasing voltage may compensate for the inhibitory effects of RyR inhibitors on peak release. In summary, release at the whole cell level cannot strictly be reconstructed by superposition of identical sparks. However, superposition works to some extent, indicating that many of the factors that determine the features of release at the cellular level can be found in the spark-generating unit.

Ca^{2+} sparks can be simulated as the result of Ca^{2+} release from a small source into a fig.1-like medium containing mobile and fixed buffers. Such models show that the *open time* of the release source is about the same as the *rise time* of the spark. Simulated sparks are about 1 μm wide, rather than 1.5-2 μm, as usually observed. Their peak intensity scales roughly linearly with release current (for constant release time), as does "signal mass", the integral of signal intensity over the volume of the spark. Depending on the assumed concentrations and properties of cellular buffers, the current intensity required to simulate a spark of average amplitude ranges from 2 to 20 pA (Smith et al., 1998; Ríos et al., 1999; Jiang et al., 2000). This result is relevant to our next question.

2.6.4. Number of Channels in a Spark Generator

It matters whether the spark generator is a single channel, or several channels in interaction. The one-channel model was formulated first (Cheng et al., 1993), but questioned later based on many indications of the involvement of multiple channels. 1) The current necessary for producing a spark, 2 - 20 pA, is substantially greater than the unitary RyR current (0.3-0.6 pA, Mejía-Alvarez et al., 1999). 2) The observation of release not constituted by sparks, causing local fluorescence increases smaller than the amplitude of average sparks (Lipp and Niggli, 1996; Shirokova and Ríos, 1997). 3) Aspects of the sparks spatial shape, including their large width and the observation of asymmetric, elongated sparks (Brum et al., 2000) or coupled fluorescence events (Parker et al., 1996), inconsistent with a point source. 4) Release agonists like caffeine (or antagonists like high $[Mg^{2+}]$), increase (reduce) spatial width of sparks, independently of the changes that these ligands induce in rise time, as if increasing (reducing) the number of open channels, rather than their open time (figure 11, and González et al., 2000a). 5) High affinity ligands that open single channels produce fluorescence events that start with a spark, then decay to a phase of much lower intensity and width (González et al., 2000b; Shtifman et al., 2000). 6) The distribution of spark open time (González et al., 2000b) has a mode, which is inconsistent with a Markovian channel gating reversibly, but is readily explained as a result of cooperative gating of several channels (Stern et al., 1997). 7) the distribution of spark amplitudes (Bridge et al., 1999; Ríos et al., 2001), which may also be modal, is also inconsistent with single Markovian channels. 8) The observed inverse correlation between spark amplitude and rise time (Lacampagne et al., 1998) is opposite to the expectation for a one-channel mechanism. 9) Sparks may migrate in space (Brum et al., 2000), which requires activity propagating along an array of sources.

Other observations favor the view that single channels produce sparks. As mentioned above, the estimates of time course of release indicate that, if multiple, channels should open and close essentially simultaneously to produce the spark (Lacampagne et al., 1998). Additionally, as illustrated in fig. 9 D, a single voltage sensor may cause a spark to terminate upon repolarization (Lacampagne et al., 2000). These diverging observations could be reconciled if sparks were caused by variable numbers of channels, normally between 2 and 6, but more under special stimuli (as in the presence of caffeine). Coupling between channels (possibly mediated by Ca^{2+}, or perhaps by allosteric interactions between neighboring RyRs, as proposed by Marx et al., 1998) should be tight --effective, fast, perhaps mutual. Quantitative models of channel gating coupled by Ca^{2+} or by Ca^{2+} and mutual allosterics, have been presented (Stern et al., 1997, 1999).

2.6.5. Possible Role of RyR3 in Sparks

Shirokova et al. (1998) could not observe sparks in adult rat fibers, while Conklin et al. (1999) found them in the mouse, albeit infrequently. This, and the fact that sparks were readily observable in embryonic mammalian muscle (where RyR3 is present), led Shirokova et al. (1998) to hypothesize that isoform 3 had a critical role in CICR and spark production. The proposal is consistent with observations at the whole cell level (Shirokova et al., 1996), with single channel properties (greater sensitivity of RyR3 to activation by Ca^{2+}; Murayama et al., 2000; Fessenden et al., 2000), and with the presence of parajunctional RyR3 channels in the frog but not the mammal (Franzini-Armstrong, personal communication and 2001). When this hypothesis was tested by examining discrete events in single-isoform systems, however, the results were contradictory (Ward et al., 1999; Shirokova et al., 1999; Conklin et al., 1999, 2000). Whether RyR3 or β have a specific role in the mechanism of sparks, for instance making them briefer and of greater amplitude, remains to be established.

In a related observation, mammalian myotubes produce sparks only in spotty regions, which do not respond to membrane depolarization (Shirokova et al., 1999). It

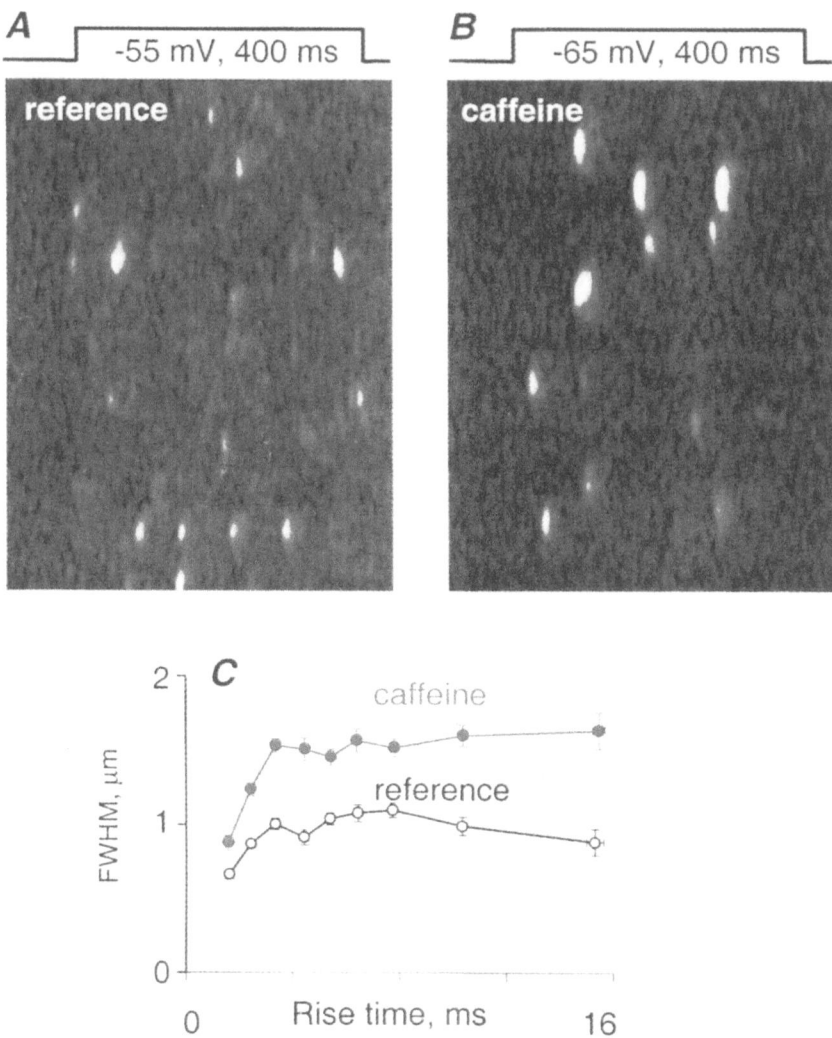

Fig. 11. *Caffeine increases spatial width of* Ca^{2+} *sparks. A, line scan image of a frog fiber, with sparks elicited by pulse depolarization. B, the same fiber, in the presence of 1 mM caffeine. Note that a lower voltage pulse was applied, to maintain the frequency of events. C, average spatial width of events (Full Width at Half Magnitude) grouped according to their rise time. The effect of caffeine on width is independent of an increase in average rise time, not shown in the plot. Modified from González et al. (2000a). Color version on page 449.*

is as if the voltage-sensing mechanism (presumably the DHPR itself) inhibited the release channels, suppressing their activation by Ca^{2+}.

3. STRUCTURE AND INTERACTIONS OF THE EC COUPLING MOLECULES

The cloning and sequencing of the main EC coupling molecules, together with experimental paradigms in muscle cells engineered for deletion of some of these molecules, have made it possible to show that specific epitopes of their primary sequence are crucial to a given function (i.e. membrane targeting, interaction with another protein, sensitivity or binding to a ligand, etc.)

3.1. The Dihydropyridine Receptor

The skeletal muscle DHP receptor (voltage sensor) comprises five subunits: α_1, $\alpha_2\delta$, β, and γ, arranged spatially as shown in fig.12. α_1 contains the pore and carries out the main functions. The other subunits have auxiliary roles, which may be important for EC coupling.

3.1.1. Alpha 1 Subunit

The skeletal muscle α_1 subunit alone may operate as a voltage sensor and a Ca^{2+} channel. It is also the target for several drugs. Initially termed α_{1S}, it is now $Ca_v1.1$, ("v" for voltage dependence, a "1" for the gene family and the other to designate the first subunit discovered within family 1).

$Ca_v1.1$ has 1600 amino acids in four domains of homology, each with 6 transmembrane segments connected by alternating cytoplasmic and extracellular loops. Such secondary structure is therefore identical to that of the voltage-dependent Na and K channels of nerve and muscle. The identity extends to the function of some portions of the molecule. The S4 segments are positively charged and participate in sensing the voltage changes, while the topologically extracellular "P" loop connecting S5 to S6 constitutes part of the pore lining.

Same as with other channels, a program of mutation/chimerization and expression, which for this subunit was largely carried out in the $Ca_v1.1$-less *mdg* mouse (Beam et al., 1986), proved successful, in the sense that changes of one or a few residues resulted in cDNAs that expressed working proteins, targeted correctly to the triads at near-normal density, but with subtly changed functional properties that could therefore be assigned to a specific epitope in the primary sequence. As with other voltage sensitive molecules, the intramembranous charge movement turned out to be a robust function (envisioned as more "primitive" and difficult to disrupt than other functions), useful to monitor density of DHPR protein insertion in the membrane. With these techniques, structural or functional roles could be assigned to many regions. What follows is a sample of a broad field.

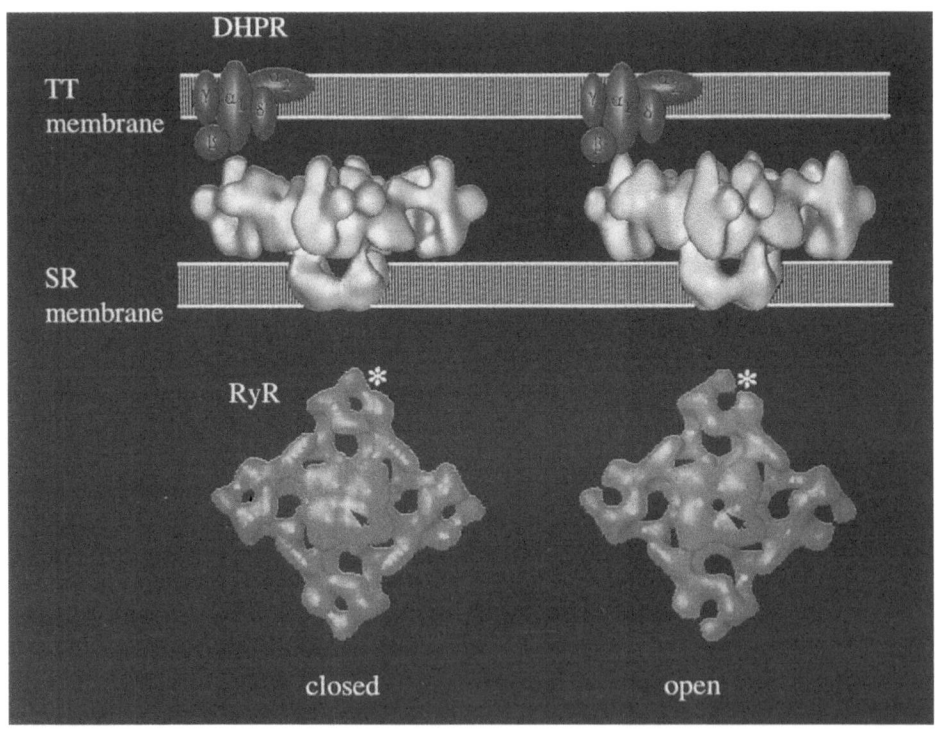

Fig. 12. Structure of the main EC coupling molecules. One of four DHPRs (purple, schematic) and the underlying RyR (from the 3D reconstruction of Orlova et al., 1996) in the reference conformation ("closed", left) or the conformation (believed to be open) reached in the presence of activating Ca^{2+} and ryanodine (3 hours in 100 µM Ca^{2+}, 100 nM ryanodine). Note the increase in height of the RyR in the lateral view. In the view from the SR lumen (bottom), the open conformation features a visible "pore" and modified "clamp" region (*), indicative of long range allosteric coupling within the channel molecule. Redrawn by I. Serysheva (Baylor College, Houston, Texas) from Orlova et al. (1996). Color version on page 450.

The intracellular loop between domains I and II is the binding site for the β subunits. The intracellular loop between II and III participates in coupling between voltage sensor and release channel. Residues 720 to 765 within the II-III loop are critical for the interaction (Nakai et al., 1998a). Oligopeptides homologous to epitopes in the II-III loop can activate or inhibit Ca^{2+} release. "Peptide A", which starts 4 residues into the N-terminal end of the II-III loop, increases the p_o of Ca^{2+} release channels in bilayers and SR vesicles (El-Hayek et al 1995). Its effects are similar to those of Imperatoxin A, a 33-aminoacid scorpion toxin of similar sequence, and have been localized further to a cluster of basic amino acids in the toxin (Gurrola et al., 1999). The homologous portion in II-III is unlikely to work physiologically, however, as $\alpha_1 s$ with this region scrambled or deleted still rescued mdg cells (Proenza et al 2000a), and the putative binding site for Imperatoxin A appears to be located too far from the TT-facing "top" of the release channel (Samsó et al., 1999, and fig. 14) to permit its direct binding with the A epitope in the normal complex. A 20 amino-acid peptide from the C-terminal part of the II-III loop is an antagonist of peptide A (El-Hayek et al., 1995). It is postulated that the homologous DHPR epitope may have a physiologic (in this case basal inhibitory) interaction with the release channel. Even if these regions do not have a functional role, peptides and toxin are interesting tools for functional (González et al., 2000b) or structural studies (Samsó et al., 1999).

The cytoplasmic loop between domains of homology III and IV, may have an auxiliary role in EC coupling. An Arg to His mutation within the loop (Monnier, 1997) occurs in some cases of Malignant Hyperthermia (described below).

A polypeptide homologous to segment 1487 - 1506 in the C terminus, inhibits Ca^{2+} release through the skeletal RyR (Slavik, 1997). The multiplicity of inhibitory segments suggests a basal closing influence of the DHPR on the RyR, which would be relieved upon conformational change of the voltage sensor. Other functional evidences, like the ability of a large depolarization to elicit release in the presence of highly inhibitory $[Mg^{2+}]$ of up to 7 mM (González et al., 2000a), point instead to a direct opening effect of the activated voltage sensor.

The C terminus has important structural roles. It has the signal to target and anchor the $Ca_v 1.1$ subunit into the surface membrane (Proenza et al., 2000b; Flucher et al., 2000). The C terminus appears to have binding sites for calmodulin, the Ca^{2+} release channel, and several other proteins.

As stated earlier, $Ca_v 1.1$ is the target for the calcium antagonist drugs. The binding site for DHPs seems to be localized at the connecting loop between IIIS5 and IIIS6 (Nakayama et al 1991; Yamaguchi et al 2000). Benzothiazepines (such as diltiazem) and phenylalkylamines (verapamil) interact with transmembrane segments IIIS6 and IVS6 (Berjukow et al 1999) as well as the pore regions of the same domains (Hockerman et al 1997).

3.1.2. β subunits

They are ~55-65-kDa proteins coded by five genes, each producing multiple isoforms by alternative splicing. In skeletal muscle the predominant isoform is β_{1a}. β subunits bind with 1:1 stoichiometry (Leung et al., 1987) and high affinity to $Ca_v 1$ at the I - II linker (Pragnell et al., 1994), via a conserved 30-amino acid motif (De Waard et al., 1994).

The β subunit modifies the density, kinetics and voltage dependence of I_{Ca}. All β isoforms increase the expression or insertion of L-type channels into the membrane, while the effects on kinetics and voltage dependence are intrinsic to each isoform.

In a β-knockout mouse the expression of $Ca_v1.1$ was drastically reduced, I_{Ca} densities and charge movement were low, and the Ca^{2+} transient was absent (Beurg et al., 1997). While transfection with $β_1$ cDNA restored the wild type phenotype, $β_{2a}$ restored I_{Ca} and charge movement but only part of the Ca^{2+} transient. To fully restore EC coupling, $β_{1a}$ required an intact C-terminus (Beurg et al., 1999). An isoform-specific C-terminal region is therefore involved in EC coupling.

3.1.3. γ Subunit

The skeletal muscle γ subunit has 223 amino acids, moderately conserved between the human, mouse, rat, and rabbit proteins. Five γ subunits have been cloned from brain (Burgesss et al., 1999). γ subunits favor inactivation of Ca^{2+} channels (Sipos et al., 2000). In skeletal $γ_1$-deficient myotubes the amplitude of peak $(Ca_v1.1)$ I_{Ca} increases, probably a consequence of a slowed down inactivation (Freise et al., 2000). These results suggest that $γ_1$ decreases the amount of Ca^{2+} entry during skeletal muscle activity, an important function for Ca^{2+} homeostasis during prolonged or high frequency activity.

3.1.4. Alpha2-delta

$α_2δ$ was first isolated in skeletal muscle and later identified in other tissues like brain, where it is part of the skeletal and cardiac L-type and neuronal N-type channels. Three genes encode the subunit; $α_2δ-1$, the skeletal muscle gene, produces 5 isoforms (Klugbauer et al., 1999). The 175-kDa protein product is post-translationally cleaved to form disulfide-linked $α_2$ and δ peptides, both of which are heavily glycosylated. A single transmembrane domain in the δ subunit anchors the $α_2δ$ protein to the membrane. $α_2$ is responsible for the modulation functions. Coexpression of $α_2δ-1$ with different $α_1$ subunits facilitates the assembly of channels in the plasma membrane. $α_2δ-1$ affects inactivation of L-type channels expressed in mammalian cells (Shirokov et al., 1998), but its modulatory roles are less clear-cut than those of the β subunit (Angelotti and Hofmann, 1996). An EC coupling role for $α_2δ$ has not been identified.

3.2. The Ryanodine Receptor

The Ca^{2+} release channel forms a supramolecular complex, constituted by the ryanodine receptor and other proteins. Mammalian tissues express three RyR isoforms, encoded by different genes. RyR1 and RyR2 are dominant in skeletal and cardiac muscle, respectively. RyR3 occurs in embryonic skeletal muscle, remaining present only in some adult muscles (especially masseter and diaphragm). In non-mammalian skeletal muscle isoforms α and β are homologous respectively to RyR1 and RyR3, but remain at the same density in the adult (Tarroni et al 1997). While RyR1 is known to interact with the DHPR functionally and structurally, RyR3 is not able to do so. Because the RyR3 homolog β isoform is present at high density in adult muscle of lower vertebrates, it must have an important role. In mammalian muscle, instead, its EC coupling function is not known. It may be minimal, because RyR3-knockout mice show no gross structural or functional changes in skeletal muscle (Takeshima et al., 1996; Clancy et al., 1999).

The difference in prevalence of the two isoforms is one of many interesting differences in EC coupling among taxonomic classes. While poikilotherms have one set of T tubules at every Z disk, the mammalian sarcomere has two sets of T tubules, right at the points where thick and thin myofilaments start to overlap. While the mammal has a ratio of ryanodine to DHP binding sites consistent with the stoichiometric pattern in fig. 4 (2 RyR per 4 DHPR), the ratio in frog muscle is roughly double (more RyRs appear to be devoid of voltage sensors, and belong to the "C" class in the nomenclature of Ríos and Pizarro, 1988). Shirokova et al. (1996) concluded from comparisons of total release, that the flux per unit area of membrane was greater in the frog than in the rat. Moreover, the peak/steady ratio of release flux was greater in the frog, and reached a maximum at intermediate values of voltage (*Time course of Ca^{2+} release*, above). This was taken as indication that the peak of release was due to "C" class channels, activated by CICR. The extra C channels in the frog appear to be RyR3 (or β), located parajunctionally, in additional rows outside the junctional double row (Franzini-Armstrong, personal communication and 2001).

The RyRs consist of four large identical subunits, each of 565 kDa, or ~ 5000 amino acids, which in the native state are associated to four small FK506-binding proteins. The large protomer, whose primary sequence (Takeshima et al., 1989) is schematically represented in fig. 13, consists of a cytoplasmic region (the "foot") and a transmembrane "pore" region (orange segment). The latter comprises the ~ 1000 amino acid COOH-terminus and appears to contain 4 membrane spanning segments, M1 to M4 (red blocks). When this portion is expressed alone, it forms a functional Ca^{2+} channel, with conductance similar to that of the entire molecule, which can be activated but not inhibited by Ca^{2+} and is sensitive to ryanodine and caffeine (Bhat et al 1997; Du and MacLennan 1999).

3.2.1. Three-dimensional Structure of the Release Channel

The tertiary/quaternary structure of the release channel is being reconstructed at increasing resolution by averaging electron microscopic images of multiple receptors, from detergent-solubilized, frozen-hydrated preparations. Fig. 12 (redrawn from Orlova et al., 1996) is a reconstruction of the release channel. A slightly different technique yields a reassuringly similar result (Fig. 14, from Stokes and Wagenknecht, 2000). Although the final aspect of the reconstruction depends in good measure on the

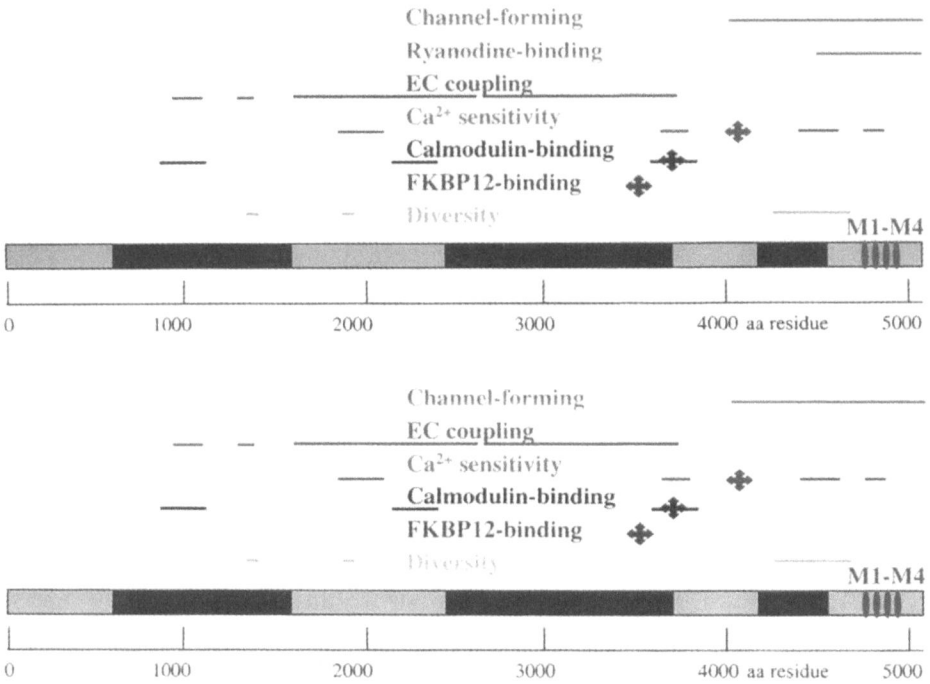

Fig. 13. Primary sequence of the Ca^{2+} release channel. Bar, primary sequence of RyR1, with spans in black depicting regions not present in the IP_3 sequence. M1-M4, transmembrane segments (Takeshima et al., 1989, but see Zorzato et al., 1990, for a different model). Horizontal green lines, areas (D1-D3) of high diversity among isoforms. Cross in purple, FKBP12-binding val-pro epitope (Cameron et al., 1997). Lines in black, identified CAM-binding regions (Menegazzi et al. 1994) with crucial cysteine marked (Moore et al. 1999). Red segments, candidate Ca^{2+}-binding regions, with crucial glutamate marked (references in text). Note that high diversity areas D2 and D3 overlap candidate regions for inhibitory Ca^{2+} binding. In blue, important areas of interaction with the voltage sensor (see text). Orange, channel-forming region (Bhat et al., 1997). Diagram modified from fig. 1 in Leong and MacLennan, 1998. Color version on page 451.

density threshold used to render solid surfaces, the results indicate intricate, loosely packed domains in the cytoplasmic foot. Some help in establishing their correspondence with tracts in the primary sequence comes from the identification of loci of interaction with ligands, visible in the reconstructions and indicated by red circles in fig. 14.

A major advance was the 3D reconstruction in the (presumably open) state induced by exposure to Ca^{2+} and ryanodine (Orlova et al., 1996). In addition to the obvious "pore" visible in fig. 12, the "open" configuration is notable for differences in the foot, including thickening (in the T-SR direction) and the apparent opening of the "clamp" between domains 6, 9 and 10. The clamps are candidate regions of interaction with the DHPRs. The correspondence of these structural changes is consistent with the idea

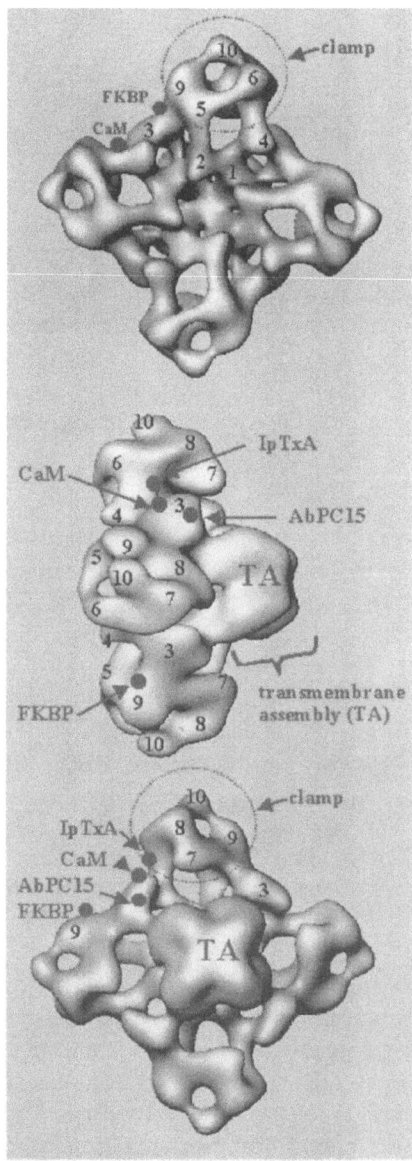

Fig. 14. Interaction sites in the Ca^{2+} release channel. Numbers designate reproducible globular sructures in the foot region. Red circles mark locations determined from 3D reconstructions of RyR complexes with the ligands: imperatoxin A (IpTxA), FKBP12, calmodulin (CaM) and a monoclonal antibody to epitope 4425-4621 (AbPC15). Reproduced from Stokes and Wagenknecht (2000). Color version on page 452.

that movements in one or more DHPRs favor RyR opening via long range allosteric effects (fig. 8).

3.2.2. The RyR and Ca^{2+}

All RyR isoforms are activated by Ca^{2+}. Candidate regions of interaction were identified (Fill et al., 1991, Chen et al., 1992; Treves et al., 1993). A putative sensor for activation is located in the pore region, comprising an essential glutamate at position 4032 for rabbit RyR1 in each protomer (Chen et al 1998; red cross in fig. 13).

Functional differences between RyR1 and RyR2 made it possible to use a chimerization approach, similar to the one applied with DHPRs, to locate the structural substrate of isoform-specific functional properties. While for the skeletal channel the half-inhibition $[Ca^{2+}]$ is near 1 mM, for RyR2 it is much higher. Even though this susceptibility might depend on degrees of thiol oxidation, the search for the site(s) involved has focused on regions where both channels diverge (Hayek et al., 2000; Bhat et al., 1997; Nakai et al., 1998b; Du and MacLennan, 1999).

3.2.3. Interactions with the DHPR

When expressed in dyspedic muscle cells, RyR2 supports Ca^{2+} release by "cardiac" activation, that is, induced by Ca^{2+} entry, but not release directly controlled by voltage. RyR1 is believed to interact mechanically with the DHPR. This contact, proposed first on a functional basis (Schneider & Chandler, 1973), was supported by observations of structural alignment between particles (Block et al., 1988), co-immunoprecipitation (Marty et al., 1994) and cross-linking (Murray & Ohlendieck, 1997). Functional observations of effects of the release channel on the voltage sensor, interpreted as the Newtonian "reaction" in a mechanical contact, strengthened the conclusion (see *Mechanical or allosteric control*, above).

In chimeras of RyR1 and RyR2, the 1000 aa region between 1635 and 2636 was important for both the forward (from DHPR to RyR) and the reverse interaction, while the next 1000 residues were able to promote I_{Ca} but did not support forward release control (Nakai et al, 1998b). Other segments could also interact with the DHPR (Yamazawa et al., 1997; Leong and MacLennan, 1998).

Important regions are suggested by naturally occurring mutations associated with inheritable diseases. Malignant hyperthermia susceptibility (MHS) features episodes of hyperthermia and muscle rigidity (by failure to terminate Ca^{2+} release) on exposure to volatile anaesthetics and relaxants. Central core disease (CCD) is a myopathy showing MH-like response to relaxants, diffuse muscular weakness, and central fiter areas (cores) of altered structure. Most mutations associated with MHS and/or CCD, are located in the foot. Mutation I4898T causes the most severe form of CCD. It is unique for being in the transmembrane domain, within the linker between segments M3 and M4 (Lynch et al., 1999).

The best *in-vitro* reproduction of these dominantly inherited diseases is achieved by co-expression of mutants with wild type RyR1. These presumably heterotetrameric channels exhibit a greater sensitivity to activation by Ca^{2+}, and cause the SR Ca^{2+} content to decrease, resting $[Ca^{2+}]_i$ to increase, and peak release flux to decrease. SR depletion probably accounts for the associated weakness, while the increased $[Ca^{2+}]_i$ may cause the structural alterations, by promoting proteolysis. Based on the conservation of Ile-4898 among RyRs, the conservation of the surrounding motif GGI(or V)G among intracellular channels, and its analogy with VGYG in the P loop of the prototypical K channel, Balshaw et al. (1999) speculated that the lumenal loop between M3 and M4 folds into the membrane to form the pore lining. A similar conclusion was reached by Zhao et al. (1999) upon showing that the mutation G4824A in mouse RyR2 (G4824 corresponds to G4895 in RyR1) drastically reduced single channel conductance

A hint of the complexity of interactions and behavior of the RyR is that MHS is also linked to a mutation in the III-IV loop of $Ca_v1.1$. The hypothesis that this loop could be involved in interactions that close the RyR1 (Leong and MacLennan, 1998) is supported by indications that the open state of the release channels is not just triggered, but to a degree maintained by the voltage sensors (Suda, 1995; Lacampagne et al., 2000; fig. 9D).

In sum, mechanical interactions between voltage sensor and Ca channel appear to be multiple, but subtle and dynamic, affecting allosterically, back and forth, the functional state of both proteins. Because there is no evidence of strong bonding, it follows that the mutual positioning of DHPRs and RyR must be maintained by their interaction with other molecules.

3.2.4. Interaction with Calmodulin

Ca^{2+}-free calmodulin (CAM) is an activator of the release channel, while Ca:CAM is inhibitory. Three possible CAM-binding regions were identified in RyR1 (Menegazzi et al., 1994; fig. 13), but one of those sites appears most important. A peptide homologous to epitope 3614 - 3643 binds CAM and Ca:CAM with high affinity. The binding site appears to shift in the N-terminal direction upon binding of Ca^{2+} to CAM (Rodney et al., 2000). Sulfhydryl oxidation of the RyR produces intersubunit disulfide bonds and prevents interactions of the RyR with CAM (Zhang et al., 1999). Cys-3635 appears critical to these interactions (Moore et al., 1999). It is believed that the critical thiols are involved in mutually exclusive interactions, either with CAM or with the adjacent subunit. Whether this dynamic role of thiols is related or not to their involvement on inhibition by Ca^{2+} and Mg^{2+} (see *Reactive sulfhydryls as modulation sites*, below) remains a worthwhile question. Fig. 14 shows the location of bound CAM in the 3D reconstruction (Wagenknecht et al., 1997). Note (especially in the bottom view) that it is on domain 3, at or near the interface between two subunits.

3.2.5. Interaction with FKBP12

Up to four copies of a small immunophilin, FKBP12 in skeletal and FKBP12.6 in cardiac muscle, are integral components of RyRs. These proteins bind the immunosupressants FK506 and rapamycin, and have peptidyl-prolyl cis-trans isomerase --rotamase-- action (Marks, 1996). FKBP12 also binds to IP_3 receptors, at a leucyl-prolyl dipeptide epitope, which structurally resembles FK506 (Cameron et al., 1997). By homology, the binding site on RyR1 has been located at the val-2462 pro-

2463 epitope (Cameron et al., 1997). Peptide substrates of the enzymatic action also require these dipeptide epitopes, but rotamase activity is not involved in the effects on channels (Timerman et al., 1995). In skinned skeletal muscle, FK506 or rapamycin, which presumably displace FKBP12, potentiated Ca^{2+} release, and eventually decoupled it from depolarization of the resealed T tubules (Lamb and Stephenson, 1996). In cardiac muscle, the immunosupressants caused spark duration to increase (Xiao et al., 1997). On channels reconstituted in bilayers, the immunosupressants increased p_o and induced subconductance openings, while FKBPs reduced p_o and the frequency of substates (Brillantes et al 1994 ; Barg et al 1997), which suggests their involvement in inter-protomer interactions allowing subunits to work cooperatively. Pairs of RyR channels were shown to exhibit coupled gating in bilayers, provided that FKBP12 was present (Marx et al., 1998). This suggests participation of the immunophilin in inter-receptor interactions. Accordingly, in a 3D reconstruction of frozen receptors (fig. 14) FKBP12 was near the region where feet would contact one another. Allosteric multi-receptor interactions could provide a mechanism for the rapid start and termination of multi-channel sparks. The observation of coupled gating in bilayers, however, has not been independently confirmed.

4. MODULATION AND MODIFICATION OF EC COUPLING

In skeletal muscle, EC coupling is viewed as a mechanism operating always at optimal speed and effectiveness, therefore not subject to much modulation. This concept is being changed by mounting evidence of redox-related modifications in the RyR. Such changes can be demonstrated in vitro, upon applying thiol reductive or oxidative interventions. Even though the physiological significance of such effects remains unknown, analogous changes may occur upon strenuous exercise, and could be part of the events of muscle fatigue. It is not known whether these changes represent an adaptive, dynamic modulatory response, or simply consist of unwanted modifications imposed on the system by the redox stress.

4.1. Reactive Sulfhydryls as Modulation Sites

The redox state of the channel protein has a marked effect on the activity of RyR channels from skeletal and cardiac muscle. Oxidation of thiol groups induces the release of Ca^{2+} from SR vesicles (Trimm et al., 1986) and activates RyR channels in bilayers (Abramson et al., 1995). Reactive thiols participate in intra-channel interactions and interactions with CAM (Porter Moore et al., 1999; Moore et al., 1999). The progressive oxidation of SH groups appears to switch the channels along a sequence of three functional modes, one with low p_o, one, named SM (skeletal muscle), which is activated by micromolar and inhibited by mM Ca^{2+}, and a third, named C (cardiac), which has somewhat higher maximal p_o and is not inhibited by Ca^{2+} (Marengo et al., 1998). Thiol oxidation changes the modulatory properties of the RyR in many ways, including suppression of the inhibitory effect of Mg^{2+} (Donoso et al., 2000). These results support the interesting idea that the details of Ca^{2+}

sensitivity and other functional properties depend less on isoform-specific features than on the redox state of the molecule.

The molecule reaction to changes in redox potential may involve the ubiquitous messenger nitric oxide, through nitrosylation of thiols. Indeed, inhibition of nitric oxide synthase (NOS) affects contractility (Kobzik et al., 1994) and RyR function (Meszaros et al., 1996). S-nitrosylation activates reversibly the channel at low $[Ca^{2+}]$(while oxidation, forming disulfides, does not) and inhibits it at $[Ca^{2+}]>100$ μM. According to Eu et al. (2000) an "O2 sensor", comprising 6-8 cysteines, determines whether a critical cysteine --possibly outside the sensor— is nitrosylated, thus providing a mechanism through which the redox potential would control dynamically the release channel.

As discussed earlier, calmodulin regulation of the RyR and cysteine oxidation are mutually interfering. CAM protects a thiol in RyR1 from oxidation, while oxidation of RyR alters CAM binding. A similar antagonism appears to exist between S-nitrosylation and inhibition by CAM (Eu et al. 2000). Such is just one instance of redox-related metabolic controls present in many cells.

4.2. ΓC Coupling Changes in Muscle Fatigue

The subject of muscle fatigue is extremely complex, with events at cell, tissue, and system levels, in addition to CNS implications (Fitts, 1994). It has been known for many years that the application of caffeine, which in muscle stimulates Ca^{2+} release, is capable of reversing the effects of fatigue (Eberstein and Sandow, 1976). Although this has long been taken as an indication of an involvement of EC coupling in fatigue, the effect of caffeine could simply be compensating a contractile deficit.

It is now clear, however, that Ca^{2+} release is diminished upon fatigue (Westerblad and Allen, 1993). This reduction in Ca^{2+} release is not paralleled by a reduction in movement of voltage sensor charge (Gyorke, 1993), which rules out a primary voltage sensor "fatiguing" mechanism.

Other possible explanations include a failure of the transmission between voltage sensor and Ca^{2+} channel, a reduction in SR Ca^{2+} load, or a change in the gating properties of the release channel itself, affecting for instance the inactivation process that terminates release after its peak.

Two effects of sustained activity are likely candidates to explain the reduction in Ca^{2+} release. One is the decline in ATP, a late consequence of strenuous exercise, which in turn may reduce the release channel's p_o independently of other modulators (Westerblad et al., 1998). Equally important may be the increase in inorganic phosphate (Fryer et al. 1995; Westerblad and Allen, 1996), which may rise to 40 mM in the active muscle cell (23). Inside the SR, Pi will form an insoluble complex with Ca^{2+} (Fryer et al., 1997) which both reduces free $[Ca^{2+}]_{SR}$ and draws additional Pi into the SR.

Other factors could have an effect: as Mg^{2+} is displaced from parvalbumin its free concentration will increase, inhibiting the RyR. Decrease in pH is a known inhibitor of release channels, but may only be a minor factor in fatigue. Local sustained increase in $[Ca^{2+}]_i$ could have an inhibitory effect, directly on the RyR, or indirectly and irreversibly, through the activation of proteases. A detailed consideration of possibilities can be found in Favero (1999).

5. ACKNOWLEDGMENTS

We are grateful to Steve Baylor, Knox Chandler, Clara Franzini-Armstrong, Susan Hamilton, Gerhard Meissner, Martin Schneider, Irina Serysheva, Natalia Shirokova and Terence Wagenknecht, for supplying illustrations used in this chapter. We thank Cecilia Hidalgo for detailed consultations and Clara Franzini-Armstrong for sharing unpublished data. We were supported by grants from the National Institutes of Health, USA.

Department of Molecular Biophysics and Physiology
Rush University
Chicago, IL 60612

6. REFERENCES

Abramson JJ, Zable AC, Favero TG, Salama G. Thimerosal interacts with the Ca^{2+} release channel ryanodine receptor from skeletal muscle sarcoplasmic reticulum. J Biol Chem. 1995 Dec 15;270(50):29644-7.

Adams, BA, T Tanabe, A Mikami, S Numa and KG Beam. 1990. Intramembrane charge movement restored in dysgenic skeletal muscle by injection of dihydropyridine receptor cDNAs. Nature 346:569-572.

Alderton JM, Steinhardt RA 2000   Calcium influx through calcium leak channels is responsible for the elevated levels of calcium-dependent proteolysis in dystrophic myotubes. J Biol Chem 275:9452-60

Almers W.   Gating currents and charge movements in excitable membranes. Rev Physiol Biochem Pharmacol. 1978;82:96-190.

Angelotti, T. and F. Hofmann (1996). "Tissue-specific expression of splice variants of the mouse voltage-gated calcium channel alpha2/delta subunit." FEBS Lett 397(2-3): 331-7.

Barg S, Copello JA, Fleischer S.   Different interactions of cardiac and skeletal muscle ryanodine receptors with FK-506 binding protein isoforms. Am J Physiol. 1997 May;272(5 Pt 1):C1726-33.

Baylor, SM and S. Hollingworth. 2000. Measurement and interpretation of cytoplasmic $[Ca^{2+}]$ signals from calcium-indicator dyes. News Physiol. Sci. 15:19-26.

Beam KG, Knudson CM, Powell JA. 1986. A lethal mutation in mice eliminates the slow calcium current in skeletal muscle cells. Nature. 320:168-70.

Berjukow, S., F. Gapp, S. Aczel, M.J. Sinnegger, J. Mitterdorfer, H. Glossmann, S. Hering (1999). Sequence differences between alpha1C and alpha1S Ca2+ channel subunits reveal structural determinants of a guarded and modulated benzothiazepine receptor. J Biol Chem 274: 6154-60.

Beurg M, Ahern CA, Vallejo P, Conklin MW, Powers PA, Gregg RG, Coronado R. 1999. Involvement of the carboxy-terminus region of the dihydropyridine receptor beta1a subunit in excitation-contraction coupling of skeletal muscle. Biophys J.77:2953-67.

Beurg M, Sukhareva M, Strube C, Powers PA, Gregg RG, Coronado R. 1997. Recovery of Ca^{2+} current, charge movements, and Ca^{2+} transients in myotubes deficient in dihydropyridine receptor beta 1 subunit transfected with beta 1 cDNA. Biophys J.73:807-18.

Bezanilla F. The voltage sensor in voltage-dependent ion channels. Physiol Rev. 2000 Apr;80(2):555-92.

Bezanilla, F, RE Taylor and JM Fernández. 1982. Distribution and kinetics of membrane polarization. I. Longterm inactivation of gating currents. J. Gen. Physiol. 79:21-40.

Bhat MB, Zhao J, Takeshima H, Ma J. Functional calcium release channel formed by the carboxyl-terminal portion of ryanodine receptor. Biophys J. 1997 Sep;73(3):1329-36.

M P. Blaustein and W. J Lederer Physiological Reviews, 79 763-854.1999. Sodium/Calcium Exchange: Its Physiological Implications.

Brandl CJ, deLeon S, Martin DR, MacLennan DH. 1987. Adult forms of the Ca^{2+}ATPase of sarcoplasmic reticulum. Expression in developing skeletal muscle. J Biol Chem.262:3768-74.

Bridge JH, Ershler PR, Cannell MB. Properties of Ca^{2+} sparks evoked by action potentials in mouse ventricular myocytes. J Physiol. 1999 Jul 15;518 (Pt 2):469-78.

Brillantes AB, Ondrias K, Scott A, Kobrinsky E, Ondriasova E, Moschella MC, Jayaraman T, Landers M, Ehrlich BE, Marks AR. 1994. Stabilization of calcium release channel (ryanodine receptor) function by FK506-binding protein. Cell. 77:513-23.

Brum G, Gonzalez A, Rengifo J, Shirokova N, Rios E. Fast imaging in two dimensions resolves extensive sources of Ca^{2+} sparks in frog skeletal muscle. J Physiol. 2000 Nov 1;528(Pt 3):419-433.

Brum G, Fitts R, Pizarro G, Rios E. Voltage sensors of the frog skeletal muscle membrane require calcium to function in excitation-contraction coupling. J Physiol. 1988 Apr;398:475-505.

Brum G, Rios E. Intramembrane charge movement in frog skeletal muscle fibres. Properties of charge 2. J Physiol. 1987 Jun;387:489-517.

Burgess, D. L., C. F. Davis, L.A. Gefrides, J.L. Noebels (1999). Identification of three novel Ca^{2+} channelgamma subunit genes reveals molecular diversification by tandem and chromosome duplication. Genome Res 9(12): 1204-13.

Caswell AH, Brandt NR, Brunschwig JP, Purkerson S. 1991. Localization and partial characterization of the oligomeric disulfide-linked molecular weight 95,000 protein (triadin) which binds the ryanodine and dihydropyridine receptors in skeletal muscle triadic vesicles. Biochemistry 30:7507-13.

Cameron AM, Nucifora FC Jr, Fung ET, Livingston DJ, Aldape RA, Ross CA, Snyder SH. 1997. FKBP12 binds the inositol 1,4,5-trisphosphate receptor at leucine-proline (1400-1401) and anchors calcineurin to this FK506-like domain. J Biol Chem272:27582-8.

Chandler WK, Rakowski RF, Schneider MF. Effects of glycerol treatment and maintained depolarization on charge movement in skeletal muscle. J Physiol. 1976 Jan;254(2):285-316.

Chen SR, Ebisawa K, Li X, Zhang L. Molecular identification of the ryanodine receptor Ca^{2+} sensor. J Biol Chem. 1998 Jun 12;273(24):14675-8.

Chen SRW, Li X, Ebisawa K, Zhang L. 1997. Functional characterization of the recombinant type 3 Ca^{2+} release channel (ryanodine receptor) expressed in HEK293 cells. J Biol Chem. 272:24234-46.

Chen W, Steenbergen C, Levy LA, Vance J, London RE, Murphy E. 1996. Measurement of free Ca^{2+} in sarcoplasmic reticulum in perfused rabbit heart loaded with 1,2-bis(2-amino-5,6 –difluorophenoxy)ethane-N,N,N',N'-tetraacetic acid by 19F NMR. J Biol Chem. 271:7398-403.

Cheng H, Lederer WJ, Cannell MB. Calcium sparks: elementary events underlying excitation-contraction coupling in heart muscle. Science. 1993 Oct 29;262(5134):740-4.

Chien, A. J., Zhao, X. L., Shirokov, R. E., Puri, T. S., Chang, C. F., Sun, D., Rios, E. & Hosey, M. M. (1995). Roles of a membrane-localized subunit in the formation and targeting of functional L-type Ca^{2+} channels. Journal of Biological Chemistry 270, 30036-30044.

Cifuentes F, Vergara J, Hidalgo C. Sodium/calcium exchange in amphibian skeletal muscle fibers and isolated transverse tubules. Am J Physiol Cell Physiol 2000 Jul;279(1):C89-97.

Clancy JS, Takeshima H, Hamilton SL, Reid MB. 1999. Contractile function is unaltered in diaphragm from mice lacking calcium release channel isoform 3. Am J Physiol 277:R1205-9

Conklin MW, Ahern CA, Vallejo P, Sorrentino V, Takeshima H, Coronado R. Comparison of Ca^{2+} sparks produced independently by two ryanodine receptor isoforms (type 1 or type 3). Biophys J. 2000 Apr;78(4):1777-85.

Conklin MW, Barone V, Sorrentino V, Coronado R. Contribution of ryanodine receptor type 3 to Ca^{2+} sparks in embryonic mouse skeletal muscle. Biophys J. 1999 Sep;77(3):1394-403.

De Waard M, Pragnell M, Campbell KP. 1994. Ca^{2+} channel regulation by a conserved beta subunit domain. Neuron. 13:495-503.

Donoso P, Aracena P, Hidalgo C. 2000. Sulfhydryl oxidation overrides Mg^{2+} inhibition of calcium-induced calcium release in skeletal muscle triads. Biophys J 79:279-86

Du, G. G. and D. H. MacLennan (1999). "Ca^{2+} inactivation sites are located in the COOH-terminal quarter of recombinant rabbit skeletal muscle Ca^{2+} release channels (ryanodine receptors)." J Biol Chem 274(37): 26120-6.

Ebashi S, Endo M. 1968. Calcium ion and muscle contraction. Prog Biophys Mol Biol.; 18:123-83.

Eberstein, A., and A. Sandow. 1976. Fatigue mechanisms in muscle fibers. In: The Effects of Use and Disuse in Neuromuscular Function, edited by E.Guttman, and P. Hnik. Amsterdam: Elsevier, 1976, p. 515-526.

El-Hayek, R., B. Antoniu, J.Wang, S.L. Hamilton, N. Ikemoto (1995). Identification of calcium release-triggering and blocking regions of the II-III loop of the skeletal muscle dihydropyridine receptor. J Biol Chem 270(38): 22116-8.

Endo, M, M Tanaka and Y Ogawa. 1970. Calcium-induced release of calcium from the sarcoplasmic reticulum of skinned skeletal muscle fibers. Nature 225:34-36

Eu, JP, J Sun, L Xu, JS Stamler and G Meissner. 2000. The skeletal muscle calcium release channel: coupled O_2 sensor and NO signaling functions. Cell 102:499-509.

Favero, TG. Sarcoplasmic reticulum Ca^{2+} release and muscle fatigue. 1999. J. App. Physiol. 87: 471-483.

Fessenden, J. D., Y. Wang, R.A. Moore, S.R. Chen, P.D. Allen, I..N. Pessah (2000). "Divergent Functional Properties of Ryanodine Receptor Types 1 and 3 Expressed in a Myogenic Cell Line." Biophys J 79(5): 2509-2525.

Fitts, RH. 1994. Cellular mechanisms of muscle fatigue. Physiol. Rev. 74:49-94.

Flucher, B. E., N. Kasielke, M. Grabner (2000). The triad targeting signal of the skeletal muscle calcium channel is localized in the COOH terminus of the alpha(1S) subunit. J Cell Biol 151: 467-78.

Franzini-Armstrong C. The sarcoplasmic reticulum and the control of muscle contraction. FASEB J. 1999 Dec;13 Suppl 2:S266-70.

Franzini-Armstrong C, Jorgensen AO. Structure and development of E-C coupling units in skeletal muscle. Annu Rev Physiol. 1994;56:509-34.

Freise, D., B. Held, U. Wissenbach, A. Pfeifer, C. Trost, N. Himmerkus, U. Schweig, M. Freichel, M. Biel, F. Hofmann, M. Hoth, V. Flockerzi (2000). Absence of the gamma subunit of the skeletal muscle dihydropyridine receptor increases L-type Ca^{2+} currents and alters channel inactivation properties. J Biol Chem 275(19): 14476-81.

Fryer, M. W., V. J. Owen, G. D. Lamb, and D. G. Stephenson. 1995. Effects of creatine phosphate and Pi on force and development of intracellular calcium movements in rat skinned skeletal muscle fibers. J. Physiol. 416: 435-454.

Fryer, M. W., J. M. West, and D. G. Stephenson. 1997. Phosphate transport into the sarcoplsmic reticulum of skinned fibres from rat skeletal muscle. J. Muscle Res. Cell Motil. 18: 161-167,

Gao, T., M. Bunemann, B.L. Gerhardstein, H. Ma, M.M. Hosey (2000). Role of the C terminus of the alpha 1C (CaV1.2) subunit in membrane targeting of cardiac L-type calcium channels. J Biol Chem 275: 25436-44.

Gao L, Balshaw D, Xu L, Tripathy A, Xin C, Meissner G. 2000. Evidence for a role of the lumenal M3-M4 loop in skeletal muscle Ca^{2+} release channel (ryanodine receptor) activity and conductance. Biophys J. 79:828-40.

Gerday, Ch. Soluble Ca-binding proteins from fish and invertebrate muscle. Molecular Physiology 2: 63-87, 1982.

Gillis, J.M. The biological significance of muscle parvalbumins, in Calcium-Binding Proteins: Structure And Function. F.L. Siegel, E Carafoli, RH Kretsinger, DH MacLennan and RH Wasserman, eds. Elsevier North Holland, 1980, pp 309-311.

Gonzalez A, Kirsch WG, Shirokova N, Pizarro G, Stern MD, Rios E. 2000a. The spark and its ember: separately gated local components of Ca^{2+} release in skeletal muscle. J Gen Physiol.115:139-58.

Gonzalez A, Kirsch WG, Shirokova N, Pizarro G, Brum G, Pessah IN, Stern MD, Cheng H, Rios E. 2000b. Involvement of multiple intracellular release channels in calcium sparks of skeletal muscle. Proc Natl Acad Sci U S A.97:4380-5.

Gonzalez, A, G Pizarro, N Shirokova, WG Kirsch, E Rios and MD Stern. 2000c. Local voltage-activated Ca release in frog skeletal muscle cells containing BAPTA. Biophys. J. 78:437A.

Gonzalez A, Rios E. Perchlorate enhances transmission in skeletal muscle excitation-contraction coupling. J Gen Physiol. 1993 Sep;102(3):373-421.

Green DR and Reed JC. 1998. Mitochondria and apoptosis. Science 281:1309-1312.

Gurrola GB, Arevalo C, Sreekumar R, Lokuta AJ, Walker JW, Valdivia HH. 1999. Activation of ryanodine receptors by imperatoxin A and a peptide segment of the II-III loop of the dihydropyridine receptor.J Biol Chem 274:7879-86.

Gyorke, S. 1993 Effects of repeated tetanic stimulation on excitation-contraction coupling in cut muscle Fibres of the frog. J. Physiol. 464: 699-710

Hasselbach, W. 1964. Relaxing factor and the relaxation of muscle. Prog. Biophys. Molec. Biol. 14:167-222.

Hayek SM, Zhu X, Bhat MB, Zhao J, Takeshima H, Valdivia HH, Ma J. 2000. Characterization of a calcium-regulation domain of the skeletal-muscle ryanodine receptor. Biochem J. 351: 57-65.

Heilbrunn, LV and FJ Wiercinski. 1947. The action of various cations on muscle protoplasm. Cell. Comp. Physiol. 29:15-32.

Hill, AV. 1948. On the time required for diffusion and its relation to processes in muscle. Proc. Roy. Soc. Ser. B 135:446-453.

Hirota A, Chandler WK, Southwick PL, Waggoner AS. 1989. Calcium signals recorded from two new purpurate indicators inside frog cut twitch fibers. J Gen Physiol. 94:597-631.

Hockerman, G. H., B. D. Johnson, M.R. Abbott, T. Scheuer, W.A. Catterall (1997). Molecular determinants of high affinity phenylalkylamine block of L-type calcium channels in transmembrane segment IIIS6 and the pore region of the alpha1 subunit. J Biol Chem 272(30): 18759-65.

Hodgkin, AL and P Horowicz. 1960. Potassium contractures in single muscle fibersl. J. Physiol. 153:386-403.

Hollingworth S, Zhao M, Baylor SM. 1996. The amplitude and time course of the myoplasmic free $[Ca^{2+}]$ transient in fast-twitch fibers of mouse muscle. J Gen Physiol.108:455-69.

Hoshi T, Zagotta WN, Aldrich RW (1990) Biophysical and molecular mechanisms of *Shaker* potassium channel inactivation Science 250:533-538.

Hoshi, T, WN Zagotta and RW Aldrich. 1991. Two types of inactivation in Shaker K channels: effects of alterations in the carboxy-terminal region. Neuron 7:548-556.

Howlett, SE, Zhu, J.-Q. and GR Ferrier. 1998. ontribution of a voltage-sensitive calcium release mechanism to contraction in cardiac ventricular myocytes. Am. J. Physiol. 274: H155-170.

Huxley, AF and RE Taylor. 1958. Local activation of striated muscle fibers. J. Physiol. 144:426-441.

Ikemoto, N, B Antoniu, JJ Kang, LG Meszaros and M Ronjat. Intravesicular calcium transient during calcium release from sarcoplasmic reticulum. Biochem. 30:5230-5237, 1991.

Inesi G.1994. Teaching active transport at the turn of the twenty-first century: recent discoveries and conceptual changes.Biophys J. 66:554-60.

Jiang YH, Klein MG, Schneider MF. Numerical simulation of Ca^{2+} "Sparks" in skeletal muscle. Biophys J. 1999 Nov;77(5):2333-57.

Jones LR, Zhang L, Sanborn K, Jorgensen AO, Kelley J. 1995. Purification, primary structure, and immunological characterization of the 26-kDa calsequestrin binding protein (junctin) from cardiac junctional sarcoplasmic reticulum. J Biol Chem. 270:30787-96.

Kamp F, Donoso P, Hidalgo C.1998. Changes in luminal pH caused by calcium release in sarcoplasmic reticulum vesicles. Biophys J. 74:290-6.

Kepplinger, K. J., H. Kahr, G. Forstner, M. Sonnleitner, H. Schindler, T. Schmidt, K. Groschner, N.M. Soldatov, C. Romanin (2000). A sequence in the carboxy-terminus of the alpha(1C) subunit important for targeting, conductance and open probability of L-type Ca^{2+} channels. FEBS Lett 477: 161-9.

Klein MG, Lacampagne A, Schneider MF. Voltage dependence of the pattern and frequency of discrete Ca^{2+} release events after brief repriming in frog skeletal muscle. Proc Natl Acad Sci U S A. 1997 Sep 30;94(20):11061-6.

Klein MG, Cheng H, Santana LF, Jiang YH, Lederer WJ, Schneider MF. Two mechanisms of quantized calcium release in skeletal muscle. Nature. 1996 Feb 1;379(6564):455-8.

Klugbauer, N., S. Dai,V. Specht, L. Lacinova, E. Marais, G. Bohn, F. Hofmann (2000). A family of gamma-like calcium channel subunits. FEBS Lett 470(2): 189-97.

Klugbauer, N., L. Lacinova, E.Marais, M. Hobom, F. Hofmann (1999). Molecular diversity of the calcium channel alpha2delta subunit. J Neurosci 19(2): 684-91.

Knudson, CM, Stang, KK, Jorgensen, AO and Campbell, KP 1993. Biochemical characterization of ultrastructural localization of a major junctional sarcoplasmic reticulum glycoprotein (triadin). J. Biol. Chem. 268:12637-12645.

Kobzik L, Reid MB, Bredt DS, Stamler JS. Nitric oxide in skeletal muscle. Nature. 1994 Dec 8;372(6506):546-8.

Lacampagne A, Klein MG, Ward CW, Schneider MF. Two mechanisms for termination of individual Ca^{2+} sparks in skeletal muscle. Proc Natl Acad Sci U S A. 2000 Jul 5;97(14):7823-8.

Lacampagne A, Ward CW, Klein MG, Schneider MF. Time course of individual Ca^{2+} sparks in frog Skeletal muscle recorded at high time resolution. J Gen Physiol. 1999 Feb;113(2):187-98.

Lamb GD, Stephenson DG. Effects of FK506 and rapamycin on excitation-contraction coupling in skeletal muscle fibres of the rat. J Physiol. 1996 Jul 15;494 (Pt 2):569-76.

Leong P, MacLennan DH. 1998. Complex interactions between skeletal muscle ryanodine receptor and dihydropyridine receptor proteins. Biochem Cell Biol.76:681-94.

Leung, A. T., T. Imagawa, K.P. Campbell (1987). Structural characterization of the 1,4-dihydropyridine receptor of the voltage-dependent Ca^{2+} channel from rabbit skeletal muscle. Evidence for two distinct high molecular weight subunits. J Biol Chem 262: 7943-6.

Lipp P, Niggli E. Submicroscopic calcium signals as fundamental events of excitation--contraction coupling in guinea-pig cardiac myocytes. J Physiol. 1996 Apr 1;492 (Pt 1):31-8.

Lynch PJ, Tong J, Lehane M, Mallet A, Giblin L, Heffron JJ, Vaughan P, Zafra G, MacLennan DH, McCarthy TV. 1999. A mutation in the transmembrane/luminal domain of the ryanodine receptor is associated with abnormal Ca^{2+} release channel function and severe central core disease. Proc Natl Acad Sci U S A 96:4164-9.

Ma J, Anderson K, Shirokov R, Levis R, Gonzalez A, Karhanek M, Hosey MM, Meissner G, Rios E. Effects of perchlorate on the molecules of excitation-contraction coupling of skeletal and cardiac muscle. J Gen Physiol. 1993 Sep;102(3):423-48.

MacLennan DH, Wong PT. 1971. Isolation of a calcium-sequestering protein from sarcoplasmic reticulum. Proc Natl Acad Sci U S A. 68:1231-5.

Marengo JJ, C Hidalgo and R Bull. 1998. Sulfhydryl Oxidation Modifies the Calcium Dependence of Ryanodine-Sensitive Calcium Channels of Excitable Cells. Biophys J 74 1263-1277.

Marks AR. 1996. Cellular functions of immunophilins. Physiol Rev 76:631-49.

Marx SO, Ondrias K, Marks AR. 1998. Coupled gating between individual skeletal muscle Ca^{2+} release channels. Science. 281:818-21.

Marty I, Robert M, Villaz M, De Jongh K, Lai Y, Catterall WA, Ronjat M. 1994 Biochemical evidence for a complex involving dihydropyridine receptor and ryanodine receptor in triad junctions of skeletal muscle. Proc Natl Acad Sci U S A 91:2270-4.

Meissner G. 1994. Ryanodine receptor/Ca^{2+} release channels and their regulation by endogenous effectors. Annu Rev Physiol. 56:485-508.

Meissner G. 1975. Isolation and characterization of two types of sarcoplasmic reticulum vesicles. Biochim. Biophys. Acta 389:51-68

Mejia-Alvarez R, Kettlun C, Rios E, Stern M, Fill M. Unitary Ca^{2+} current through cardiac ryanodine receptor channels under quasi-physiological ionic conditions. J Gen Physiol. 1999 Feb;113(2):177-86.

Melzer W, Rios E, Schneider MF. 1987. A general procedure for determining the rate of calcium release from the sarcoplasmic reticulum in skeletal muscle fibers. Biophys J. 51:849-63.

Menegazzi P, Larini F, Treves S, Guerrini R, Quadroni M, Zorzato F. Identification and characterization of three calmodulin binding sites of the skeletal muscle ryanodine receptor. Biochemistry. 1994 Aug 9;33(31):9078-84.

Meszaros LG, Minarovic I, Zahradnikova A. Inhibition of the skeletal muscle ryanodine receptor calcium release channel by nitric oxide. FEBS Lett. 1996 Feb 12;380(1-2):49-52.

Mobley, BA and B Eisenberg. 1975. Sizes of components in frog skeletal muscle measured by methods of stereology. J. Gen. Phsiol. 66:31-45.

Monnier, N., V. Procaccio, P. Stieglitz, J. Lunardi. 1997. "Malignant-hyperthermia susceptibility is associated with a mutation of the alpha 1-subunit of the human dihydropyridine-sensitive L-type voltage-dependent calcium-channel receptor in skeletal muscle [see comments]." Am J Hum Genet 60: 1316-25.

Monod, J, J Wyman and J-P Changeux. 1965. On the nature of allosteric transitions: a plausible model. J. Molec. Biol. 12:88-118.

Moore CP, Rodney G, Zhang JZ, Santacruz-Toloza L, Strasburg G, Hamilton SL. 1999 Apocalmodulin and Ca^{2+} calmodulin bind to the same region on the skeletal muscle Ca^{2+} release channel. Biochemistry. 38:8532-7.

Murayama, T, T Oba, E Katayama, H Oyamada, K Oguchi, M Kobayashi, K Otsuka and Y Ogawa. 1999. Further characterization of the type 3 ryanodine receptor (RyR3) purified from rabbit diaphragm. J. Biol. Chem. 274: 17297-17308.

Murray BE, Ohlendieck K. 1997. Cross-linking analysis of the ryanodine receptor and alpha1-dihydropyridine receptor in rabbit skeletal muscle triads. Biochem J. 324:689-96.

Nakai, J., T. Tanabe, T. Konno, B. Adams, K.G. Beam (1998). Localization in the II-III loop of the dihydropyridine receptor of a sequence critical for excitation-contraction coupling. J Biol Chem 273: 24983-6.

Nakai J, Sekiguchi N, Rando TA, Allen PD, Beam KG. Two regions of the ryanodine receptor involved coupling with L-type v channels. J Biol Chem. 1998 May 29;273(22):13403-6.

Nakai J, Ogura T, Protasi F, Franzini-Armstrong C, Allen PD, Beam KG 1997. Functional nonequality of the cardiac and skeletal ryanodine receptors. Proc Natl Acad Sci U S A 94:1019-22

Nakai J, Dirksen RT, Nguyen HT, Pessah IN, Beam KG, Allen PD. 1996 Enhanced dihydropyridine receptor channel activity in the presence of ryanodine receptor. Nature. 380:72-5.

Nakayama, H., M. Taki, J. Striessnig, H. Glossmann, W.A. Catterall, Y. Kanaoka (1991). Identification of 1,4-dihydropyridine binding regions within the alpha 1 subunit of skeletal muscle Ca^{2+} channels by photoaffinity labeling with diazipine. Proc Natl Acad Sci U S A 88: 9203-7.

Nelson MT, Cheng H, Rubart M, Santana LF, Bonev AD, Knot HJ, Lederer WJ. Relaxation of arterial smooth muscle by calcium sparks. Science. 1995 Oct 27;270(5236):633-7.

Noceti, F, P Baldelli, X Wei, N Qin, L Toro, L Birnbaumer and E Stefani. 1996. Effective gating charges per channel in voltage[dependent K^+ and Ca^{2+} channels. J. Gen. Physiol. 108:143-155.

Orlova EV, Serysheva II, van Heel M, Hamilton SL, Chiu W. 1996. Two structural configurations of the skeletal muscle calcium release channel. Nat Struct Biol.3:547-52.

Pape, PC, D-S Jong and WK Chandler, 1995. Calcium release and its voltage dependence in frog cut muscle fibers equilibrated with 20 mM EGTA. J. Gen. Physiol. 106:259-336)

Parker I, Zang WJ, Wier WG. Ca^{2+} sparks involving multiple Ca^{2+} release sites along Z-lines in rat heart cells. J Physiol. 1996 Nov 15;497 (Pt 1):31-8.

Penniston JT, Enyedi A, J Membr Biol 1998 165:101-9 Modulation of the plasma membrane Ca^{2+} pump.

Pizarro G, Shirokova N, Tsugorka A, Rios E. 'Quantal' calcium release operated by membrane voltage in frog skeletal muscle. J Physiol. 1997 Jun 1;501 (Pt 2):289-303.

Pizarro G, Fitts R, Uribe I, Rios E. The voltage sensor of excitation-contraction coupling in skeletal muscle. Ion dependence and selectivity. J Gen Physiol. 1989 Sep;94(3):405-28.

Porter Moore C, Zhang JZ, Hamilton SL. 1999. A role for cysteine 3635 of RYR1 in redox modulation and calmodulin binding. J Biol Chem. 274:36831-4.

Pragnell M, De Waard M, Mori Y, Tanabe T, Snutch TP, Campbell KP. 1994. Calcium channel beta-subunit binds to a conserved motif in the I-II cytoplasmic linker of the alpha 1-subunit. Nature 368:67-70.

Proenza, C., C. M. Wilkens, K.G. Beam (2000a). Excitation-contraction coupling is not affected by scrambled sequence in residues 681-690 of the dihydropyridine receptor II-III loop [In Process Citation]. J Biol Chem 275: 29935-7.

Proenza, C., C. Wilkens, N.M. Lorenzon, K.G Beam (2000b). "A carboxyl-terminal region important for the expression and targeting of the skeletal muscle dihydropyridine receptor." J Biol Chem 275(30): 23169-74.

Protasi F, Takekura H, Wang Y, Chen SR, Meissner G, Allen PD, Franzini-Armstrong C. 2000. RYR1 and RYR3 have different roles in the assembly of calcium release units of skeletal muscle. Biophys J. 79:2494-508.

Protasi, F., C. Franzini-Armstrong and P.D. Allen. 1998. Role of ryanodine receptors in the assembly of calcium release units in skeletal muscle. J. Cell. Biol. 140:831-842.

Ríos, E, N Shirokova, WG Kirsch, G Pizarro, MD Stern, H Cheng and A Gonzalez. 2001. A preferred amplitude of calcium sparks in skeletal muscle. Biophys. J. 80. In Press.

Rios E, Stern MD, Gonzalez A, Pizarro G, Shirokova N. Calcium release flux underlying Ca^{2+} sparks of frog skeletal muscle. J Gen Physiol. 1999 Jul;114(1):31-48.

Rios E, Karhanek M, Ma J, Gonzalez A. An allosteric model of the molecular interactions of excitation-contraction coupling in skeletal muscle. J Gen Physiol. 1993 Sep;102(3):449-81.

Rios E, Pizarro G. 1991. Voltage sensor of excitation-contraction coupling in skeletal muscle. Physiol Rev. 71:849-908.

Rios, E and G Pizarro, 1988. The voltage sensors and calcium channels of excitation-contraction coupling. News In Physiol. Sci. 3:223-227.

Rios E, Brum G. 1987. Involvement of dihydropyridine receptors in excitation-contraction coupling in skeletal muscle. Nature.325:717-20.

Robertson, SP, JD Johnson and JD Potter, 1981. The time course of Ca exchange with calmodulin, troponin, parvalbmin and myosin in response to transient increases in Ca. Biophys. J. 34, 559-59.

Rodney GG, Moore CP, Williams BY, Zhang JZ, Krol J, Pedersen SE, Hamilton SL; 2000. Calcium Binding to Calmodulin Leads to an N-Terminal Shift in its Binding Site on the Ryanodine Receptor. J Biol Chem In Press.

Rome LC, Syme DA, Hollingworth S, Lindstedt SL, Baylor SM. 1996. The whistle and the rattle: the design of sound producing muscles. Proc Natl Acad Sci U S A. 93:8095-100.

Rosenberg RL, Hess P, Tsien RW. 1988. Cardiac calcium channels in planar lipid bilayers. L-type channels and calcium-permeable channels open at negative membrane potentials. J Gen Physiol. 92:27-54.

Samsó M, Trujillo R, Gurrola GB, Valdivia HH, Wagenknecht T. 1999. Three-dimensional location of the imperatoxin A binding site on the ryanodine receptor. J Cell Biol. 146:493-9.

Schatzmann HJ. 1989. The calcium pump of the surface membrane and of the sarcoplasmic reticulum. Annu Rev Physiol. 51:473-85.

Schneider MF and BJ Simon, Inactivation of calcium release form the sarcoplasmic reticulum in frog skeletal muscle. J. Physiol. 405:727-745, 1988.

Schneider MF, Chandler WK. Voltage dependent charge movement of skeletal muscle: a possible step in excitation-contraction coupling. Nature. 1973 Mar 23;242(5395):244-6.

Seoh SA, Sigg D, Papazian DM, Bezanilla F. 1996. Voltage-sensing residues in the S2 and S4 segments of the Shaker K^{+} channel. Neuron. 16:1159-67.

Shirokov R, Ferreira G, Yi J, Rios E. Inactivation of gating currents of L-type calcium channels. Specific role of the alpha 2 delta subunit. J Gen Physiol. 1998 Jun;111(6):807-23.

Shirokova N, Shirokov R, Rossi D, Gonzalez A, Kirsch WG, Garcia J, Sorrentino V, Rios E. Spatially segregated control of Ca^{2+} release in developing skeletal muscle of mice. J Physiol. 1999 Dec 1;521 Pt 2:483-95.

Shirokova N, Garcia J, Rios E. Local calcium release in mammalian skeletal muscle. J Physiol. 1998 Oct 15;512 (Pt 2):377-84.

Shirokova N, Rios E. Small event Ca^{2+} release: a probable precursor of Ca^{2+} sparks in frog skeletal muscle. J Physiol. 1997 Jul 1;502 (Pt 1):3-11.

Shirokova N, Garcia J, Pizarro G, Rios E. Ca2+ release from the sarcoplasmic reticulum compared in amphibian and mammalian skeletal muscle. J Gen Physiol. 1996 Jan;107(1):1-18.

Shtifman A, Ward CW, Wang J, Valdivia HH, Schneider MF. Effects of imperatoxin A on local sarcoplasmic reticulum Ca^{2+} release in frog skeletal muscle. Biophys J. 2000 79(2):814-27.

Simon BJ, Hill DA. Charge movement and SR calcium release in frog skeletal muscle can be related by a Hodgkin-Huxley model with four gating particles. Biophys J. 1992 May;61(5):1109-16.

Sipos, I., U. Pika-Hartlaub, F. Hofmann, B.E. Flucher, W. Melzer (2000). Effects of the dihydropyridine receptor subunits gamma and alpha2delta on the kinetics of heterologously expressed L-type Ca^{2+} channels. Pflugers Arch 439(6): 691-9.

Slavik, K. J., J. P. Wang, B.Aghdasi, J.Z Zhang, F. Mandel, N. Malouf, S.L. Hamilton. (1997). A carboxy-terminal peptide of the alpha 1-subunit of the dihydropyridine receptor inhibits Ca^{2+}-release channels. Am J Physiol 272: C1475-81.

Smith GD, Keizer JE, Stern MD, Lederer WJ, Cheng H. A simple numerical model of calcium spark formation and detection in cardiac myocytes. Biophys J. 1998 Jul;75(1):15-32.

Smith JS, R Coronado and G Meissner. 1985. Sarcoplasmic reticulum contains adenine nucleotide-activated calcium channels. Nature 316:446-449.)

Stern MD, Song LS, Cheng H, Sham JS, Yang HT, Boheler KR, Rios E Local control models of cardiac excitation-contraction coupling. A possible role for allosteric interactions between ryanodine receptors. J Gen Physiol. 1999 Mar;113(3):469-89.

Stokes, DL and T Wagenknecht. 2000. Calcium transport across the sarcoplasmic reticulum. Structure and function of Ca^{2+}-ATPase and the ryanodine receptor. Eur. J. Biochem. 267: 5274-5279.

Suda N. Involvement of dihydropyridine receptors in terminating Ca^{2+} release in rat skeletal myotubes. J Physiol. 1995 Jul 1;486 (Pt 1):105-12.

Sutko, JL, JA Airey, W Welch and L Ruest. 1997. The pharmacology of ryanodine and related compounds. Pharmacol. Rev. 49:53-98

Sutko JL, Airey JA. Ryanodine receptor Ca^{2+} release channels: does diversity in form equal diversity in function? Physiol Rev. 1996 76:1027-71.

Sham JS, Song LS, Chen Y, Deng LH, Stern MD, Lakatta EG, Cheng H. 1998. Proc Natl Acad Sci U S A. 95:15096-101. Termination of Ca^{2+} release by a local inactivation of ryanodine receptors in cardiac myocytes.

Stokes DL, Wagenknecht T. 2000. Calcium transport across the sarcoplasmic reticulum structure and function of Ca^{2+}-ATPase and the ryanodine receptor. Eur J Biochem. Sep;267(17):5274-9.

C Szegedi, S Sárközi, A Herzog, I Jóna and M Varsányi. Calsequestrin: more than 'only' a luminal Ca^{2+} buffer inside the sarcoplasmic reticulum. Biochem. J. (1999) 337, 19–22.

Szentesi P, Kovacs L, Csernoch L. Deterministic inactivation of calcium release channels in mammalian skeletal muscle. J Physiol. 2000 Nov 1;528(Pt 3):447-456.

Takekura, H., L. Bennett, T Tanabe, KG Beam and C Franzini-Armstrong. 1994. Restoration of junctional tetrads in dysgenic myotubes by dihydropyridine receptor cDNA. Biophysical J. 67:793-803.

Takekura H, Franzini-Armstrong C. Correct targeting of dihydropyridine receptors and triadin in dyspedic mouse skeletal muscle in vivo. Dev Dyn. 1999 Apr;214(4):372-80.

Takeshima H, Ikemoto T, Nishi M, Nishiyama N, Shimuta M, Sugitani Y, Kuno J, Saito I, Saito H, Endo M, Iino M, Noda T. 1996. Generation and characterization of mutant mice lacking ryanodine receptor type 3. J Biol Chem. 271:19649-52.

Tanabe, T, A Mikami, S Numa and KG Beam. 1990. Cardiac type excitation-contraction coupling in dysgenic skeletal muscle injected with cardiac dihydropyridine receptor cDNA. Nature 344:451-453.

Tanabe, T, KG Beam, JA Powell and S Numa. 1988. Restoration of excitation-contraction coupling and slow calcium current in dysgenic muscle by dihydropyridine receptor complementary DNA. Nature 336:134-139.

Tarroni, P., D. Rossi, A.Conti, V. Sorrentino (1997). Expression of the ryanodine receptor type 3 calcium release channel during development and differentiation of mammalian skeletal muscle cells. J Biol Chem 272(32): 19808-13.

Timerman, A. P., Wiederrecht, G., Marcy, A., and Fleischer, S. (1995) Characterization of an Exchange Reaction between Soluble FKBP-12 and the FKBP-Ryanodine Receptor Complex . Modulation by FKBP mutants deficient in peptidyl-prolyl isomerase activity. J. Biol. Chem. 270, 2451-2459

Trimm JL, Salama G, Abramson JJ. Sulfhydryl oxidation induces rapid calcium release from sarcoplasmic reticulum vesicles. J Biol Chem. 1986 Dec 5;261(34):16092-8.

Tsugorka A, Rios E, Blatter LA. Imaging elementary events of calcium release in skeletal muscle cells. Science. 1995 Sep 22;269(5231):1723-6.

Veratti, E. 1961. Investigations on the fine structure of striated muscle fiber. J. Biophys. Biochem. Cytol. 10:1-60.

Wagenknecht T, Radermacher M, Grassucci R, Berkowitz J, Xin HB, Fleischer S. 1997. Locations of calmodulin and FK506-binding protein on the three-dimensional architecture of the skeletal muscle ryanodine receptor. J Biol Chem. 272:32463-71.

Wang S, Trumble WR, Liao H, Wesson CR, Dunker AK, Kang CH. 1998. Crystal structure of calsequestrin from rabbit skeletal muscle sarcoplasmic reticulum. Nat Struct Biol 5:476-83

Ward, CW, D Castillo, F Protasi, Y Wang, S Chen, M Schneider and P Allen. 1999. Expression of RyR3, but not RyR1, produces Ca^{2+} sparks in dyspedic myotubes. Biophys. J. 76:A386.

Westerblad H, Allen DG, Bruton JD, Andrade FH, Lannergren J 1998. Mechanisms underlying the reduction of isometric force in skeletal muscle fatigue. Acta Physiol Scand 162:253-60.

Westerblad H, Allen DG. 1996. The effects of intracellular injections of phosphate on intracellular Calcium and force in single fibres of mouse skeletal muscle.Pflugers Arch 431:964-70.

Westerblad H, Duty S, Allen DG 1993. Intracellular calcium concentration during low-frequency fatigue in isolated single fibers of mouse skeletal muscle. J Appl Physiol 75:382-8.

Xiao RP, Valdivia HH, Bogdanov K, Valdivia C, Lakatta EG, Cheng H. The immunophilin FK506-binding protein modulates Ca^{2+} release channel closure in rat heart. J Physiol. 1997 Apr 15;500 (Pt 2):343-54.

Xu L, Eu JP, Meissner G, Stamler JS. Activation of the cardiac calcium release channel (ryanodine receptor) by poly-S-nitrosylation. Science. 1998 Jan 9;279(5348):234-7.

Xu, L, A Tripathy, DA Pasek and G Meissner. 1998. Potential for pharmacology of ryanodine receptor/calcium release channels. Ann. NY Acad Sci 853:130-148.

Yamaguchi, S., Y. Okamura, T.Nagao, S. Adachi-Akahane S. (2000). Serine residue in IIIS5-S6 linker of L-type Ca^{2+} channel α_{1C} subunit is the critical determinant of the action of dihydropyridine Ca^{2+} channel agonists. J Biol Chem. 275: 41504-11.

Zhang JZ, Wu Y, Williams BY, Rodney G, Mandel F, Strasburg GM, Hamilton SL. Oxidation of the skeletal muscle Ca^{2+} release channel alters calmodulin binding. Am J Physiol. 1999 Jan;276(1 Pt 1):C46-53.

L. Zhang , J Kelley , G Schmeisser , Y M. Kobayashi and L R. Jones. Complex Formation between Junctin, Triadin, Calsequestrin, and the Ryanodine Receptor. Proteins of the cardiac junctional sarcoplasmic reticulum membrane. J. Biol. Chem. 272: 23389-23397, 1997.

Zhao M, Li P, Li X, Zhang L, Winkfein RJ, Chen SR. 1999. Molecular identification of the ryanodine receptor pore-forming segment. J Biol Chem.274:25971-4.

A.W. TRAFFORD & D.A. EISNER

EXCITATION-CONTRACTION COUPLING IN CARDIAC MUSCLE

1. INTRODUCTION

Excitation-contraction coupling concerns the processes linking electrical excitation of the surface membrane to contraction. As shown in Fig. 1A, the action potential produces a systolic increase of calcium that then activates contraction. This chapter will focus on the control of calcium and will largely ignore the contractile protein mechanisms that follow the increase of Ca. These events are described in detail in other chapters. Equally, on grounds of space, we will largely ignore the whole field of pharmacological modulation of e-c coupling. The reader is directed to the following books for recent summaries of the broad field of cardiac electrophysiology and contraction (Bers, 2001;Zipes & Jalife, 2000;Sperelakis et al., 2001). An overview of cellular Ca handling is shown in Fig. 1A.

2. CONTROL OF CALCIUM BY THE SURFACE MEMBRANE

The resting or diastolic Ca concentration is of the order of 100 nM, a value 4 orders of magnitude less than the extracellular concentration. This and the negative inside membrane potential of the cell results in a large electrochemical gradient for Ca entry into the cell which must be opposed by energy dependent active transport of Ca out of the cell.

2.1. Calcium Removal from the Cell

Two processes have been identified that remove Ca ions from the cell: Na-Ca exchange and the sarcolemmal Ca-ATPase.

2.1.1. Na-Ca Exchange
This area has been comprehensively reviewed (Blaustein & Lederer, 1999;Philipson & Nicoll, 2000;Shigekawa & Iwamoto, 2001). The Na-Ca exchange uses the energy provided by the entry of sodium ions into the cell to fuel extrusion of calcium from the cell. If n Na ions exchange for each Ca then thermodynamics predicts that there

R.J. Solaro and R.L. Moss (eds), Molecular Control Mechanisms in Striated Muscle Contraction, 49-89
©2002 Kluwer Academic Publishers, Printed in the Netherlands

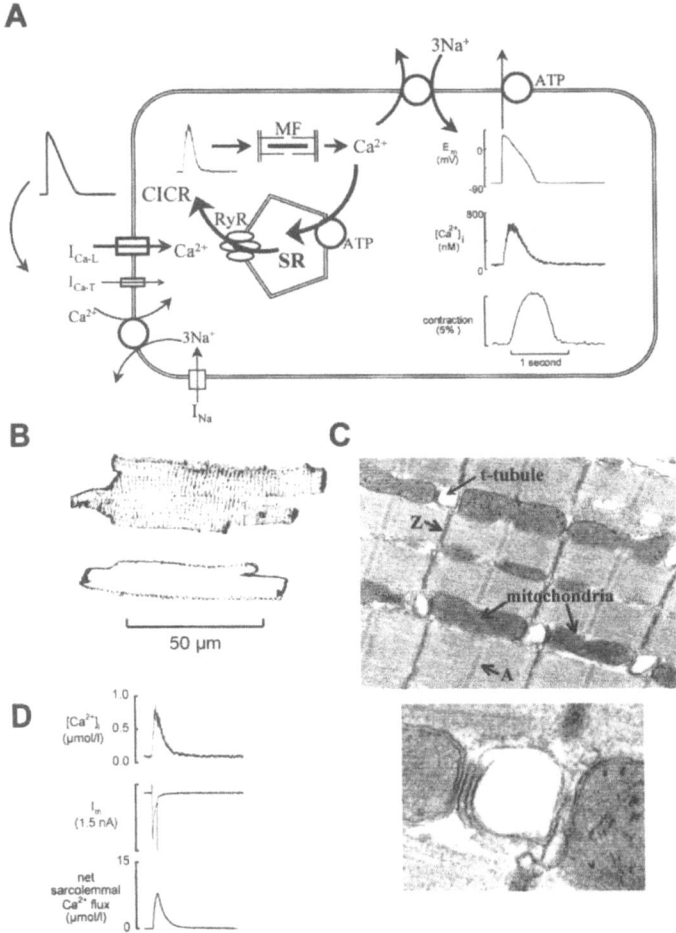

Fig. 1. Overview of e-c coupling. A. *Summary of Ca handling pathways in the cell. This shows Ca entry via the L- and T- type Ca channels and Na-Ca exchange. Ca efflux occurs on Na-Ca exchange and Ca-ATPase. Na entry occurs via the Na channel. Entry of Ca into the cell stimulates release of Ca from the sarcoplasmic reticulum (SR) via calcium induced Ca release (CICR) resulting in a calcium transient that activates the myofilaments (MF). Relaxation is initiated by Ca being taken up into the SR (Ca-ATPase) or pumped out of the cell. The data traces show: top, action potential; middle, [Ca²⁺]ᵢ; bottom, cell shortening all recorded from a ferret ventricular myocyte.* **B.** *t-tubular distribution. Confocal images of cells stained with the membrane dye di-8-ANEPPS. The cell on the top is from the ventricle and that below from the atrium. High intensity is black in these images.* **C.** *Electron micrographs. The top image shows the sarcomeric structure indicating the A & Z lines, mitochondria, t-tubule. The bottom one shows a higher magnification view of the t-tubule with adjacent SR.* **D.** *Ca flux balance. The top two traces show a Ca transient produced by depolarisation and the accompanying membrane current. The bottom traces shows the calculated changes of total cell Ca content. The depolarising pulse has a duration of 100 ms.*

will be net ionic flux at membrane potentials positive or negative to the reversal potential for the exchanger (E_{Na-Ca}) given by:

$$E_{Na-Ca} = 3E_{Na} - 2E_{Ca} \tag{1}$$

Where E_{Na} and E_{Ca} have their usual meanings:

$$E_{Na} = \frac{nRT}{F} ln \frac{[Na^+]_o}{[Na^+]_I} \qquad E_{Ca} = \frac{RT}{2F} ln \frac{[Ca^{2+}]_o}{[Ca^{2+}]_i} \tag{2}$$

Such an exchanger can produce a net Ca efflux so long as the membrane potential is more negative than E_{Na-Ca}. This means as shown in Fig. 2B that at potentials negative to E_{Na-Ca} there is net Ca efflux and Na influx whereas, at more positive potentials, net Ca influx and Na efflux are observed. If we take n=3 (see below) and typical diastolic values for intra- and extracellular concentrations: $[Ca^{2+}]_o$. 1 mM; $[Ca^{2+}]_i$, 100 nM; $[Na^+]_o$, 145 mM; $[Na^+]_i$, 10 mM then one obtains a value for E_{Na-Ca} of -32 mV. With a systolic $[Ca^{2+}]_i$ of 1 μM, a value for E_{Na-Ca} of +29 mV is obtained. Therefore, at diastole, the Na-Ca exchange will produce net Ca influx if the membrane is depolarised to potentials positive to –32 mV whereas with the systolic values depolarisation to greater than +29 mV will be required to produce Ca influx. This rather simple approach ignores the complicating effects of local Na and Ca gradients immediately adjacent to the cytoplasmic face of the sarcolemma that may occur during SR Ca release or as a result of Na influx via I_{Na}. In the case of Ca, it has been suggested that during SR Ca release the sub sarcolemmal Ca concentration (Trafford et al., 1998b;Lewartowski et al., 1996;Langer & Peskoff, 1996), which will determine E_{Na-Ca} may be of the order of several tens of μM and as such will make Ca influx via Na-Ca exchange more difficult to achieve given the action potential profile.

Most work suggests that 3 Na^+ are exchanged with each Ca^{2+}, i.e. n=3. This conclusion is based on measurements of reversal potential as well as radioactive flux measurements (Reeves & Hale, 1984;Ehara et al., 1989;Matsuoka & Hilgemann, 1992). However recent studies have suggested that, at least under some conditions, n=4 (Fujioka et al., 2000). This difference of stoichiometry will have two consequences. (1) If n =4 then the transport of each Ca^{2+} will be accompanied by movement of two positive charges and thus the current generated by Na-Ca exchange will be twice as large than if 3 Na^+ move. (2) The reversal potential will be more positive (about +40 mV even with the assumed diastolic concentrations above) and net Ca^{2+} influx will be even less likely to occur.

The activity of the Na-Ca exchange is controlled by the concentration of Na and Ca on both sides of the membrane. As shown in Fig. 2A, an increase of intracellular Ca concentration increases the inward current (equivalent to Ca efflux) in giant patches (Matsuoka & Hilgemann, 1992). In the intact cell (Fig. 2C), the larger the

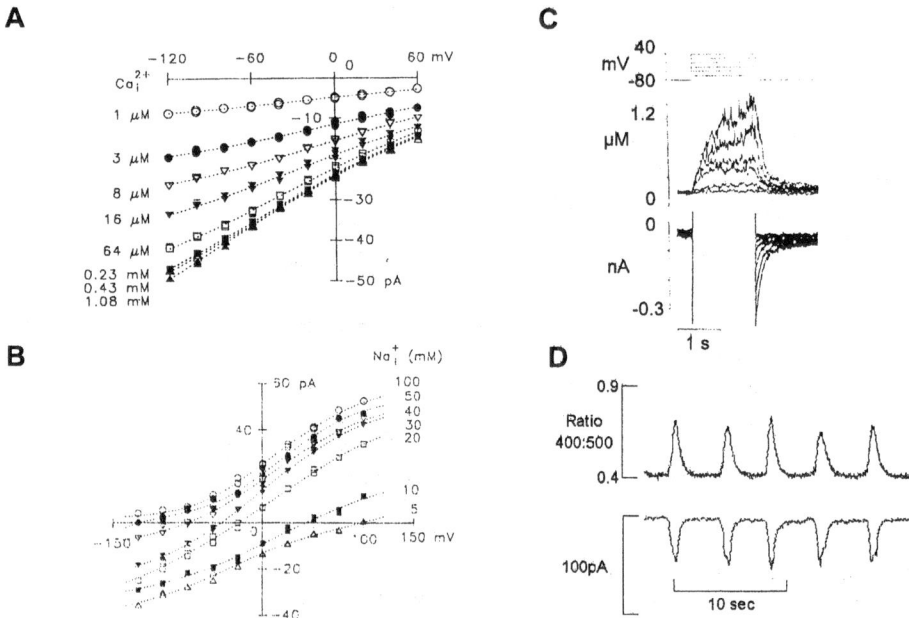

Fig. 2. Sodium Calcium exchange in cardiac muscle. *A. Na-Ca exchange current recorded from a giant patch. Traces show current voltage (I-V) relationships at various [Ca²⁺]ᵢ. The experiment was performed in the absence of cytoplasmic Na so only inward currents (equivalent to Ca efflux) are seen (Matsuoka & Hilgemann, 1992). B. I-V relations as a function of [Na⁺]ᵢ (1 µM [Ca²⁺]ᵢ). Note that the reversal potential shifts positive as [Na⁺]ᵢ is decreased (Matsuoka & Hilgemann, 1992). C. Na-Ca exchange currents and Ca movements. Pulses were applied to different potentials (SR disabled with ryanodine) resulting in Ca influx on Na-Ca exchange. On repolarization the decrease of [Ca²⁺]ᵢ is accompanied by inward Na-Ca current (Barcenas-Ruiz et al., 1987). D. Na-Ca exchange activated by spontaneous Ca release. The membrane potential was held at –80 mV. Spontaneous release of Ca from the SR (top) activates Na-Ca exchange currents (bottom). (Diaz et al., 1997)*

increase of $[Ca^{2+}]_i$ produced by depolarisation, the greater the inward Na-Ca current on repolarization. Ca efflux is a reasonably linear function of $[Ca^{2+}]_i$ over the diastolic to systolic range of $[Ca^{2+}]_i$ (Beuckelmann & Wier, 1989). This can also be seen by observing (Fig. 2D) the inward Na-Ca currents produced by spontaneous Ca release from the SR (Díaz *et al.*, 1997). Efflux of Ca is inhibited and influx stimulated by an increase of intracellular Na concentration. As shown in Fig. 2B, an increase of cytoplasmic Na concentration decreases the inward current or Ca efflux and increases outward current/ Ca influx. Intracellular Ca has an additional effect on the exchanger. As Ca is decreased to levels less than 1 μM, not only does Ca efflux decrease but so, also, does Ca influx. A K_D of about 120- 300 nM has been found for this activating effect (Hilgemann, 1990;Weber *et al.*, 2001). Activity of the exchange may also be stimulated by phosphorylation (Perchenet *et al.*, 2000) and in addition to these effects of the transported ions, the activity of the exchange can be modified by inositol lipids such as PIP_2 (phosphatidylinositol 4,5-bisphosphate) (He *et al.*, 2000) and agonists leading to increases in protein kinase C (Ballard & Schaffer, 1996;Ballard *et al.*, 1994).

2.1.2. Sarcolemmal Ca-ATPase

In the absence of a Na gradient, the cardiac cell can still regulate its resting Ca concentration. Thus, while abrupt removal of external Na results in an increase of $[Ca^{2+}]_i$ this decreases to a level which may be only slightly greater than control (Allen *et al.*, 1983). There are two possible mechanisms for this regulation of Ca. (1) At least in the short term, it could be due to sequestration of Ca by mitochondria or (2) it could reflect the activity of a sacolemmal Ca-ATPase. Distinguishing between these alternatives is not straightforward (Barry *et al.*, 1986). Unlike the Na-Ca exchange, the Ca-ATPase does not appear to generate an electric current so its activity cannot be followed electrophysiologically. There is also a lack of specific inhibitors for the ATPase. Work has, however, shown that, in the absence of Na-Ca exchange and SR, Ca regulation is inhibited by the sarcolemmal Ca-ATPase inhibitor carboxyeosin (Bassani *et al.*, 1995b;Choi & Eisner, 1999). This suggests a significant role for a Ca-ATPase. An example of this is shown in Fig. 3.

Application of caffeine produces an initial increase of $[Ca^{2+}]_i$ due to SR Ca release which, under control conditions, relaxes as Ca^{2+} is pumped out of the cell. In the absence of Na^+ the decay is slowed but not abolished showing that something, in addition to Na-Ca exchange can remove Ca from the cell. However the subsequent addition of carboxyeosin to inhibit the sarcolemmal Ca-ATPase results in complete inhibition of the relaxation of $[Ca^{2+}]_i$.

Several splice variants of the plasma membrane Ca-ATPase (PMCA) have been identified (see Strehler & Zacharias, 2001 for a review). If there is a significant role for the sarcolemmal Ca-ATPase in cardiac muscle, one must consider why the cardiac cell has two separate mechanisms for Ca extrusion. One suggestion which has been made for both axons (DiPolo & Beaugé, 1979) and the heart (Carafoli, 1985) is that the Na-Ca exchange is a low affinity high capacity system designed to deal with Ca loads whereas the Ca-ATPase, as a high affinity, low capacity system regulates resting Ca levels. However the fractional effect of Na removal on the rate of decay of $[Ca^{2+}]_i$ is the same over a wide range of $[Ca^{2+}]_i$ (Lamont & Eisner, 1996). In recent work we have measured the absolute Ca efflux from the cell by measuring the rate of decay of $[Ca^{2+}]_i$ and knowing the Ca buffering properties of

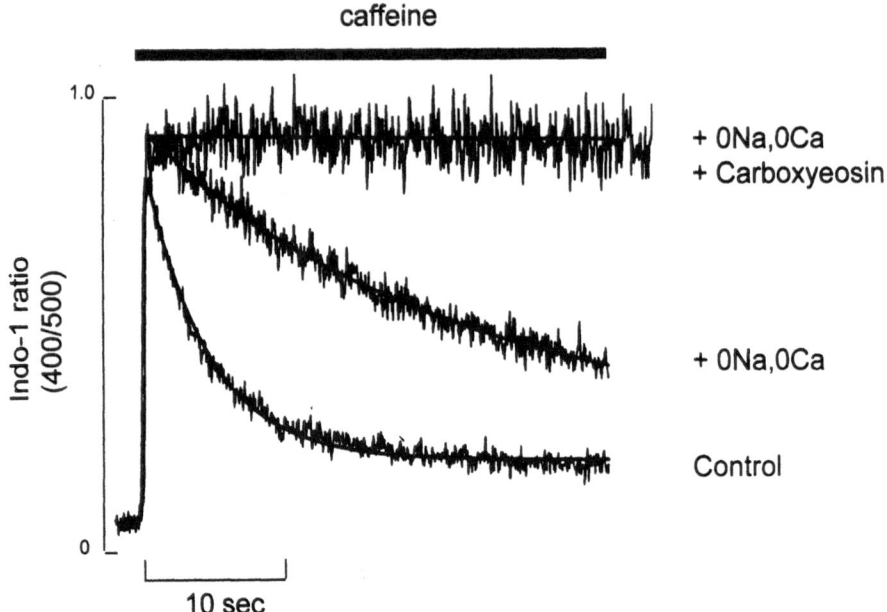

Fig. 3. Ca regulation in the absence of Na-Ca exchange. *The trace shows the effects on [Ca²⁺]ᵢ (measured as the Indo-1 fluorescence ratio) of applying caffeine (10 mM) to release Ca from the SR. Caffeine was applied in three conditions: control; 0-Na, 0-Ca; 0-Na,0-Ca plus carboxyeosin. From (Choi & Eisner, 1999).*

the cell. The results showed that the fraction of Ca efflux which was via the Ca-ATPase compared to Na-Ca exchange was independent of $[Ca^{2+}]_i$ suggesting that both mechanisms contribute over the whole range of $[Ca^{2+}]_i$ (Choi *et al.*, 2000a). Finally, it should be noted that work in which the sarcolemmal Ca-ATPase was overexpressed found no effect on Ca regulation (Hammes *et al.*, 1998). The role of the sarcolemmal Ca-ATPase is therefore still incompletely explained. One possibility is that it acts as a final defence when the activity of Na-Ca exchange is compromised by elevated intracellular Na.

2.2. Ca Influx

The best characterized route of Ca entry into the ventricular cell is via the L-type Ca current. For reviews of the properties of this channel see (Hess & Tsien, 1984;Tsien, 1983;Hofmann *et al.*, 1999). For the purposes of the present article, the following properties of the L-type channel are important. (1) The channel is activated on depolarisation with the activation range occurring over –40 to 0 mV and is therefore known as a high voltage activated channel. Further depolarisation decreases Ca entry due to a decreased driving force. (2) Following activation, the channel inactivates, partly as a direct result of depolarisation but largely as a consequence of the increase of Ca (Imredy & Yue, 1992). (3) The open probability, rather than the unitary conductance of the channel is increased by phosphorylation (Osterrieder *et al.*, 1982;Cachelin *et al.*, 1983).

In addition to L-type Ca channels, T-type channels are also present. These activate at more negative potentials (-70 mV) than the L-type and are often referred to as low voltage activated Ca channels. They are most abundant in regions of the heart other than the ventricle and may be important in pacemaker activity (Vassort & Alvarez, 1994).

Most of the Ca entry into the cell on depolarisation occurs via the L-type Ca current with possibly a contribution from "reverse" Na-Ca exchange (see below). However since in an unstimulated cell $[Ca^{2+}]_i$ does not fall to zero despite the ongoing removal of Ca from the cell, this suggests that there must be some continuous Ca entry. Recent measurements estimate that the influx is of the order of 4 $\mu mol.l^{-1}s^{-1}$ (Choi *et al.*, 2000a), however, these experiments were performed at membrane potential of –40mV and as such the actual 'background' Ca entry may be expected to be somewhat less at the more usual resting membrane potential of approximately -80mV where the P_o of the L-type Ca channel should be even lower although Ca entry via other routes would be thermodynamically favoured at more negative potentials. Despite this, the exact identity of this resting Ca entry is uncertain. Among the various possibilities are. (1) A low probability of opening of the L- or T-type Ca channels. That the former may contribute is suggested by the fact that the calculated Ca influx into a resting cell is decreased by nifedipine. However the reduction is only about 20% (Choi *et al.*, 2000a) suggesting that other routes of Ca entry must also exists. (2) Background (B type) channels have been identified from single channel recordings (Coulombe *et al.*, 1989;Lefevre *et al.*, 1994) although the lack of specific pharmacology makes it difficult to investigate these. Partly on the evidence that inhibitors of the PMCA affect B-type channels, it has recently been suggested that these channels may represent a form of the PMCA (Antoine *et al.*, 2001). This exciting possibility deserves further study. (3) It might be that an equilibrium is reached at which Na-Ca exchange produces no net flux. In this case Na-Ca exchange would be producing both Ca influx and efflux. Such an explanation is difficult to reconcile with the estimates for reversal potential of the Na-Ca exchange presented above.

2.3. Ca Flux Balance

It should be noted that, in the steady state, over one cardiac cycle calcium influx must equal efflux. That this does, indeed, occur is shown by the records of Fig 1D. The top trace shows the intracellular Ca transient. The current record (middle trace) shows Ca entry via the L-type Ca current and, on repolarization, a Na-Ca exchange "tail" current. These Ca fluxes are integrated in the bottom trace which demonstrates that about 8 μmol of Ca enter per litre cell volume via the Ca current and an equal amount is pumped out of the cell on repolarization. The importance of this flux balance to the control of systolic Ca will be emphasised throughout the course of this review.

3. CA BUFFERING

The amplitude of the systolic Ca transient depends not only on the fluxes of Ca into and out of the cytoplasm but, in addition, on the buffering properties of the cell. As is the case in other cells, most of the Ca in the cytoplasm is complexed to buffers rather than being free. In order to measure the degree of Ca buffering in the cell one must add a known amount of total Ca to the cytoplasm and then measure the resulting change of $[Ca^{2+}]_i$. For this to be valid, the measurement of the amount of Ca added must either be unaffected by or corrected for any Ca removal from the cell. Two approaches have been used. (1) Depolarizations are used to activate Ca entry via the L-type Ca current. The integral of these currents then gives the amount of total Ca entry. So long as Ca removal by both the SR and surface membrane are inhibited, this can be compared with the measured change of free Ca and the buffering characteristics obtained (Berlin *et al.*, 1994). (2) An alternative method is to make use of caffeine. As shown in Fig. 4A (top trace), the application of caffeine

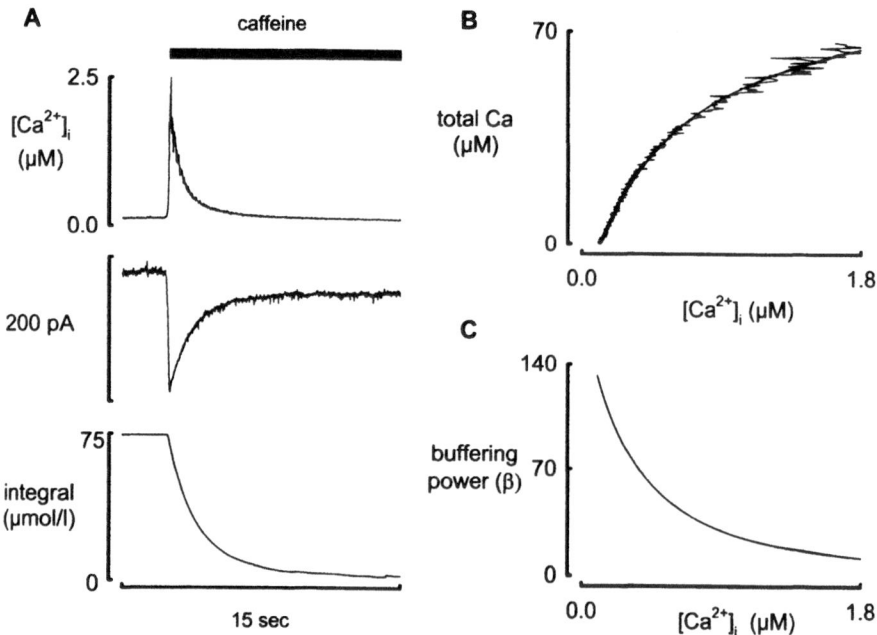

Fig. 4. **Ca buffering.** *A. Measurement. Caffeine (10 mM) was applied for the period shown above resulting in a transient increase of $[Ca^{2+}]_i$ accompanied by the activation of Na-Ca exchange current (middle). The total amount of Ca removed from the cell (expressed per litre cell volume) is shown in the bottom trace. **B.** Buffer curve obtained by plotting total Ca as a function of free. **C.** Buffering power (from the slope of the graph in B) as a function of $[Ca^{2+}]_i$.*

results in an increase of $[Ca^{2+}]_i$ which is due to Ca release from the SR. This increase of $[Ca^{2+}]_i$ decays within a few seconds due to Ca pumping out of the cell. This Ca pumping is largely on Na-Ca exchange and generates the electrogenic current seen in the middle trace of Fig. 4A. If this current is integrated it gives the amount of Ca removed from the cell by Na-Ca exchange. Correction for the fact that some Ca is removed by other mechanisms gives the total amount of Ca removed from the cell as a function of time and this is shown in the lower (integral) trace of Fig. 4A. Fig. 4B shows the total Ca (Ca_T) record plotted as a function of the free ($[Ca^{2+}]_i$). It can be seen that this is a saturating function of free Ca. The buffer power, defined as the slope of this relationship, is plotted as a function of $[Ca^{2+}]_i$ in Fig. 4C. It is clear that the buffer power decreases as $[Ca^{2+}]_i$ increases.

The slope (β) can be defined as:

$$\beta = d(Ca_T)/d[Ca^{2+}]_i \qquad (3)$$

For a simple buffer

$$Ca_T = B_{max.}[Ca^{2+}]_i /([Ca^{2+}]_i + K_d) \qquad (4)$$

Where B_{max} is the maximum binding capacity; then

$$\beta = B_{max.}K_d /([Ca^{2+}]_i + K_d)^2 \qquad (5)$$

This means that the buffer power will decrease as $[Ca^{2+}]_i$ increases. With a K_d of 0.6 µM and B_{max} of 175 µmol.l^{-1} then the value of β will increase from 214 at a diastolic $[Ca^{2+}]_i$ of 100 nM to 41 at a systolic $[Ca^{2+}]_i$ of 1 µM and 16 at 2 µM. In order to increase $[Ca^{2+}]_i$ by 400 nM from resting of 100 to 500 nM requires 55 µmol.l^{-1}. A further increase of 400 nM (to 900 nM) requires only 25 µmol.l^{-1} and further increase to 1300 nM needs only 14 µmol.l^{-1}. These buffer curves presumably reflect a lumped average of several different buffers which will include contributions from troponin, calmodulin and various membrane binding sites (Fabiato, 1983). This tendency to saturation of the buffers has significant effects on systolic $[Ca^{2+}]_i$ (Trafford et al., 2000;Díaz et al., 2001). Another method has been used to measure Ca buffering properties in the intact heart. This makes use of the fact that exogenous buffers decrease systolic and increase diastolic $[Ca^{2+}]_i$ (Kirschenlohr et al., 2000) . These effects could be modelled by a cooperative model of Ca buffers in which at least two Ca^{2+} ions are bound (Smith et al., 2000). This sort of buffer will show, over a certain range, an increase of buffering power with increasing $[Ca^{2+}]_i$ and in future work it will be important to compare the predictions of this model with those described above.

4. THE SARCOPLASMIC RETICULUM

There is abundant evidence that, in many mammalian species, the bulk of the Ca which activates contraction comes from the SR rather than entering directly across the surface membrane. Such evidence includes the fact that the changes of contraction or $[Ca^{2+}]_i$ which occur during the "staircase" are not accompanied by parallel changes of the L-type Ca current (Beeler & Reuter, 1970;Isenberg, 1982). In addition, systolic $[Ca^{2+}]_i$ is greatly decreased by drugs which interfere with SR Ca handling (Allen *et al.*, 1983;Hess & Wier, 1984;Marban & Wier, 1985;Kirby *et al.*, 1992;Negretti *et al.*, 1993). It should, however, be noted that the exact contribution from the SR is species dependent (Bers, 1985).

Some of the relevant ultrastructure is shown in Figs 1B & C. Fig. 1B shows the surface membrane of both ventricular and atrial cells. The presence of transverse (t) tubules in ventricular cells is shown by the banding pattern in the ventricular cell and, equivalently, the lack of banding in the atrial cells shows the absence of t-tubules. In mammalian ventricular myocytes electron microscopy studies have shown that the regions of the SR containing the Ca release channels (RyR) are very closely apposed to the surface membrane within the t-tubule network forming diads (Fig. 1C). Further, it is also within this region of close apposition that the L-type Ca channels appear to be localised (Sun *et al.*, 1995). This close anatomical relationship between the source of trigger Ca (the L-type Ca channel) and the site of SR Ca release (the RyR) has been argued to be of great functional significance in enabling tight control over the process of CICR (Stern, 1992;Soeller & Cannell, 1997;Cannell *et al.*, 1995). In cells without t-tubules (avian myocytes, mammalian atria and Purkinje fibres) an equivalent relationship exists between the SR and the surface membrane (Sommer, 1995;Sommer & Johnson, 1968;Junker *et al.*, 1994). In addition in all cardiac cells some of the SR (so called "corbular" SR) does not make contact with the surface membrane). Calcium is accumulated into the SR via a Ca-ATPase (SERCA) and released through a Ca release channel, the ryanodine receptor (RyR). Both these Ca transporters are discussed extensively below. There are also IP_3 receptors although their exact function is unclear (Lipp *et al.*, 2000;Verma *et al.*, 1992). The amount of Ca stored in the SR is an important determinant of contractility (see below). SR Ca content can be measured qualitatively by releasing Ca from the SR into the cytoplasm and measuring the resulting increase of either $[Ca^{2+}]_i$ or the consequent contraction. Both rapid cooling (Kurihara & Sakai, 1985;Bers & Bridge, 1988) and application of caffeine (Chapman & Léoty, 1976;Smith *et al.*, 1988;O'Neill *et al.*, 1990) have been used to release calcium. A quantitative measurement can be obtained in one of two ways. (1) If the cytoplasmic Ca buffering properties are known then the total amount of Ca released from the SR can be calculated (Bassani *et al.*, 1995a). (2) Since most of the calcium is pumped out of the cell by the electrogenic Na-Ca exchange, the amount of Ca pumped by this exchange can be calculated from the integral of the exchange current (Varro *et al.*, 1993). After correction for Ca removal by other mechanisms, the total amount of Ca released from the SR can be obtained (Fig.4). Both methods give Ca contents of up to 100 μmoles per litre cell volume. It should be noted that these values are of the total Ca content of the SR. The SR contains the Ca buffer calsequestrin which has a K_d for Ca of about 500 μM (Mitchell *et al.*, 1988). Measurements on isolated SR vesicles have characterized the intra-SR Ca buffering and found a K_d of 0.6 mM in fair agreement with that expected for calsequestrin (Shannon & Bers, 1997). Interestingly at 100 nM Ca outside the vesicles the intra-SR total Ca was about one half the maximum level. As pointed out by the original authors, it may be unwise to extrapolate measurements from isolated vesicles to the intact beating heart. Nevertheless, the buffering properties of the SR will mean that as the buffers become saturated a given increase of total SR Ca will produce a larger fractional increase of free SR Ca. This (see later) may have important consequences for the relationship between SR content and Ca release. Finally, in addition to calsequestrin, there is another Ca binding protein, calreticulin, which is expressed in the embryo and repressed postnatally. Overexpression of this compound results in sudden death (Nakamura *et al.*, 2001).

4.1. Ca Release from the SR

Ca release from the SR occurs through a specialized release channel known as the ryanodine receptor (RyR). This is a homotetramer (200 kD in molecular weight). The RyR is closely associated with other proteins. At the cytoplasmic surface it binds the FK506 binding protein (FKBP12.6) (Timerman *et al.*, 1996) and sorcin (Meyers *et al.*, 1995b). Interestingly sorcin can also bind to the L-type Ca channel and this has been suggested to be of significance in linking the L-type channel to the RyR (Meyers *et al.*, 1998). At the lumenal surface the RyR binds the Ca buffer calsequestrin (for a brief review see (Valdivia, 1998)). The FK binding protein stabilizes the RyR and its experimental removal results in the appearance of long lasting subconductance states resulting in an increased Ca leak flux (Brillantes *et al.*, 1994). The dissociation of FKBP12.6 from the RyR is increased by phosphorylation of the RyR (Marx *et al.*, 2000). Furthermore, overexpression of FKBP12 decreases the leak efflux of Ca from the SR (Prestle *et al.*, 2001). Sorcin has been found to decrease the open probability of the isolated SR and this effect is decreased by phosphorylation of sorcin (Lokuta *et al.*, 1997). Interestingly, an increase of $[Ca^{2+}]_i$ has been reported to translocate sorcin to membranes and this could therefore contribute to a Ca-dependent inactivation of Ca release (Meyers *et al.*, 1995a;Valdivia, 1998).

At least in skeletal muscle, the RyR shows the phenomenon of coupled gating (Marx et al., 1998) where two channels open and close simultaneously. In order to see this coupled gating the accessory protein FKBP12 also had to be present.

Some of the properties of the RyR which are important for this review are shown in Fig. 5, of particular relevance are the increase of open probability (p_o) of an increase of Ca^{2+} at either the cytosolic or lumenal faces and the opposing effects on channel p_o of caffeine and the local anaesthetic tetracaine.

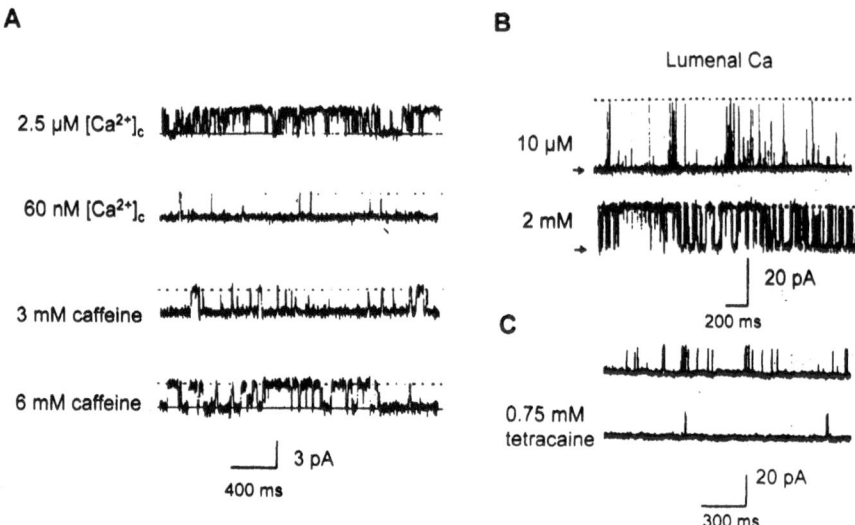

Fig. 5 Properties of isolated RyRs. *All records are taken from experiments on isolated RyRs reconstituted into lipid bilayers. A. Increasing [Ca²⁺] on the cytoplasmic surface or adding caffeine increase open probability (Rousseau & Meissner, 1989). B. Increasing [Ca²⁺] at the lumenal surface increases open probability (Sitsapesan & Williams, 1997). C. Tetracaine decreases the open probability (Györke et al., 1997). Where necessary, traces have been modified from the original such that channel opening corresponds to an upward deflection.*

4.1.1. Ca Induced Ca Release Triggered by the L-type Ca Current

The open probability of the RyR is increased by an increase of Ca concentration on the cytoplasmic surface (Rousseau *et al.*, 1986). The idea that Ca release from the SR could be induced via Ca influx into the cell (so called calcium induced Ca release or CICR) was championed by Fabiato using experiments on cells which had been "skinned" to remove their surface membrane (Fabiato 1985a,b). A variety of approaches then implicated CICR in intact cells and, furthermore, suggested that Ca entry via the L-type current triggered Ca release. (1) Removal of external Ca abruptly abolished contraction and the Ca transient (Fig. 6C) despite the fact that the SR still contained Ca (Smith *et al.*, 1988;Näbauer *et al.*, 1989). (2) Flash photolysis of "caged" Ca buffers to release a small amount of Ca produced a contraction (Fig. 6B) which required a functional SR (Valdeolmillos *et al.*, 1989;Näbauer & Morad, 1990;Niggli & Lederer, 1990). (3) The voltage dependence of the amplitude of the Ca transient can be similar to that of the L-type Ca current (Fig. 6A) and, in particular, both increase at first with depolarisation and then decrease as the Ca equilibrium potential is reached.   Importantly, following very strong depolarisations, Ca is released on *repolarization*, presumably because of entry of Ca ions into the cell (London & Krueger, 1986;Beuckelmann & Wier, 1988;Cannell *et al.*, 1987). It should, however, be noted that other studies have found more complicated effects of membrane potential (see later).

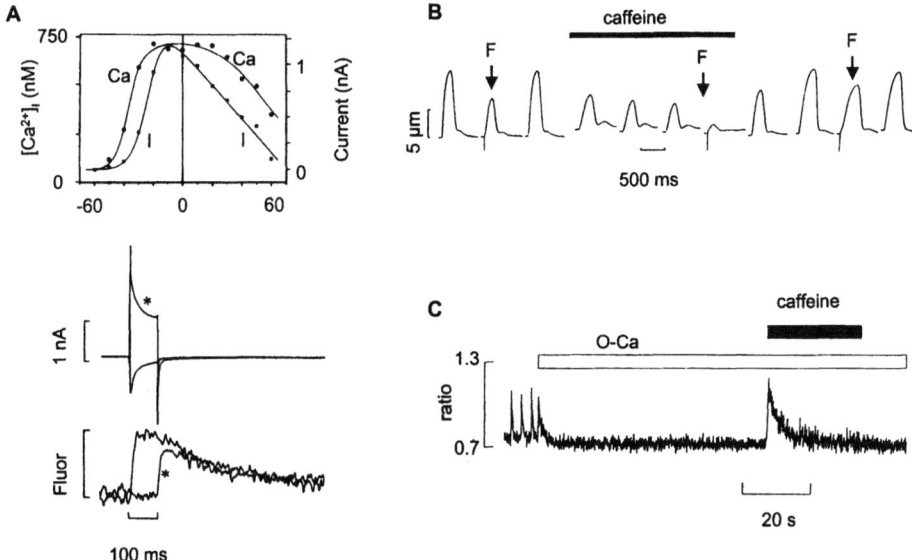

Fig. 6. Evidence showing CICR. A. Voltage dependence of the Ca transient. The top panel shows a graph of the amplitude of the L-type Ca current (I) and Ca transient (Ca) on depolarisation.   The bottom panel shows current and $[Ca^{2+}]_i$ (fluorescence) records produced by depolarisations to +10 mV and +100 mV (*). Note that depolarisation to +100 abolishes the Ca transient on depolarisation and elicits a transient on repolarization (Cannell et al., 1987). B. Flash photolysis of "caged" Ca results in an SR dependent contraction. The contractions were elicited by depolarising pulses except for those shown as (F) which were produced by photolytic release of Ca from DM-nitrophen. The application of caffeine (2 mM) decreases both depolarisation and photolytically- evoked responses (Näbauer & Morad, 1990). C. Removal of extracellular Ca abolishes the systolic Ca transient although the SR still contains Ca (Eisner, 1990).

4.1.2. Other Sources of Trigger Ca

In principle, any sarcolemmal Ca influx should be able to trigger Ca release from the SR. The physiologically interesting question is whether mechanisms in addition to Ca entry via the L-type Ca current contribute to Ca release under (patho)physiological conditions. Specifically can other mechanisms contribute on the timescale of the observed contraction, rather than simply resulting in a very slow response (Sipido *et al.*, 1997;Sipido *et al.*, 1998)? In order to answer this question it is important to measure the rate of Ca release or development of tension. At least two factors will determine the efficacy of a given total cell Ca entry to trigger release from the SR. (1) The anatomical relationship between the putative trigger and the RyR. Thus recent work has shown that dyadic region contains L-type Ca channels and RyR but not Na-Ca exchange or Na^+ channels (Scriven *et al.*, 2000). (2) The flux per site. Thus, for a given Ca entry into the cell, a trigger where entry occurs at a few points (situated near RyR) will be more effective than a case where entry is via many sites. In this context it is worth noting that the number of ions flowing per second through a channel such as the L-type Ca channel will be much greater than that through the Na-Ca exchange.

In addition to the L-type Ca current, other sources for trigger Ca have been suggested. Recent data have shown that the T-type current is considerably less effective than the L-type at triggering Ca release (Sipido *et al.*, 1998;Zhou & January, 1998). Another potential trigger is the Na-Ca exchange. This has been suggested to contribute Ca in two ways. (1) Depolarization will shift the driving force on Na-Ca exchange such that net Ca entry occurs (Fig. 2B). Evidence in favour of this is provided by studies which show that SR release can be triggered by depolarisation even in the apparent absence of the L-type Ca current (Nuss & Houser, 1992;Levi *et al.*, 1993;Kohmoto *et al.*, 1994;Wasserstrom & Vites, 1997). This effect of depolarisation was inhibited either by reduction of intracellular Na (Vornanen *et al.*, 1994;Levi *et al.*, 1996;Litwin *et al.*, 1998) or addition of the Na-Ca exchange inhibitor XIP (Kohmoto *et al.*, 1994). However other work has found that Na-Ca exchange (compared to the L-type Ca current) is a weak activator of Ca release from the SR (Sipido *et al.*, 1997;Sham *et al.*, 1995). Fig. 7 shows that, athough a Ca transient can still be observed when the L-type Ca current is inhibited, at least over the plateau range of potentials, its rate of rise is very slow suggesting that it will not contribute significantly to normal excitation-contraction coupling. This weak effect of Na-Ca exchange could be due to the fact that the Na-Ca exchange may be situated further away from the RyR than is the L-type Ca channel (Scriven *et al.*, 2000).

Another way in which Na-Ca exchange may produce a trigger Ca influx is in response to an increase of intracellular Na concentration. Thus it has been shown that, when the L-type Ca current is inhibited, there was SR release that was sensitive to inhibition by either addition of TTX or removal of external Ca. This was explained as due to initial Na entry through the TTX-sensitive Na channel increasing the local Na concentration near the membrane followed by Na efflux and Ca influx on Na-Ca exchange (Leblanc & Hume, 1990). Although similar results have been obtained by others (Lipp & Niggli, 1994) it has been argued that voltage clamp problems and changes of SR content may account for at least some of the observations (Bouchard *et al.*, 1993;Sham *et al.*, 1992;Sipido *et al.*, 1995b;Evans & Cannell, 1997).

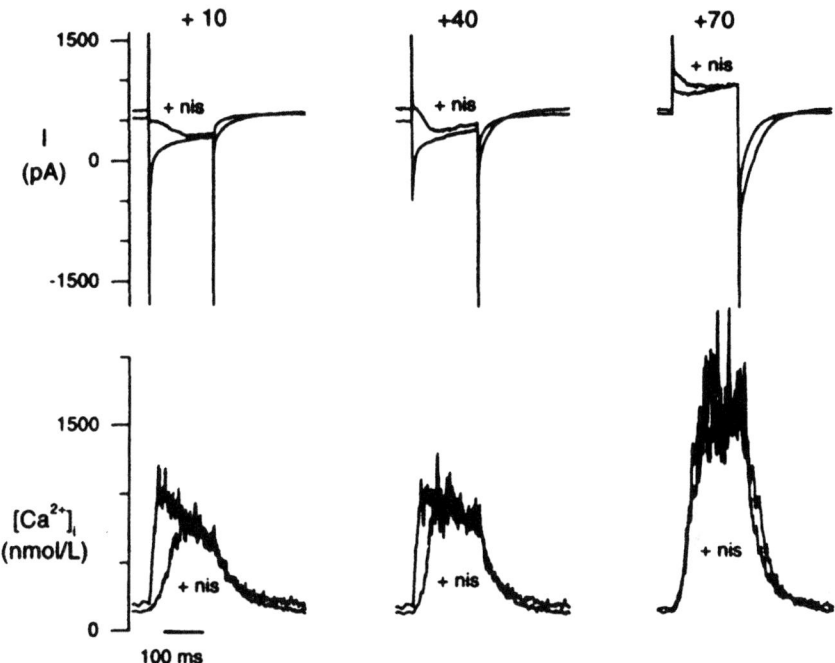

Fig. 7. Na-Ca exchange is a weak trigger for the Na-Ca exchange. *The traces show membrane current (top) and [Ca²⁺]ᵢ (bottom) in response to depolarising pulses to the potentials indicated above. Two traces are superimposed: control and with the L-type current blocked by 20 µM nisoldipine (+nis). (Sipido et al., 1997)*

The overall consensus today is that Ca entry via the Na-Ca exchange is not the major trigger for SR Ca release but it may serve to modify the effects of the L-type Ca current (Litwin *et al.*, 1998).

Yet another source for trigger Ca has been suggested to be Ca influx through Na channels. It has been reported that cAMP or ouabain can make classical Na channels become more selective to Ca^{2+} ions (or "promiscuous") such that the resulting Ca influx triggers Ca release from the SR (Santana *et al.*, 1998). The evidence for the increase of Ca permeability was that the reversal potential shifted in the positive direction. This result has been challenged by a later study investigating TTX-sensitive Na channels expressed in CHO cells where no evidence for such an effect was observed (Nuss & Marban, 1999). However another study by the original authors found that, in a different expression system (HEK cells) a shift of reversal potential was observed (Cruz *et al.*, 1999). Another example of a putative flux of Ca through Na channels is provided by $I_{Ca(TTX)}$ which is a Na current also blockable by TTX which has apparently distinct properties from the conventional TTX-sensitive Na channel and allows Ca permeation (Aggarwal *et al.*, 1997). A further complication in this area comes from work on expressed cardiac Na channels showing that, under appropriate conditions, they can produce a current component resembling $I_{Ca(TTX)}$ suggesting that this component may simply be a different mode of behaviour of the classical Na channel (Guatimosim et al., 2001). At this time, perhaps the best summary of this confusing area is that it has not yet been shown that Ca entry through any of the potential Na channels can produce Ca release equivalent to that produced by entry through the L-type channel.

4.1.3. Voltage Sensitive Ca Release

The last section raised the possibility that sources other than the L-type Ca current may contribute to the Ca trigger that activates Ca release from the SR. An even more radical suggestion is that a Ca trigger may not always be needed and that sarcolemmal depolarisation per se (rather than any induced Ca current) may trigger Ca release from the SR. Such a voltage dependent mechanism has long been known to exist in skeletal muscle (González & Ríos, 2001). Work in cardiac muscle has also suggested that when the Ca entry into the cell is inhibited, depolarisation still causes contraction (Ferrier & Howlett, 1995;Hobai et al., 1997). This mechanism requires conditions suggested to produce phosphorylation (Ferrier et al., 1998) and can be observed at membrane potentials more negative than those which activate the L-type current (see review (Ferrier & Howlett, 2001)). It has, however, been suggested that the apparent demonstration of Ca release in the absence of Ca entry into the cell could be due to inadequate voltage clamp control (Piacentino III et al., 2000). In summary, it is not yet clear that a very small fraction of uninhibited L-type Ca current cannot account for the observed e-c coupling in the presence of Ca channel blockers. If a voltage activated mechanism is important in the heart it is difficult to understand the results of expression studies in which either the cardiac or skeletal isoform of the DHP receptor (Ca channel) was expressed (García et al., 1994). When the cardiac type was expressed then a cardiac type of e-c coupling was observed with the amplitude of the Ca transient changing in parallel with that of the L-type Ca current. In contrast when the skeletal type was expressed contraction increased with depolarisation.

4.2. Ca Sparks

The use of confocal microscopy showed small (less than 1 μm) brief (decaying within a few tens of milliseconds) increases of $[Ca^{2+}]_i$ termed Ca "sparks" (Cheng et al., 1993) – see Fig. 8A. These Ca sparks were inhibited by blockers of SR function and shown to arise near the transverse tubules (Shacklock et al., 1995;Cheng et al., 1996) as would be expected if they are due to Ca release from the SR (Fig. 8B). The frequency of occurrence of sparks is increased by depolarisation (Shacklock et al., 1995;Cannell et al., 1994). Importantly, as shown in Fig. 8C, the voltage-dependence of spark occurrence was what would be expected if they were activated by the opening of single Ca channels (López-López et al., 1995;Cannell et al., 1995;Cleemann et al., 1998;Collier et al., 1999). Perhaps the most direct evidence linking Ca entry through L-type Ca channels to Ca sparks comes from

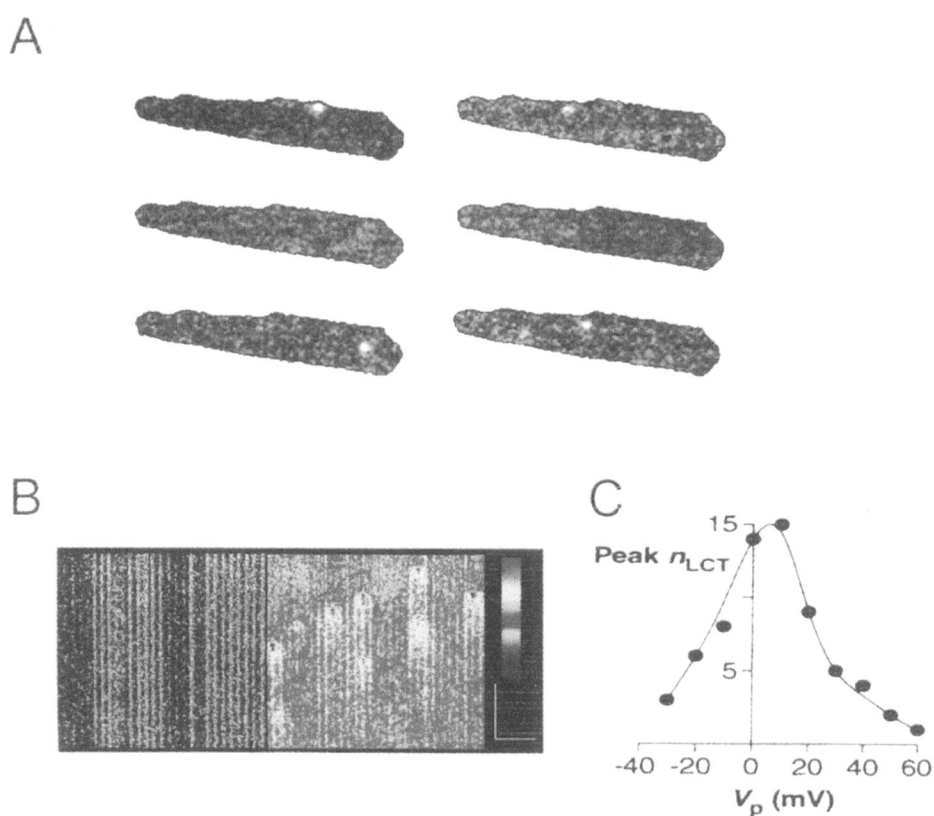

Fig. 8. Existence and properties of Ca sparks in ventricular cells. A. Confocal (XY) images of a rat ventricular myocyte (redrawn from Cheng et al., 1993). The images were obtained every 0.5 sec. B. Spatial origin of sparks. Both panels show linescans (time runs from top to bottom). The left hand traces show fluorescence of di-8 and therefore the bright lines correspond to the positions of the t-tubules. The right hand traces were recorded with fluo-3 and show Ca sparks occurring at t-tubules (Shacklock et al., 1995). C. Voltage dependence of spark occurrence. The graph shows spark frequency (referred to here as local Ca transients –LCT) as a function of membrane potential (López-López et al., 1995) Color version on page 453.

simultaneous cell-attached patch and confocal microscopy (Wang *et al.*, 2001). Wang et al first disabled the SR with caffeine and showed that they could measure small rises of $[Ca^{2+}]_i$ (termed sparklets) which were correlated with the opening of *single* L-type Ca channels as measured directly with the patch clamp. Importantly on some occasions, when the SR was functional (Fig. 9), the opening of a single L-type channel could be seen to activate a spark due to SR release thereby confirming directly the link between L-type Ca channel opening and SR Ca release.

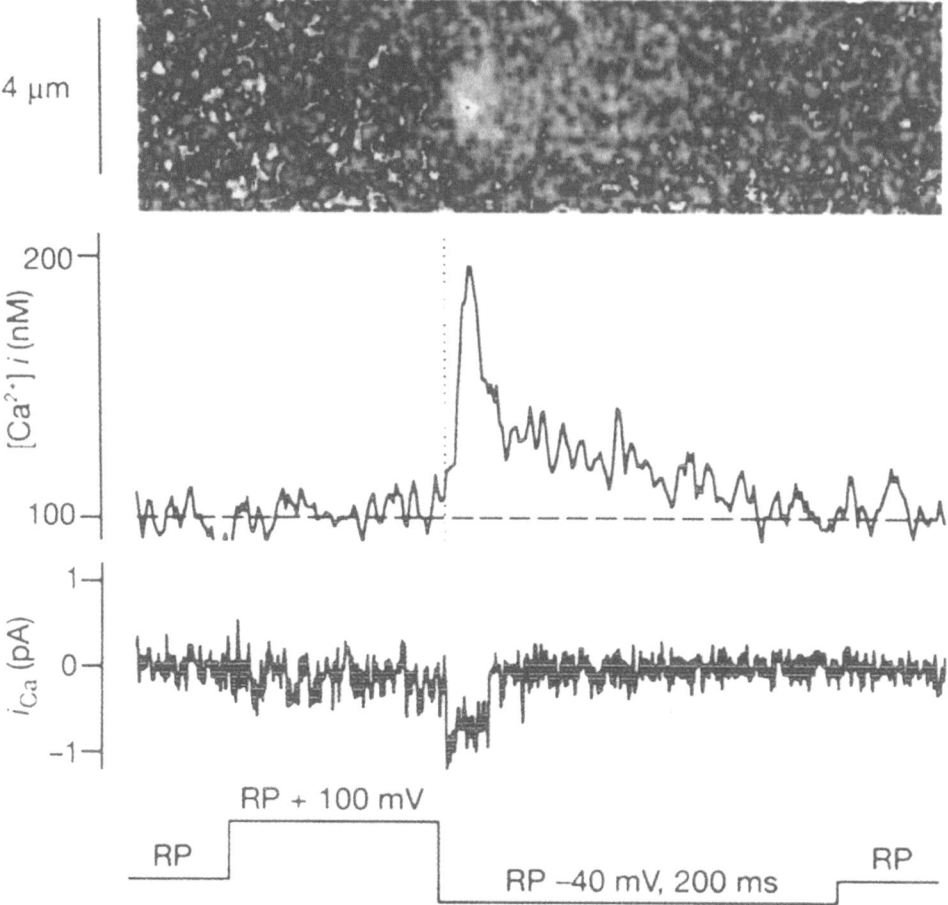

Fig. 9. Correlation of the opening of an L-type channel with a Ca spark. The cell-attached patch clamp technique was used. The membrane was initially depolarised by 100 mV. Repolariation then allows Ca entry through L-type channels accompanied by a spark (Wang et al., 2001). Color version on page 454.

4.2.1. How Many RyR Per Spark?

A very important question concerns the number of RyR that must open to produce a spark. While the original work was consistent with the opening of one RyR resulting in a spark (Cheng *et al.*, 1993), more recent work has questioned this. Work using flash photolysis of caged Ca compounds showed that the Ca release from the SR triggered by a photolytically-generated increase of $[Ca^{2+}]_i$ was uniform and did not result from measurable sparks (Lipp & Niggli, 1996). It was suggested, therefore, that the spark represents the opening of several RyR. A more recent study used two photon photolysis of caged Ca to produce local SR Ca release of smaller amplitude than the Ca spark perhaps reflecting Ca release from a single RyR (Lipp

& Niggli, 1998) again suggesting that the spark is due to the opening of more than one RyR. The recent work combining single channel measurements of L-type Ca channels and confocal microscopy allowed the magnitude of the spark (due to RyR release) to be compared with the sparklet resulting from an electrophysiologically measured Ca entry through the L-type channel. This gave an estimate of the Ca flux responsible for a spark as equivalent to a current of 2.1 pA which was estimated to result from the opening of 4 to 6 RyRs (Wang et al., 2001),

4.2.2. What Terminates Ca Release from the SR?

In order to obtain useful, graded Ca release from the SR, there must exist mechanism(s) to turn off Ca release. The release flux has been calculated from the measured $[Ca^{2+}]_i$ during a transient and calculated to be over within less than 40 msec (Sipido & Wier, 1991). More recently, Ca release during a spark was shown to decay with a time constant of 6-12 ms (Lukyanenko et al., 1998). Another and perhaps more direct way to measure the Ca release flux involves using a high concentration of EGTA (a slow buffer) in the presence of a fast, low affinity Ca indicator (Oregon Green). Under these conditions Ca released from the SR binds to the indicator and the resulting brief Ca transient or "spike" provides an index of Ca release from the SR (Song et al., 1998). The release flux decayed with a time constant of the order of 10 ms. Various explanations have been proposed for the decay of this release. (1) Inactivation of the Ca release process. It was originally suggested that an increase of $[Ca^{2+}]_i$ might initially increase Ca release but then result in inactivation (Fabiato, 1985b). This hypothesis came from experiments on skinned cells and subsequent work on intact cells has questioned its relevance (Näbauer & Morad, 1990). However, more recent work is supportive of a role for inactivation. Thus measurements of Ca spikes using the EGTA/Orgeon Green method (see above) showed that Ca release occurred on the initial depolarisation. Subsequent Ca entry through L-type Ca channels (produced by hyperpolarization) did not result in Ca release (Sham et al., 1998). Most convincingly, at any one release site, if release occurred on depolarisation it did not occur on repolarization whereas if no release had occurred on depolarisation (and therefore there was presumably no inactivation) then release could be seen on repolarization. (2) A related mechanism is termed "adaptation". This was discovered in experiments on RyR reconstituted into lipid membranes where a step increase of $[Ca^{2+}]$ produced an increase of open probability that "adapted" to a lower level. A further step increase of $[Ca^{2+}]$ resulted in a further adapting increase of open probability (Györke & Fill, 1993;Györke et al., 1994). It is not clear that the rate of this adaptation is sufficient to account for the turn off of release in the intact cell but this problem may be resolved by the fact that adaptation depends on accessory proteins such as FKBP and sorcin (Valdivia, 1998). (3) Another suggestion is that of "stochastic attrition". The idea is that once the L-type Ca channels have closed, $[Ca^{2+}]_i$ near the RyR depends on the number of RyR open. Each RyR will close randomly. If all RyR in this region close than $[Ca^{2+}]_i$ will fall to levels below that which activates RyR opening (Stern, 1992). Of course the probability that all the local RyRs will be closed at one time depends inversely on the number of RyR and therefore this hypothesis alone can only account for termination of Ca release if there are only a small number of RyR. Another argument against stochastic attrition is that larger Ca release fluxes did not inactivate more slowly as might be expected for this mechanism (Lukyanenko *et al.*, 1998). (4) A further possibility is that Ca release leads to depletion of the SR and this will decrease the release flux . This effect might be expected to be particularly important given the steep dependence between SR Ca content and Ca release (see below)(Lukyanenko *et al.*, 1996). However the fact that SR release terminates even in response to a small depolarisation which produces only a small Ca transient and therefore presumably does not deplete the SR is hard to reconcile with this hypothesis. Furthermore in experiments where apparent inactivation was observed, Ca could still be released by caffeine (Sham *et al.*, 1998). Work using photolysis of caged calcium in the whole cell found that as the interval between flashes was decreased the response decreased with a time constant of about 300 msec. Interestingly this behaviour was not seen if only local photolysis was used (DelPrincipe *et al.*, 1999). As the authors pointed out, the difference between the responses to local and whole cell photolysis is unlikely to be due to differences in $[Ca^{2+}]_i$ and therefore is hard to explain by inactivation. Rather, they suggest, it could be due to the fact that whole cell photolysis depletes the whole SR whereas the local depletion due to local photolysis may be quickly replenished by diffusion of Ca from adjacent regions of the SR network.

4.2.3. Synchrony of Ca Release

The original demonstration of Ca sparks showed that Ca release occurred from discrete sites. The discrete nature of these release sites was best seen under resting conditions. On depolarisation the frequency of sparks increased and the idea was therefore proposed that e-c coupling occurred by the recruitment of many sparks (Cannell *et al.*, 1995). Under conditions where the L-type Ca current was partially inhibited and therefore the number of sparks decreased then individual sparks could be seen even during depolarisation (Cheng *et al.*, 1995;López-López *et al.*, 1995). Interestingly along a single line scan there can be variability in the time of occurrence of each spark (Cannell *et al.*, 1994). A study using the high buffer technique has shown that Ca release is not always synchronized between different release sites (Song *et al.*, 1998). Furthermore, a recent study has shown that this lack of synchronization is much more evident in cells taken from infarcted hearts compared to those form control ones (Litwin *et al.*, 2000). Conversely, synchronization is increased during beta stimulation leading to the suggestion that an increased synchrony can be another way to increase the amplitude of the Ca transient (Song *et al.*, 2001). Future work will presumably elucidate the importance of synchrony/dyssynchrony in normal and abnormal e-c coupling.

4.3. Sub-Sarcolemmal Ca

The above discussion of Ca sparks makes it clear that near the RyR, the Ca concentration is much higher than in the bulk of the cytoplasm. Given the close proximity of the RyR to the sarcolemma, it is not therefore surprising that Ca near the sarcolemma can be different from that in the bulk. This means that the Ca concentration that affects, for example, the properties of sarcolemmal carriers and channels will be different from that in the bulk. For example the kinetics of inactivation of the Ca current behave as if the Ca concentration which is affecting the channel changes more quickly than the bulk cytoplasmic Ca (Adachi-Akahane *et al.*, 1996;Sipido *et al.*, 1995a). Similarly the caffeine-evoked change of both the Na-Ca exchange current (Trafford *et al.*, 1995) and the Ca-activated chloride current (Papp *et al.*, 1995;Trafford *et al.*, 1998a) appear to be controlled by faster changing Ca than that measured in the bulk cytoplasm. It is also worth noting that Ca is not the only ion that may have a different concentration near the sarcolemma as local Na gradients have also been suggested (Carmeliet, 1992;Bielen *et al.*, 1991;Fujioka *et al.*, 1998;Terracciano, 2001) and, in one case, measured directly (Wendt-Gallitelli *et al.*, 1993).

4.4. Heterogeneity of Excitation-Contraction Coupling Within the Heart

The bulk of the discussion in this review and, indeed, of experimental work has been performed on ventricular cells perhaps mainly because the ventricle makes the major contribution to cardiac output. Brief mention should, however, be made of other regions of the heart. A major difference between atrial and ventricular cells is that, in the majority of species, the former do not have t-tubules e.g. Figure 1B (Hüser *et al.*, 1996;Mitcheson *et al.*, 1997). A consequence of this is that the systolic Ca transient arises initially near the surface of the cell, where there is close apposition of the SR and sarcolemma and propagates into the interior (Hüser *et al.*, 1996). Even within the periphery of the cell where $[Ca^{2+}]_i$ rises first it has been shown that there are 'eager sites' where the SR seems to release Ca^{2+} preferentially before recruiting adjacent regions of SR and forming a wave of Ca^{2+} release propagating towards the cell interior (Mackenzie *et al.*, 2001). Similar peripheral Ca^{2+} release upon depolarisation and propagation into the interior is observed in Purkinje cells which, again, are devoid of t-tubules (Cordeiro *et al.*, 2001;Boyden *et al.*, 2000). The exact consequences of this type of propagated Ca^{2+} release on contractile performance and cellular electrophysiology remain to be determined in both the atrial and purkinge systems.

5. HOW IS CICR REGULATED?

As shown in Fig. 1A, there are at least three sites at which CICR can be regulated. (1) The Ca content of the SR: the more Ca there is in the SR, the greater the Ca release produced by the opening of a certain number of RyR. (2) The properties of the RyR, in particular the number of channels expressed and the relationship between open probability and $[Ca^{2+}]_i$. (3) The amplitude and kinetics of the sarcolemmal L-type Ca current. It will become clear from the following discussion that these control points cannot be considered independently.

5.1. Regulation of SR Ca Content and Effects of Systolic Ca

It should be obvious that the Ca content of the SR depends on the balance between Ca uptake into the SR (mediated by SERCA) and Ca release through the RyR. Much of the Ca in the SR is bound to the buffer calsequestrin (Campbell et al., 1983). It should, therefore, be remembered that free Ca (which presumably determines the flux of Ca through the RyR) may not change in exact proportion to the total Ca content of the SR. Overexpression of calsequestrin decreases SR Ca release despite an increase of SR Ca content (Jones et al., 1998;Wang et al., 2000).

5.1.1. Regulation of SERCA

The rate of Ca pumping by SERCA is increased by cytoplasmic Ca concentration and decreased by intra-SR Ca (Inesi & De Meis, 1989). The dependence on cytoplasmic Ca concentration will mean that manoeuvres that increase cytoplasmic Ca will increase SERCA activity and thence SR Ca content. This explains why decreasing Ca extrusion from the cell on Na-Ca exchange (e.g. by increasing intracellular (Smith et al., 1988;Bers & Bridge, 1988) or decreasing extracellular Na concentration (Meme et al., 2001) increase SR content. SERCA is also regulated by the inhibitory accessory protein phospholamban. The activity of SERCA is increased by phosphorylation of phospholamban (Hicks et al., 1979;Tada et al., 1974) and this contributes to the faster rate of decay of the systolic Ca transient during stimulation with beta adrenergic agonists (McIvor et al., 1988;Endoh & Blinks, 1988). An extreme equivalent of beta stimulation is seen in animals that have been bred to be deficient in phospholamban. Such animals have a smaller response than controls to beta stimulation as well as faster decaying systolic Ca transients (Li et al., 2000;Wolska et al., 1996) and increased SR Ca contents (Li et al., 1998). However, it should be remembered in these animals that are deficient in phospholamban there will still be positive inotropic effects of beta adrenergic stimulation thorough phosphorylation of other Ca^{2+} regulatory proteins e.g. the L-type Ca^{2+} channel and RyR.

Ultimately the extent to which SERCA can accumulate Ca into the SR will be determined by the free energy available from ATP hydrolysis. It has been suggested that this thermodynamic limits sets the Ca content of the SR (Shannon et al., 2000). While this effect may contribute it cannot be the only explanation since tetracaine and protons that act by decreasing the leak of Ca^{2+} from the SR lead to a large increase in SR Ca^{2+} content (Overend et al., 1997;Györke et al., 1997;Overend et al., 1997;Choi et al., 2000b).

5.1.2. Autoregulation of Sarcolemmal Fluxes and SR Ca Content

An important mechanism for controlling SR Ca content arises from the fact that Ca release from the SR modulates *sarcolemmal* Ca fluxes (Trafford *et al.*, 1997). Specifically, an increase of the amplitude of the systolic Ca transient increases Ca efflux from the cell on Na-Ca exchange (Barcenas-Ruiz *et al.*, 1987) and decreases Ca entry on the L-type Ca current by increasing the rate and extent of Ca-dependent inactivation (Adachi-Akahane *et al.*, 1996;Sipido *et al.*, 1995a;Grantham & Cannell, 1996). The interactions between these phenomena are shown in Fig. 10. In this

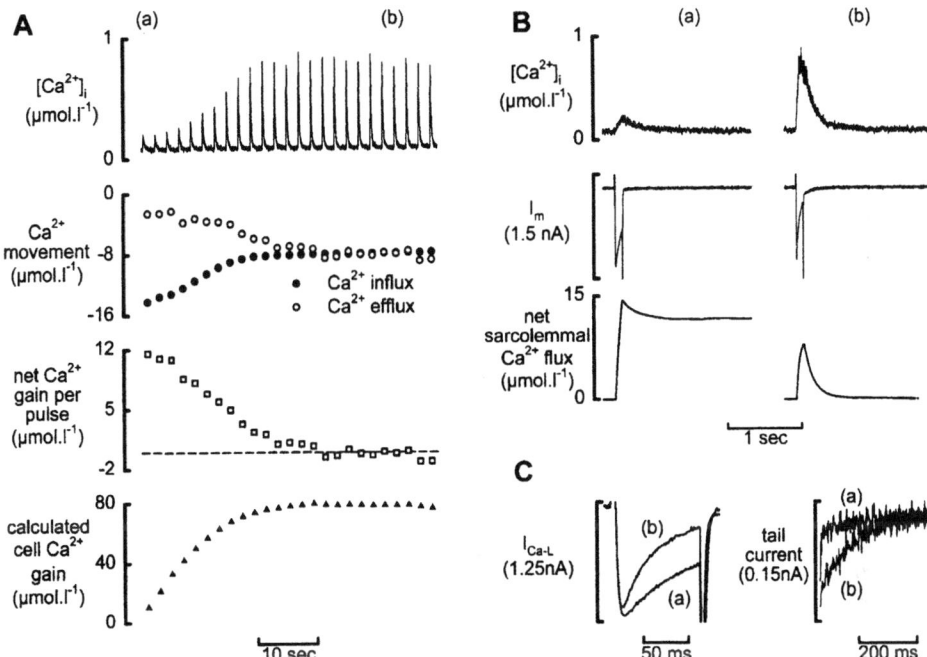

Fig. 10. Coordinated control of sarcolemmal Ca fluxes and SR Ca loading.
A. Before this record begins the SR Ca store had been depleted by application of 10mM caffeine. On removal of caffeine voltage clamp pulse were applied (-40 - 0mV, 100ms, 0.5 Hz). The panels show (top to bottom) $[Ca^{2+}]_i$; Sarcolemmal Ca movements obtained by integrating the L-type Ca current on depolarisation (filled symbols) and the Na-Ca exchange tail current on repolarisation (open symbols); Net Ca gain on each pulse (influx on I_{Ca-L} – efflux on Na-Ca exchange); Calculated cell Ca gain obtained from the cumulative net Ca fluxes on each pulse. B. Speciment records on expanded timescale showing (top to bottom) $[Ca^{2+}]_i$; Membrane current; Net sarcolemmal Ca movement. (a) represents the first pulse on resuming stimulation after depleting the SR Ca store and (b) once a steady state Ca transient has been reached. C. Expanded membrane current records obtained from the pulses in B. Records show (left) I_{Ca-L} and Na-Ca exchange tail current (right). Note the increased rate of inactivation of I_{Ca-L} on the steady state pulse (b) compared to the first pulse (a) and the increased Na-Ca exchange tail current (b) due to the increased Ca transient amplitude. Adapted from Trafford et al (1997).

experiment the SR had been emptied by application of 10 mM caffeine. Fig. 10A shows that when stimulation was recommenced the Ca transient was initially very small, as a result of the low SR Ca content. However, over the timecourse of a few stimuli, the amplitude of the Ca transient increases. Specimen membrane currents are shown in Fig. 10B. Trace *b* was recorded in the steady state. The systolic Ca transient is accompanied by Ca entry on the L-type Ca current accompanied by efflux on Na-Ca exchange (best seen from the expanded tail current of Fig. 10C. The integrated traces show that (as in Fig. 1D) Ca influx = efflux. A very different situation is seen for the first Ca transient (trace *a*). Here the small Ca transient is accompanied by a larger Ca current which inactivates more slowly than in *b* (left hand panel Fig. 10C). The Ca efflux on Na-Ca exchange is smaller than in *b* (right hand panel Fig. 10C). This has the effect that, rather than influx and efflux being equal, on this first pulse there is net Ca entry. Similar calculations can be made for each pulse and the results are shown in Fig. 10A. It can be seen that, initially, Ca entry is greater than exit. However, with time, the entry decreases and the exit increases until the two are in balance. This is emphasised by the third panel of Fig. 10C which shows the difference between entry and exit i.e. the net entrance per pulse. These net entries can be summed in order to calculate the net gain by the cell of Ca. In this particular experiment the cell had gained 80 µmol of Ca per litre cell volume.

The fact that the magnitude of the systolic Ca transient influences sarcolemmal Ca fluxes allows the cell to control its and therefore the SR Ca content. If it was not for this mechanism it is difficult to see how SR Ca content could be controlled and steady state contractions could be achieved (Eisner *et al.*, 1998;Eisner *et al.*, 2000)

5.1.3. The Relationship Between SR Ca Content and Systolic Ca
The reason that it is essential to regulate SR content is that the amplitude of the systolic Ca transient is a steep function of the content. Evidence supporting this can be seen in Fig. 10: by pulse number 4 the SR Ca content has recovered to 50% of its final value yet there is only a small recovery of the amplitude of the Ca transient. Such a steep relationship was shown first in previous studies (Bassani *et al.*, 1995a;Spencer & Berlin, 1995). At least three factors may contribute to this relationship. (1) The higher the SR Ca content, the larger the driving force for Ca efflux from the SR. Any tendency towards saturation of the Ca buffer calsequestrin will mean that a given increase of total SR Ca will lead to a fractionally greater change of free Ca and thence of driving force. (2) The larger the Ca efflux from the SR, the more the activation of adjacent RyRs (Spencer & Berlin, 1995;Spencer & Berlin, 1997). (3) It may originate from the fact that an increase of $[Ca^{2+}]$ on the lumenal face of the RyR increases the open probability (Sitsapesan & Williams, 1994;Sitsapesan & Williams, 1997;Lukyanenko *et al.*, 1999). Related to this, the frequency of Ca sparks has been reported to increase with an increase of SR content (Satoh *et al.*, 1997) although another study has not found this result (Song *et al.*, 1997).

The fact that Ca release from the SR is a steep function of SR content and the dependence of Ca influx and efflux on the Ca transient means that even small changes of SR content may have significant effects on sarcolemmal fluxes. We have found, for example, that Ca efflux is proportional to (SR content)$^{3.3}$ (Trafford et al., 2001). This steep relationship will allow tight control of SR content but, at least in principle, it may make the control unstable. If the SR content increases then the extra efflux from the cell may decrease the content to below the steady state value. At this lower content the efflux may be so small that, on the next stimulus the content increases to a higher level resulting in alternation of SR content and systolic Ca from beat to beat (Eisner et al., 2000). Although other mechanisms may also contribute, this feedback may contribute to the phenomenon of alternans where alternate large and small contractions are produced.

5.2. Modulation of the RyR

As mentioned above, the open probability of the RyR is increased by an increase of $[Ca^{2+}]_i$ and this is the basis of CICR. In addition, other ligands can change the properties of the RyR. In fact, the RyR has a very broad pharmacology, a fact that may be related to its very large size. Recent reviews have catalogued many of the substances which affect RyR function (Zucchi & Ronca-Testoni, 1997;Meissner, 1994). Of particular relevance to this review, many substances including caffeine (Rousseau & Meissner, 1989), 2,3-butanedione monoxime (BDM) (Tripathy et al., 1999) and cyclic ADP ribose (Sitsapesan et al., 1994;Sitsapesan & Williams, 1995) increase the open probability of the RyR. Among the manoeuvres which decrease the open probability are the local anaesthetic tetracaine (Tinker & Williams, 1993;Xu et al., 1993;Györke et al., 1997), or a decrease of either pH or [ATP] (Xu et al., 1996).

Given the above, it was thought that altering the open probability of the RyR would affect the amplitude of the systolic Ca transient. Thus a manoeuvre that decreases the open probability would be expected to decrease systolic Ca. Equally decreasing the number of RyRs might decrease systolic Ca. However, as shown in Fig. 11, the experimental observations are very different from these expectations. This figure shows that agents which increase the open probability such as caffeine (A) (O'Neill & Eisner, 1990;Trafford et al., 2000) or BDM (B) (Adams et al., 1998) produce only a transient increase of systolic $[Ca^{2+}]_i$ whereas those that decrease it such as acidification by the weak acid butyrate (C) (Choi et al., 2000b) or the addition of tetracaine (D) (Overend et al., 1998) produce only a transient decrease of systolic $[Ca^{2+}]_i$. The mechanism of these transient effects is shown in Fig. 12. Fig. 12Aa shows the timecourse of changes of systolic $[Ca^{2+}]_i$ produced by adding a low concentration of caffeine (500 µM). This results in a transient increase of systolic $[Ca^{2+}]_i$ followed by an undershoot on removal of caffeine. The amplitude and timecourse of the transient in the steady state in caffeine are identical to that in control (Trafford et al., 2000). Fig. 12B shows specimen records for the control (i) and second transient in caffeine (ii). As was shown in Fig. 1, in control the Ca entry via the L-type Ca current balances the efflux on Na-Ca exchange (Bb &Bc). Trace

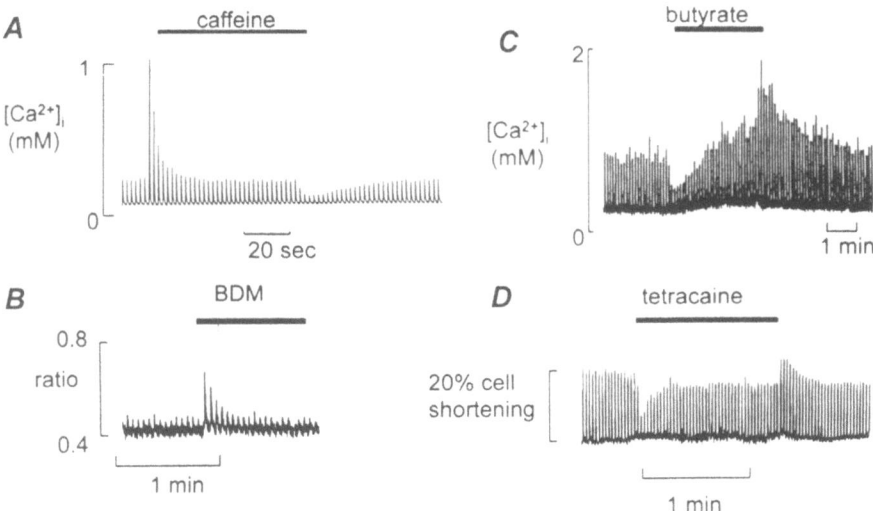

Fig. 11 *Effects of various modifiers of RyR function on systolic $[Ca^{2+}]_i$ or contraction.*
*Panels show the effects of : **A**. caffeine (0.5 mM); **B**. BDM (butane dione monoxime);*
***C**. Butyrate (30 mM at constant external pH); 5 mM; **D**. tetracaine (100 µM).*

ii shows that the large transient in caffeine is accompanied by an increased efflux on
Na-Ca exchange such that, rather than being in balance, the cell is predicted to lose
Ca. Measurements of net sarcolemmal Ca fluxes for all pulses are shown in Fig.
Ab. It is clear that Ca^{2+} influx = Ca^{2+} efflux in both the control and steady state in
caffeine. However, following application of caffeine Ca^{2+} efflux is transiently
greater than Ca^{2+} influx. When these measurements are used to calculate changes of
total cell Ca, it can be seen that the fade of the Ca transient in caffeine is associated
with a decrease of SR Ca content (Ac). Trace Ad shows the calculated fractional
release of Ca from the SR. This increases initially due to the direct effect of caffeine
before decaying to a level which is still greater than control. This secondary
decrease is due to loss of SR Ca decreasing fractional release. This decreased SR
content has been measured directly, by applying 10mM caffeine to discharge the SR
Ca^{2+} store and integrating the resulting Na-Ca exchange current (Trafford *et al.*,
2000). Similarly potentiation of CICR by BDM is also accompanied by a measured
decrease of SR content (Adams *et al.*, 1998) whereas the depression of CICR
produced by acidosis (Choi *et al.*, 2000b) or tetracaine (Overend *et al.*, 1998)

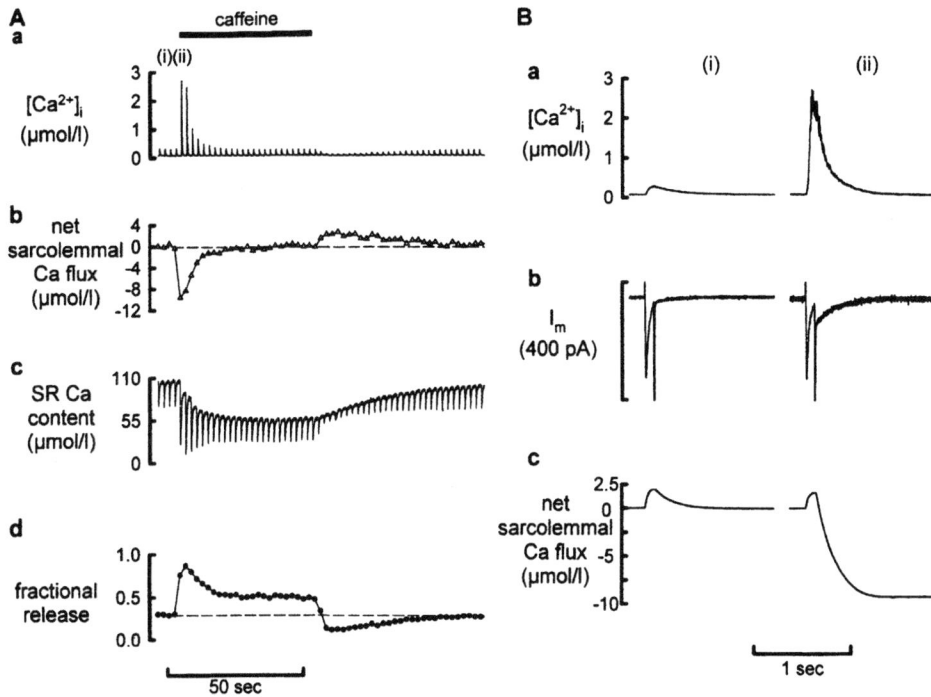

***Fig. 12. Changes of sarcolemmal fluxes and SR Ca²⁺ release during potentiation of CICR
by caffeine (500 μM). A. Time course.*** *Traces show (from top to bottom): a, [Ca²⁺]ᵢ; b, net
sarcolemmal flux calculated as Ca entry on Iₑₐ minus efflux on Na-Ca exchange; c, SR Ca
content; d, fractional release, the fraction of the SR Ca content which is released.* ***B.***
*Specimen records from a control transient (i) and the first in caffeine (ii). For each transient
records show (from top to bottom): [Ca²⁺]ᵢ, membrane current; calculated movement of Ca
across the surface membrane. Note that for transient (i), influx = efflux whereas, for (ii) the
efflux is greater than the influx. Data from a rat myocyte. Taken from Eisner et al. (2000)
modified from Trafford et al. (2000).*

produce transient decreases in systolic [Ca²⁺]ᵢ but this is accompanied by increases
of SR content and recovery of the systolic Ca²⁺ transient back to control levels.

The transient nature of the changes of systolic Ca content produced by agents
which only affect the RyR result from the effects of systolic Ca on sarcolemmal
fluxes reviewed above in association with Fig. 10. Thus, in the steady state, Ca
influx = Ca efflux. If the influx does not change then the efflux must be constant.
Therefore, as long as the properties of the Na-Ca exchange do not alter, this
requirement for a constant Ca efflux requires that the amplitude of the Ca transient
be identical to control in the steady state presence of the modulator of the RyR.

The above arguments apply equally to situations in which the number rather than
the properties of the RyR are altered. Thus it has been suggested that in heart failure
the degree of expression of the RyR may be decreased and that this may result in a

decreased amplitude of the systolic Ca transient (Vatner et al., 1994;Hittinger et al., 1999;Yamamoto et al., 1999;Teshima et al., 2000). The argument above would simply predict that, if the number of RyR decreases, the SR Ca content will increase until the combination of an increased SR content and a decreased number of RyR restores the Ca efflux to control levels.

This analysis assumes that the RyR only alters during systole. It is, however, well known that a sufficient increase of the RyR open probability will make the SR leaky to Ca even during diastole. Thus high concentrations of caffeine deplete the SR of calcium even in the absence of stimulation (Trafford et al., 1997) and therefore increased opening of the RyR results in a decreased systolic Ca transient. A similar mechanism has been suggested to account for the decreased systolic transient in heart failure (Marx et al., 2000) where it was found that, in failing hearts, the RyR was hyperphosphorylated leading to increased Ca leak through the presence of long lasting subconductance states of the RyR .

5.2.1. The Observed Effects of Natural Substances which Affect the RyR P_o

The above analysis runs counter to much work suggesting that agents which increase the open probability of the RyR will increase the amplitude of the systolic Ca transient. As an example, we will consider one such compound cyclic ADP Ribose (cADPR) which releases Ca from intracellular stores in a variety of cell types (for review see (Galione, 1993)). Studies in cardiac muscle have found that cADPR increases and its antagonists decrease the amplitude of the systolic Ca transient (Rakovic et al., 1996;Iino et al., 1997;Cui et al., 1999). Furthermore cADPR increases the open probability of the RyR (Mészáros et al., 1993). If the only action of cADPR is to increase the open probability of the RyR then these experimental results do not support the analysis of the previous section. It is interesting, therefore, to note that a recent paper has reported that the increase of spark frequency produced by cADPR is *not* due to a direct effect on the RyR but, rather, is secondary to a stimulation of SERCA by cADPR leading to an increase of SR content (Lukyanenko et al., 2001).

5.3. Control of Ca Entry Into the Cell

It has been known for many years that manoeuvres which increase Ca entry into the cell increase the amplitude of the Ca transient and/or the force of contraction (Beeler & Reuter, 1970;Barcenas-Ruiz & Wier, 1987;Cannell et al., 1987). In terms of the above discussion, this could occur via two mechanisms: (1) increased trigger Ca resulting in the opening of more RyR; (2) Increased loading of the cell with calcium and thereby an increased SR Ca content.

A recent study has shown that increasing external Ca concentration (Ca_o) from 1 to 2 mM produces a significant increase of systolic $[Ca^{2+}]_i$ but has *no* effect on SR Ca content. Lowering Ca_o to 0.2 mM produced a marked decrease of systolic Ca_o accompanied by a modest *increase* of SR content. Over the whole range studied (0.2 to 2 mM Ca_o) there was a ~10 fold *increase* in the amplitude of systolic $[Ca^{2+}]_i$

accompanied by a ~15% *decrease* of SR content (Fig. 13). In other words, when external Ca and therefore Ca entry via the L-type Ca current is changed, there is

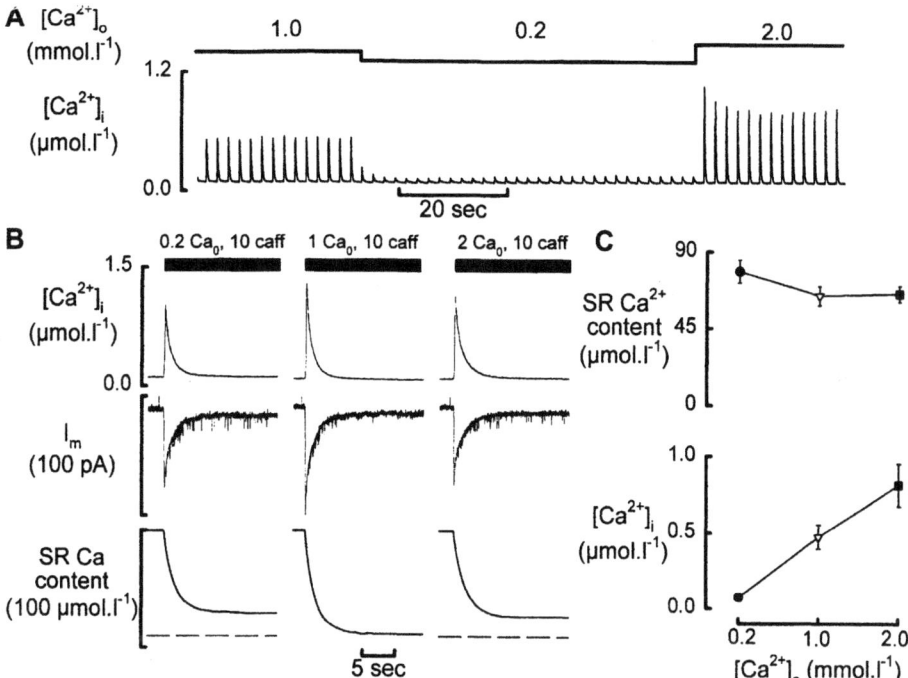

*Fig.13 Coordinated control of Ca release and loading in response to changes of Ca entry. A. Timecourse of change of systolic $[Ca^{2+}]_i$ in response to the changes of external Ca concentration indicated above. **B.** SR Ca contents following stimulation in the various external Ca concentrations. SR content was measured by applying 10 mM caffeine and integrating the resulting Na-Ca exchange current. In each panel traces show: top, $[Ca^{2+}]_i$; middle, current; bottom, integrated current. **C.** Mean data showing systolic $[Ca^{2+}]_i$ (top) and SR content (bottom) as a function of external Ca (Trafford et al. 2001).*

very little effect on SR content (Trafford *et al.*, 2001). The explanation of this observation is that, as far as the SR content is concerned, increasing Ca entry has two opposing effects. (1) It increases the net Ca entry into the cell. (2) It triggers Ca release from the SR thereby resulting in the loss of Ca from the cell via Na-Ca exchange. The relative constancy of SR content over a wide range of Ca_o must mean that these two mechanisms balance each other. In other words the triggering and loading effects of the L-type Ca current are coordinated to keep SR Ca content constant. This has two important consequences (Eisner & Trafford, 2000). (1) It allows an increase of L-type Ca current to increase contraction without the need to increase SR content. This is potentially advantageous since an increase of SR content both requires increased energy dependent uptake of Ca into the SR and predisposes to spontaneous and arrhythmogenic Ca release from the SR (Orchard *et al.*, 1983;Ferrier *et al.*, 1973;Wier *et al.*, 1983). (2) The lack of an increase of SR content means that changing Ca entry into the cell can produce an immediate effect on systolic $[Ca^{2+}]_i$. In contrast, an increase of SR content takes several beats to develop (see Fig. 10 & 13A).

The fact that the L-type Ca current has two roles inasmuch as it triggers Ca release from the SR and loads the cell with Ca is important and, as reviewed above, allows the SR Ca content to be maintained in the face of a change of trigger. This can be compared with the situation resulting from a voltage-dependent release mechanism. If such a mechanism is potentiated then Ca will be lost from the SR and a transient increase of contractility expected. It is worth noting that in skeletal muscle where a voltage dependent mechanism is well-established, efflux of Ca across the surface membrane is much less important than in the heart and these considerations do not apply.

The properties of the L-type Ca channel itself are also fundamental to e-c coupling. The trigger role depends on the amount of Ca^{2+} that enters initially on depolarisation and therefore on the peak current. In contrast the loading function depends on the integrated Ca entry. If we approximate the action potential as a square pulse so that the L-type Ca current decays exponentially then the total Ca entry = peak current * time constant of decay. Therefore for a given peak current and therefore trigger, the greater the time constant of decay of the Ca current, the larger the loading and thence the greater the SR content.

5.4. Physiological Inotropy: How is E-C Coupling Regulated?

In the previous section we have mentioned the effect of increasing the L-type Ca current and pointed out that the coordination of the increased Ca loading and increased trigger results in a larger systolic Ca transient with no change of SR content. An increase of the Ca current is one component of the β receptor mediated effects of catecholamines. However β-stimulation also has other effects. Phosphorylation of phospholamban will increase the activity of SERCA . This effect alone accounts for the observed increase of the rate of decay of systolic $[Ca^{2+}]_i$ (Allen & Kurihara, 1980;McIvor *et al.*, 1988;Spurgeon *et al.*, 1990). Stimulation of SERCA should increase the Ca content of the SR because SERCA now competes more effectively with sarcolemmal extrusion mechanisms. It is difficult to predict exactly how much a given degree of SERCA stimulation will increase the SR Ca content. In the steady state the Ca efflux from the cell must be unaffected by stimulation of SERCA. The decrease of the duration of the Ca transient will decrease the amount of Ca pumped out of the cell on Na-Ca exchange and this must be compensated for by an increase in the amplitude of the systolic Ca transient as is seen physiologically during ß-adrenergic stimulation (Hussain & Orchard, 1997). If we assume that the rate constant of decay of the Ca transient is doubled then (assuming that efflux is proportional to $[Ca^{2+}]_i$) the amplitude must also be doubled. Because the amplitude of the Ca transient is a steep function of SR content then only a small increase of content may be required to achieve this magnitude of increase of the amplitude of the Ca transient. Specifically since Ca efflux varies as $\sim SR^3$ then one would expect that only a 25% increase of SR content would be observed.

5.5. Where Next?

Although an enormous amount has been learnt about the mechanisms regulating various steps of excitation-contraction coupling, it is important to note that there is very little data on the magnitude of the SR content and systolic Ca under conditions of normal temperature, heart rate and autonomic stimulation. Equally, much more information is required concerning the regulation of these parameters in major disease conditions such as heart failure.

6. ACKNOWLDEGEMENTS

Work from the authors laboratory was supported by grants from the British Heart Foundation and Wellcome Trust.

Unit of Cardiac Physiology
The University of Manchester
1.524 Stopford Building
Oxford Road
Manchester M13 9PT
Email: eisner@man.ac.uk
Tel: +44 161 275 2702
Fax: +44 161 275 2703

7. REFERENCES

Adachi-Akahane, S., Cleemann, L., & Morad, M. (1996). Cross-signaling between L-type Ca^{2+} channels and ryanodine receptors in rat ventricular myocytes. *Journal of General Physiology* **108**, 435-454.

Adams, W. A., Trafford, A. W., & Eisner, D. A. (1998). 2,3-butanedione monoxime (BDM) decreases sarcoplasmic reticulum Ca content by stimulating Ca release in isolated rat ventricular myocytes. *Pflügers Archiv* **436**, 776-781.

Aggarwal, R., Shorofsky, S. R., Goldman, L., & Balke, C. W. (1997). Tetrodotoxin-blockable calcium currents in rat ventricular myocytes; a third type of cardiac cell sodium current. *Journal of Physiology* **505**, 353-369.

Allen, D. G., Eisner, D. A., Lab, M. J., & Orchard, C. H. (1983). The effects of low sodium solutions on intracellular calcium concentration and tension in ferret ventricular muscle. *Journal of Physiology* **345**, 391-407.

Allen, D. G. & Kurihara, S. (1980). Calcium transients in mammalian ventricular muscle. *European Heart Journal* **1**, 5-15.

Antoine, S., Pinet, C., & Coulombe, A. (2001). Are B-type Ca^{2+} channels of cardiac myocytes akin to the passive ion channel in the plasma membrane Ca^{2+} pump? *Journal of Membrane Biology* **179**, 37-50.

Ballard, C., Mozaffain, M., & Schaffer, S. (1994). Signal transduction mechanism for the stimulation of the Na^+/Ca^{2+} exchanger by insulin. *Molecular and Cellular Biochemistry* **135**, 113-119.

Ballard, C. & Schaffer, S. (1996). Stimulation of Na^+/Ca^{2+} exhanger by phenylephrine, angiotensin II and endothelin 1. *Journal of Molecular and Cellular Cardiology* **28**, 11-17.

Barcenas-Ruiz, L., Beuckelmann, D. J., & Wier, W. G. (1987). Sodium-calcium exchange in heart: Membrane currents and changes in $[Ca^{2+}]_i$. *Science* **238**, 1720-1722.

Barcenas-Ruiz, L. & Wier, W. G. (1987). Voltage dependence of intracellular $(Ca^{2+})_i$ transients in guinea pig ventricular myocytes. *Circulation Research* **61**, 148-154.

Barry, W. H., Rasmussen, C. A., Jr., Ishida, H., & Bridge, J. H. (1986). External Na-independent Ca extrusion in cultured ventricular cells Magnitude and functional significance. *Journal of General Physiology* **88**, 393-411.

Bassani, J. W. M., Yuan, W., & Bers, D. M. (1995a). Fractional SR Ca release is regulated by trigger Ca and SR Ca content in cardiac myocytes. *American Journal of Physiology* **268**, C1313-C1329.

Bassani, R. A., Bassani, J. W. M., & Bers, D. M. (1995b). Relaxation in ferret ventricular myocytes: role of the sarcolemmal Ca ATPase. *Pflügers Archiv* **430**, 573-578.

Beeler, G. W. & Reuter, H. (1970). The relation between membrane potential, membrane currents and activation of contraction in ventricular myocardial fibres. *Journal of Physiology* **207**, 211-229.

Berlin, J. R., Bassani, J. W. M., & Bers, D. M. (1994). Intrinsic cytosolic calcium buffering properties of single rat cardiac myocytes. *Biophysical Journal* **67**, 1775-1787.

Bers, D. M. (1985). Ca influx and sarcoplasmic reticulum Ca release in cardiac muscle activation during postrest recovery. *American Journal of Physiology* **248**, H366-H381.

Bers, D. M. (2001). *Excitation-Contraction Coupling and Cardiac Contractile Force.*, 2 ed. Kluwer Academic Publishers, Dordrecht/Boston/London.

Bers, D. M. & Bridge, J. H. B. (1988). Effect of acetylstrophanthidin on twitches, microscopic tension fluctuations and cooling contractures in rabbit ventricle. *Journal of Physiology* **404**, 53-69.

Beuckelmann, D. J. & Wier, W. G. (1988). Mechanism of release of calcium from sarcoplasmic reticulum of guinea-pig cardiac cells. *Journal of Physiology* **405**, 233-255.

Beuckelmann, D. J. & Wier, W. G. (1989). Sodium-calcium exchange in guinea-pig cardiac cells: exchange current and changes in intracellular Ca^{2+}. *Journal of Physiology* **414**, 499-520.

Bielen, F. V., Glitsch, H. G., & Verdonck, F. (1991). Changes of the subsarcolemmal Na^+ concentration in internally perfused cardiac cells. *Biochimica et Biophysica Acta* **1065**, 269-271.

Blaustein, M. P. & Lederer, W. J. (1999). Sodium/calcium exchange: its physiological implications. *Physiological Reviews* **79**, 763-854.

Bouchard, R. A., Clark, R. B., & Giles, W. R. (1993). Role of sodium-calcium exchange in activation of contraction in rat ventricle. *Journal of Physiology* **472**, 391-413.

Boyden, P. A., Pu, J., Pinto, J., & ter Keurs, H. E. D. J. (2000). Ca^{2+} transients and Ca^{2+} waves in Purkinje cells. Role in action potential initiation. *Circulation Research* **86**, 448-455.

Brillantes, A.-M. B., Ondrias, K., Scott, A., Kobrinsky, E., Ondriasova, E., Moschella, M. C., Jayaraman, T., Landers, M., Ehrlich, B. E., & Marks, A. R. (1994). Stabilization of calcium release channel (ryanodine receptor) function by FK506-binding protein. *Cell* 77, 513-523.

Cachelin, A. B., DePeyer, J. E., Kokubun, S., & Reuter, H. (1983). Ca^{2+} channel modulation by 8-bromocyclic AMP in cultured heart cells. *Nature* 304, 462-464.

Campbell, K. P., MacLennan, D. H., Jorgensen, A. O., & Mintzer, M. C. (1983). Purification and characterization of calsequestrin from canine cardiac sarcoplasmic reticulum and identification of the 53,000 dalton glycoprotein. *Journal of Biological Chemistry* 258, 1197-1204.

Cannell, M. B., Berlin, J. R., & Lederer, W. J. (1987). Effect of membrane potential changes on the calcium transient in single rat cardiac muscle cells. *Science* 238, 1419-1423.

Cannell, M. B., Cheng, H., & Lederer, W. J. (1994). Spatial Non-Uniformities of $[Ca^{2+}]_i$ during excitation-Contraction Coupling in Cardiac Myocytes. *Biophysical Journal* 67, 1942-1956.

Cannell, M. B., Cheng, H., & Lederer, W. J. (1995). The control of calcium release in heart muscle. *Science* 268, 1045-1049.

Carafoli, E. (1985). The homeostasis of calcium in heart cells. *Journal of Molecular and CellularCardiology* 17, 203-212.

Carmeliet, E. (1992). A fuzzy subsarcolemmal space for intracellular Na^+ in cardiac cells? *Cardiovascular Research* 26, 433-442.

Chapman, R. A. & Léoty, C. (1976). The time-dependent and dose-dependent effects of caffeine on the contraction of the ferret heart. *Journal of Physiology* 256, 287-314.

Cheng, H., Cannell, M. B., & Lederer, W. J. (1995). Partial inhibition of Ca^{2+} current by methoxyverapamil (D600) reveals spatial nonuniformities in $[Ca^{2+}]_i$ during excitation contraction coupling in cardiac myocytes. *Circulation Research* 76, 236-241.

Cheng, H., Lederer, M. R., Lederer, W. J., & Cannell, M. B. (1996). Calcium sparks and $[Ca^{2+}]_i$ waves in cardiac myocytes. *American Journal of Physiology* 270, C148-C159.

Cheng, H., Lederer, W. J., & Cannell, M. B. (1993). Calcium sparks: elementary events underlying excitation-contraction coupling in heart muscle. *Science* 262, 740-744.

Choi, H. S. & Eisner, D. A. (1999). The role of the sarcolemmal Ca-ATPase in the regulation of resting calcium concentration in rat ventricular myocytes. *Journal of Physiology* 515, 109-118.

Choi, H. S., Trafford, A. W., & Eisner, D. A. (2000a). Measurement of calcium entry and exit in quiescent rat ventricular myocytes. *Pflügers Archiv* 440, 600-608.

Choi, H. S., Trafford, A. W., Orchard, C. H., & Eisner, D. A. (2000b). The effect of acidosis on systolic Ca and sarcoplasmic reticulum Ca content in isolated rat ventricular myocyes. *Journal of Physiology*529, 661-668.

Cleemann, L., Wang, W., & Morad, M. (1998). Two-dimentional confocal images of organisation, density and gating of focal Ca^{2+} release sites in rat cardiac myocytes. *Proceedings of the National Academy of Sciences, USA.* 95, 10984-10989.

Collier, M. L., Thomas, A. P., & Berlin, J. R. (1999). Relationship between L-type Ca^{2+} current and unitary sarcoplasmic reticulum Ca^{2+} release events in rat ventricular myocytes. *Journal of Physiology* 516, 117-128.

Cordeiro, J. M., Spitzer, K. W., Giles, W. R., Ershler, P. E., Cannell, M. B., & Bridge, J. H. B. (2001). Location of the initiation site of calcium transients and sparks in rabbit heart Purkinje cells. *The Journal of Physiology* 531, 301-314.

Coulombe, A., Lefèvre, I. A., Baró, I., & Coraboeuf, E. (1989). Barium- and calcium-permeable channels open at negative membrane potentials in rat ventricular myocytes. *Journal of MembraneBiology* 111, 57-67.

Cruz, J. S., Santana, L. F., Frederick, C. A., Isom, L. L., Malhotra, J. D., Mattei, L. N., Kass, R. S., Xia, J., An, R.-H., & Lederer, W. J. (1999). Whether "slip-mode conductance" occurs. *Science* 284, 711a.

Cui, Y., Galione, A., & Terrar, D. A. (1999). Effects of photoreleased cADP-ribose on calcium transients and calcium sparks in myocytes isolated from guinea-pig and rat ventricle. *Biochemical Journal* 342, 269-273.

DelPrincipe, F., Egger, M., & Niggli, E. (1999). Calcium signalling in cardiac muscle: refractoriness revealed by coherent activation. *Nature Cell Biology* 1, 323-329.

DiPolo, R. & Beaugé, L. (1979). Physiological role of ATP-driven calcium pump in squid axon. *Nature* 278, 271-273.

Díaz, M. E., Trafford, A. W., & Eisner, D. A. (2001). The role of intracellular Ca buffers in determining the shape of the systolic Ca transient in cardiac ventricular myocytes. *Pflügers Archiv* **442**, 96-100.

Díaz, M. E., Trafford, A. W., O'Neill, S. C., & Eisner, D. A. (1997). Measurement of sarcoplasmic reticulum Ca^{2+} content and sarcolemmal Ca^{2+} fluxes in isolated rat ventricular myocytes during spontaneous Ca^{2+} release. *Journal of Physiology* **501**, 3-16.

Ehara, T., Matsuoka, S., & Noma, A. (1989). Measurement of reversal potential of Na^+-Ca^{++} exchange current in single guinea-pig ventricular cells. *Journal of Physiology* **410**, 227-249.

Eisner, D. A. (1990). Intracellular sodium in cardiac muscle: effects on contraction. *Experimental Physiology* **75**, 437-457.

Eisner, D. A., Choi, H. S., Díaz, M. E., O'Neill, S. C., & Trafford, A. W. (2000). Integrative analysis of calcium cycling in cardiac muscle. *Circulation Research* **87**, 1087-1094.

Eisner, D. A. & Trafford, A. W. (2000). No role for the ryanodine receptor in regulating cardiac contraction? *News in Physiological Sciences* **15**, 275-279.

Eisner, D. A., Trafford, A. W., Díaz, M. E., Overend, C. L., & O'Neill, S. C. (1998). The control of Ca release from the cardiac sarcoplasmic reticulum: regulation versus autoregulation. *Cardiovascular Research* **38**, 589-604.

Endoh, M. & Blinks, J. R. (1988). Actions of sympathomimetic amines on the Ca^{2+} transient and contraction of rabbit myocardium: Reciprocal changes in myofibrillar responsiveness to Ca^{2+} mediated through α- and β-adrenoceptors. *Circulation Research* **62**, 247-265.

Evans, A. M. & Cannell, M. B. (1997). The role of L-type Ca^{2+} current and Na^+ current-stimulated Na/Ca exchange in triggering SR calcium release in guinea-pig cardiac ventricular myocyte. *Cardiovascular Research* **35**, 294-302.

Fabiato, A. (1983). Calcium-induced release of calcium from the cardiac sarcoplasmic reticulum. *American Journal of Physiology* **245**, C1-C14.

Fabiato, A. (1985a). Simulated calcium current can both cause calcium loading in and trigger calcium release from the sarcoplasmic reticulum of a skinned canine cardiac purkinje cell. *Journal of General Physiology* **85**, 291-320.

Fabiato, A. (1985b). Time and calcium dependence of activation and inactivation of calcium-induced release of calcium from the sarcoplasmic reticulum of a skinned canine cardiac Purkinje cell. *Journal of General Physiology* **85**, 247-289.

Ferrier, G. R. & Howlett, S. E. (1995). Contractions in guinea-pig ventricular myocytes triggered by a calcium-release mechanism separate from Na^+ and L-currents. *Journal of Physiology* **484**, 107-122.

Ferrier, G. R., Saunders, J. H., & Mendez, C. (1973). A cellular mechanism for the generation of ventricular arrhythmias by acetylstrophanthidin. *Circulation Research* **32**, 600-609.

Ferrier, G. R., Zhu, J., Redondo, I. M., & Howlett, S. E. (1998). Role of c-AMP-dependent protein kinase A in activation of a voltage-sensitive release mechanism for cardiac contraction in guinea-pig myocytes. *Journal of Physiology* **513**, 185-201.

Ferrier, G. R. & Howlett, S. E. (2001). Cardiac excitation-contraction coupling: role of membrane potential in regulation of contraction. *American Journal of Physiology* **280**, H1928-H1944.

Fujioka, Y., Komeda, M., & Matsuoka, S. (2000). Stoichiometry of Na^+-Ca^{2+} exchange in inside-out patches excised from guinea-pig ventricular myocytes. *Journal of Physiology* **523**, 339-351.

Fujioka, Y., Matsuoka, S., Ban, T., & Noma, A. (1998). Interaction of the Na^+-K^+ pump and Na^+-Ca^{2+} exchange via $[Na^+]i$ in a restricted space of guinea-pig ventricular cells. *The Journal of Physiology* **509**, 457-470.

Galione, A. (1993). Cyclic ADP-ribose: a new way to control calcium. *Science* **259**, 325-326.

García, J., Tanabe, T., & Beam, K. G. (1994). Relationship of calcium transients to calcium currents and charge movements in myotubes expressing skeletal and cardiac dihydropyridine receptors. *Journal of General Physiology* **103**, 125-147.

González, A. & Ríos, E. (2001). Excitation-contraction coupling in skeletal muscle., eds. Moss, R. L. & Solaro, R. J., Kluwer.Grantham, C. J. & Cannell, M. B. (1996). Ca^{2+} influx during the cardiac action potential in guinea pig ventricular myocytes. *Circulation Research* **79** , 194-200.

Guatimosim, S., Sobie, E. A., dos Santos Cruz, J., Martin, L. A., & Lederer, W. J. (2001). Molecular identification of a TTX-sensitive Ca^{2+} current. *American Journal of Physiology* **280**, C1327-C1339.

Györke, S. & Fill, M. (1993). Ryanodine receptor adaptation: control mechanism of Ca^{2+}-induced Ca^{2+} release in heart. *Science* **260**, 807-809.

Györke, S., Lukyanenko, V., & Györke, I. (1997). Dual effects of tetracaine on spontaneous calcium release in rat ventricular myocytes. *Journal of Physiology* **500**, 297-309.

Györke, S., Vélez, P., Suáre-Isla, B., & Fill, M. (1994). Activation of single cardiac and skeletal ryanodine receptor channels by flash photolysis of caged Ca^{2+}. *Biophysical Journal* **66**, 1879-1886.

Hammes, A., Oberdorff-Maass, S., Rother, T., Nething, K., Gollnick, F., Linz, K. W., Meyer, R., Hu, K., Han, H., Gaudron, P., Ertl, G., Hoffmann, S., Gantan, U., Vetter, R., Schuh, K., Benkwitz, C., Zimmer, H. G., & Neyses, L. (1998). Overexpression of the sarcolemmal calcium pump in the myocardium of transgenic rats. *Circulation Research* **83**, 877-888.

He, Z., Feng, S., Tong, Q., Hilgemann, D. W., & Philipson, K. D. (2000). Interaction of PIP_2 with the XIP region of the cardiac Na/Ca exchanger. *American Journal of Physiology* **278**, C661-C666.

Hess, P. & Tsien, R. W. (1984). Mechanism of ion permeation through calcium channels. *Nature* **309**, 453-456.

Hess, P. & Wier, W. G. (1984). Excitation-contraction coupling in cardiac Purkinje fibers. Effects of caffeine on the intracellular $[Ca^{2+}]$ transient, membrane currents, and contraction. *Journal of General Physiology* **83**, 417-433.

Hicks, M. J., Shigekawa, M., & Katz, A. M. (1979). Mechanism by which cyclic adenosine 3':5'-monophosphate-dependent protein kinase stimulates calcium transport in cardiac sarcoplasmic reticulum. *Circulation Research* **44**, 384-391.

Hilgemann, D. W. (1990). Regulation and deregulation of cardiac Na-Ca exchange in giant excised sarcolemmal membrane patches. *Nature* **344**, 242-245.

Hittinger, L., Ghalen, B., Chen, J., Edwards, J. G., Kudej, R. K., Iwase, M., Kim, S., Vatner, D. E. (1999). Reduced subendocardial ryanodine receptors and consequent effects on cardiac function in conscious dogs with left ventricular hypertrophy. *Circulation Research* **84**, 999-1006.

Hobai, I. A., Howarth, F. C., Pabbathi, V. K., Dalton, G. R., Hancox, J. C., Zhu, J.-Q., Howlett, S. E., Ferrier, G. R., & Levi, A. J. (1997). "Voltage-activated Ca release" in rabbit, rat and guinea-pig ardiac myocytes, and modulation by internal cAMP. *Pflügers Archiv* **435**, 164-173.

Hofmann, F., Lacinova, L., & Klugbauer, N. (1999). Voltage-dependent calcium channels: from structure to function. *Reviews of Physiology Biochemistry and Pharmacology* **139**, 33-87.

Hussain, M. & Orchard, C. H. (1997). Sarcoplasmic reticulum Ca^{2+} content, L-type Ca^{2+} current and the Ca^{2+} transient in rat myocytes during b-adrenergic stimulation. *Journal of Physiology* **505**, 385-402.

Hüser, J., Lipsius, S. L., & Blatter, L. A. (1996). Calcium Gradients during excitation-contraction coupling in cat atrial myocytes. *Journal of Physiology* **494**, 641-651.

Iino, S., Cui, Y., Galione, A., & Terrar, D. A. (1997). Actions of cADP-Ribose and its antagonists on contraction in guinea pig isolated ventricular myocytes: Influence of temperature. *Circulation Research* **81**, 879-884.

Imredy, J. P. & Yue, D. T. (1992). Submicroscopic Ca^{2+} diffusion mediates inhibitory coupling between individual Ca^{2+} channels. *Neuron* **9**, 197-207.

Inesi, G. & De Meis, L. (1989). Regulation of steady state filling in sarcoplasmic reticulum. Roles of back-inhibition, leakage and slippage of the calcium pump. *Journal of Biological Chemistry* **264**, 5929-5936.

Isenberg, G. (1982). Ca entry and contraction as studied in isolated bovine ventricaular myocytes. *Zeitschrift Fur Naturforschung* **37**, 502-512.

Jones, L. R., Suzuki, Y. J., Wang, W., Kobayashi, Y. M., Ranesh, V., Franzini-Armstrong, C., Cleemann, L., & Morad, M. (1998). Regulation of Ca^{2+} signaling in transgenic mouse cardiac myocytes overexpressing calsequestrin. *Journal of Clinical Investigation* **101**, 1385-1393.

Junker, J., Sommer, J. R., Sar, M., & Meissner, G. (1994). Extended junctional sarcoplasmic reticulum of avian cardiac muscle contains functional ryanodine receptors. *Journal of Biological Chemistry* **269**, 1627-1634.

Kirby, M. S., Sagara, Y., Gaa, S., Inesi, G., Lederer, W. J., & Rogers, T. B. (1992). Thapsigargin inhibits contraction and Ca^{2+} transient in cardiac cells by specific inhibition of the sarcoplasmic reticulum Ca^{2+} pump. *Journal of Biological Chemistry* **267**, 12545-12551.

Kirschenlohr, H. L., Grace, A. A., Vandenberg, J. I., Metcalfe, J. C., & Smith, G. A. (2000). Estimation of systolic and diastolic free intracellular Ca^{2+} by titration of Ca^{2+} buffering in the ferret heart. *Biochemical Journal* **346**, 385-391.

Kohmoto, O., Levi, A. J., & Bridge, J. H. B. (1994). Relation between reverse sodium-calcium exchange and sarcoplasmic reticulum calcium˙release in guineà pig ventricular cells. *Circulation Research* **74**, 550-554.

Kurihara, S. & Sakai, T. (1985). Effects of rapid cooling on mechanical and electrical responses in ventricular muscle of guinea-pig. *Journal of Physiology* **361**, 361-378.

Lamont, C. & Eisner, D. A. (1996). The sarcolemmal mechanisms involved in the control of diastolic intracellular calcium in isolated rat cardiac trabeculae. *Pflügers Archiv* **432**, 961-969.

Langer, G. A. & Peskoff, A. (1996). Calcium concentration and movement in the diadic cleft space of the cardiac ventricular cell. *Biophysical Journal* **70**, 1169-1182.

Leblanc, N. & Hume, J. R. (1990). Sodium current-induced release of calcium from cardiac sarcoplasmic reticulum. *Science* **248**, 372-376.

Lefevre, T., Coulombe, A., & Coraboeuf, E. (1994). Tonically active (background) calcium channels unmasked by phenothiazines in rat ventricular myocytes. *Biophysical Journal* **66**, a321.

Levi, A. J., Brooksby, P., & Hancox, J. C. (1993). A role for depolarisation induced calcium entry on the Na-Ca exchange in triggering intracellular calcium release and contraction in rat ventricular myocytes. *Cardiovascular Research* **27**, 1677-1690.

Levi, A. J., Li, J., Litwin, S. E., & Spitzer, K. W. (1996). Effect of internal sodium and cellular calcium load on voltage-dependence of the Indo-1 transient in guinea-pig ventricular myocytes. *Cardiovascular Research* **32**, 534-550.

Lewartowski, B., Janiak, R., & Langer, G. A. (1996). Effect of sarcoplasmic reticulum Ca release into diadic region on Na/Ca exchange in cardiac myocytes. *Journal of Physiology and Pharmacology* **47**, 577-590.

Li, L., Chu, G., Kranias, E. G., & Bers, D. M. (1998). Cardiac myocyte calcium transport in phospholamban knockout mouse: relaxation and endogenous CaMKII effects. *American Journal of Physiology* **274**, H1335-H1347.

Li, L., Desantiago, J., Chu, G., Kranias, E. G., & Bers, D. M. (2000). Phosphorylation of phopholamban and troponin I in b-adrenergic-induced acceleration of cardiac relaxation. *American Journal of Physiology* **278**, H769-H779.

Lipp, P., Laine, M., Tovey, S. C., Burrell, K. M., Berridge, M. J., Li, W., & Bootman, M. D. (2000). Functional InsP3 receptors that may modulate excitation-contraction coupling in the heart. *Curr.Biol.* **10**, 939-942.

Lipp, P. & Niggli, E. (1994). Sodium current-induced calcium signals in isolated guinea-pig ventricular myocytes. *Journal of Physiology* **474**, 439-446.

Lipp, P. & Niggli, E. (1996). Submicroscopic calcium signals as fundamental events of excitation-contraction coupling in guinea-pig cardiac myocytes. *Journal of Physiology* **492**, 31-38.

Lipp, P. & Niggli, E. (1998). Fundamental calcium release events revealed by two photon excitation photolysis of caged calcium in guinea pig cardiac myocytes. *Journal of Physiology* **508**, 801-809.

Litwin, S. E., Li, J., & Bridge, J. H. B. (1998). Na-Ca exchange and the trigger for sarcoplasmic reticulum Ca release: studies in adult rabbit ventricular myocytes. *Biophysical Journal* **75**, 359-371.

Litwin, S. E., Zhang, D., & Bridge, J. H. (2000). Dyssynchronous Ca^{2+} sparks in myocytes from infarcted hearts. *Circulation Research* **87**, 1040-1047.

Lokuta, A. J., Meyers, M. B., Sander, P. R., Fishman, G. I., & Valdivia, H. H. (1997). Modulation of Cardiac Ryanodine Receptors by Sorcin. *Journal of Biological Chemistry* **272** , 25333-25338.

London, B. & Krueger, J. W. (1986). Contraction in voltage-clamped, internally perfused single heart cells. *Journal of General Physiology* **88**, 475-505.

López-López, J. R., Shacklock, P. S., Balke, C. W., & Wier, W. G. (1994). Local, stochastic release of Ca^{2+} in voltage-clamped rat heart cells: visualization with confocal microscopy. *Journal of Physiology* **480**, 21-29.

López-López, J. R., Shacklock, P. S., Balke, C. W., & Wier, W. G. (1995). Local calcium transients triggered by single L-type calcium channel currents in cardiac cells. *Science* **268**, 1042-1045.

Lukyanenko, V., Györke, I., & Györke, S. (1996). Regulation of calcium release by calcium inside the sarcoplasmic reticulum in ventricular myocytes. *Pflügers Archiv* **432**, 1047-1054.

Lukyanenko, V., Györke, I., Wiesner, T. F. & Györke, S. (2001). Potentiation of Ca^{2+} release by cADP-ribose in the heart is mediated by enhanced SR Ca^{2+} uptake into the sarcoplasmic reticulum. *Circulation Research* **89**, 614-622.

Lukyanenko, V., Subramanian, S., Györke, I., Wiesner, T. F., & Györke, S. (1999). The role of luminal Ca in the generation of Ca waves in rat ventricular myocytes. *Journal of Physiology* **518**, 173-186.

Lukyanenko, V., Wiesner, T. F., & Gyorke, S. (1998). Termination of Ca^{2+} release during Ca^{2+} sparks in rat ventricular myocytes. *Journal of Physiology* **507**, 667-677.

Mackenzie, L., Bootman, M. D., Berridge, M. J., & Lipp, P. (2001). Predetermined recruitment of calcium release sites underlies excitation- contraction coupling in rat atrial myocytes. *Journal of Physiology* **530**, 417-429.

Marban, E. & Wier, W. G. (1985). Ryanodine as a tool to determine the contributions of calcium entry and calcium release to the calcium transient and contraction of cardiac Purkinje fibers. *Circulation Research* **56**, 133-138.

Marx, S. O., Ondrias, K., & Marks, A. R. (1998). Coupled gating between individual skeletal muscle Ca^{2+} release channels (ryanodine receptors). *Science* **281**, 818-821.

Marx, S. O., Reiken, S., Hisamatsu, Y., Jayaraman, T., Burkhoff, D., Rosemblit, N., & Marks, A. R. (2000). PKA phosphorylation dissociates FKBP12.6 from the calcium release channel (ryanodine receptor): defective regulation in failing hearts. *Cell* **101**, 365-376.

Matsuoka, S. & Hilgemann, D. W. (1992). Steady-state and dynamic properties of the cardiac Na/Ca exchange cycle. Ion and voltage dependences of the transport cycle. *Journal of General Physiology* **100**, 963-1001.

McIvor, M. E., Orchard, C. H., & Lakatta, E. G. (1988). Dissociation of changes in apparent myofibrillar Ca^{2+} sensitivity and twitch relaxation induced by adrenergic and cholinergic stimulation in isolated ferret cardiac muscle. *Journal of General Physiology* **92**, 509-529.

Meissner, G. (1994). Ryanodine receptor/Ca^{2+} release channels and their regulation by endogenous effectors. *Annual Review of Physiology* **56**, 485-508.

Meme, W., O'Neill, S. C., & Eisner, D. A. (2001). Low sodium inotropy is accompanied by diastolic Ca^{2+} gain and systolic loss in isolated guinea-pig ventricular myocytes. *Journal of Physiology* **530**, 487-495.

Meyers, M. B., Puri, T. S., Chien, A. J., Gao, T., Hsu, P. H., Hosey, M. M., & Fishman, G. I. (1998). Sorcin Associates with the Pore-forming Subunit of Voltage-dependent L-type Ca^{2+} Channels. *Journal of Biological Chemistry* **273**, 18930-18935.

Meyers, M. B., Zamparelli, C., Verzili, D., Dicker, A. P., Blanck, T. J., & Chiang, B. N. (1995a). Calcium-dependent translocation of sorcin to membranes: functional relevance in contractile tissue. *FEBS Letters* **357**, 230-4.

Meyers, M. B., Pickel, V. M., Sheu, S. S., Sharma, V. K., Scotto, K. W., & Fishman, G. I. (1995b). Association of Sorcin With the Cardiac Ryanodine Receptor. *Journal of Biological Chemistry* **270**, 26411-26418.

Mészáros, L. G., Bak, J., & Chu, A. (1993). Cyclic ADP-ribose as an endogenous regulator of the non-skeletal type ryanodine receptor Ca^{2+} channel. *Nature* **364**, 76-79.

Mitchell, R. D., Simmerman, H. K. B., & Jones, L. R. (1988). Ca^{2+} binding effects on protein conformation and protein interactions of canine cardiac calsequestrin. *Journal of Biological Chemistry* **263**, 1376-1381.

Mitcheson, J. S., Hancox, J. C., & Levi, A. J. (1997). Voltage dependence of the Fura-2 transient in rabbit left atrial myocytes at 37 degrees C. *Pflügers Archiv* **433**, 817-826.

Nakamura, K., Robertson, M., Liu, G., Dickie, P., Nakamura, K., Guo, J. Q., Duff, . J., Opas, M., Kavanagh, K., & Michalak, M. (2001). Complete heart block and sudden death in mice overexpressing calreticulin. *Journal of Clinical Investigation* **107**, 1245-1253.

Näbauer, M., Callewaert, G., Cleemann, L., & Morad, M. (1989). Regulation of calcium release is gated by calcium current, not gating charge, in cardiac myocytes. *Science* **244**, 800-803.

Näbauer, M. & Morad, M. (1990). Ca^{2+}-induced Ca^{2+} release as examined by photolysis of caged Ca^{2+} in single ventricular myocytes. *American Journal of Physiology* **258**, C189-C193.

Negretti, N., O'Neill, S. C., & Eisner, D. A. (1993). The effects of inhibitors of sarcoplasmic reticulum function on the systolic Ca^{2+} transient in rat ventricular myocytes. *Journal of Physiology* **468**, 35-52.

Niggli, E. & Lederer, W. J. (1990). Voltage-independent calcium release in heart muscle. *Science* **250**, 565-568.

Nuss, H. B. & Houser, S. R. (1992). Sodium-calcium exchange-mediated contractions in feline ventricular myocytes. *American Journal of Physiology* **263**, H1161-H1169.

Nuss, H. B. & Marban, E. (1999). Whether "Slip-mode conductance" occurs. *Science* **284**, 711a.

O'Neill, S. C., Donoso, P., & Eisner, D. A. (1990). The role of $[Ca^{2+}]_i$ and $[Ca^{2+}]_i$-sensitization in the caffeine contracture of rat myocytes: measurement of $[Ca^{2+}]_i$ and $[caffeine]_i$. *Journal of Physiology* **425**, 55-70.

O'Neill, S. C. & Eisner, D. A. (1990). A mechanism for the effects of caffeine on Ca^{2+} release during diastole and systole in isolated rat ventricular myocytes. *Journal of Physiology* **430**, 519-536.

Orchard, C. H., Eisner, D. A., & Allen, D. G. (1983). Oscillations of intracellular Ca^{2+} in mammalian cardiac muscle. *Nature* **304**, 735-738.

Osterrieder, W., Brum, G., Hescheler, J., & Trautwein, W. (1982). Injection of subunits of cyclic AMP-dependent protein kinase into cardiac myocytes modulates Ca^{2+} current. *Nature* **298**, 576-578.

Overend, C. L., Eisner, D. A., & O'Neill, S. C. (1997). The effect of tetracaine on spontaneous Ca release and sarcoplasmic reticulum calcium content in rat ventricular myocytes. *Journal of Physiology* **502**, 471-479.

Overend, C. L., O'Neill, S. C., & Eisner, D. A. (1998). The effect of tetracaine on stimulated contractions, sarcoplasmic reticulum Ca^{2+} content and membrane current in isolated rat ventricular myocytes. *Journal of Physiology* **507**, 759-769.

Papp, Z., Sipido, K. R., Callewaert, G., & Carmeliet, E. (1995). Two components of $[Ca^{2+}]_i$ activated chloride current during large $[Ca^{2+}]_i$ transients in single rabbit heart purkinje cells. *Journal of Physiology* **483**, 319-330.

Perchenet, L., Hinde, A. K., Patel, K. C., Hancox, J. C., & Levi, A. J. (2000). Stimulation of Na/Ca exchange by the beta-adrenergic/protein kinase A pathway in guinea-pig ventricular myocytes at 37 degrees C. *Pflügers Archiv* **439**, 822-   828.

Philipson, K. D. & Nicoll, D. A. (2000). Sodium-calcium exchange: a molecular perspective. *AnnualReview of Physiology* **62**, 111-133.

Piacentino III, V., Dipla, K., Gaughan, J. P., & Houser, S. R. (2000). Voltage-dependent Ca^{2+} release from the SR of feline ventricular myocytes is explained by Ca^{2+}-induced Ca^{2+} release. *Journal of Physiology* **523**, 533-548.

Prestle, J., Janssen, P. M. L., Janssen, A. P., Zeitz, O., Lehnart, S. E., Bruce, L., Smith, G. L., & Hasenfuss, G. (2001). Overexpression of FK506-Binding Protein FKBP12.6 in Cardiomyocytes Reduces Ryanodine Receptor-Mediated Ca^{2+} Leak From the Sarcoplasmic Reticulum and Increases Contractility. *Circulation Research* **88**, 188-194.

Rakovic, S., Galione, A., Ashamu, G. A., Potter, B. V. L., & Terrar, D. A. (1996). A specific cyclic ADP-ribose antagonist inhibits cardiac excitation-contraction coupling. *Current Biology* **6**, 989-996.

Reeves, J. P. & Hale, C. C. (1984). The stoichiometry of the cardiac sodium-calcium exchange system. *Journal of Biological Chemistry* **259**, 7733-7739.

Rousseau, E. & Meissner, G. (1989). Single cardiac sarcoplasmic reticulum Ca^{2+}-release channel: a activation by caffeine. *American Journal of Physiology* **256**, H328-H333.

Rousseau, E., Smith, J. S., Henderson, J. S., & Meissner, G. (1986). Single channel and $^{45}Ca^{2+}$ flux measurements of cardiac sarcoplasmic reticulum calcium channel. *Biophysical Journal* **50**, 1009-1014.

Santana, L. F., Gómez, A. M., & Lederer, W. J. (1998). Ca^{2+} flux through promiscuous cardiac Na^+ channels: slip-mode conductance. *Science* **279**, 1027-1033.

Satoh, H., Blatter, L. A., & Bers, D. M. (1997). Effects of $[Ca^{2+}]_i$, SR Ca^{2+} load, and rest on Ca^{2+} spark frequency in ventricular myocytes. *American Journal of Physiology* **272**, H657-H668.

Scriven, D. R. L., Dan, P., & Moore, E. D. W. (2000). Distribution of proteins implicated in excitation-contraction coupling in rat ventricular myocytes. *Biophysical Journal* **79**, 2682-2691.

Shacklock, P. S., Wier, W. G., & Balke, C. W. (1995). Local Ca^{2+} transients (Ca^{2+} sparks) originate at transverse tubules in rat heart cells. *Journal of Physiology* **487**, 601-608.

Sham, J. S. K., Cleemann, L., & Morad, M. (1992). Gating of the cardiac Ca^{2+} release channel: the role of Na^+ current and Na^+-Ca^{2+} exchange. *Science* **255**, 850-853.

Sham, J. S. K., Cleemann, L., & Morad, M. (1995). Functional coupling of Ca^{2+} channels and ryanodine receptors in cardiac myocytes. *Proceedings of the National Academy of Sciences, USA.* **92**, 121-125.

Sham, S. K., Song, L.-S., Chen, Y., Deng, L.-H., Stern, M. D., Lakatta, E. G., & Cheng, H. (1998). Termination of Ca^{2+} release by a local inactivation of ryanodine receptors in cardiac myocytes. *Proceedings of the National Academy of Sciences, USA.* **95**, 15096-15101.

Shannon, T. R. & Bers, D. M. (1997). Assessment of intra-SR free [Ca] and buffering in rat heart. *Biophysical Journal* **73**, 1524-1531.

Shannon, T. R., Ginsburg, K. S., & Bers, D. M. (2000). Reverse mode of the sarcoplasmic reticulum calcium-pump and load-dependant cytosolic calcium decline in voltage-clamped cardiac ventricular myocytes. *Biophysical Journal* **78**, 322-333.

Shigekawa, M. & Iwamoto, T. (2001). Cardiac Na^+-Ca^{2+} exchange. *Circulation Research* **88**, 864-876.

Sipido, K. R., Callewaert, G., & Carmeliet, E. (1995a). Inhibition and rapid recovery of Ca^{2+} current during Ca^{2+} release from sarcoplasmic reticulum in guinea pig ventricular myocytes. *CirculationResearch* **76**, 102-109.

Sipido, K. R., Carmeliet, E., & Pappano, A. (1995b). Na^+ current and Ca^{2+} release from the sarcoplasmic reticulum during action potentials in guinea-pig ventricular myocytes. *Journal of Physiology* **489**, 1-17.

Sipido, K. R., Carmeliet, E., & Van de Werf, F. (1998). T-type Ca^{2+} current as a trigger for Ca^{2+} release from the sarcoplasmic reticulum in guinea-pig ventricular myocytes. *Journal of Physiology* **508**, 439-451.

Sipido, K. R., Maes, M., & Van de Werf, F. (1997). Low efficiency of Ca^{2+} entry through the Na^+-Ca^{2+} exchanger as trigger for Ca^{2+} release from the sarcoplasmic reticulum. *Circulation Research* **81**, 1034-1044.

Sipido, K. R. & Wier, W. G. (1991). Flux of Ca^{2+} across the sarcoplasmic reticulum of guinea-pig cardiac cells during excitation-contraction coupling. *Journal of Physiology* **435**, 605-630.

Sitsapesan, R., McGarry, S. J., & Williams, A. J. (1994). Cyclic ADP-ribose competes with ATP for the adenine nucleotide sites on the cardiac ryanodine receptor Ca^{2+} release channel. *Circulation Research* **75**, 596-600.

Sitsapesan, R. & Williams, A. J. (1994). Regulation of the gating of the sheep cardiac sarcoplasmic reticulum Ca^{2+}- release channel by luminal Ca^{2+}. *Journal of Membrane Biology* **137**, 215-226.

Sitsapesan, R. & Williams, A. J. (1995). Cyclic ADP-ribose and related compounds activate sheep skeletal sarcoplasmic reticulum Ca^{2+} release channel. *American Journal of Physiology* **268**, C1235-C1240.

Sitsapesan, R. & Williams, A. J. (1997). Regulation of current flow through ryanodine receptors by luminal Ca^{2+}. *Journal of Membrane Biology* **159**, 179-185.

Smith, G. A., Dixon, H. B. F., Kirschenlohr, H. L., Grace, A. A., Metcalfe, J. C., & Vandenberg, J. I. (2000). Ca^{2+} buffering in the heart: Ca^{2+} binding to and activation of cardiac myofibrils. *Biochemical Journal* **346**, 393-402.

Smith, G. L., Valdeolmillos, M., Eisner, D. A., & Allen, D. G. (1988). Effects of rapid application of caffeine on intracellular calcium concentration in ferret papillary muscles. *Journal of General Physiology* **92**, 351-368.

Soeller, C. & Cannell, M. B. (1997). Numerical simulation of local calcium movements during L-type caclium channel gating in the cardiac diad. *Biophysical Journal* **73**, 97-111.

Sommer, J. R. (1995). Comparative anatomy: In praise of a powerful approach to elucidate mechanisms translating cardiac excitation into purposeful contraction. *Journal of Molecular and Cellular Cardiology* **27**, 19-35.

Sommer, J. R. & Johnson, E. A. (1968). Cardiac muscle. A comparative study of purkinje fibres and ventricular fibres. *Journal of Cell Biology* **36**, 497-526.

Song, L.-S., Sham, J. S. K., Stern, M. D., Lakatta, E. G., & Cheng, H. (1998). Direct measurement of SR release flux by tracking Ca^{2+} spikes in rat cardiac myocytes. *Journal of Physiology* **512**, 677-691.

Song, L.-S., Stern, M. D., Lakatta, E. G., & Cheng, H. (1997). Partial depletion of sarcoplasmic reticulum calcium does not prevent calcium sparks in rat ventricular myocytes. *Journal of Physiology* **505**, 665-675.

Song, L. S., Wang, S. Q., Xiao, R. P., Spurgeon, H., Lakatta, E. G., & Cheng, H. (2001). beta-Adrenergic Stimulation Synchronizes Intracellular Ca^{2+} Release During Excitation-Contraction Coupling in Cardiac Myocytes. *Circulation Research* **88**, 794-801.

Spencer, C. I. & Berlin, J. R. (1995). Control of sarcoplasmic reticulum calcium release during calcium loading in isolated rat ventricular myocytes. *Journal of Physiology* **488**, 267-279.

Spencer, C. I. & Berlin, J. R. (1997). Calcium-induced release of strontium from the sarcoplasmicreticulum of rat cardiac ventricular myocytes. *Journal of Physiology* **504**, 565-578.

Sperelakis, N., Kurachi, Y., Terzic, A., & Cohen, M. V. (2001). *Heart Physiology and Pathophysiology*, 4 ed., pp. 1-1261. Academic, San Diego.

Spurgeon, H. A., Stern, D. M., Baartz, G., Raffaeli, S., Hansford, R. G., Talo, A., Lakatta, E. G., & Capogrossi, M. C. (1990). Simultaneous measurements of Ca^{2+}, contraction and potential in cardiac myocytes. *American Journal of Physiology* **258**, h574-h586.

Stern, M. D. (1992). Theory of excitation-contraction coupling in cardiac muscle. *Biophysical Journal* **63**, 497-517.

Strehler, E. E. & Zacharias, D. A. (2001). Role of Alternative Splicing in Generating Isoform Diversity Among Plasma Membrane Calcium Pumps. *Physiological Reviews* **81**, 21-50.

Sun, X., Protasi, F., Takahashi, M., Takeshima, H., Ferguson, D. G., & Franzini-Armstrong, C. (1995). Molecular Architecture of Membranes Involved in Excitation-Contraction coupling of Cardiac Muscle. *Journal of Cell Biology* **129**, 659-671.

Tada, M., Kirchberger, M. A., Repke, D. I., & Katz, A. M. (1974). The stimulation of calcium transport in cardiac sarcoplasmic reticulum by adenosine 3':5'-monophosphate-dependent protein kinase. *Journal of Biological Chemistry* **249**, 6174-6180.

Terracciano, C. M. (2001). Rapid inhibition of the Na^+-K^+ pump affects Na^+-Ca^{2+} exchanger-mediated relaxation in rabbit ventricular myocytes. *Journal of Physiology* **533**, 165-173.

Teshima, Y., Takahashi, N., Saikawa, T., Hara, M., Yasunaga, S., Hidaka, S., & Sakata, T. (2000). Diminished expression of sarcoplasmic reticulum Ca^{2+}-ATPase and ryanodine sensitive Ca^{2+} channel mRNA in streptozotocin-induced diabetic rat heart. *Journal of Molecular and Cellular Cardiology* **32**, 655-664.

Timerman, A. P., Onoue, H., Xin, H.-B., Barg, S., Copello, J., Wiederrecht, G., & Fleischer, S. (1996). Selective binding of FKBP12.6 by the cardiac ryanodine receptor. *Journal of Biological Chemistry* **271**, 20385-20391.

Tinker, A. & Williams, A. J. (1993). Charged local anesthetics block ionic conduction in the sheep cardiac sarcoplasmic reticulum calcium release channel. *Biophysical Journal* **65**, 852-864.

Trafford, A. W., Díaz, M. E., & Eisner, D. A. (1998a). Ca-activated chloride current and Na-Ca exchange have different timecourses during sarcoplasmic reticulum Ca release in ferret ventricular myocytes. *Pflügers Archiv* **435**, 743-745.

Trafford, A. W., Díaz, M. E., & Eisner, D. A. (1998b). Ca-activated chloride current and Na-Ca exchange have different timecourses during sarcoplasmic reticulum Ca release in ferret ventricular myocytes. *Pflügers Archiv* **435**, 743-745.

Trafford, A. W., Díaz, M. E., & Eisner, D. A. (2001). Coordinated control of cell Ca^{2+} loading and triggered release from the sarcoplasmic reticulum underlies the rapid inotropic response to increased L-type Ca^{2+} current. *Circulation Research* **88**, 195-201.

Trafford, A. W., Díaz, M. E., Negretti, N., & Eisner, D. A. (1997). Enhanced calcium current and decreased calcium efflux restore sarcoplasmic reticulum Ca content following depletion. *CirculationResearch* **81**, 477-484.

Trafford, A. W., Díaz, M. E., O'Neill, S. C., & Eisner, D. A. (1995). Comparison of subsarcolemmal and bulk calcium concentration during spontaneous calcium release in rat ventricular myocytes. *Journal ofPhysiology* **488**, 577-586.

Trafford, A. W., Díaz, M. E., Sibbring, G. C., & Eisner, D. A. (2000). Modulation of CICR has no maintained effect on systolic Ca^{2+}: simultaneous measurements of sarcoplasmic reticulum and sarcolemmal Ca^{2+} fluxes in rat ventricular myocytes. *Journal of Physiology* **522**, 259-270.

Tripathy, A., Xu, L., Pasek, D. A., & Meissner, G. (1999). Effects of 2,3-butanedione 2-monoxime on Ca^{2+} release channels (ryanodine receptors) of cardiac and skeletal muscle. *Journal of Membrane Biology* **169**, 189-198.

Tsien, R. W. (1983). Calcium channels in excitable membranes. *Annual Review of Physiology* **45**, 341-358.

Valdeolmillos, M., O'Neill, S. C., Smith, G. L., & Eisner, D. A. (1989). Calcium-induced calcium releaseactivates contraction in intact cardiac cells. *Pflügers Archiv* **413**, 676-678.

Valdivia, H. H. (1998). Modulation of intracellular Ca^{2+} levels in the heart by sorcin and FKBP12, two accessory proteins of ryanodine receptors. *Trends in Pharmacological Sciences* **19**, 479-482.

Varro, A., Negretti, N., Hester, S. B., & Eisner, D. A. (1993). An estimate of the calcium content of the sarcoplasmic reticulum in rat ventricular myocytes. *Pflügers Archiv* **423**, 158-160.

Vassort, G. & Alvarez, J. (1994). Cardiac T-type calcium current: pharmacology and roles in cardiac tissues. *Journal of Cardiovascular Electrophysiology* **5**, 376-393.

Vatner, D. E., Sato, N., Kiuchi, K., Shannon, R. P., & Vatner, S. F. (1994). Decrease in myocardial ryanodine receptors and altered excitation-contraction coupling early in the development of heart failure. *Circulation* **90**, 1423-1430.

Verma, A., Hirsch, D. J., & Snyder, S. H. (1992). Calcium pools mobilized by calcium or inositol 1,4,5-triposphate are differentially localized in rat heart and brain. *Molecular Biology of the Cell* **3**, 621-631.

Vornanen, M., Shepherd, N., & Isenberg, G. (1994). Tension-voltage relations of single myocytes reflect Ca release triggered by Na/Ca exchange at 35°C but not at 23°C. *American Journal of Physiology* **267**, C623-C632.

Wang, S. Q., Song, L. S., Lakatta, E. G., & Cheng, H. (2001). Ca^{2+} signalling between single L-type Ca^{2+} channels and ryanodine receptors in heart cells. *Nature* **410**, 592-596.

Wang, W., Cleemann, L., Jones, L. R., & Morad, M. (2000). Modulation of focal and global Ca^{2+} release in calsequestrin-overexpressing mouse cardiomyocytes. *The Journal of Physiology* **524**, 399-414.

Wasserstrom, J. A. & Vites, A. M. (1997). The role of Na^+-Ca^{2+} exchange in activation of excitation-contraction coupling in rat ventricular myocytes. *Journal of Physiology* **493**, 529-542.

Weber, C. R., Ginsburg, K. S., Philipson, K. D., Shannon, T. R., & Bers, D. M. (2001). Allosteric regulation of Na/Ca exchange current by cytosolic Ca in intact cardiac myocytes. *Journal of General Physiology* **117**, 119-132.

Wendt-Gallitelli, M. F., Voigt, T., & Isenberg, G. (1993). Microheterogeneity of subsarcolemmal sodium gradients, electron probe microanalysis in guinea-pig ventricular myocytes. *Journal of Physiology* **472**, 33-44.

Wier, W. G., Kort, A. A., Stern, M. D., Lakatta, E. G., & Marban, E. (1983). Cellular calcium fluctuations in mammalian heart: direct evidence from noise analysis of aequorin signals in Purkinje fibers. *Proceedings of the National Academy of Sciences, USA.* **80**, 7367-7371.

Wolska, B. M., Stojanovic, M. O., Luo, W., Kranias, E. G., & Solaro, R. J. (1996). Effect of ablation of phospholamban on dynamics of cardiac myocyte contraction and intracellular Ca^{2+}. *AmericanJournal of Physiology* **271**, C391-C397.

Xu, L., Jones, R., & Meissner, G. (1993). Effects of local anesthetics on single channel behavior of skeletal muscle release channel. *Journal of General Physiology* **101**, 207-233.

Xu, L., Mann, G., & Meissner, G. (1996). Regulation of cardiac Ca^{2+} release channel (Ryanodine receptor) by Ca^{2+}, H^+, Mg^{2+}, and adenine nucleotides under normal and simulated ischemicconditions. *Circulation Research* **79**, 1100-1109.

Yamamoto, T., Yano, M., Kohno, M., Hisaoka, T., Ono, K., Tanigawa, T., Saiki, Y., Hisamatsu, Y., Ohkusa, T., & Matsuzaki, M. (1999). Abnormal Ca^{2+} release from cardiac sarcoplasmic reticulum in tachycardia-induced heart failure. *Cardiovascular Research* **44**, 146-155.

Zhou, Z. & January, C. T. (1998). Both T- and L-Type Ca^{2+} Channels Can Contribute to Excitation-Contraction Coupling in Cardiac Purkinje Cells. *Biophysical Journal* **74**, 1830-1839.

Zipes, D. P. & Jalife, J. (2000). *Cardiac Electrophysiology: from cell to bedside*, 3 ed., pp. 1-1111. W.B. Saunders, Philadelphia.

Zucchi, R. & Ronca-Testoni, S. (1997). The sarcoplasmic reticulum Ca^{2+} channel/ryanodine receptor: modulation by endogenous effectors, drugs and disease states. *Pharmacological Reviews* **49**, 1-51.

RHEA J.C. LEVINE & ROBERT W. KENSLER

THE THICK FILAMENT OF VERTEBRATE STRIATED MUSCLE

The Structure at Rest and at Work

1. INTRODUCTION: THE IMPORTANCE OF THE STRIATED MUSCLE THICK FILAMENT

Contraction in mammalian skeletal muscle and cardiac muscle occurs through a sliding filament mechanism involving two sets of interdigitating filaments: the myosin containing thick filaments and the actin-containing thin filaments (Huxley and Hanson, 1954; Huxley and Niedergerke, 1954; Davies, 1963; Squire, 1981). Interaction between these filaments to produce force or shortening involves a cyclic binding of the myosin heads on the thick filament to the actin of the thin filament in the presence of calcium and ATP (Huxley, 1969; Huxley and Simmons, 1971; Brenner, 1987; Schoenberg, 1988). Both the biochemistry and mechanics of this mechanism have been extensively studied (Taylor, 1979; Cooke, 1986; Hibberd and Trentham, 1986; Goldman and Brenner, 1987; Schoenberg, 1988; Ford et al., 1977; Brenner, 1987; Brenner, 1990) in skeletal muscle, and understanding of the interaction between myosin and actin has been facilitated by recent x-ray crystallographic determinations of the structure of the S-1 subunit portion of the chicken skeletal muscle myosin head (Rayment et al., 1993) and of actin (Holmes et al., 1990a; Holmes et al., 1990b).

These data have led to the proposal that ATP hydrolysis is coupled to a small conformational change in the motor domain of the head that is amplified by movement of the lever arm region of the head to produce force with the adjacent actin-containing thin filaments (Rayment et al., 1993). Support for this hypothesis has come from cryoelectron microscope studies of thin filaments decorated with smooth muscle myosin which show a rotation of the lever arm by 3 - 3.5 nm upon release of ADP (Whittaker et al., 1995); from in vitro motility assays of mutant Dictyostelium myosins with different lever arm lengths (Uyeda et al., 1996); and from the recent determination of the crystal structure of a vertebrate smooth muscle motor domain complexed with various nucleotide analogs (Dominguez et al., 1998; Houdusse, 1999).

Although these studies have significantly increased our knowledge of myosin and actin interactions, a number of questions about muscle function and regulation

91

R.J. Solaro and R.L. Moss (eds), Molecular Control Mechanisms in Striated Muscle
Contraction, 91-141
©2002 Kluwer Academic Publishers, Printed in the Netherlands

remain (Millman, 1998). Many relate to the structure of the thick filament and the constraints placed by its structure on the interaction of the myosin head with actin. Information on thick filament structure is important for interpreting results from a variety of studies, including the changes in intensity of the myosin reflections in the x-ray diffraction pattern during contraction (Huxley *et al.*, 1994), the structure and orientation of weakly-binding crossbridges (Brenner *et al.*, 1984; Squire *et al.*, 1987; Xu *et al.*, 1987; Bagni *et al.*, 1992; Harford and Squire, 1992; Pollard *et al.*, 1993; Walker *et al.*, 1994), the role of phosphorylation of the thick filaments (Sweeney and Stull, 1986; Craig *et al.*, 1987; Levine *et al.*, 1991; 1995; 1996) and the role of the various accessory proteins in the thick filament (Craig and Offer, 1976; Bennett *et al.*, 1985a; Starr *et al.*, 1985; Weisberg and Winegrad, 1998).

This is particularly true for cardiac muscle. Although similar to skeletal muscle in the arrangement of thick and thin filaments into sarcomeres, cardiac muscle differs from skeletal muscle in a number of physiological properties, including its high resting tension, steep relationship between active force and sarcomere length (Frank-Starling relation) (Fuchs and Wang, 1997) and modulatory regulation by phosphorylation (Jeacocke and England, 1980; Garvey *et al.*, 1988; Schlender and Bean, 1991; Hartzell and Titus, 1982). Evidence suggests that at least some of these differences may relate to the structure of the cardiac thick filament and its crossbridges (Matsubara and Millman, 1974; Weisberg and Winegrad, 1998) or accessory proteins (Weisberg and Winegrad, 1998; Jeacocke and England, 1980; Garvey *et al.*, 1988; Schlender and Bean, 1991; Hartzell and Titus, 1982). The significance of this is highlighted by recent evidence that mutations in the β-myosin heavy-chain gene (Fananapazir *et al.*, 1993) and the gene for myosin-binding protein-C (Yu *et al.*, 1998; Niimura *et al.*, 1998; Freiburg and Gautel, 1995) may both be linked to certain types of familial hypertrophic cardiomyopathy, an important cause of sudden death in young adults.

A full understanding of contraction in mammalian striated muscle necessitates knowing the detailed molecular structure of the thick filament; and the changes that occur in its structure as it changes from the non-active relaxed state to the contractile state. Although the precise arrangement of myosin and accessory proteins in the mammalian striated muscle thick filament and their possible regulatory role is still not completely understood, a number of studies on the vertebrate thick filament have shed light on its structure.

2. THE THICK FILAMENT UNDER RELAXING CONDITIONS

2.1. General Structure of the Vertebrate Thick Filament

Biochemical and electron microscopic studies (Huxley, 1963; Starr and Offer, 1971; Craig and Offer, 1976; Ip and Heuser, 1983; Kensler and Stewart, 1983; 1986; 1989; 1993; Variano-Marston *et al.*, 1984; Bennett *et al.*, 1985b; Cantino and Squire, 1986; Padron and Craig, 1989; Craig *et al.*, 1992; Sosa *et al.*, 1994; Kensler and Woodhead, 1995) have demonstrated that the vertebrate thick filament is a bipolar structure ~1.5 microns in length, composed predominantly of myosin II with smaller amounts of at least six to ten accessory proteins.

2.1.1. Myosin Arrangement

Myosin II is a large asymmetrically shaped molecule composed of two heavy chains and two pairs of associated light chains. The two heavy chains fold to form two globular domains (the heads or S1 subunits, Fig. 1a) at one end of the molecule and a long rod-like coiled-coil domain (the myosin rod) extending to the other end of the molecule. An essential light chain and a regulatory light chain (Fig. 1a) are associated with the heavy chain in each of the two heads.

The myosins (Fig. 1a) are thought to be arranged so that the long rod domains of the molecules aggregate to form the filament shaft or backbone, while the globular heads lie on the surface (Figs. 1b and c) in a position to form crossbridges with the adjacent thin filaments (Huxley, 1963). At the center of the filament the myosin rods are packed in an antiparallel arrangement to give the myosin head-free bare zone region of the filament, while on either side of this region the myosin rods pack together in a parallel arrangement to give the crossbridge (myosin head) studded arms of the filament; thus establishing the bipolar structure of the filament (Fig. 1b).

X-ray diffraction studies of relaxed vertebrate skeletal muscle (Huxley and Brown, 1967; Haselgrove, 1975; Haselgrove, 1980; Haselgrove and Rodger, 1980; Rome, 1972; Millman, 1979) have established that the myosin heads on the filament are arranged approximately helically (Fig. 1c) with an axial repeat of 42.9nm and an average axial rise of 14.3nm between adjacent levels of heads. A series of meridional reflections, not expected from ideal helical symmetry, on the non-meridional layer lines (those which are not multiples of three) (Huxley and Brown, 1967; Haselgrove, 1975; Haselgrove, 1980; Haselgrove and Rodger, 1980; Harford and Squire, 1986) of the patterns have been interpreted as indicating a perturbation of the crossbridge array from ideal helical symmetry (Huxley and Brown, 1967; Yagi et al., 1981; Squire et al., 1982).

2.1.2. Accessory Proteins

In addition to myosin, at least six to ten other accessory proteins are associated with myosin in the vertebrate striated muscle thick filament and are thought to be responsible for a series of eleven transverse stripes (Fig. 1c) at 43 nm axial spacing across each half of the A-band (Starr and Offer, 1971; Craig and Offer, 1976; Yamamoto and Moos, 1981; Starr and Offer, 1983; Bennett et al., 1985a; b). Of these, myosin binding protein-C (MyBP-C) and myosin binding protein-X (MyBP-X) appear to be present in the largest amount, whereas myosin binding protein-H (MyBP-H) and the other accessory proteins are present in smaller amounts (Starr and Offer, 1971; Starr and Offer, 1983; Bennett et al., 1985a; b). In addition to these

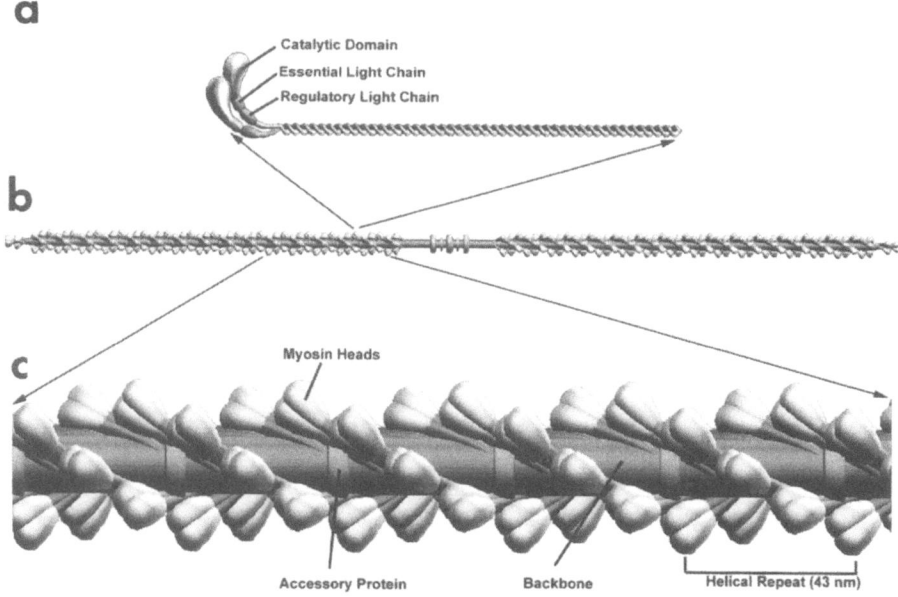

Figure 1. *A diagram of the arrangement of myosin and accessory proteins in the vertebrate striated muscle thick filament. (a) Drawing of myosin II molecule with its long rod-like coiled-coil domain and two globular heads (S-1 subunits) each with a regulatory light chain and essential light chain. (b) Drawing of arrangement of myosin into a bipolar thick filament with a central bare zone region and two flanking myosin head-studded (crossbridge-bearing) arms. (c) Enlarged segment of the filament arm showing the helical arrangement of the myosin heads on the surface of the filament backbone. The 43 nm helical repeat (every 3rd crossbridge level) is indicated. The presence of bands of accessory proteins at 43-44 nm intervals is also shown.*

proteins, the M-band proteins (Luther *et al.*, 1981; Pask *et al.*, 1994) and the giant elastic protein titin (Wang *et al.*, 1979; 1993; Maruyama, 1996; Labeit *et al.*, 1992) are associated with the filament and may play an important role in maintaining the integrity of the sarcomere.

The MyBP-C family comprises isoforms specific for slow, fast and cardiac muscles (Yamamoto and Moos, 1983), while MyBP-H is typically absent from cardiac muscle, but may be present in the conducting fibers of the heart (Vaughan *et al.*, 1993a; Alyonycheva *et al.*, 1997). Cloning has shown that most of these proteins share a conserved domain pattern and belong to the intracellular immunoglobulin superfamily (Einheber and Fischman, 1990; Furst *et al.*, 1992;

Vaughan *et al.*, 1992; Vaughan *et al.*, 1993b). MyBP-C is known to bind to myosin by a C-terminal myosin binding IgI domain which is conserved within the whole family (Okagaki *et al.*, 1993).

Attempts to form synthetic filaments with the native structure by adding stoichiometric amounts of myosin and various accessory proteins have failed (see Davis, 1988 for a discussion), implying that the native structure is relatively unique and that it may be important to understand the exact arrangement of these accessory proteins relative to the myosin. This emphasizes the importance of establishing the precise location and arrangement of the accessory proteins relative to the myosin in the filament backbone.

Unfortunately, as has been recently noted (Chew and Squire, 1995; Squire *et al.*, 1998), very little information is available on the detailed packing and orientation of the myosin rods and accessory proteins in the native filament backbone. Electron microscopy and computer image analysis of cross sections of thick filaments in thin sections have shown that there are probably nine subfilaments in transverse section, but have not elucidated the detailed packing of myosin within the subunits. Analysis of the sequences of charge groups (Parry, 1981; McLachlan and Karn, 1982) along the myosin rod, and recent evidence that the solubility and packing properties of myosin reside in a 29 residue region near the C-terminus of the rod (Sohn *et al.*, 1997) provide a basis for understanding the 14.3 nm and 43 nm periodicities seen in the vertebrate thick filament and for modeling of the myosin packing (Chew and Squire, 1995), but have not established the precise packing of the myosin in the backbone.

In addition to providing a basis for better understanding the interaction between the myosin backbone and accessory proteins, data on the backbone structure is important for better understanding the factors involved in maintaining the highly ordered helical arrangement of the myosin heads in the relaxed thick filament, and how factors such as phosphorylation of myosin light chain 2 or of MyBP-C may contribute to release of the heads from the backbone and activation (Levine *et al.*, 1995; Levine *et al.*, 1996).

Although the role of the accessory proteins in the thick filament may be primarily structural, recent evidence from studies of the proteins in cardiac muscle suggest that they may also have a regulatory role (Garvey *et al.*, 1988; Jeacocke and England, 1980; Hartzell and Titus, 1982; Schlender and Bean, 1991; Weisberg and Winegrad, 1996; Weisberg and Winegrad, 1998). Recent experimental evidence supporting this role for the protein MyBP-C in cardiac muscle will be described in greater detail below.

2.2 Electron Microscopic Approach to Thick Filament Study

Although extremely powerful, x-ray diffraction studies, largely because of the loss of the phase information in the patterns and the problem of lattice sampling of the diffraction maxima, have neither allowed a definitive determination of the filament structure nor determined the changes which take place in it during contraction. Many questions such as the number of myosin heads at each crossbridge level; their precise orientation and location; and the changes which occur in filament structure during contraction could not be definitively answered. For this reason, we and others several years ago turned to the alternative (supplementary) approach of using electron microscopic and computer image analysis techniques to determine the structure of thick filaments. In the case of vertebrate thick filaments (Kensler and Woodhead, 1995; Kensler *et al.*, 1994; Kensler and Stewart, 1983; 1986; 1989; 1993; Stewart and Kensler, 1986), filaments in their relaxed state can be isolated with their native helical arrangement of myosin heads preserved. The structure is amenable to determination by a combination of electron microscopy (negative staining and platinum shadowing) and Fourier based computer image analysis. Using this approach, the thick filaments from frog (Kensler and Stewart, 1983; Kensler and Stewart, 1986; Stewart and Kensler, 1986), fish (Kensler and Stewart, 1989), rabbit (Kensler *et al.*, 1994; Kensler and Stewart, 1993), and chicken skeletal muscles (Kensler and Woodhead, 1995) have been shown to have a three-stranded helical arrangement of myosin heads. In addition, these isolated filaments have been used as model systems to explore the effects of such environmental factors as temperature (Kensler *et al.*, 1994; Kensler and Woodhead, 1995), ionic strength (Kensler *et al.*, 1994), phosphorylation of myosin light chain (Levine *et al.*, 1995; 1996) and MyBP-C (Weisberg and Winegrad, 1998), and crosslinking of the myosin heads (Levine, 1993; 1997) on the structure of the filaments and their interactions with actin.

2.3 Analysis of Isolated Mammalian Thick Filaments

Many recent studies of the effects of changes in parameters such as temperature, ionic strength, and phosphorylation of myosin light chain 2 and MyBP-C on crossbridge arrangement have been done using isolated mammalian skeletal muscle thick filaments. The successful isolation of these filaments with the preservation of their relaxed helical arrangement of crossbridges depended upon the finding by Wray (1987) that the helical order is lost in x-ray diffraction patterns at low temperatures. Thus to preserve the helical order it is necessary to maintain the filaments at warmer temperatures, such as the 20-25° C temperature used in our filament isolation procedures. The relaxed crossbridge arrangement in these filaments is extremely sensitive to small changes in factors such as temperature or phosphorylation; and the resulting changes in thick filament structure as will be discussed later may provide insight into the changes in filament structure occurring during contraction. A key to the success of these studies is the well ordered structure of these filaments under relaxing conditions. This ordered structure facilitates the detection of any changes in the crossbridge arrangement occurring as a result of the experimental manipulations.

2.3.1. Electron Microscopic Appearance
Fig. 2 shows a gallery of such isolated relaxed mammalian (rabbit) skeletal muscle thick filaments with negative staining at medium (a-d) and high magnifications (e-f).

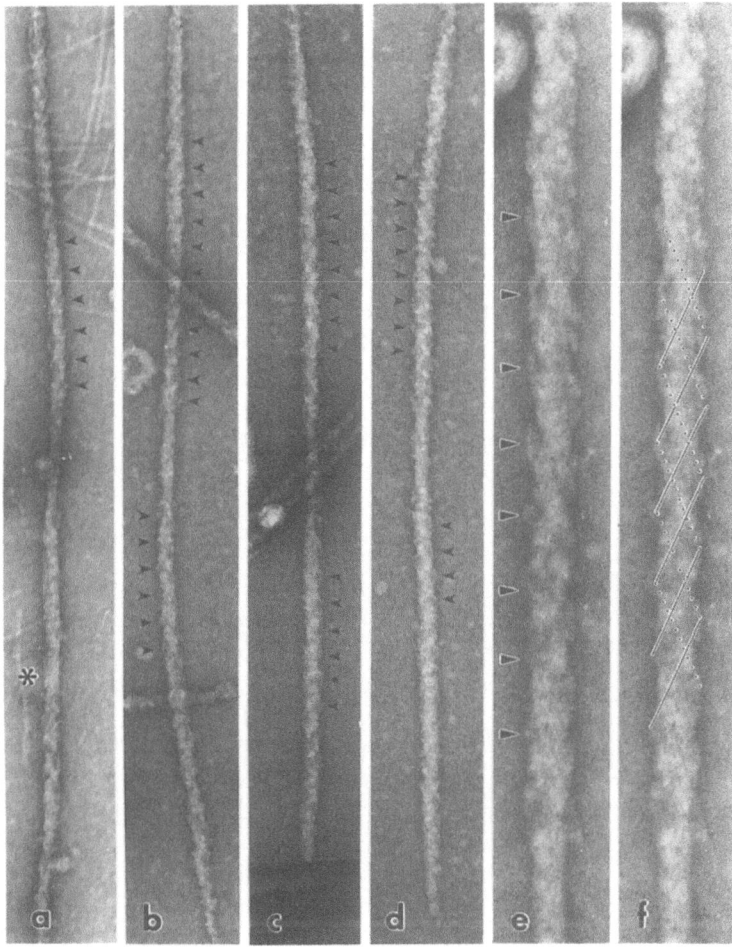

Figure 2. *A gallery of medium (a-d) and high (e-f) magnification images of negatively stained rabbit psoas muscle thick filaments. The arrowheads indicate the 43 nm axial periodicity, which can be seen along the filament arms in most regions, except for occasional small regions (asterisks) in which the filaments appear to lose order. Note that the crossbridge array appears to lack the bilaterally symmetrical appearance that would be expected if the crossbridges lay along an even number of helical or near-helical strands. (e) and (f) show two very high magnification images of the same filament differentially labelled to show both the distinct 43 nm periodicity in the crossbridge array (e) and the helical paths along which the crossbridges appear to lie (f). The arrowheads in (e) indicate where crossbridges appear to project from the backbone of the filament at regular axial repeats of 43 nm. Note that crossbridges related by mirror plane symmetry do not project from the other side of the filament. In (f) the solid lines and dotted lines illustrate the helical paths as seen on both the front (solid lines) and back (dotted lines) of the filament image. Note that the helical tracks appear to delineate a three-stranded helical pattern.*

Fig. 3 shows a similar gallery of the filaments with unidirectional platinum shadowing at low (a) and medium (b-e).   Both the negatively stained images and the platinum shadowed images clearly show the well ordered 43 nm near-helical periodicity expected from x-ray diffraction studies (Huxley and Brown, 1967; Haselgrove, 1975; Haselgrove, 1980; Haselgrove and Rodger, 1980; Rome, 1972; Millman, 1979).   In high magnification images of the negatively stained filaments (Fig. 2e) this periodicity can clearly be seen to result from an axial repeat in the crossbridge pattern every third crossbridge level.   Careful inspection of high magnification images of the isolated rabbit filaments (Fig. 2e) also reveals that these images lack the mirror plane symmetry of the crossbridges at each crossbridge level that would be expected if the myosin heads lie along an even number of helical strands. This can be seen at the arrowheads in Fig. 2e where crossbridges at 43 nm intervals   project from the left side of the filament, without a symmetrically equivalent crossbridge projecting from the other side at that crossbridge level. That the myosin heads lie along three helical strands is suggested in fig. 2f by fitting lines to the helical paths of crossbridge density on the upper (solid lines) and lower (dotted lines) sufaces of the filament.

2.3.2. Computer Image Analysis

The periodicity of the filaments was confirmed by computer image analysis of the filament images. Fourier transforms of both the negatively stained and platinum shadowed filament images showed a series of layer lines indexing close to the expected orders of a 43 nm near-helical repeat (Fig. 4a); thus consistent with the x-ray diffraction results (Huxley and Brown, 1967; Haselgrove, 1975; Haselgrove, 1980; Haselgrove and Rodger, 1980; Rome, 1972; Millman, 1979) for the vertebrate thick filament. The similarity in appearance between the computed transforms of the isolated rabbit thick filaments and the x-ray diffraction pattern from living vertebrate striated muscle is significant in suggesting that the electron microscope images reflect well the *in vivo* structure of the thick filaments.

In the case of the negatively stained filaments, the patterns often extended to at least the sixth layer line. In addition to the meridional reflections on the 3rd and 6th layer lines which come from the axial rise of 14.3 nm between crossbridge levels, the transforms frequently showed additional meridional reflections on the 2nd, 4th, and 5th layer lines. These appear to correspond to the "forbidden" meridional reflections also seen in the vertebrate x-ray diffraction pattern (Huxley and Brown, 1967). As previously noted, the presence of these reflections in the vertebrate x-ray diffraction pattern has been interpreted as indicating a perturbation from ideal helical symmetry. This conclusion has been confirmed by our reconstruction of the isolated frog thick filament (Stewart and Kensler, 1986). Thus the presence of these reflections in the transforms of the isolated mammalian thick filaments suggests that the mammalian filaments, like frog thick filaments, are perturbed from ideal helical symmetry.

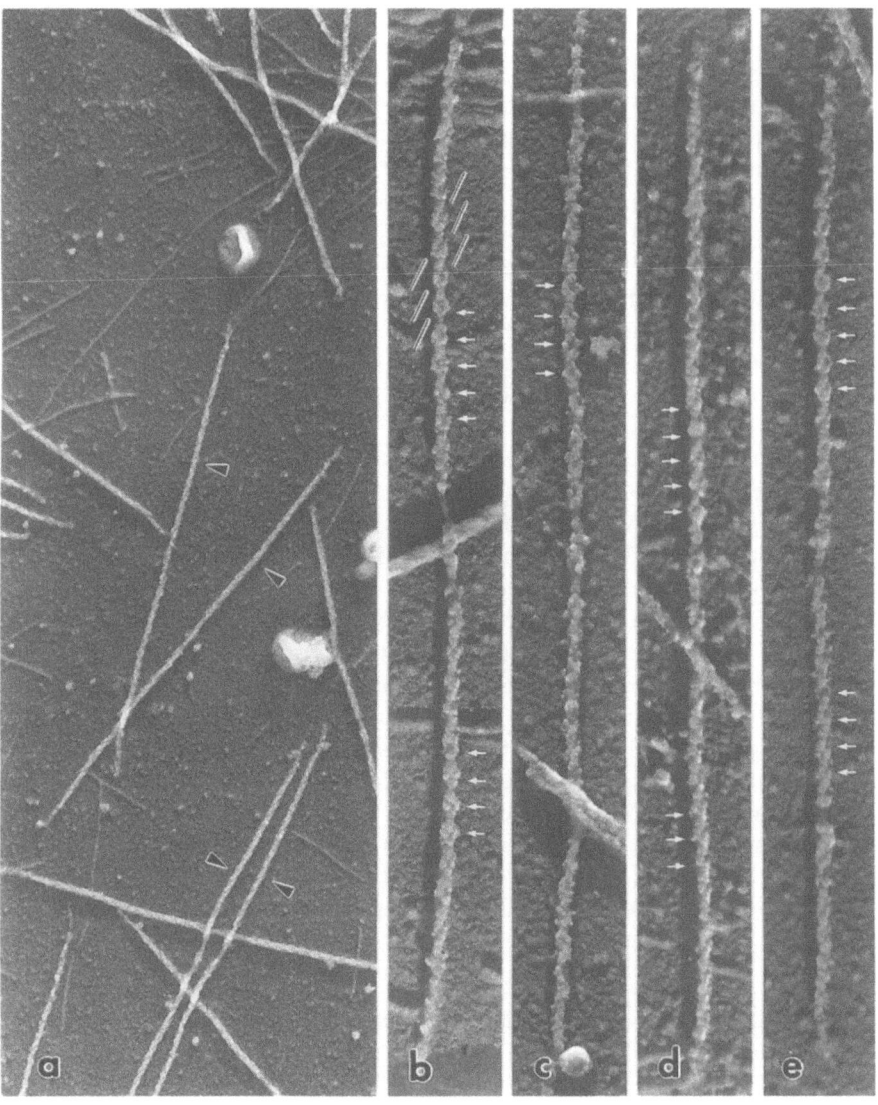

Figure 3. *Low (a) and medium (b-e) magnification images of rabbit psoas muscle thick filaments unidirectionally shadowed with platinum. Note the periodic right-handed near-helical appearance of the shadowed filaments The arrowheads in (a) point to filament arms showing a distinct periodicity. This appearance can be seen to result (b-e) from subunits lying along a series of near-helical strands, which occur every 43 nm axially, except at the bare zone. The sets of diagonal lines in (b) illustrate a region in which the right-handed near-helical arrangement of the subunits can be clearly seen, while the arrows in (b-e) illustrate the 43 nm axial periodicity of the helical strands along the filament.*

Analysis of the phase data along the first layer line in the transforms (Fig. 4b) of the images of the negatively stained filaments revealed that the phases of the primary maxima were close to 180° out of phase; consistent with the myosin heads lying along an odd number of helical strands (Moody, 1967). Consistent with this, filtered images of the filaments obtained using the data along the first six layer lines (Fig. 4c-d) showed the "saw-tooth" pattern of stain-excluding densities previously shown in the frog and demonstrated to be consistent with a three-stranded arrangement of the myosin heads (Stewart and Kensler, 1986). These filtered images show more clearly the odd-stranded arrangement of the crossbridges evident in the high-magnification images of the negatively stained filaments (Fig. 2e). Consistent with this, filtered images of the platinum shadowed filaments (Fig. 4e-g) clearly show the helical arrangement of the myosin heads, with 4 to 5 crossbridge levels per half- turn of each helix. This is consistent with a three-stranded arrangement of myosin heads with the nine crossbridge levels per full turn expected for a filament with 3 myosins per crossbridge level, and a helical repeat every third crossbridge level.

Unfortunately, in the case of the mammalian skeletal muscle thick filament, these studies have not yet been extended to a three-dimensional reconstruction because of a technical problem. Analysis shows that, although well ordered, the filaments are flattened greater than 20%; thus precluding the calculation of a three-dimensional reconstruction using the helical reconstruction technique previously used for isolated frog thick filaments (Stewart and Kensler, 1986). A similar problem has been seen for isolated fish filaments and is illustrated in fig. 5 (b-c), where the flattening can be seen in the decreased diameter at high tilt angles. Although this problem has had little effect on studies using these filaments to examine the effects of temperature, ionic strength, and phosphorylation, it means that information from a three-dimensional reconstruction of the mammalian thick filament is not available. Information on the three-dimensional structure of the mammalian thick filament, therefore, must be obtained by extrapolation from the amphibian results. At the current level of resolution of the reconstruction, however, it is likely that most of the features of the reconstruction will also be applicable with small differences to the mammalian thick filament.

2.4. 3-D Reconstruction of the Vertebrate Thick Filament

The frog thick filament was the first vertebrate thick filament examined by computer image analysis (Kensler and Stewart, 1983; 1986; Stewart and Kensler, 1986). Like the mammalian filaments it shows a distinct 43 nm near-helical periodicity (Fig. 5a). From an analysis of images of filaments tilted at defined angles around their axes (with a goniometer tilt stage) in the electron microscope, (Stewart and Kensler (1986) computed a three-dimensional reconstruction of these filaments (Fig. 5d-g). This provides much more detail about the shape and arrangement of the myosin heads, and about the probable location of accessory proteins.

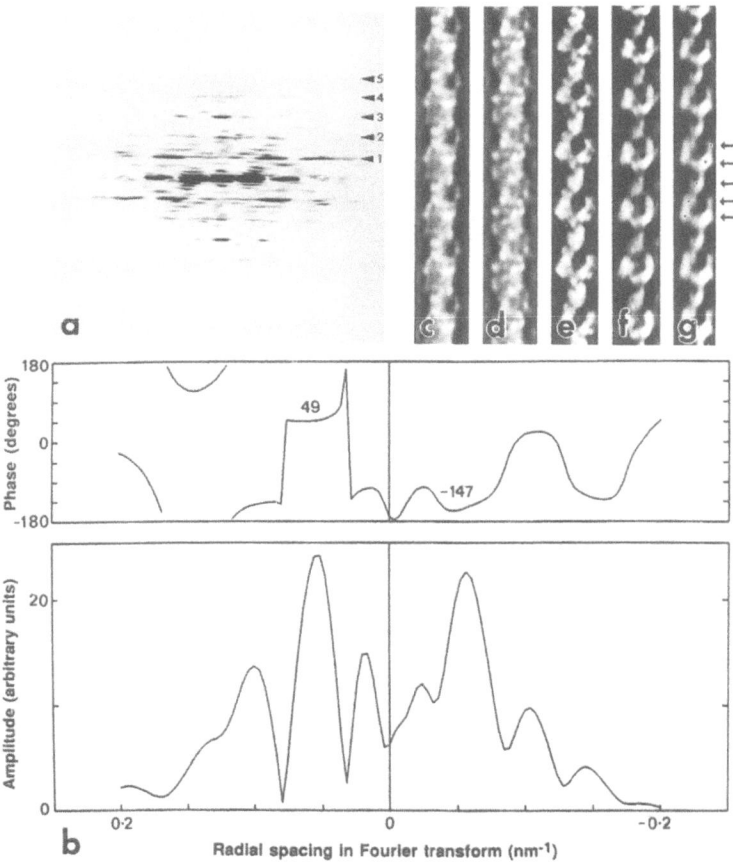

Figure 4. *(a) illustrates a computed transform of a negatively stained rabbit psoas muscle thick filament. The transform shows a series of layer lines (indicated by the arrowheads) indexing close to the expected orders of a 43 nm helical (or near-helical) periodicity with a meridional reflection on the 3rd layer line corresponding to the average axial rise of 14.3 nm per crossbridge level, and additional meridional reflections not expected from ideal helical symmetry on the 2nd, 4th, and 5th layer lines. (b) Phase and amplitude data along the 1st layer line as a function of the radial spacing from the meridian for a transform similar to (a) is shown. Note that the phase difference for the primary maxima on the layer line is close to 180°, consistent with an odd-stranded arrangement of the crossbridges. (c-d) show images from two different negtively stained rabbit psoas thick filaments filtered using the data along the first six layer lines. The helical repeat every third crossbridge level (43 nm) is clearly seen. Note also the well defined "sawtooth" pattern of the crossbridges which we have previously shown is consistent with a three-stranded arrangement of the crossbridges in other species. (e-g) show filtrations using the 1st four layer lines for three different shadowed psoas thick filaments. In each image four to five subunits can be seen along each of a set of helical (or near-helical) tracks. This is illustrated for a single track in (g) by the black dots at the center of each subunit, which indicate that each track extends over 5 crossbridge levels per half-turn of the helix. A similar pattern can be recognized for the other images.*

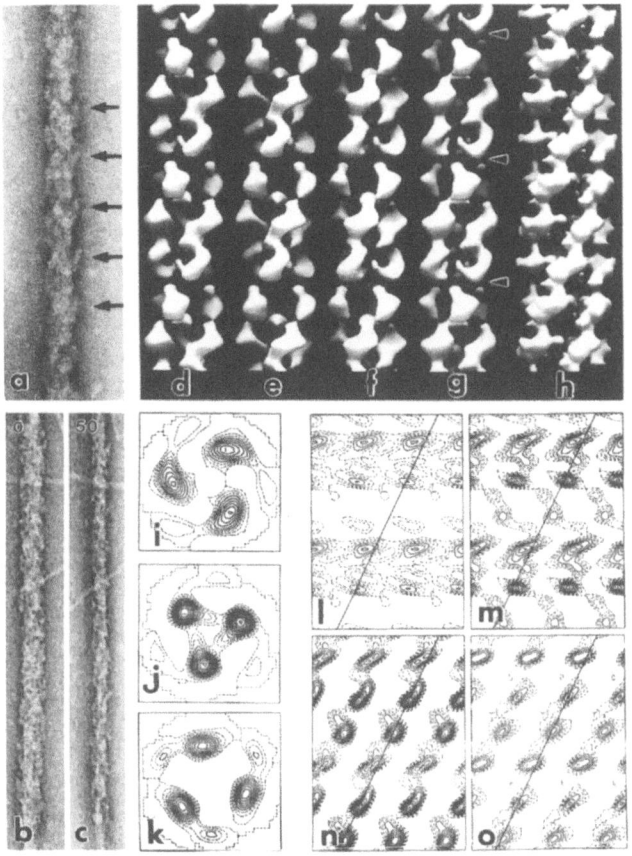

Figure 5. (a) High magnification image of a negatively stained frog skeletal muscle thick
filament. The arrows indicate the 43 nm near-helical periodicity of the filament. Note the
general similarity of the appearance of the crossbridge arrangement in the frog thick filament
to that seen in the mammalian filaments in Fig. 2. (b-c) A negatively stained fish thick
filament tilted at angles of 0° and 50° around its axis. Note the much smaller diameter at 50°
tilt, which is consistent with flattening of the filament. (d-g) Pseudosolid model of the frog
filament reconstruction. Views (d-g) show the reconstruction rotated successively by 30°,
while view (h) is a scorpion thick filament for comparison. The arrowheads along side (g)
indicate the wider gap separating the groups of three crossbridge levels which form a
repeating unit. Compare this to the evenly spaced crossbridge levels in the scorpion thick
filament.   (i-k) Sections of the reconstruction perpendicular to the filament axis through the
center of mass of the three different crossbridge levels within the 43 nm repeat. Note the
different radius for the crossbridge masses at each level. (l-o) Cylindrical sections through
the reconstruction at radii of 5 nm, 8 nm, 11 nm, and 14 nm, respectively. Note in (m-o) that
the crossbridge masses deviate by ~2 nm from the helical tracks indicated by the lines.

2.4.1. The Crossbridge Arrangement

The most significant results from the reconstruction were that it clearly showed the crossbridges to be arranged along three strands as a series of golf club shaped subunits (Fig. 5d-g) similar to those previously observed in such invertebrate filaments as *Limulus* (Fig. 5h) and scorpion (Stewart *et al.*, 1985). Within each of these subunits are two densities which correspond in size to that expected for the myosin heads. Although it was not possible from the reconstruction to determine if the heads came from the same myosin or from two myosins at adjacent crossbridge levels, crosslinking studies with $Bis_{22}ATP$ support a "splayed" arrangement of the myosin heads (Levine, 1993; 1997). In this arrangement, the heads of each myosin would be splayed with one head pointing up along the helical track and one pointing down along the helical track. Heads from myosins at adjacent crossbridge levels would overlap along the helical strand to form the densities seen in the reconstruction.

The frog reconstruction, in addition, provided evidence of a periodic perturbation in the arrangement of the crossbridges from ideal helical symmetry. This perturbation is not a random disordering of the crossbridges, but rather exists as a repeating axial, radial, and azimuthal change in the positions of the crossbridges from that expected for ideal helical symmetry. The axial perturbation can be seen in the reconstruction (Figs. 5d-g) as the grouping of crossbridges into repeating units of three crossbridge levels separated by a slightly wider space (arrowheads in Fig. 5g). The axial spacings between the three crossbridge levels making up the repeat are 13.5 nm, 16.5 nm, and 12.9 nm respectively, rather than the constant 14.3 nm expected for an ideal helix. The radial perturbation (Figs. 5i-k) can be seen in the slightly different radius of the center of mass of the myosin head densities for each of the three crossbridge levels within each repeating unit of three crossbridge levels. The azimuthial perturbation (Figs. 5l-o) is revealed in the crossbridge densities not all lying exactly along the helical path (line) in the cylindrical projections.

Evidence from x-ray diffraction patterns has been interpreted for many years as consistent with the presence of such a perturbation (Huxley and Brown, 1967), and speculations as to its origin had been put forth (Yagi *et al.*, 1981; Squire *et al.*, 1982) but the exact nature of the perturbation could not be resolved. The results from the three-dimensional reconstruction of the frog filaments provide some of the first evidence about the nature of this perturbation. At present, it is not clear if this perturbation results from the nature of the packing of the myosin rods in the shaft; the presence of the accessory proteins; or other factors.

2.4.2. Accessory Protein Location

A final feature of interest in the reconstruction is the presence of a series of densities along the shaft of the filament tentatively ascribed to the accessory protein (Stewart and Kensler, 1986). These densities occurred at a radius consistent with a position on the shaft of the filament, and did not appear to be a part of the crossbridges that could be seen at higher radii in the reconstruction. These bands of density occurred at axial spacings of approximately 43 nm (Fig. 5l), which is thought to be the axial spacing of several of the accessory proteins. Of these, MyBp-C would be the most likely candidate, since the areas of the filaments used for the reconstruction correspond to the regions in which MyBP-C is thought to be present.

2.4.3. Similarity of Mammalian Striated Muscle Thick Filament to the Frog Filament

Although it is not possible to state that the mammalian skeletal muscle thick filament has the same structure, it is clear that it, like the frog thick filament, has a three-stranded arrangement of the myosin crossbridges. In addition, the presence of relatively strong meridional reflections not expected from ideal helical symmetry (i.e "forbidden" meridional reflections – Huxley and Brown, 1967) in its transform clearly shows that it has also has a crossbridge arrangement perturbed from ideal helical symmetry.

In addition, recent results reported for the isolated rabbit cardiac muscle thick filament show that it also has a well ordered helical arrangement of crossbridges consistent with a three-stranded arrangement of the crossbridges (Kensler, 1997; 1998; 2001). Initial analysis suggests that its crossbridge arrangement is as perturbed, if not more so, as the rabbit skeletal muscle thick filament.

2.5. Maintenance of the Relaxed Crossbridge Arrangement

Evidence from both X-ray diffraction (Huxley and Brown, 1967; Haselgrove, 1975; Haselgrove, 1980; Haselgrove and Rodger, 1980; Rome, 1972; Millman, 1979) and electron microscopy (Kensler and Woodhead, 1995; Kensler et al., 1994; Kensler and Stewart, 1983; 1986; 1989; 1993; Stewart and Kensler, 1986) of vertebrate thick filaments, as well as several invertebrate thick filaments; have clearly shown that under relaxing conditions the majority of the myosin heads at any one time remain close to the filament backbone in a well ordered helical or near-helical arrangement. This suggests that there are factors which help to maintain the helical array.

2.5.1. Factors Stabilizing the Crossbridge Arrangement

Among the factors which have been suggested as helping to stabilize the helical array are: 1) an increase in stiffness of the head correlated with the predominant nucleotide hydrolysis state present during relaxation (Wray, 1987; Dominguez et al., 1998; Houdusse et al., 1999; Xu et al., 1999); 2) overlap and weak binding of myosin heads from adjacent crossbridge levels to each other along the helical tract (Crowther et al., 1985; Stewart et al., 1985; Levine, 1993; 1997; Padron et al., 1998); and 3) charge dependent interactions between positively charged regions of the regulatory light chain and negatively charged regions of the backbone (Levine et al., 1995; 1996; 1998). Evidence supports all of these possibilities, and it seems probable that all of these factors play a coordinated role in maintaining the helical order of the heads in the relaxed filament. The differences observed in the lability of the crossbridge helical arrangement for thick filaments from different species to factors such as temperature and ionic strength (Wray, 1987; Wakabayashi *et al.,* 1988; Kensler *et al.,* 1994; Kensler and Woodhead, 1995; Xu *et al.,* 1999) most likely reflects species dependent differences in the influence of these factors in maintaining the relaxed crossbridge arrangement.

2.5.2. Strength of the Factors stabilizing Crossbridge Arrangement

By their nature, the forces helping to stabilize the helical array of heads must be relatively weak in order to allow the heads to be released from the filament backbone and to interact with the adjacent actin filaments. This is consistent with data from a number of experiments suggesting that an equilibrium exists between the the ordered helical arrangement of the heads and the more disordered state involved in the attachment of the heads to actin and contraction (see Xu *et al.,* 1999). As will be detailed in the next several sections, a number of environment factors can influence thick filament structure and provide information on the probable changes in the thick filament under activation conditions.

3. THE THICK FILAMENT IN ACTIVATED MUSCLE

Some thick filament preparations contain more disordered filaments than those that retain the near-helical array of surface myosin described as characteristic of the relaxed state. We, and others have found that filament structure is related to the conditions under which the sample is prepared. This implies that a variety of environmental conditions affect the surface array of myosin heads. Examination of the effects of these factors may clarify those of activation phenomena on thick filament structure. Although some of the perturbing factors may be specific to filaments from the muscles of a particular class or species, they, too, may elucidate responses of the myosin head to discrete biochemical steps in the crossbridge cycle of intact muscle.

3.1. Factors InfluencingTthick Filament Structure

Before the molecular model for the myosin head (Rayment *et al.,* 1993) became available, it was difficult to ascribe any changes in myosin head attitude to specific internal conformational shifts between domains. It was apparent, however, that the myosin head array on mammalian and other vertebrate thick filaments was exquisitely sensitive to such environmental manipulations as nucleotide depletion and changes in temperature (Wray *et al.,* 1986), and fairly sensitive to changes in ionic strength. Similar to the situation noted earlier with scallop (Vibert and Craig, 1985) and *Limulus* (Levine *et al.,* 1986) thick filaments, nucleotide-depleted vertebrate filaments become completely disordered (Schlicting and Wray, 1986; Xu *et al.,* 1999). Myosin heads extend randomly to varying distances from the filaments' backbones. If thin filaments lie alongside thick filaments within 18 nm of their surfaces in ATP-free solutions, myosin heads become attached to these thin filaments in an angled conformation with regular spacing of about 37-40 nm. The attitudes of the bound myosin heads are similar, if not identical, to that of subfragments 1 (S1s) producing "decorated" thin filaments.

Temperature sensitivity is more of a class-specific phenomenon and is especially the case with thick filaments from warm-blooded species: birds and mammals. It was noted, both by x-ray diffraction of muscle bundles (Wray *et al.,* 1986; Lowy *et al.,* 1991; Malinchick *et al.,* 1997; Xu *et al.,* 1997; 1999) and by electron microscopy of negatively-stained preparations of thick filaments (Kensler *et al.,* 1994; Kensler and Woodhead, 1995; Levine *et al.,* 1996), that lowering of the ambient temperature below 18° C perturbed the near-helical surface array (Fig. 6). Within the thick-thin filament lattice, the mass of the myosin heads moved away from the thick filament backbone and closer to the thin filaments (Xu *et al.,* 1997). One rationale suggested for this effect is that at the colder temperatures the ATP is not hydrolyzed in the active site of the myosin head (Wray *et al.,* 1986; Schlicting and Wray, 1986; Rapp et al. 1991) This implies that myosin heads with intact ATP in their active sites may have conformations different than those with ADP.Pi or an intermediate, which was confirmed by x-ray crystallographic (Wakabayashi, 1992; Fisher *et al.,* 1995) and electron spin resonance (Ostap *et al.,* 1995; Roopnarine and Thomas, 1996; Adhikari *et al.,* 1997; Baker *et al.,* 1998) studies. It was suggested that the disordering head conformation was "straight", which extended further away from the filament backbone than a "bent" head with ADP.Pi in the active site. Only the latter form the ordered array of relaxed thick filaments (Xu *et al.,* 1999).

Another factor producing disorder of the surface array of myosin heads on thick filaments, although to a lesser extent than temperature, is the ionic strength of the surrounding medium. Several investigators noted that at low ionic strengths (≤ 50 mM) the myosin head array also lost its near-helical, relaxed order and myosin heads were weakly bound to actin (Brenner *et al.,* 1982, 1984; Xu *et al.,* 1997, 1999). A comparative electron microscopic study of fish and rabbit filaments, (Kensler *et al.,* 1994) provided evidence that the disordering effects of temperature and ionic strength are most pronounced on the mammalian thick filaments and less so on those of fish.

3.2. Activation Effects: Myosin Regulatory Light Chains

It is important to determine if biochemical events in the muscle fiber during activation have noticeable effects on thick filament structure in vertebrate and specifically mammalian striated muscle and if such effects reflect changes in physiological parameters. In these tissues, contraction is directly regulated by calcium released from the terminal cisternae of sarcoplasmic reticulum binding to troponin C (TnC), which is part of the regulatory complex on the thin filaments. The ultimate result of such binding is exposure of the myosin-binding site on the actin molecules. However, calcium also has other effects and binds to other sites in the sarcoplasm, including calmodulin and at higher concentrations, the high affinity single divalent cation binding sites on the calmodulin-like regulatory light chains of myosin (Metzger and Moss, 1992; Diffee et al. 1995; 1996).

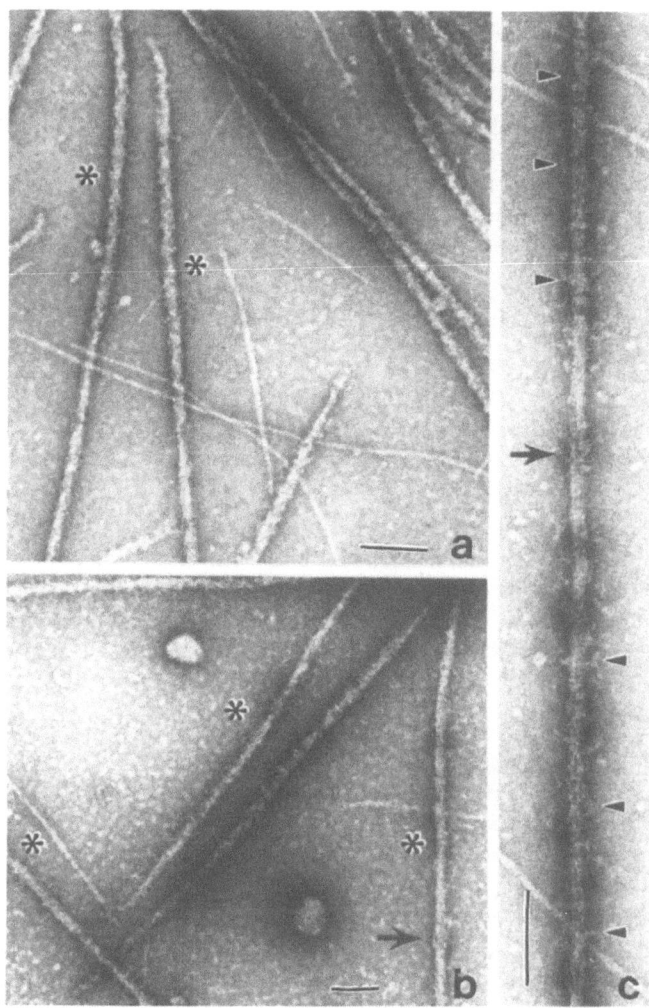

Figure 6. *Effects of temperature on mammalian thick filament structure. Electron micrographs of rabbit thick filaments maintained at 25 oC (a) and 4 oC (b and c) prior to staining. The typical 43 nm repeat periodicity of the relaxed pattern of myosin heads is indicated by asterisks in image 6a, which shows a preparation made at 25 oC. The surface helical periodicity is absent from the field of filaments shown in image 6b, and the single filament shown in image 6c, both from a preparation made at 4 oC. The asterisks in b indicate the disordered appearance of the arrangement of myosin heads on the surface of the low-temperature thick filaments; the black arrow points to a bare zone region on one filament. In c, an image at higher magnification, the central bare zone is again indicated by the black arrow. Arrowheads point to myosin heads that extend to high radii from the filament backbone in this disordered structure. Bars = 100nm.*

The calcium-calmodulin complex activates a series of cellular responses. One of these, phosphorylation of the regulatory light chains (light chain 2, LC2) of myosin II by calmodulin-activated, tissue-specific myosin light chain kinase (MLCK), is a nearly ubiquitous phenomenon, with varying effects in different muscle and tissue types. It directly regulates motor activity in vertebrate smooth muscles and in non-muscle cells. In the latter it may also affect the aggregation of myosin II into "thick filaments" by altering the binding affinity of regions of the molecule for other myosin sites. Actin-myosin interaction in these tissues requires phosphorylation of a serine residue (Ser 19 in smooth muscle myosin) in the regulatory light chain N-terminus; dephosphorylation abolishes such interaction (Sellers and Adelstein, 1987; Ito and Hartshorne, 1990). LC2 phosphorylation is also required, although not sufficient, for activation of dually-regulated invertebrate striated muscles; calcium must also bind to troponin C on the thin filaments (Lehman, *et al.*, 1973; Sellers, 1981; Kerrick and Bolles, 1981; Wang *et al.*, 1993). In tarantula and *Limulus* muscles, thick filaments respond to phosphorylation of their myosin regulatory light chains (one of the two regulatory light chains in *Limulus* myosin) by becoming disordered, with myosin heads extending away from the filaments' surfaces (Craig *et al.*, 1987; Levine *et al.*, 1991; Padron *et al.*, 1991). This response is similar to that of scallop filaments incubated with calcium (which directly binds to myosin in this muscle) and has been interpreted as a direct effect of activation (Vibert and Craig, 1985). In vertebrate striated muscles, with thin filament-based regulation, contraction does not depend on LC2 phosphorylation. However, under normal physiological conditions, MLCK is activated during contractile activity, resulting in phosphorylation of the regulatory light chains (Perrie *et al.*, 1973; Stull and High, 1977; Barany and Barany, 1979).

3.2.1. Physiological Expression of LC2 Phosphorylation
The physiological observation of potentiation of twitch tension following a tetanus in intact muscle was ascribed to phosphorylation of the myosin regulatory light chains that most likely occurred during the prolonged contraction. Studies on the effects of myosin light chain phosphorylation on single permeabilized fibers of rabbit psoas also demonstrated potentiation of tension development at low levels of sarcoplasmic calcium. There was an increase in the rate of rise of tension and increased calcium sensitivity at calcium levels between pCa 6.2 and pCa 5.3 (Manning and Stull, 1982; Metzger *et al.*, 1989; Sweeney and Stull, 1990; Sweeney *et al.*, 1993). These effects were ascribed to an increase in the apparent crossbridge attachment rate when LC2 was phosphorylated (Sweeney and Stull, 1990; Metzger *et al.*, 1989). More recently, it was suggested that the relaxation rate might also be slowed (Patel *et al.*, 1998). All of these findings could be due to myosin heads moving close to actin on thin filaments. In mammalian striated muscle the effects of myosin light chain phosphorylation are relatively long-lived, due to the low level of activity of the tissue-specific phosphatase that dephosphorylates the regulatory light chains (Sweeney *et al.*, 1993).

3.2.2. Structural Expression of LC2 phosphorylation

Electron microscopic and optical diffraction, as well as some x-ray studies, show that indeed LC2 phosphorylation reversibly perturbs the helical or near-helical order of the relaxed arrangement of myosin heads on the surfaces of native thick filaments in fibers from both dually-regulated arthropod and thin filament-regulated vertebrate muscles (Craig *et al.*, 1987; Padron *et al.*, 1991; Levine *et al.*, 1991; 1995; 1996). In negatively-stained preparations of thick filaments separated from permeabilized rabbit psoas fibers, phosphorylated myosin heads appear randomly positioned around the filament backbone and some reach to high radii away from the filament surface, similar to the situation in thick filaments from chelicerate arthropod muscles. The relaxed order is restored following dephosphorylation with tissue-specific phosphatase. Fourier transforms confirm the reappearance of myosin layer lines associated with the relaxed state after dephosphorylation (Levine *et al.*, 1996) (Fig. 7). Structural changes associated with LC2 phosphorylation may be interpreted in at least two ways. One is that these changes reflect an increase in the mobility of myosin heads, which move to different attitudes and reach higher radii from the filament backbone than they are able to in the relaxed, unphosphorylated state. Another interpretation is that LC2 phosphorylation affects the state of the nucleotide in the active site of the motor domain of the myosin head, which produces a conformational change of the head, or a shift in the rotation of the regulatory domain about the motor domain, with a result similar to that of lowering the temperature (Wray *et al.*, 1986; Xu *et al.*, 1999). In either case, mobile or conformationally changed heads become "caught" in random positions by the negative-staining procedure and are seen as "disordered". Most studies have concentrated on the first of these interpretations, as it has proved readily testable, although both interpretations may be operative.

Disordering thick filament structure by increasing myosin head mobility. The hypothesis of increased head mobility is consistent with results of crosslinking subfragments 2 (S2) to the backbones of synthetic filaments formed from vertebrate skeletal muscle myosin. These studies also indicated that LC2 phosphorylation promoted partial release of myosin heads from the filaments' surfaces (Mrakovicec-Zenic and Reisler, 1983). This interpretation of the structural change fits well with the notion that phosphorylation-induced increased mobility of myosin heads produces the observed physiological effects on force generation (Manning and Stull, 1982; Metzger *et al.*, 1989; Sweeney *et al.*, 1993; Patel *et al.*, 1998), since mobile heads are likely to spend longer times in greater proximity to their binding sites on thin filament actins than "relaxed" heads that remain close to the thick filament backbones. Such heads are in optimal positions to interact with actin, even under conditions of low sarcoplasmic calcium. This interpretation is further supported in that when unphosphorylated myosin heads of thick filaments are brought closer to thin filaments by decreasing the lattice spacing within the A-bands of permeabilized fibers (either by stretching or by osmotic compression), increases in calcium-sensitivity, seen as a leftward shift of the pCa - force relation, occur, similar to those seen following LC2 phosphorylation. Subsequent phosphorylation of the regulatory light chains, however, does not affect tension development further (Yang *et al.*,

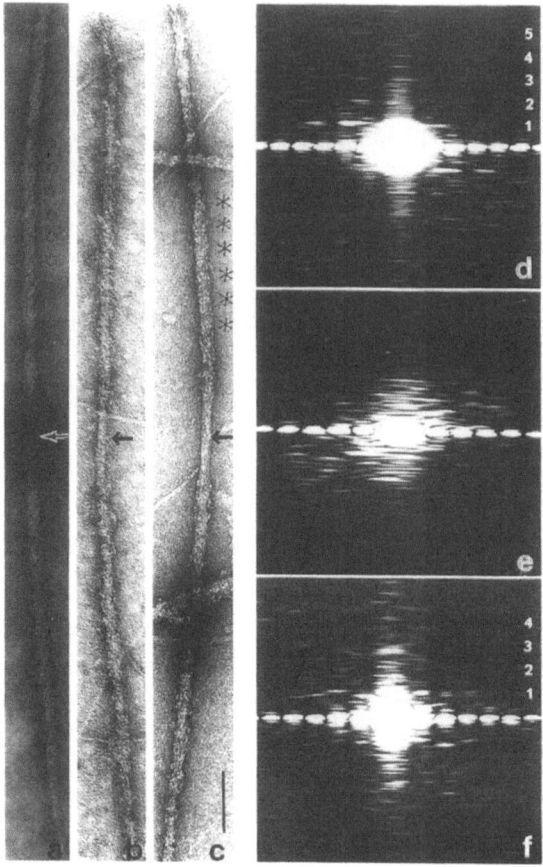

Figure 7. *Reversible structural effects of LC2 phosphorylation. Electron micrographs (EMs) (a, b, c) and corresponding optical transforms (OTs) (d, e, f) showing the reversible disordering of the relaxed arrangement of myosin heads on rabbit thick filaments by phosphorylation/dephosphorylation of LC2s. On EMs of negatively stained filaments, bare zones are indicated by arrows and the near-helical repeat periodicity of myosin heads on the filaments' surfaces, where present, is indicated by asterisks (a and c). Bar = 100 nm. Myosin –based layer-line positions (where present) are numbered in the upper right quadrant of OTs (d and f). (a and c): Image and OT of unphosphorylated native thick filament. The relaxed pattern of myosin heads is present on the filament surface and myosin-based layer-lines are present in the OT. (b and e): Image and OT of thick filament after LC2 phosphorylation by incubation of a permeabilized fiber with skeletal muscle MLCK. Note the disorder of the myosin head array on the filament surface and absence of myosin-based reflections in the OT. (d and f) Image and OT of thick filament from a permeabilized fiber that had been incubated first with MLCK to phosphorylate LC2s, and then with the catalytic subunit of skeletal muscle phosphatase. Note the reappearance of surface order to the filament and of some of the myosin-based reflections to the OT. Reproduced with permission from Yang et al., 1998, Figure 1, p. 143, Academic Press.*

1998) (Fig. 8). Thus, the effects of phosphorylation are mimicked by decreasing the distance between the myosin heads and the thin filaments, as would occur with increased head mobility.

Are changes in charge responsible for increased myosin head mobility? Phosphorylation of smooth muscle or cellular myosin LC2s is associated with a structural change at the level of individual, solubilized molecules; a shift from the "straight" 6s to the "safety-pin-bent" 10s form of the molecule. The latter conformation is interpreted as representing, in one molecule, phosphorylation-dependent interactions that occur between molecules during aggregation to form thick filaments (needed for directional motility in cells) (Craig *et al.*, 1983; Trybus, 1994). Reasonably, it is assumed that these interactions are charge-related. In vertebrate smooth muscle, phosphorylation-mediated regulation also requires intramolecular rearrangement of regulatory light chains, involving the phosphoserine in the N-terminus and one or more of the four arginines in the C-terminal half of the molecule (Yang and Sweeney, 1995), also a charge-related phenomenon. Although similar conformational changes have not been seen in individual vertebrate striated muscle myosin molecules or even LC2 moeities, one may assume that the documented structural changes of the thick filaments may be related to changes in charge associated with LC2 phosphorylation. In fact, a reduction in the net positive charge near the phosphorylation site of the vertebrate skeletal LCs mimics the effect of phosphorylation, in terms of its capacity to modulate contraction and produce a leftward shift of the pCa - force relation (Sweeney *et al.*, 1994). This change in the charge complement of the LC N-terminus occurs when phosphate binds to the phosphorylatable serine, adjacent to several positively charged lysine and arginine residues (Saraswat and Lowey, 1991; Chavanieu *et al.*, 1994). Thus, it is attractive to suggest that the structural change of thick filaments that is associated with LC2 phosphorylation is, first, indicative of increased head mobility, and second, may result from altered charge-related interactions between the N-terminal region of the regulatory light chain and the thick filament backbone. In filaments with unphosphorylated myosin, the basic sequence of this region of the LC2 may help to maintain the relaxed, "ordered" array of myosin heads by charge-based interactions with underlying regions of other myosin molecules in the filament backbone. Neutralization of the net postive charge by LC2 phosphorylation may remove such a constraint on head mobility and produce both a disordered filament surface and a leftward shift of the pCa - force relationship in the fiber (Fig. 9).

One way to explore this hypothesis is to alter the charge complement on the N-terminus of the mammalian skeletal muscle regulatory light chain and examine filament structure with and without phosphorylation. The native endogenous regulatory light chains of permeabilized rabbit psoas muscle fibers have been exchanged with: (a) expressed wild type rabbit skeletal muscle regulatory light chains, (b) expressed, wild-type, human smooth muscle regulatory light chains and (c) two mutant recombinant human smooth muscle regulatory light chains. In one of these, serine 19 (the phosphorylatable serine) is replaced with negatively-charged

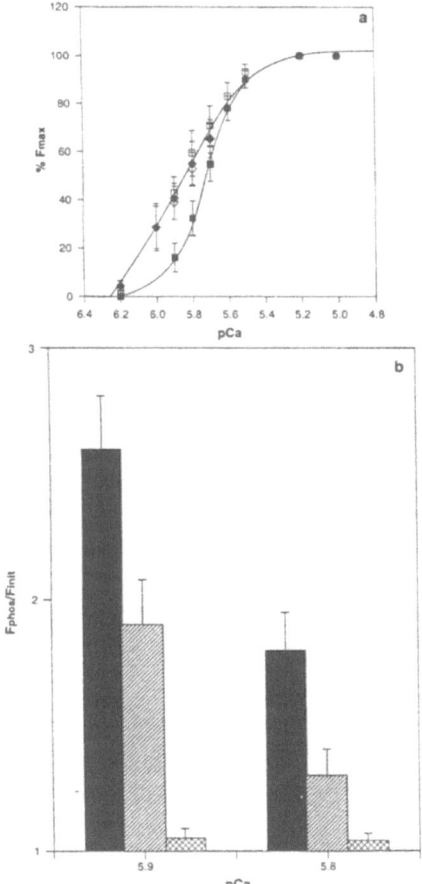

Figure 8. *Physiological effects of osmotic compression. (a) Force-pCa relationships before and after LC2 phosphorylation in mermeabilized fibers held at 2.5 μm sarcomere length, in the presence (n=7) and absence (n=6)of 8% dextran. Points are mean values ± SD. All determinations were made at 23°C. Filled squares: values for force-pCa measurements before LC2 phosphorylation in the absence of dextran; open squares: values for force-pCa measurements after LC2 phosphorylation in the absence of dextran; filled diamonds: values for force-pCa measurements before LC2 phosphorylation in the presence of 8% dextran; filled diamonds: values for force-pCa measurements after LC2 phosphorylation in the presence of 8% dextran. (b) Histograms showing the effects of osmotic compression on the amount of potentiation of force development obtained following phosphorylation of the myosin RLC. Bars represent the ratio of force developed after LC2 phosphorylation to the force developed prior to phosphorylation at either pCa 5.9 or pCa 5.8. Graphed values are the means + SD from 6 fibers in the case of 0% (filled bars) and 4% (diagonally hatched bars) and from 7 fibers in the case of 8% (cross-hatched bars) dextran. Note the lack of effect of LC2 phosphorylation on the force-pCa relation in the case of the fully compressed lattice (8% dextran), when the thick and thin filaments are already close together. Reproduced with permission from Yang et al., 1998, Figure 2, p. 144, Academic Press.*

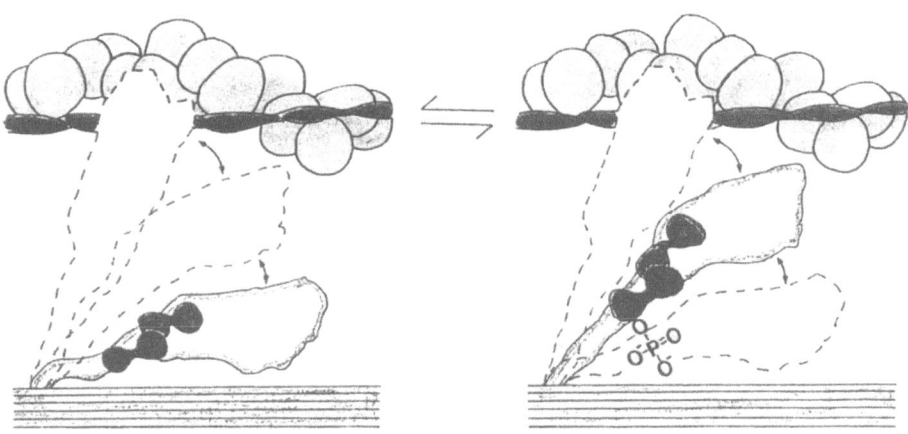

Figure 9. *Drawing of the suggested effect of LC2 phosphorylation on the average position (solid-line head) of the myosin head relative to the filament lattice. Left: LC2 unphosphorylated. Right: LC2 phosphorylated.*

glutamate. In the other, threonine 18 and serine 19 are both replaced with glutamate. The smooth muscle constructs bind to the skeletal muscle myosin heavy chain more strongly than either native or expressed skeletal muscle LC2 (Sweeney *et al.*, 1994; Trybus, 1994). 90 to 100% exchange is achieved by several cycles of exchange at low temperature (8° C) in the presence of a 10 M excess of expressed peptide. Thick filaments made hybrid with either skeletal or smooth muscle wild-type constructs display good relaxed order of the myosin surface array both on micrographs and Fourier transforms (Figs. 10 and 11). Upon incubation with myosin light chain kinase, however, they both become disordered: the filaments hybridized with the smooth muscle construct more severely so (Figs. 10 and 11). The calcium sensitivity of permeabilized fibers hybrid for either construct is increased by LC2 phosphorylation (Fig. 12) (Levine *et al.*, 1998).

Thick filaments hybrid for the mutant construct in which the phosphorylatable serine is replaced with negatively charged glutamate appear completely disordered, even under relaxing conditions (Fig. 13). Permeabilized fibers hybrid for this construct and for the mutant light chain where both threonine 18 and serine 19 are replaced by glutamate also show increased calcium sensitivity. Such findings strongly support the hypothesis that the physiological effects of myosin regulatory light chain phosphorylation are due to increased myosin head mobility and that such increased mobility is a result of a change in charge complement in the light chain N-terminus by the addition of negatively charged phosphate to the phosphorylatable serine.

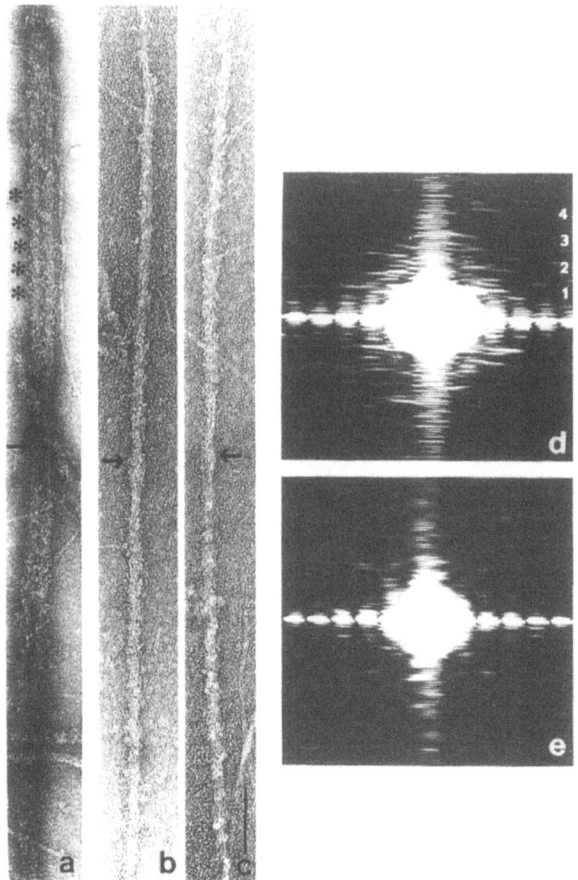

Figure 10. *Structural effects of exchange of wild-type skeletal muscle LC2s (WTskLC2) into permeabilized psoas fibers. (a, b, c) EMs and (d and e) OTs demonstrating the effect of LC2 phosphorylation on the structure of thick filaments following the exchange of endogenous LC2s for WTskLC2s expressed in E. coli (control exchange) in permeabilized fibers. In EMs, bare zones of negatively stained filaments are indicated by black arrows and the periodic surface structure (where present) by asterisks. Bar = 100 nm. The myosin-based layer-lines (when present) are numbered in the upper right-hand quadrand of the OT. (a and d): Image and OT of filament under relaxing conditions, following exchange of endogenous LC2s for (identical) expressed, unphosphorylated WTskLC2s. Note the helical periodicity characteristic of native relaxed filaments ont eh hybrid filament after exchange. This is echoed by the presence of myosin-based reflections in the OT. (b and c): Images of filaments after phosphorylation of exchanged LC2s by incubation of the permeabilized, exchanged fiber with skeletal MLCK, and (e) an OT of one such filament. Note the disorder of the filaments' surfaces on the EMs and the absence of myosin-based reflections in the OT, all of which indicate structural variability of the positions of the myosin heads due to phosphorylation, as seen in native filaments. Reproduced with permission from Levine et al., 1998, Figure 2, p. 153, Academic Press.*

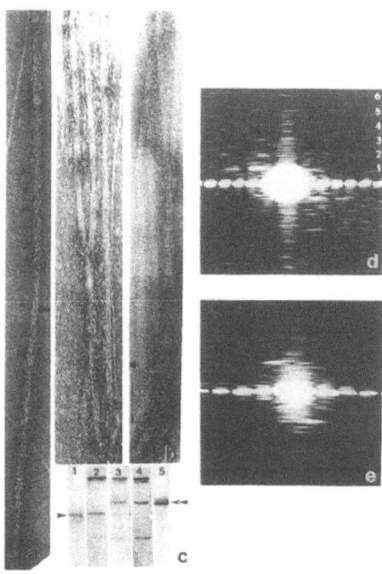

Figure 11. *Structural effects of exchange of wild-type smooth muscle LC2s (WTsmLC2) into permeabilized psoas fibers. (a and b) EMs, (c) SDS-polyacrylamide gel, and (d and e) OTs illustrating the effect of LC2 phosphorylation on the structure of thick filaments following the exchange of endogenous LC2s for WTsmLC2s expressed in E. coli (experimental exchange) in permeabilized fibers. In EMs, bare zones of filaments are indicated by black arrows and the periodic surface structure (where present), by asterisks. Bar = 100 nm. The myosin-based layer-lines (where present) are numbered in the upper right-hand quadrant of the OT. (a and d): EM and OT of filament after complete exchange of endogenous LC2s for the expressed, unphosphorylated WTsmLC2s (see gel in c). Note the maintenance of the periodic surface repeat of the myosin head array characteristic of the relaxed state and identical to that of filaments after the control exchange seen in Figure RL4). The OT shows myosin-based layer-lines consistent with the relaxed filament structure. (b and e): EMs of groups of filaments with exchanged WTsmLC2s, after phosphorylation of the exchanged, permeabilized fiber with smooth muscle MLCK and an OT from one filament from such a group. Note the total disarray of the myosin heads on the filaments after LC phosphorylation; the heads are positioned in all directions and reach high radial attitudes from the filaments' shafts. Some phosphorylated heads interact with actin on neighboring thin filaments. The OT lacks myosin-based reflections. (c): Lower half of a Coomassie brilliant blue-stained gradient SDS-polyacrylamide gel showing complete exchange of endogenous skeletal LC2 for expressed WTsmLC2. The band identified as skeletal LC2 is indicated by a single arrowhead; that of smooth muscle LC2 by double arrowheads. The unlabeled upper doublet and lower bands are, respectively, native skeletal muscle essential light chains LC1 and LC3. Lane 1, expressed WTskLC2. Lane 2, permeabilized fiber segment prior to exchange procedure. Lane 3, permeabilized fiber segment halfway through exchange procedure; approximately 70% of the endogenous LC2 is replaced by WTsmLC2. Lane 4, permeabilized fiber segment at completion of exchange procedure. All of the endogenous LC2 is replaced by the expressed WTsmLC2. There is no change in the essential LCs. Lane 5, expressed WTsmLC2. Reproduced with permission from Levine et al., 1998, Figure 3, p. 154, Academic Press.*

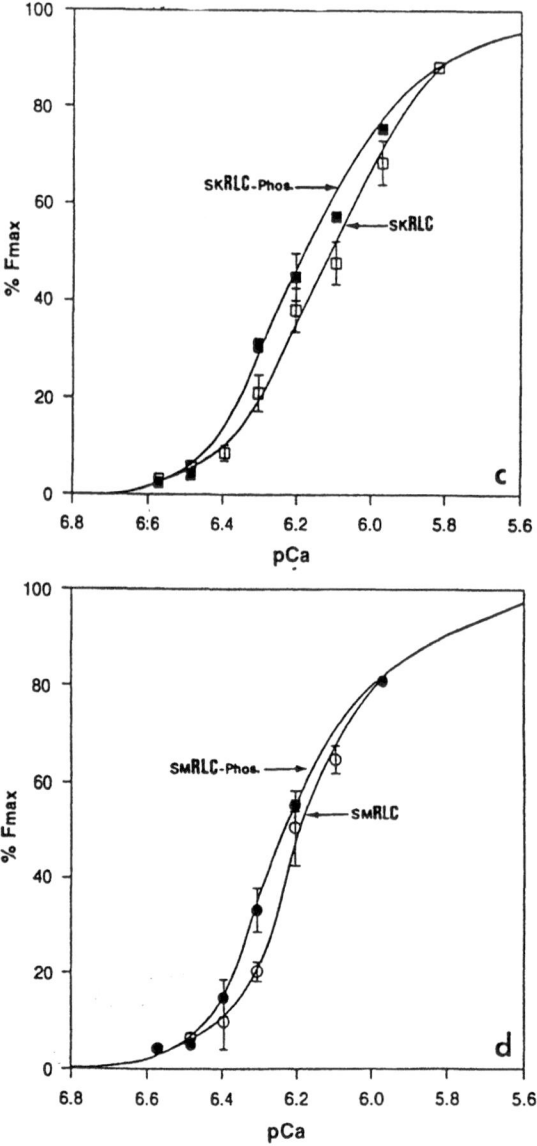

Figure 12. *Physiologic effects of phosphorylation of LC2s in hybrid fibers. Force-pCa relationships of permeabilized rabbit psoas fibers in which endogenous LC2s have been replaced by bacterially expressed WTskLC2s or WTsmLC2s, before and after phosphorylation. All force measurements were made at 25°C and are expressed as means ± SE (normalized to dephosphorylated WTLCs at pCa4.3). (a) Force-pCa relationships before (open squares) and after (filled squares) phosphorylation of exchanged WTskLC2s. (b) Force-pCa relationships before (open circles) and after (filled circles) phosphorylation of exchanged WTsmLC2s. Modified with permission from Levine et al., 1998, Figure 1, p. 152. Academic Press.*

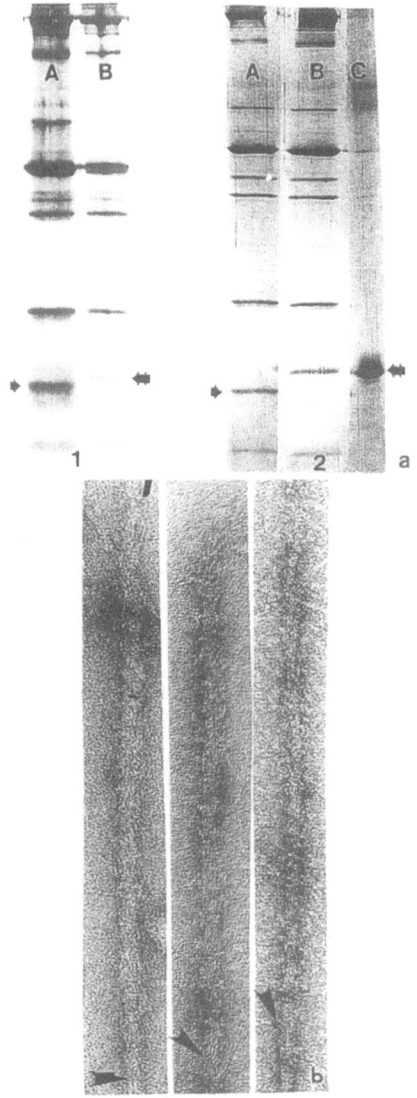

Figure 13. *Exchange with mutant smooth muscle LC2s and the structural effect on thick filaments of hybrid fibers. (a) SDS-polyacrylamide gels: (1) ca. 90% exchange with SmLC2-S19E and (2) 100% exchange with SmLC2-T18E, S19E, respectively. Single arrowheads indicate endogenous SkLC2; double arrowheads indicate SmLC2. A: fiber pre-exchange. B: fiber after exchange with mutant LCs. C: expressed mutant SmLC. (b) Thick filaments from fibers with exchanged SmLC2-S19E. Even under relaxing conditions, filaments hybrid for an LC2 with an acidic residue 19 are disordered.*

Finally, although this effect has not been studied structurally, it is likely that phosphorylation of LC2 may increase the compactness or decrease the flexibility of the light chain, since it protects against digestion of the light chain by trypsin or papain (Ritz-Gold *et al.*, 1980).

3.2.3. A Functional Rationale for LC2 Phosphorylation/Dephosphorylation in Mammalian Skeletal Muscle

The effect of LC2 phosphorylation on the calcium sensitivity of force generation and the rate of force development is most pronounced is at sarcomere lengths where thick-thin filament overlap is optimal, *i.e.* rest length. This is also the sarcomere length at which most muscles contract. It may be that phosphorylation of the myosin regulatory light chain functions as a recent "muscle memory", with the extent of activation at a specific calcium level dependent on the muscle's immediately previous contractile history. A very recently active muscle, having myosin regulatory light chains still in the phosphorylated state and "disordered" thick filaments with mobile myosin heads (probably at high radii from the filaments' backbones), would respond more rapidly and with greater force to suboptimal levels of sarcoplasmic calcium than would a long-term relaxed muscle where dephosphorylated heads are in an ordered array close to the filaments' surfaces. The response would be strongest at optimal (complete or near-complete) overlap. Such a mechanism is economical, since, when myosin is phosphorylated, the cascade of activation-associated events requires only a small increase in sarcoplasmic calcium, so less calcium has to be pumped back into the sarcoplasmic reticulum, at a decreased energy cost. The persistence of cycles of phosphorylation and dephosphorylation in mammalian skeletal muscle remains unclear, since it seems as if continuous phosphorylation may be advantageous to contractile economy. Thus, in *Drosophila melanogaster* indirect flight muscle, where LC2 phosphorylation is necessary for sufficient oscillatory work to support flight, the light chain is maintained in a phosphorylated state (Tohtong *et al.*, 1995). As suggested by Yang *et al.*, (1998), reversible LC2 phosphorylation in mammalian striated muscle may reflect the fact that many very fast muscles are active only for a small part of the time, which makes energy use by myosin, independent of actin-myosin interaction, a significant "housekeeping" expense in this tissue. If, in non-active muscle, the myosin heads are maintained in the ordered, near-helical array of the relaxed filament structure, they may be trapped in a state of lowered ATP turnover. (Increased head mobility might increase the myosin ATPase rate). Then, energy savings would occur in the dephosphorylated, relaxed state, as well as in active muscle with phosphorylated myosin. The two "economizing" influences may act together to preserve this cycle and its modulatory role in mammalian skeletal muscle.

3.2.4. Effects of calcium binding to LC2 in mammalian striated muscle

The divalent cation-binding site on the myosin regulatory light chain appears to play a role in modulation of contractile activity under experimental conditions where the light chains are not phosphorylated. Thus, when the endogenous LC2 was partially replaced with an unphosphorylated mutant light chain with a reduced affinity for divalent cations, maximum tension at pCa 4.5 declined and the rate constant of tension redevelopment was severely reduced (Diffee *et al.*, 1995). This was due to a decrease in the rate of formation of force-generating crossbridges as well as an apparent increase in the rate of crossbridge detachment (Diffee *et al.*, 1996). These effects were largely reversed after re-exchange of native LC2 for the mutant form. Increased levels of free magnesium, competing for the divalent cation-binding site with calcium, had a similar effect in fibers with native, endogenous LC2 (Metzger and Moss, 1992). This suggests that the observed effects of calcium binding occur at quite high calcium levels. The effect of an activating solution at pCa 5.0 on the structure of relaxed rabbit thick filaments has been examined under conditions where the regulatory light chains could not be phosphorylated (absence of MLCK and calmodulin). There was no evidence of disorder due to calcium and those filaments that were well separated from thin filaments on the grid retained well-ordered helical arrays of surface myosin heads. If, however, a thin filament ran closely parallel to a thick filament, crossbridges connected the two, with a regular, periodic repeat (Fig. 14). This is most likely due to derepression of the thin filaments at this level of calcium concentration. Thus, it is likely that, in the absence of LC2 phosphorylation, the divalent cation-binding site of the peptide influences crossbridge attachment and detachment rates at high levels of calcium.

3.2.5. Depletion of LC2 changes thick filament structure
Alterations in myosin regulatory light chains other than phosphorylation can affect thick filament structure, causing increased calcium sensitivity and rate of rise of tension and disorder of the myosin surface array. This is especially apparent when light chains are depleted or removed. Extraction of about 33% of LC2 from skinned rat skeletal muscle fibers decreases the maximum velocity of shortening (V_{max}) by 50% and increases active tensions in the range of pCa 6.6 – 5.7, but does not effect maximum tension (Moss *et al.*, 1982; Hofmann *et al.*, 1990). Removal of 40 – 70% of LC2 abolishes the calcium-dependence of tension development in skinned fibers (Patel *et al.*, 1996). These findings are consistent with LC removal increasing myosin head mobility. When examined structurally, it is seen that the thick filaments separated from such tissue are highly disordered. The myosin heads extend to extremely high radii from the backbone, the filaments appear "frayed" at their tapered ends and myosin layer lines disappear from the diffraction pattern (Fig. 15). This is not unexpected, since the light chains are the only support for the α-helical region of each myosin head that ultimately joins its molecular partner in forming the coiled-coil structure of the myosin rod (Rayment *et al.*, 1993). An unsupported α-helix might well have reduced stability and display increased head mobility, especially in the absence of any charge-dependent interactions that might maintain it in the "relaxed" position near the filament backbone. A similar, but more extreme situation occurs with the thick filaments from cardiac tissue of patients with end-stage idiopathic dilated cardiomyopathy (IDC). In these patients, there is activation of a neutral protease, mekratin (Margossian *et al.*, 1992), which has high specificity for ventricular LC2. These light chains are depleted by 60 to 70% (Margossian *et al.*, 1998; Levine *et al.*, 1999). The thick filaments are

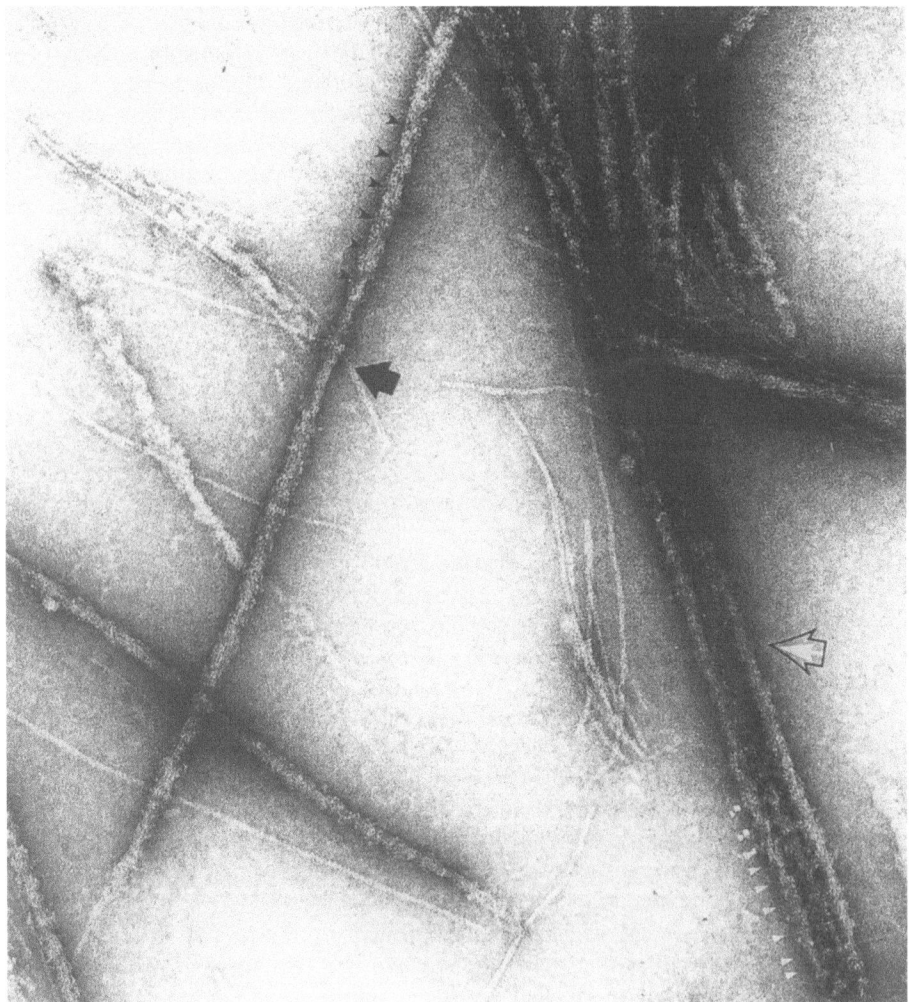

Figure 14. *EM of a field of negatively stained thick and thin filaments incubated in the presence of calcium (pCa 5.0). Large black arrowhead points to bare zone. Ordered surface structure (43 nm repeat periodicity) is indicated by small black arrowheads. Periodicity associated with attachments between neighboring filaments is indicated by small white arrowheads. Filaments involved are indicated with large white arrowhead. This image shows that calciium alone has no effect on the relaxed filament structure. Modified and reproduced with permission from Levine et al., 1996, Figure 4, p. 902, The Biophysical Society.*

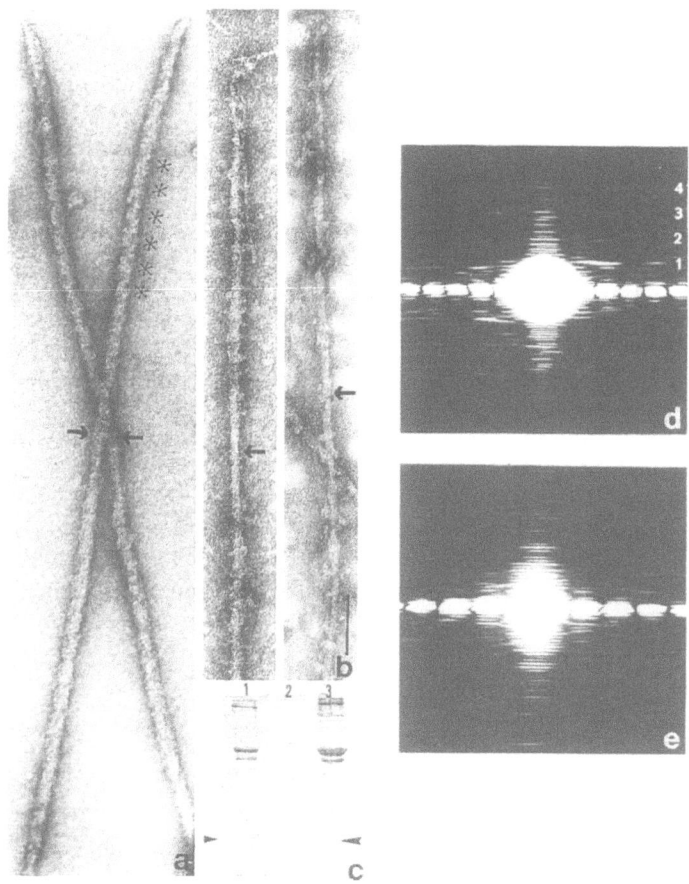

Figure 15. *Structural effects of LC2 depletion. (a and b) EMs, (c) SDS-polyacrylamide gel and (c and d) OTs, all demonstrating the effect of LC2 removal on thick filament structure. Arrows indicate bare zones in EMs of negatively stained thick filaments. The repeat periodicity on the filaments' surfaces is indicated (in one case) by asterisks. Bar=100nm. The myosin layer-lines (when present) are numbered in the upper right quadrant of the OT. (a and d) EM of thick filaments separated from permeabilized rabbit psoas fiber under relaxing conditions with endogenous LC2s intact and unphosphorylated and an OT from such a filament. Note the repeating surface pattern on the thick filament and the presence of myosin-based layer-lines in the OT. (b and e) EMs of filaments separated from a permeabilized rabbit posas fiber following removal of LC2s. Note the severe disruption of surface order on filaments lacking LC2s (some depletion of essential LCs also occurs), which is echoed by the lack of myosin-based reflections in the OT. (c) Coomassie brilliant blue-stained SDS-12% polyacrylamide gel of (lane 3) fiber segment before LC2 removal; (lane 2) supernatant remaining in chamber after removal procedure; and (lane 1) fiber segment after LC2 removal. The band identified as LC2 is indicated by the arrowheads. Even thought less protein was loaded on lane 1 than lane 3, the LC2 is most affected by the removal procedure; a faint band corresponding to LC2 is present on lane 2.*

extremely disordered with myosin heads extending to very high radii from the filament backbones. The thick filaments are very fragile and fall apart or fragment into smaller pieces (Fig. 16). Thick filaments from heart explants from patients suffering from other end-stage cardiac disease are well-ordered in relaxing solution and retain their integrity (Levine *et al.*, 1999). LC2-depleted myosin isolated from IDC hearts forms only short synthetic filaments (Margossian *et al.*, 1987), indicating that the presence of the regulatory light chains may be required to maintain filament integrity, at least in cardiac muscle.

3.3. Cardiac Thick Filaments: the Role of Myosin-Binding Protein C (MYBP-C)

Beyond the effects of nucleotide depletion, temperature, ionic strength and state of phosphorylation and/or integrity of the myosin regulatory light chains, other factors also affect thick filament structure and the physiology of striated muscle tissue. Cardiac muscle differs from skeletal muscle in that regulation of activity is more complex in ventricular myocytes than it is in skeletal muscle fibers, which are either totally activated or inactive. Cardiac cells are always active, yet exhibit constant change in functional parameters, such as time to peak force development, time to relaxation, F_{max}, etc., imposed on them by specific requirements of organismic activity and their ventricular location. Unlike skeletal muscle, which is directly innervated and activated by cholinergic endings, cardiac muscle receptors receive positive and negative input indirectly from a variety of autonomic nerves and systemically circulating hormones, all of which modify the intrinsic activity of the cells. Second messenger systems, often involving calcium, effect the necessary changes to the contractile proteins. There are also more modulating controls to regulation in cardiac than skeletal muscle (both have thin filament regulatory mechanisms); tropomyosin I (TnI) and tropomyosin T (TnT) are each phosphorylated by specific kinases, including cyclicAMP (cAMP) dependent protein kinases A and C (PKA and PKC) (Venema and Kuo, 1993; Anderson *et al.*, 1991), which affect the calcium sensitivity and contractile rate (Solaro and Van Eyk, 1996; Hofmann and Lange, 1994).

3.3.1 Cardiac MyBP-C: location and phosphorylation.
An additional protein of the immunoglobulin-fibronectin family, myosin-binding protein C (MyBP-C), is present as an accessory protein to myosin in vertebrate skeletal and cardiac muscle (Offer *et al.*, 1973; Starr and Offer, 1978; Moos *et al.*, 1975). There are two skeletal isoforms (fast and slow) and one cardiac isoform of the protein (Weber *et al.*, 1993). Cardiac MyBP-C is a different gene product, made on a different chromosome, than either of the skeletal muscle isoforms. It has an additional immunoglobulin module (101 residues) at the amino terminus (C0) and the linker between the C1 and C2 immunoglobulin domains contains three accessible phosphorylation sites: A, B and C that are not present in the skeletal MyBP-Cs (Gautel *et al.*, 1995; Gruen and Gautel, 1999; Kunst *et al.*, 2000). In both muscle types, MyBP-C lies in a series of seven stripes with 43 nm periodicity in the

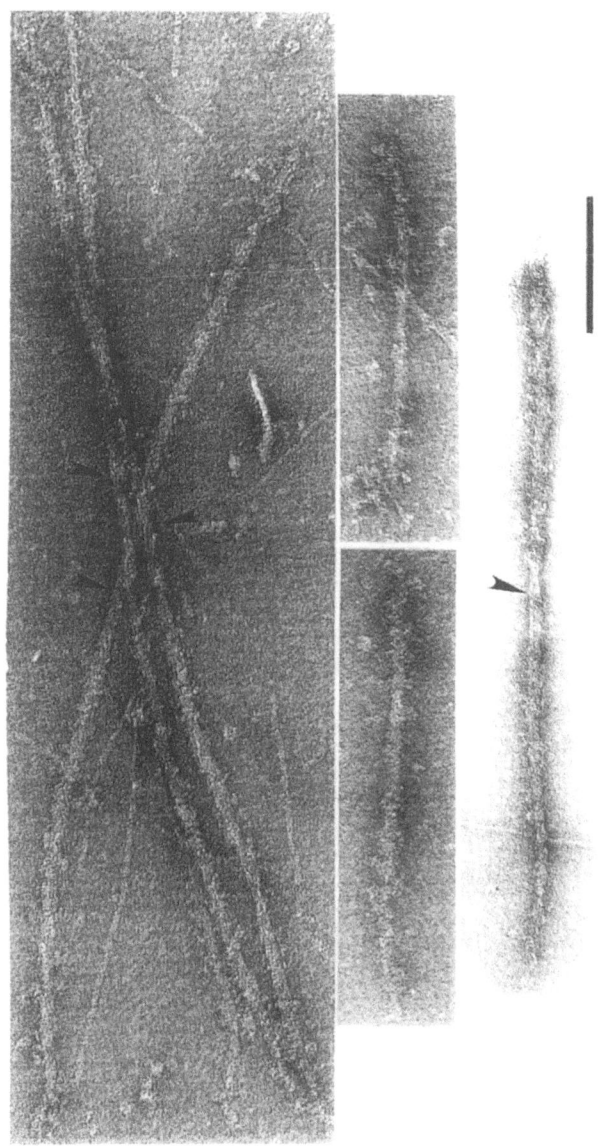

Figure 16. *EMs of thick filaments separated from a cardiac explant of end-stage IDC. Arrows indicate bare zones. Bar = 100 nm. These filaments range in length from 0.7 to 1.6 μm and filament fragments are frequently found. No remnants of the repeat periodicity of the helical arrangement of myosin heads typical of relaxed filaments are present, even after an hour in relaxing solution. Reproduced with permission from a portion of Levine et al., 1999, Figure 2, p.4, Kluwer Academic Publishers.*

middle third of each half A-band (Craig and Offer, 1976). Three molecules of MyBP-C are present per thick filament at each stripe. The N-terminus of MyBP-C binds to the α-helical region of myosin proximal to the head (S2) and the C-terminus has binding sites for titin, LMM and possibly actin (Moos *et al.*, 1975; Starr and Offer, 1978; Labeit and Kolmer, 1995; Gilbert *et al.*, 1996; Freiburg and Gautel, 1996; Kunst *et al.*, 2000).

In rats and humans, a calcium-calmodulin dependent enzyme: calmodulin kinase (CAMK) specific for cardiac MyBP-C (Hartzell and Glass, 1984; Schlender and Bean, 1991) phosphorylates only site B, the first site (temporally) to be phosphorylated. Once this occurs, sites A and C can also be phosphorylated, usually by PKA (Gautel *et al.*, 1995). In the isolated protein, all three sites can be phoshorylated by PKA (Garvey et al, 1988; Gautel *et al.*, 1995). While LC2 phosphorylation normally occurs at activating levels of calcium (Sweeney *et al.*, 1993), MyBP-C is phosphorylated at much lower calcium concentrations (Garvey *et al.*, 1988). The functional effects of full phosphorylation of MyBP-C are expressed in the presence of an intact A-band lattice and include an increase in both F_{max} and ATPase activity (Winegrad and Weisberg, 1987; Garvey *et al.*, 1988; McClellan *et al.*, 1994; *in press*; Weisberg and Winegrad, 1998).

3.3.2 External calcium affects MyBP-C phosphorylation and thick filament structure
Changes in external calcium concentration affect the state of phosphorylation of MyBP-C (Kulikovskaya *et al.*, 2000; McClellan *et al.*, *in press*) and also produce structural changes of the thick filament that are independent of the state of myosin regulatory light chain phosphorylation (Winegrad and Levine, 2000; Levine *et al.*, 2001). Immediately after dissection, during hypoxia of the intact ventricular bundles, cardiac MyBP-C is unphosphorylated, but recovers in oxygenated Krebs' solution containing 2.5 mM calcium, becoming first monophosphorylated, then multiphosphorylated. Thick filaments separated from tissue after two hours at 2.5 mM calcium largely appear ordered, with a visible 43nm helical repeat of myosin heads on their surfaces (Fig. 17a). Diffraction patterns show myosin reflections. (Fig. 17A). Some instances of interaction between myosin heads and thin filaments are seen, but these are not common.

After intact bundles are incubated at very low levels of external calcium (Krebs' sol· n containing 1.25 mM calcium) for two hours, the unphosphorylated state of My ⸲ C is favored, and the majority of thick filaments appear disordered (Fig. 17b) Myⸯ ⸲ reflections are absent from Fourier transforms (Fig. 17B). Very few interactions between myosin heads and adjacent thin filaments occur on the grids of separated filament preparations, even if the thin filaments are lying within a few nanometers of the thick filaments. If, after being at low calcium for two hours, the intact bundles are then placed for as few as ten minutes in Krebs' solution containing high calcium (7.5 mM), MyBP-C becomes rephosphorylated at multiple sites. The backbone structure of the filaments appears to loosen in the region of the myosin heads, but not in the bare zones, and the helical arrangement of the myosin heads on the filaments' surfaces is restored (Fig. 17c), as are the myosin-based reflections (Fig. 17C$_1$). Thus the thick filaments have larger diameters and are ordered. Many interactions between individual thin and thick filaments occur, with

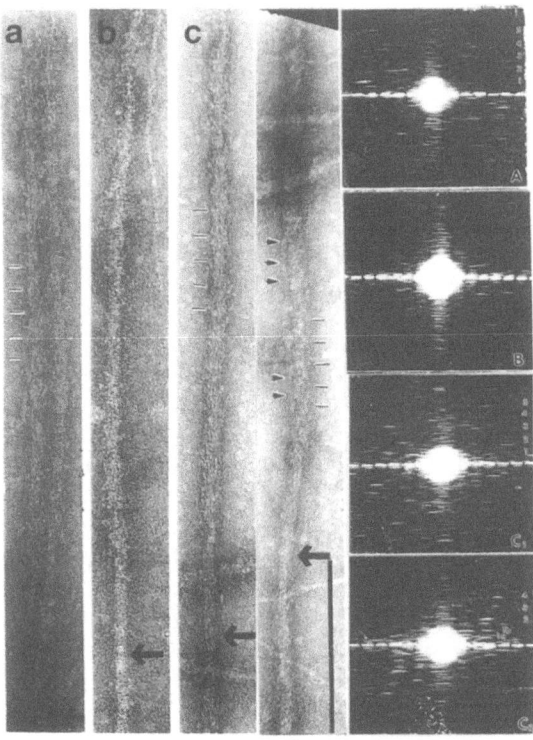

Figure 17. *Structural effects of different external calcium concentrations on thick filament structure and actin-myosin interaction in intact rat cardiac fibers. (a, b and c) EMs and (A, B, C_1 and C) OTs of thick filaments from fiber bundles after exposure to different levels of external calcium. In EMs, black arrows indicate bare zones. Bar = 200 nm. Parallel lines indicate regions of helical periodicity of surface myosin (where present). Small arrowheads indicate points of actin-myosin interaction. Myosin-based layer-lines (where present) are numbered in the upper right quadrant of the OTs (a and A) EMs of filaments separated from a fiber bundle after two hours at 2.5 mM calcium and an OT from one such filament. The filaments showed helical surface periodicity and had a mean diameter of 30.7 nm. Myosin-based reflections are present in the OT. (b and B) EM of a filament separated from fiber bundle after two hours at 1.25 mM calcium and an OT from one such filament. The filament shows a typically disordered surface structure. Such filaments had a wide range of diameters, consistent with myosin heads present in different positions and at different radial distances from the shafts. No myosin-based reflections are present on the OT. (c and C_1, C_2). EMs of filaments separated from a fiber bundle after two hours at 1.25 mM calcium followed by ten minutes at 7.5 mM calcium and OTs from such filaments. The filaments show good surface helical periodicity, but have uniformly larger diameters than those in a, with a mean of 34.4 nm. These filaments frequently show numerous sites of myosin heads, still in their helical positions, interacting with neighboring thin filaments. Myosin-based reflections are present on the OTs (C_1, C_2). On OT C_2, from a thick filament with an extensive region of interaction with a thin filament, the myosin-based first layer-line at 43 nm^{-1} is labeled 1a and a layer-line associated with the regulatory protein repeat on thin filaments at ca. 38 nm^{-1} is labeled 1b. Filaments with periodic surface arrays of myosin heads structures have phosphorylated MyBP-C (gels not shown).*

about a 40 nm spacing (Weisberg and Winegrad, 1996; Winegrad and Levine, 2000; Levine *et al.,* 2001) (Fig. 17c). X-ray diffraction shows an increase in the intensity of the 11 and decrease in intensity of the 10 equatorial reflections when intact tissue is incubated for ten minutes in high calcium (7.5mM) Krebs' after two hours in the low calcium (1.25 mM) solution (Winegrad and Levine, 2000; Levine *et al.,* 2001). This is consistent with the movement of a significant mass of the thick filaments toward the thin filaments within the A-band lattice. Myosin order has not been examined in x-ray diagrams of intact bundles, but Fourier transforms obtained from electron micrographs of thick filaments separated from similarly treated muscle show off-meridional and meridional reflections consistent with the helical organization of myosin heads (Winegrad and Levine, 2000; Levine *et al.,* 2001) and occasionally, the presence of both myosin and actin-based layerlines from regions of thin-thick filament interaction (Fig. $17C_2$). Thus, although the myosin heads approach the thin filaments and may interact with actin (in the A-band lattice) they retain the ordered "relaxed" arrangement of myosin heads on their surfaces, instead of becoming disordered. This underscores the observation that these filaments have larger diameters than the majority of those that were incubated in solutions with either in 2.5 or 1.25 mM calcium. Furthermore, the results of these studies are similar to those in which isolated thick filaments were incubated with PKA, in which the dominant filament structure appeared to be related to the degree of MyBP-C phosphorylation (Weisberg and Winegrad, 1996; 1998). Physiological studies show an increase in F_{max} (McClellan *et al.,* 2001).

At all external calcium concentrations, the level of LC2 phosphorylation, assessed on two-dimensional gels, was 5% or lower. It is important to note that the structural changes in thick filaments observed in these studies occurred at internal calcium concentrations considerably lower than those associated with the structural changes accompanying LC2 phosphorylation and activation.

3.3.3 Expressed fragments of MyBP-C: binding to myosin

Different fragments of cardiac MyBP-C have been expressed (Gruen and Gautel, 1999). The C1C2 fragment, which contains the phosphorylatable sites, binds to the A-bands in myofibrils and to myosin S2 but the flanking C0C1 or C2-C5 fragments appear not to (Gruen and Gautel, 1999; Kunst *et al.,* 2000). The C-terminal fragment binds to the light meromyosin (LMM) portion of the rod. Such binding also occurs in permeabilized skeletal (Gruen and Gautel, 1999; Kunst *et al.,* 2000) and cardiac (Winegrad, *personal communication*) fibers. It is not known if the parameters of binding of the various fragments reflect those of binding by the entire protein to the thick filament. However, the state of phosphorylation of the C1C2 fragment of MyBP-C affects its ability to bind to myosin. When phosphorylated, this N-terrminal fragment has less affinity for myosin than when it is unphosphorylated (Gruen *et al.,* 1999; Kunst *et al.,* 2000). In skeletal fibers, non-phosphorylated cardiac C1C2, when bound to myosin, decreases the calcium-activated force, increases rigor force and shifts the pCa – force curve to the left (Kulikovskaya *et al.,* 2001). None of these effects are seen with phosphorylated C1C2. Since phosphorylation of the endogenous protein has a marked effect on thick filament structure and physiological activity, including increases in ATPase activity and F_{max}, it is important to examine the effects of the phosphorylated and non-phosphorylated fragments in permeabilized cardiac muscle and on the structure of thick filaments obtained from such specimens. These studies are currently underway. Preliminary findings indicate that C0C1 produces an increase in the diameter when it binds to separated cardiac thick filaments while C1C2 binding disorders the filaments (Levine et al., 2001abstr.). This is consistent with the presence of two myosin-binding sites on the MyBP-C N-terminus; a high-affinity site on C2 (when the protein is unphosphorylated) and a lower affinity site on C0 (Flavigny et al., 1999). In the intact protein the latter might come into play when the protein is phosphorylated.

Given the extreme sensitivity of MyBP-C to small changes in intrasarcomeric calcium levels and its ability to bind with varying affinity to the S2 region of myosin, its role in the sarcomere is likely to be significant. It is apparent that it is required during development for the appropriate formation of thick filament structure, although the way in which it affects this parameter is not understood. Removal of up to 80 to 90% of MyBP-C produces severe disorder and even disaggregation of cardiac thick filaments (Levine et al., 2001abstr; McClellan et al., 2001abstr). In the absence of MyBP-C, skeletal muscle myosin fails to form thick filaments with native structure (Koretz, 1979). The results of all of the other studies discussed above infer that it probably plays a role, at least in cardiac muscle, in phosphorylation-dependent modulation of contractile activity, which is highly variable and changes on a beat-to-beat basis. Its role in maintaining filament order and integrity, which may or may not be phosphorylation-dependent, may permit turnover of myosin molecules in "less active" myocytes. This suggestion is based on the observation that thick filaments with non-phosphorylated MyBP-C are more disordered and have backbone regions well exposed so that whole myosin molecules might participate in protein turnover. The results of MyBP-C fragment binding to myosin and A-bands in permeabilized skeletal fibers requires that the situation be examined in greater detail in cardiac cells and thick filaments.

3.4 Familial Hypertrophic Cardiomyopathies and Thick Filament Structure

Single-base, missense mutations in the genes for β-myosin heavy chains (β-MHC), essential and myosin regulatory light chains and myosin-binding protein C (as well as other sarcomeric proteins, including titin, and the thin filament proteins) all have been associated with familial hypertrophic cardiomyopathies (FHC) of varying severities. (Poetter et al., 1996; Bonne et al., 1998). The defects in both light and heavy chains of myosin are predominantly residue substitutions in highly conserved regions of the molecules. In cardiac MyBP-C, however, the defects more frequently produce molecules that are truncated at the carboxy-terminus (Flavigny et al., 1999). Since ventricular myosin has the same light and heavy chain complement as slow skeletal muscle myosin, the study of structure and function of individuals with the disease and their normal relatives has mainly been performed on biopsies of soleus or deltoid muscles, both rich in slow twitch fibers (80 and 50%, respectively). Recently, however, transgenic mouse models that incorporate the genes for the abberant proteins have become available (Geisterfer-Lawrence et al., 1996; Epstein, 1998).

To date, the structures of native thick filaments have been examined from several patients and their unaffected kindred with FHC due to only two different mutations on myosin. In these studies appears that the location on the protein that produces the defective function determines, in part, the effect of the mutation on the thick filament. No structural studies of filaments from fibers with truncated MyBP-C have been performed, although differently truncated molecules appear to bind more or less specifically to the A-band regions of cardiomyocytes (Flavigny et al., 1999).

3.4.1. Thick Filament Structure in the R403Q Mutation on the β-MHC Gene
These studies were performed on soleus single fibers. Thick filaments were obtained from one-half fiber after physiological testing. The other half was used for gel electrophoretic analysis of the myosin present. The mutation decreases shortening velocity, but has no effect on F_{max}, although there is a marked increase in resting tension. In sectioned tissue many of the sarcomeres and A-bands appear shorter than those in samples from normal fibers. The thick filaments are also significantly shorter than those obtained from normal fibers (1.25 µm vs. 1.55 µm), but are otherwise unremarkable. (Dantzig et al., 1996) (Fig. 18).

3.4.2. Thick Filament Structure in the E22K Mutation on LC2
This mutation changes a neutral to a basic residue, very close to the N-terminal region of LC2. Pathology associated with it and other light chain mutations includes thickening of the mid region of the interventricular septum (Epstein, 1998). Increased calcium sensitivity has been found in permeabilized single slow skeletal fibers from biopsies of the deltoid muscle of affected persons, but not of their unaffected kin. The thick filaments from mutant fibers are always disordered under relaxing conditions, indicating increased mobility of the myosin heads (Fig. 19). Mutant fibers have permanently increased calcium sensitivity (Fig. 20). Although this mutation is not directly in the N-terminus of the LC2, it is sufficiently close to it to produce a local conformational change that may affect the ability of the light chain to engage in the normal interactions with underlying myosin regions that maintain the relaxed array of myosin heads. This and other cardiomyopathic light chain mutations have been incorporated into transgenic mice and are being studied further (Epstein, 1998).

4. SUMMARY

The relaxed structure of the mammalian thick filament has undergone detailed analyses by electron microscopic and diffraction techniques and is fairly well understood, although there are still outstanding questions. When isolated under relaxing conditions, the filaments show a distinct helical arrangement of the myosin heads with an axial repeat every third crossbridge level (43 nm). This is consistent with previous x-ray diffraction studies of the living muscle by other investigators, and is also consistent with our earlier studies from other vertebrate classes. This periodicity has been confirmed by computer image analysis of the images of the

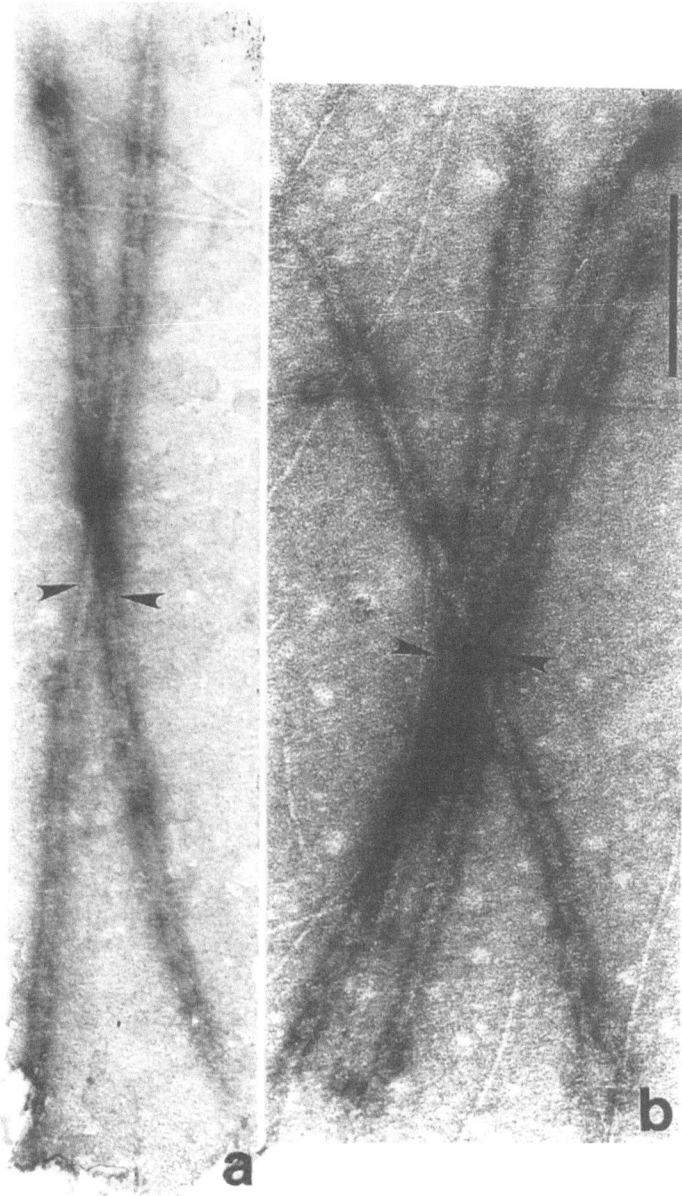

Figure 18. *Structure of thick filaments from patients with an FHC-producing mutation as compared with those of their unaffected kin. EMs of thick filaments separated from (a) normal human soleus fibers and (b) soleus fibers with the R403Q mutation, from patients with FHC. Note the difference in lengths of thick filaments in (a) vs. (b). Arrowheads indicate bare zones. Bars = 250 nm.*

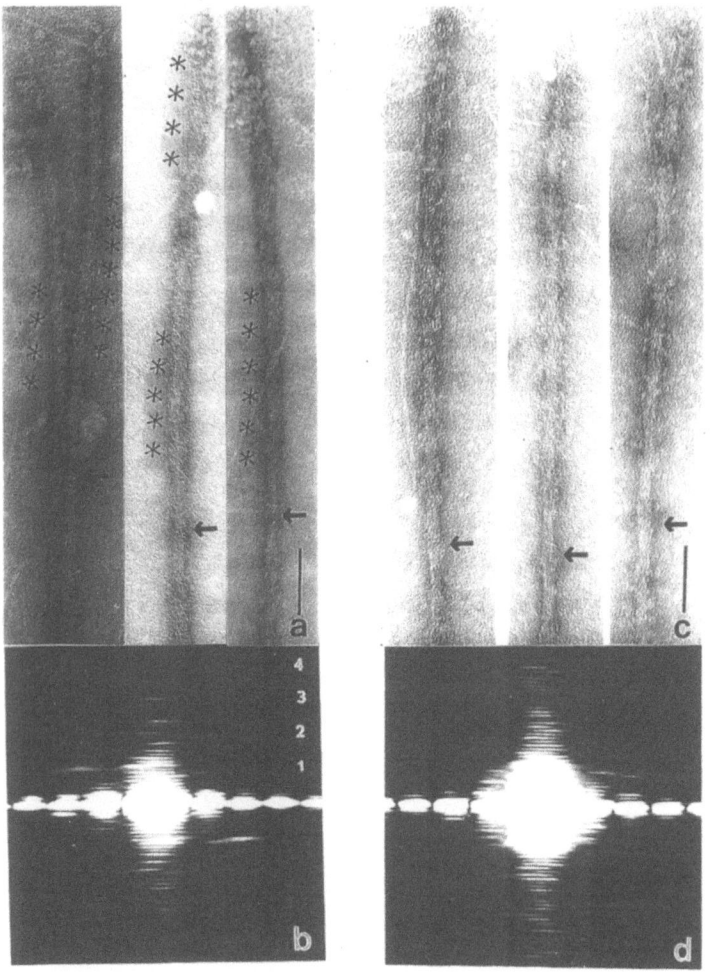

Figure 19.   *Structural effects of an FHC-producing mutation of LC2. (a and c) EMs and (b and d) OTs of thick filaments separated from permeabilized slow human deltoid fibers, showing the effect of the E22K mutation near the N-terminus of LC2, associated with FHC, on thick filament structure. In EMs, the bare zones are indicated by the black arrows and the helical periodicity of the relaxed arrangement of myosin heads on the filaments' surfaces (where present) is indicated by asterisks. Bars = 100 nm. Myosin-based layer-lines (where present) are numbered in the upper right quadrant of the OTs. (a and b) EMs of thick filaments and an OT from one such filament obtained from a normal relative of an affected individual and maintained under relaxing conditions. These filaments display the characteristic helical repeat of surface myosin heads associated with the relaxed state. (c and d) EMs of thick filaments and an OT from one such filament obtained from an individual diagnosed with a severe form of FHC. Both the images and the OT show complete disorder of the filaments' surface structure, even after one hour under relaxing conditions. Reproduced with permission from Levine et al., 1998, Figure 6, p. 158, Academic Press.*

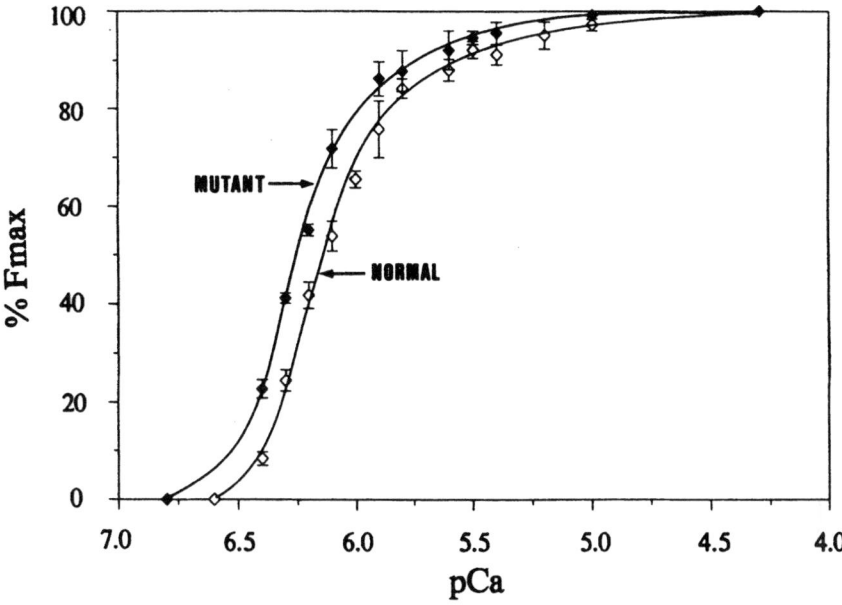

Figure 20. Physiological effects of an FHC-producing mutation of LC2. Force-pCa relationships of permeabilized deltoid fibers from a patient with the E22K mutation near the N-terminus of LC2 (filled diamonds) and from a normal relative (open circles). Measurements were made at 20°C. Points are means ± SE from 12 fibers from the mutant biopsy and 8 fibers from the normal sample. The values are normalized to maximal force in each group. Note the leftward shift of the relationship with the mutant fibers, a result to be expected if myosin heads are mobile. Reproduced with permission from Levine et al., 1998, Figure 5, p. 157, Academic Press.

isolated thick filaments, and evidence has been presented supporting a three-stranded arrangement of the myosin heads in the relaxed mammalian skeletal muscle thick filament.

Although a three-dimensional reconstruction of the mammalian skeletal muscle thick filament has not yet been achieved, a reconstruction of the isolated frog thick filament provides indications of the probable structure of the mammalian striated muscle thick filament. The three-dimensional reconstruction of the frog thick filament confirms the three-stranded arrangement of the myosin heads and provides an explanation for the long known perturbation of the myosin head array from ideal helical symmetry.

The constancy of the helical arrangement of the myosin heads in the thick filament from the different vertebrate classes under relaxing conditions suggests that there are factors helping to stabilize this helical arrangement. These factors are probably fairly weak, however, since the thick filament must be capable of reversibly releasing the myosin heads from this helical arrangement in order to interact with the adjacent actins under activating conditions. Thus the arrangement of the myosin heads is not static, but in a dynamic equilibrium between the ordered and non-ordered states.

Perturbations of the surface array of myosin heads away from their relaxed organization, seen with various environmental changes such as decreased temperature and to a lesser extent, decreased ionic strength, are hallmarks of biochemical changes to myosin that occur on activation. The most obvious of these is phosphorylation of the regulatory light chains of myosin, which produces a reversible disordering of the relaxed myosin array. The mechanism for this is most likely neutralization of a series of positively charged residues in the amino-terminus of the light chain by binding of phosphate to serine (15 in skeletal muscle myosin). This change in charge then frees the myosin head from any charge-based interaction with underlying myosin rods and increases head mobility. Mobile heads are caught, during processing for electron microscopy, in random and extended attitudes of disorder. Mobile heads come close to actin monomers on thin filaments, which accounts for the physiological findings associated with LC2 phosphorylation: increased calcium sensitivity and increased apparent binding to actin. Depletion of LC2 promotes unstability of the light chain domain of the myosin head and also leads to increased head mobility. Calcium alone seems to have no effect on the relaxed structure of the filaments.

Changes in cardiac thick filament structure are also associated with phosphorylation of MyBP-C, a thick filament accessory protein. This occurs at much lower calcium levels than LC2 phosphorylation and appears to be independent of the latter. Filaments with unphosphorylated MyBP-C are disordered, but do not interact well with thin filaments. Fibers from which such filaments are obtained show decreased actomyosin ATPase activity and F_{max}. Those filaments with multi-phosphorylated MyBP-C have backbones of increased diameter, with well-ordered helical arrays of myosin heads on their surfaces. These heads, despite being close to the backbone in attitude, interact extremely well with thin filaments. The increased filament diameter brings them closer to the thin filaments in the A-band lattice. These fibers have increased actomyosin ATPase activity and F_{max}.

Several disease states also affect thick filament structure. Loss of LC2 and fragility of thick filaments is seen in end-stage idiopathic cardiomyopathy. Activation of specific protease depletes the light chains and filament stability appears to be dependent on their presence. In several forms of familial hypertrophic cardiomyopathy thick filament structure and fiber physiology are also compromised.

Therefore, while reversible changes in thick filament structure occur during normal contractile activity in skeletal and cardiac fibers and appear to have important functions in economizing contractility and possibly permitting turnover of myosin, perpetuation of states of altered structure that occur in abnormal situations may reflect compromised function. Different types of abnormalities in structure may be diagnostic of particular molecular bases for different diseased states.

5. ACKNOWLEDGEMENTS

RWK acknowledges support by grant 9607749S from the American Heart Association, a Minorities Basic Research Support Grant S06 GM08224 from the National Institutes of Health, and, in part, by funding from a "Research Centers in Minority Institutions" Award G12RR-03051, from the National Center for Research Resources, National Institutes of Health.

Rhea J.C. Levine
Department of Neurobiology & Anatomy
MCP-Hahnemann University School of Medicine
2900 Henry Avenue
Philadelphia, PA 19129

Robert W. Kensler
Department of Anatomy
University of Puerto Rico
Medical Sciences Campus
San Juan, PR 00935

6. REFERENCES

Adhikari, B., K. Hideg and P. Fajer. (1997) Independent mobility of catalytic and regulatory domains of Myosin heads. *Proc. Natl. Acad. Sci. USA.* **94**:9643-9647.

Alyonycheva,T., L.Cohen-Gould, C. Siewert, D. A. Fischman, and T. Mikawa. (1997) Skeletal muscle-specific myosin binding protein-H is expressed in Purkinje fibers of the cardiac conduction system. *Circ. Res.* **80**:665-672.

Anderson, P., N. Malouf, A. Oakeley, E. Pagani and P. Allen. (1991) Troponin T isoform expression in the normal and failing human left ventricle: a comparison among normal and failing adult hearts, fetal hears and adult and fetal skeletal muscle. *Circ Res.* **60**:1226-1233

Bagni, M. A., G. Cecchi, F. Colomo, and P. Garzella. (1992) Are weakly binding bridges present in resting intact muscle fibers? *Biophys. J.* **63**:1412-1415.

Baker, J., I. Brust-Mascher, S. Ramachandran, L. LaConte and D. Thomas. (1998) A large and distinct rotation of the myosin light chain domain occurs upon muscle contraction. *Proc. Natl. Acad. Sci. USA.* **95**:2944-2949.

Barany, K. and M. Barany. (1979) Phosphorylation-dephosphorylation of the 18,000 dalton light chain of myosin during the contraction-relaxation cycle of frog muscle. *J. Biol. Chem.* **254**:3617-3623.

Bennett, P., R. Craig, R. Starr, and G. Offer. (1985b) The ultrastructural location of C-protein, X-protein and H-protein in rabbit muscle. *J. Muscle Res. Cell Motil.* **7**:550-567.

Bennett, P., R. Starr, A. Elliot, and G. Offer. (1985a) The structure of C-protein and X-protein molecules and a polymer of X-protein. *J. Mol. Biol.* **184**:297-309.

Bonne, G., L. Carrier, P. Richard, B. Hainque and K. Schwartz. (1998) Familal hypertrophic cardiomyopathy: from mutations to functional defects. *Circ. Res.* **83**:579-593.

Brenner, B. (1987) Mechanical and structural approaches to correlation of crossbridge action with actomyosin ATPase in solution. *Annu. Rev. Physiol.* **48**:655-672.

Brenner, B. (1990) Muscle mechanics and biochemical kinetics. *In* Molecular Mechanisms in Muscular Contraction. J.M.Squire, editor. Macmillan, New York. 77-149.

Brenner, B., M. Schoenberg, J. Chalovich, L. Greene and E. Eisenberg. (1982) Evidence for cross-bridge attachment in relaxed muscle at low ionic strength. *Proc. Natl. Acad. Sci. USA.* **79**:7288-7291.

Brenner, B., L. Yu and R. Podolsky. (1984) X-ray diffraction evidence for cross-bridge formation in relaxed muscle fibers at various ionic strengths. *Biophys. J.* **46**:299-306.

Cantino, M. and J. M.Squire. (1986) Resting myosin cross-bridge configuration in frog muscle thick filaments. *J. Cell Biol.* **102**:610-618.

Chavanieu, A., N. Keane, P. Quirk, B. Levine, B. Calas, L. Wei and L. Ellis. (1994) Phosphorylation effects on flanking charge residues: Structural implications for signal transduction in protein kinases. *Eur. J. Biochem.* **224**:115-123.

Chew, M. W. K. and J. M.Squire. (1995) Packing of a-helical coiled-coil myosin rods in vertebrate muscle thick filaments. *J. Struct. Biol.* **115**:233-249.

Cooke, R. (1986) The mechanism of muscle contraction. *CRC Crit. Rev. Biochem.* 21:53-118.

Craig, R., L. Alamo, and R. Padron. (1992) Structure of the myosin filaments of relaxed and rigor vertebrate striated muscle studied by rapid freezing electron microscopy. *J. Mol. Biol.* **228**:474-487.

Craig, R. and G. Offer. (1976) The location of C-protein in rabbit skeletal muscle. *Proc. R Soc Lond B Biol Sci* **192**:451-461.

Craig, R., R. Padron and J. Kendrick-Jones. (1987) Structural changes accompanying phosphorylation of tarantula muscle myosin filaments. *J. Cell Biol.* **105**:1319-1327.

Craig, R., R. Smith and J. Kendrick-Jones. (1983) Light-chain phosphorylation controls the conformation of vertebrate non-muscle and smooth muscle myosin molecules. *Nature (Lond.)* **302**:436-439.

Crowther, R. A., R. Padron, and R. Craig. (1985) Arrangement of the heads of myosin in relaxed thick filaments from tarantula muscle. *J. Mol. Biol.* **184**:429-439.

Dantzig, J., M. Leonard, L. Fananapazir, N. Epstein, Y. Goldman, R. Levine and H.L. Sweeney. (1996) A-band length, slack length and tension in skinned soleus fibers from patients with hypertrophic cardiomyopathy caused by a mutation in the b-myosin heavy chain. *Biophys. J.* **70**:A290.

Davies, R. E. (1963) A molecular theory of muscle contraction: calcium dependent contractions with hydrogen bond formation plus ATP-dependent extensions of part of the myosin-actin crossbridges. *Nature (Lond.)* **199**:1068-1074.

Davis, J. S. (1988) Interaction of C-protein with pH 8.0 synthetic thick filaments prepared from the myosin of vertebrate skeletal muscle. *J. Muscle Res. Cell Motil.* **9**:174-183.

Diffee, G., M. Greaser, F. Reinach and R. Moss. (1995) Effects of a non-divalent cation binding mutant of myosin regulatory light chain on tension generation in skinned skeletal muscle fibers. *Biophys. J.* **68**:1443-1452.

Diffee, G., J. Patel., F. Reinach, M. Greaser and R. Moss. (1996) Altered kinetics of contraction in skleltal muscle fibers containing a mutant myosin regulatory light chain with reduced divalent cation binding. *Biophys. J.* **71**:341-350.

Dominguez, R., Y. Freyzon, K. M. Trybus, and C. Cohen. (1998) Crystal structure of a vertebrate smooth muscle myosin motor domain and its complex with the essential light chain: visualization of the pre-power stroke state. *Cell* **94**:559-571.

Einheber, S. and D. Fischman. (1990) Isolation and characterization of a cDNA clone encoding avian skeletal C-protein: an intracellular member of the immunoglobulin superfamily. *Proc. Natl. Acad. Sci.* **87**:2157-2161.

Epstein, N. (1998) The molecular biology and pathophysiology of hypertrophic cardiomyopathy due to mutations in the beta myosin heavy chains and the essential and regulatory light chains. *Adv. Exp. Med. and Biol.* **453**:105-115.

Fananapazir, L., M. C. Dalakas, F. Cyran, G. Cohn, and N. D.Epstein. (1993) Missense mutations in the b-myosin heavy-chain gene cause central core disease in hypertrophic cardiomyopathy. *Proc. Natl. Acad. Sci.* **90**:3993-3997.

Fisher, A., C. Smith, J. Thoden, R. Smith, K. Sutoh, H. Holden and I. Rayment. (1995) Structural studies of myosin:nucleotide complexes: a revised model for the molecular basis of muscle contraction. *Biophys. J.* **68**:19S-26S.

Flavigny, J., M. Souchet, P. Sibillon, I. Berregi-Bertrand, B. Hainque, A. Mallet, A. Bril, K. Schwartz and L. Carrier. (1999) COOH-terminal truncated cardiac myosin-binding protein C mutants resulting from familial hypertrophic cardiomyopathy mutations exhibit altered expression and/or incorporation in fetal rat cardiomyocytes. *J. Mol. Biol.* **294**:443-456.

Ford, L. E., A. F. Huxley, and R. M.Simmons. (1977) Tension responses to sudden length change in stimulated frog muscle fibres near slack length. *J. Physiol.* **269**:441-515.

Freiburg, A. and M. Gautel. (1996) A molecular map of the interactions of titin and myosin-binding protein C: implications for sarcomeric assembly in familian hypertrophic cardiomyopathy. *Eur. J. Biochem.* **235**:317-323.

Fuchs, F. and Y. P. Wang. (1997) Length-dependence of actin-myosin interaction in skinned cardiac muscle fibers in rigor. *J. Mol. Cell. Cardiol.* **29**:3267-3274.

Furst, D. O., U. Vinkemeyer, and K. Weber. (1992) Mammalian skeletal muscle C-protein: purification from bovine muscle, binding to titin and the characterization of a full length cDNA. *J. Cell Sci.* **102**:769-778.

Garvey, L., E. Kranias and J. Solaro. (1988) Phosphorylation of C protein, troponin I and phospholamban in isolated rabbit hearts. *Biochem. J.* **249**:709-14.

Gautel, M., O. Zuffardi, A. Freiberg and S. Labeit. (1995) Phosphorylation switches specific for the cardiac isoforms of myosin binding protein C: a modulator of cardiac contraction? *EMBO J.* **14**:1952-1960.

Geisterfer-Lowrance, A., M. Christe, D. Conner, J. Ingwall, F. Schoen, C. Seidman and J. Seidman. (1996) A mouse model of familial hypertrophic cardiomyopathy. *Science.* **272**:731-734.

Gilbert, R, M. Kelly, T. Mikawa and D. Fischman. (1996) The carboxyl terminus of myosin binding protein-C (MyBP-C, C-protein) specifies incorporation into the A-band of striated muscle. *J. Cell Sci.* **109**:101-111.

Goldman, Y. E. and B. Brenner. (1987) Special topic: molecular mechanisms of muscle contraction. *Annu. Rev. Physiol.* **49**:629-635.

Gruen, M. and M. Gautel. (1999) Mutations in b-myosin S2 that cause familial hypertrophic cardiomyopathy (FHC) abolish the interaction with the regulatory domain of myosin-binding protein-C. *J. Mol. Biol.* **286**;933-949.

Harford, J. and J. M. Squire. (1986) "Crystalline" myosin crossbridge array in relaxed bony fish muscle. Low angle X-ray diffraction from plaice fin muscle and its interpretation. *Biophys. J.* **50**:145-155.

Harford, J. J. and J. M. Squire. (1992) Evidence for structurally different attached states of myosin crossbridges on actin during contraction of fish muscle. *Biophys. J.* **63**:387-396.

Hartzell, H.C. and D. Glass. (1984) Phosphorylation of purified cardiac muscle protein by purified cAMP-dependent and endogenous Ca-calmodulin-dependent protein kinases. *J. Biol. Chem.* **259**:15587-15596.

Hartzell, H.C. and L.Titus. 1982. Effects of cholinergic and adrenergic agonists on phosphorylation of a 165,000-dalton myofibrillar protein in intact cardiac muscle. *J. Biol. Chem.* **257**:2111-2120.

Haselgrove, J. C. (1975) X-ray evidence for confomational changes in the myosin filaments of vertebrate striated muscle. *J. Mol. Biol.* **92**:113-143.

Haselgrove, J. C. (1980) A model of myosin crossbridge structure consistent with the low-angle X-ray diffraction pattern of vertebrate muscle. *J. Muscle Res. Cell Motil.* **1**:177-191.

Haselgrove, J. C. and C. D. Rodger. (1980) The interpretation of X-ray diffraction patterns from vertebrate striated muscle. *J. Muscle Res. Cell Motil.* **1**:371-390.

Hibberd, M. G. and D. R. Trentham. (1986) Relationships between chemical and mechanical events during muscle contraction. *Annu. Rev. Biophys. Biophys. Chem.* **15**:119-161.

Hofmann, P.A. and J.H. Lange,III. (1994) Effects of phosphorylation of troponin I and C protein on isometric tension and velocity of unloaded shortening in skinned single ventricular myocytes from rats. *Circ. Res.* **74**:718-726.

Hofmann, P., J. Metzger, M. Greaser and R. Moss. (1990) Effects of partial extraction of light chain 2 on the Ca^{2+} sensitivities of isometric tension, stiffness and the velocity of shortening in skinned skeletal muscle fibers. *J. Gen. Physiol.* **95**:477-498.

Holmes, K. C., D. Popp, W. Gebhard, and W. Kabasch. (1990a) Atomic model of the actin filament. *Nature* **347**:44-49.

Holmes, K. C., D. Popp, W. Gebhard, and W. Kabasch. (1990b) The structure of F-actin calculated from X-ray diffraction diagrams and the 0.6 nm crystal structure. *In* Molecular Mechanisms in Muscular Contraction. J.M. Squire, editor. CRC Press, Inc., Boca Raton. 65-75.

Houdusse, A., V. N. Kalabokis, D. Himmel, A. G. Szent-Gyorgyi, and C. Cohen. (1999) Atomic structure of scallop myosin subfragment S1 complexed with MgADP: a novel conformation of the myosin head. *Cell* **97**:459-470.

Huxley, A. F. and R. Niedergerke. (1954) Structural changes in muscle during contraction. Interference microscopy of living muscle fibers. *Nature* **173**:971-972.

Huxley, A. F. and R. M. Simmons. (1971) Proposed mechanism of force generation in striated muscle. *Nature (Lond.)* **233**:533-538.

Huxley, H. E. (1963) Electron microscope studies of the structure of natural and synthetic protein filaments from muscle. *J. Mol. Biol.* **7**:281-308.

Huxley, H. E. (1969) The mechanism of muscle contraction. *Science* **164**:1356-1366.

Huxley, H. E. and W. Brown. (1967) The low angle X-ray diagram of vertebrate striated muscle and its behaviour during contraction and vigor. *J. Mol. Biol.* **30**:383-434.

Huxley, H. E. and J. Hanson. (1954) Changes in the cross-striations of muscle during contraction and stretch and their structural interpretation. *Nature (Lond.)* **173**:973-976.

Huxley, H. E., A. Stewart, H. Sosa, and T. Irving. (1994) X-ray diffraction measurements of the extensibility of actin and myosin filaments in contracting muscle. *Biophys. J.* **67**:2411-2421.

Ip, W. and J. Heuser. (1983) Direct visualization of the myosin crossbridge helices on relaxedrabbit psoas thick filaments. *J. Mol. Biol.* **171**:105-109.

Ito, M. and D. Hartshorne. (1990) Phosphorylation of myosin as a regulatory mechnanism in smooth muscle. *Prog. Clin Biol. Res.* **327**:57-72.

Jeacocke, S. and P. England. (1980) Phosphorylation of a myofibrillar protein of Mr 150,000 in perfused rat heart, and the tentative identification of this as C-protein. *FEBS Letters* **122**:129-132.

Kensler, R.W. (1997). Isolation and analysis of rabbit cardiac muscle thick filaments. Puerto Rico Health Science Journal **16**:226.

Kensler, R. W. (1998) Analysis of Rabbit Cardiac Muscle Thick Filaments. *Biophysics J.* **74**:A363.

Kensler, R. W. (2001) Mammalian cardiac muscle thick filaments: Their periodicity and interactions with actin. Submitted to Biophys. J.

Kensler, R., S. Peterson and M. Norberg. (1994) The effects of changes in temperature or ionic strength on isolated rabbit and fish skeletal muscle thick filaments. *J. Mus. Res. Cell Motil.* **15**:69-79.

Kensler, R. W. and M. Stewart. (1983) Frog skeletal muscle thick filaments are three-stranded. *J. Cell Biol.* **96**:1797-1802.

Kensler, R. W. and M. Stewart. (1986) An ultrastructural study of crossbridge arrangement in the frog thigh muscle thick filament. *Biophys. J.* **49**:343-351.

Kensler, R. W. and M. Stewart. (1989) An ultrastructural study of crossbridge arrangement in the fish skeletal muscle thick filament. *J. Cell Sci.* **94**:391-401.

Kensler, R. W. and M. Stewart. (1993) The relaxed crossbridge pattern in isolated rabbit psoas muscle thick filaments. *J. Cell Sci.* **105**:841-848.

Kensler, R. and J. Woodhead. (1995) The chicken thick filament: Temperature and the relaxed crossbridge arrangement. *J. Mus. Res. Cell Motil.* **16**: -90

Kerrick, W. and L. Bolles. (1981) Regulation of Ca^{2+}-activated tension in *Limulus* striated muscle. *Pflug. Arch.* **392**:121-124.

Koretz, J. (1979) Effects of C-protein on synthetic myosin filament structure. *Biophys. J.* **27**:433-446.

Kulikovskaya, I., G. McClellan, M. Gautel and S. Winegrad. (2001) Interaction of N-terminus of cardiac binding protein C with myosin and the effect on contractility. *Biophys. J.* **80**:261A.

Kulikovskaya, I., G. McClellan, and S. Winegrad. (2000) Effect of extracellular Ca on the phosphorylation of myosin-binding protein (MyBP-C) in cardiac muscle. *Biophys. J.***78**:118A (abstr.)

Kunst, G., K. Kress, M. Gruen, D. Uttenweiler, M. Gautel and R. Fink. (2000) MyBP-C (C-Protein)- a phosphorylation dependent force regulator in muscle that controls the attachment of myosin heads by interaction with myosin – S2. *Circ. Res.* **86**:51-58.

Labeit, S. and B. Kolmer. (1995) Titins: giant proteins in charge of muscle ultrastructure and elasticity. *Science* **270**:293-296.

Labeit, S., M. Gautel, A. lakey, and J. Trinick. (1992) Towards a molecular understanding of titin. EMBO J. **11**:1711-1716.

Lehman, W., A.G. Szent-Gyorgyi and J. Kendrick-Jones. Lehman, W., A.G. Szent-Gyorgyi and J. Kendrick-Jones. (1973) Myosin-linked regulatory systems. *Cold Spring Harbor Symp. Quant. Biol.* **37**:310-30.

Levine, R. J. C. (1993) Evidence for overlapping myosin heads on relaxed thick filaments of fish, frog, and scallop striated muscle. *J. Struct. Biol.* **110**:99-110.

Levine, R. J. C. (1997) Differences in myosin head arrangement on relaxed thick filaments from *Lethocerus* and rabbit muscles. J. Muscle Res. Cell Motil. **18**:529-543.

Levine, R., J. Caulfield, P. Norton, P. Chantler, M. Deziel, H. Slayter and S. Margossian. (1999) Myofibrillar protein structure and assembly during idiopathic dilated cardiomyopathy. *Molec. Cell. Biochem.* **195**:1-10.

Levine, R., P. Chantler, R. Kensler and J. Woodhead. (1991) Effects of phosphorylation by myosin light chain kinase on the structure of *Limulus* thick filaments. *J. Cell Biol.* **113**:563-572.

Levine, R., R. Kensler and P. Levitt. (1986) Crossbridge and backbone structure of invertebrate thick filaments. *Biophys. J.* **49**:135-8.

Levine, R. J. C., R. W. Kensler, H. L. Sweeney, and Z. Yang. (1992) Phosphorylation of myosin light chain produces changes in thick filaments from rabbit skeletal muscle. *Molecular Biology of the Cell* **3**:363a (abstract).

Levine, R., R. Kensler, Z. Yang, J. Stull and H.L. Sweeney. (1996) Myosin regulatory light chain phosphorylation affects the structure of rabbit thick filaments. *Biophys. J.* **71**:898-907.

Levine, R., R. Kensler, Z. Yang, and H. L. Sweeney. (1995) Myosin regulatory light chain phosphorylation and the production of functionally significant changes in myosin head arrangement on striated muscle thick filaments. *Biophys. J.* **68**:224S.

Levine, R., I. Kulikovskaya, G. McClellan, M. Gautel and S. Winegrad. (2001) Cardiac MyBP-C: removal or addition of defined fragments affects thick filament structure. *Biophys. J.* **80**:261A.

Levine, R., A. Weisberg, B. Millman, I. Kulikovskaya, G. McClellan and S. Winegrad. (2001) Multiple structures of thick filaments in resting cardiac muscle and their influence on cross bridge interactions. *Biophys. J.* **81**:1070-1082.

Levine, R., Z Yang, N. Epstein, L. Fananapazir, J. Stull and H. L. Sweeney. (1998) Structural and functional reflections of alterations in the myosin regulatory light chains of mammalian skeletal muscle. *J. Struct. Biol.* **121**:149-161.

Lowy, J., D. Popp and A. Stewart. (1991) X-ray studies of order-disorder transitions in the myosin heads of skinned rabbit psoas muscles. *Biophys. J.* **60**:812-824.

Luther, P. K., P. M. G. Munro, and J. M. Squire. (1981) Three-dimensional structure of the vertebrate Muscle A-band. III. M-region structure and myosin filament symmetry. J. Mol. Biol. **151**:703-730.

Malinchik, S., S. Xu and L. Yu. (1997) Temperature-induced structural changes in the myosin thick filament of skinned rabbit psoas muscle. *Biophys. J.* **73**:2304-2312.

Manning, D. and J. Stull. (1982) Myosin light chain phosphorylation-dephosphorylation in mammalian skeletal muscle. *Am. J. Physiol.* **242**:C234-241.

Margossian, S., H. White, J. Caulfield, P. Norton, S. Taylor and H. Slayter. (1992) Light chain 2 profile and activity of human ventricular myosin during dilated cardiomyopathy. *Circulation* **85**:1720-1733.

Margossian, S., P. Anderson, P. Chantler, H. Patel, M. Deziel, W. Stafford, P. Norton, A. Malhotra, F. Yang, J. Caulfield and H. Slayter. (1998) Calcium regulation in the human myocardium during dilated cardiomyopathy: a structural basis for impaired Ca^{2+}-sensitivity. *Molec. Cell. Biochem.* **194**:301-313.

Margossian, S., T. Huiatt and H. Slayter. (1987) Control of filament length by the regulatory light chains in skeletal and cardiac myosins. *J. Biol. Chem.* **262**:5791-5796.

Maruyama, K. (1986) Connectin, an elastic filamentous protein of striated muscle. Int. rev. Cytol. **104**:81-114.

Matsubara, I. and B. Millman. (1974) X-ray diffraction patterns from mammalian heart muscle. *J. Mol. Biol.* **82**:527-536.

McClellan, G., I. Kulikovskaya and S. Winegrad. (2001) Cardiac myosin: use it or lose it or get help from cMyBP-C. *Biophys. J.* **80**:261A.

McClellan, G., I. Kulikovskaya I and S. Winegrad. (2001) Changes in cardiac contractilitiy related to calcium-mediated changes in phosphorylation of myosin binding protein C (MyBP-C). *Biophys J.* **81**:1082-1093.

McClellan, G., A. Weisberg and S. Winegrad. (1994) Cyclic AMP can raise or lower cardiac actomyosin ATPase activity depending on alpha adrenergic activity. *Am. J. Physiol.* **267**:H431-H442.

McLachlan, A. D. and J. Karn. (1982) Periodic charge distributions in the myosin rod amino acid sequence match cross-bridge spacings in muscle. *Nature* **299**:226-230.

Metzger, J., M. Greaser and R. Moss. (1989) Variations in cross-bridge attachment rate and tension with phosphorylation of myosin in mammalian skinned skeletal muscle fibers. *J. Gen. Physiol.* **93**:855-883.

Metzger, J. and R. Moss. (1992) Myosin light chain 2 modulates calcium-sensitive cross-bridge transitions in vertebrate skeletal muscle. *Biophys. J.* **63**:460-468.

Millman, B. (1979) X-ray diffraction from chicken skeletal muscle. *In* Proceedings of the First John M. Marshall Symposium in Cell Biology. F.A.Pepe, J.M.Sanger, and V.Nachmias, editors. Academic Press, New York. 351-4.

Millman, B. M. (1998) The filament lattice of striated muscle. *Physiological reviews* **78**:359-391.

Moody, M. F. (1967) Structure of the sheath of bacteriophage T4. I. Structure of the contracted sheath and polysheath. *J. Mol. Biol.* **25**:167-200.

Moos, C., G. Offer, R. Starr and P. Bennett. (1975) Interaction of C-protein with myosin, myosin rod and light meromyosin. *J. Mol Biol.* **97**:1-9.

Moss, R., G. Giulian and M. Greaser. (1982) Physiological effects accompanying the removal of myosin LC2 from skinned skeletal muscle fibers. *J. Biol. Chem.* **257**:8588-8591.

Mrakovcic-Zenic, A. and E. Reisler. (1983) Light-chain phosphorylation and cross-bridge conformation in myosin from vertebrate skeletal muscle. *Biochemistry* **22**:525-529.

Niimura, H., L. L. Bachinski, S. Sangwatanaroj, H. Watkins, A. E. Chudley, W. McKenna, A. Kristinsson, R. Roberts, M. Sole, B .J. Maron, J. G. Seidman, and C. E.Seidman. (1998) Mutations in the gene for cardiac myosin-binding protein C and late-onset familial hypertrophic cardiomyopathy. *New England J. Medicine* **338**:1248-1257.

Offer, G., C. Moos and R. Starr. (1973) A new protein of the thick filaments of vertebrate skeletal myofibrils: extraction, purification and characterization. J. Mol. Biol. **74**:653-676.

Okagaki, T., F. E. Weber, D. A. Fischman, K. T. Vaughan, T. Mikawa, and F. C. Reinach. (1993) The major myosin-binding domain of skeletal muscle MyBP-C (C-protein) resides in the COOH-terminal immunoglobulin C2 motif. *J. Cell Biol.* **123**:619-626.

Ostap, E., V. Barnett and D. Thomas. (1995) Resolution of three structural states of spin-labeled myosin in contracting muscle. *Biophys. J.* **69**:177-188.

Padron, R., L. Alamo, J. Murgich, and R. Craig. (1998) Towards an atomic model of the thick filaments of muscle. *J. Mol. Biol.* **275**:35-41.

Padron, R. and R. Craig. (1989) Disorder induced in non-overlap myosin crossbridges by loss of ATP. *Biophys. J.* **56**: 927-935.

Padron, R., N. Pante, H. Sosa and J. Kendrick-Jones. (1991) X-ray diffraction study of the strctrual changes accompanying phosphorylation of tarantula muscle. *J. Mus. Res. Cell Motil.* **12**:235-241.

Parry, D. A. (1981) Structure of rabbit skeletal myosin: analysis of the amino acid sequences of two fragments from the rod region. *J. Mol. Biol.* **153**:459-464.

Pask, H. T., K. L. Jones, P. K. Luther, and J. M. Squire. (1994) M-band structure, M-band interactions and contraction speed in vertebrate cardiac muscles. J. Muscle res. Cell Motil. **15**:633-645.

Patel, J., G. Diffee, X. Huang and R. Moss. (1998) Phosphorylation of myosin regulatory light chain eliminates force-dependent changes in relaxation rates in skeletal muscle. *Biophys. J.* **74**:360-368.

Patel, J., G. Diffee and R. Moss. (1996) Myosin regulatory light chain modulates the Ca^{2+} dependence of the kinetics of tension development in skeletal muscle fibers. *Biophys. J.* **70**:2333-2340.

Perrie, W., L. Smillie and S. Perry. (1973) A phosphorylated light chain component of myosin from skeletal muscle. *Biochem. J.* **135**:151-164.

Poetter, K., H. Jiang, S. Hassanzadeh, S. Master, A. Chang, M. Dalakas, I. Rayment, J. Sellers, L. Fananapazir and N. Epstein. (1996) Mutations in either the essential or regulatory light chains of myosin are associated with a rare myopathy in human heart and skeletal muscle. *Nature Genetics.* **13**:63-69.

Pollard, T. D., D. Bhandari, P. Maupin, D. Wachsstock, A. G.Weeds, and H. G. Zot. (1993) Direct visualization by electron microscopy of the weakly bound intermediates in the actomyosin adenosine triphosphatase cycle. *Biophys. J.* **64**:454-471.

Rapp, G., M. Schrumpf and J. Wray. (1991) Kinetics of the structural change in the myosin filaments of relaxed psoas fibres after a millisecond temperature-jump. *Biophys. J.* **59**:35a (abstr.)

Rayment, I., W. Rypniewski, K. Schmidt-Base, R. Smith, D. Tomchick, M. Benning, D. Winkelmann, G. Wesenberg and H. Holden. (1993) Three-dimensional structure of myosin subfragment-1: a molecular motor. *Science (Wash.)*: **261**:50-58.

Ritz-Gold, C., R. Cooke, D. Blumenthal and J. Stull. (1980) Light chain phosphorylation alters the conformation of skeletal muscle myosin. *Biochem. Biophys. Res. Comm.* **93**:209-214.

Rome, E. (1972) Relaxation of glycerinated muscles: low angle x-ray diffraction studies. *J. Mol. Biol.* **65**:331-345.

Roopnarine, O. and D. Thomas. (1996) Orientation of intermediate nucleotide states of indane dion spin-labeled myosin heads in muscle fibers. *Biophys. J.* **70**:9795-2806.

Saraswat, L. and S. Lowey. (1991) Engineered cysteine mutants of myosin light chain 2. *J. Biol. Chem.* **266**:19777-19785.

Schlender, K and L. Bean. (1991) Phosphorylation of chicken cardiac C protein by calcium clamodulin-dependent protein kinase II. *J. Biol. Chem.* **266**:2811-2817.

Schlichting, I. and J. Wray. (1986) Behaviour of crossbrdiges in non-overlap frlg muscle in the presence and absence of ATP. *J. Mus. Res. Cell Motil.* 7:9 (abstr.)

Schoenberg, M. (1988) Characterization of the myosin adenosine triphosphate (M.ATP) crossbridge in rabbit and frog skeletal muscle fibers. *Biophys. J.* **54**:135-148.

Sellers, J. (1981) Phosphorylation-dependent regulation of *Limulus* muscle. *J. Biol. Chem.* **256**:9274-9278.

Sellers, J. and R.S. Adelstein. Regulation of contractile activity. (1987) In: *The Enzymes*, Eds: P.D. Boyer and E.G. Krebs. Academic Press, Orlando. pp. 381-418.

Sohn, R. L., K. L. Vikstrom, M. Strauss, C. Cohen, A. G. Szent-Gyorgyi, and L. A. Leinwand. (1997) A 29 residue region of the sarcomeric myosin rod is necessary for filament formation. *J. Mol. Biol.* **266**:317-330.

Solaro, R. J. and J. Van Eyk. (1996) Altered interactions among thin filament proteins modulate cardiac function. *J. Mol. Cell. Cardiol.* **28**:217-230.

Sosa, H., D. Popp, G. Ouyang, and H. E. Huxley. (1994) Ultrastructure of skeletal muscle fibers studied by a plunge quick freezing method: myofilament lengths. *Biophys. J.* **67**:283-292.

Squire, J., M. Cantino, M. Chew, R. Denny, J. Harford, L. Hudson, and P. Luther. (1998) Myosin rod-packing schemes in vertebrate muscle thick filaments. *J. Struct. Biol.* **122**:128-138.

Squire, J. M. (1981) The Structural Basis of Molecular Contraction. Plenum Press, New York. 157-264.

Squire, J. M., J. J. Harford, A. C. Edmun, and M. Sjostrom. (1982) Fine structure of the A-band in cryo-sections. III. Crossbridge distributionand the axial structure of the human C-zone. *J. Mol. Biol.* **155**:467-494.

Squire, J. M., R. J. Podolsky, L. C. Yu, and B. Brenner. (1987) Equatorial X-ray diffraction from resting skinned single fibres of fish muscle: little evidence for crossbridge attachment at low ionic strength. *J. Muscle Res. Cell Motil.* **8**:66.

Starr, R., R. Almond, and G. Offer. (1985) Location of C-protein, H-protein, and X-proteinin rabbit skeletal muscle fibre types. *J. Muscle Res. Cell Motil.* **6**:227-256.

Starr, R. and G. Offer. (1971) Polypeptide chains of intermediate molecular weight in myosin preparations. *FEBS Letters* **15**:40-44.

Starr, R., and G. Offer. (1978) The interaction of C-protein with heavy meromyosin and subfragment-2. *Biochem. J.* **171**:813-816.

Starr, R. and G. Offer. (1983) H-protein and X-protein. Two new components of the thick filaments of vertebrate skeletal muscle. *J. Mol. Biol.* **170**:675-698.

Stewart, M. and R. W. Kensler. (1986) Arrangement of heads in relaxed thick filaments from frog skeletal muscle. *J. Mol. Biol.* **192**:831-851.

Stewart, M., R. W. Kensler, and R. J. C. Levine. (1985) Three-dimensional reconstructionof thick filaments from *Limulus* and scorpion muscle. *J. Cell Biol.* **101**:402-411.

Stull, J. and C. High. (1977) Phosphorylation of skeletal muscle contractile proteins *in vivo*. *Biochem Biophys. Res. Comm.* **77**:1078-1083.

Sweeney, H.L., B. Bowman and J. Stull. (1993) Myosin light chain phosphorylation in vertebrate striated muscle: regulation and function. *Am J. Physiol.* **264**:C1085-1095.

Sweeney, H.L. and J. Stull. (1986) Phosphorylation of myosin in permeabilized mammalian cardiac and skeletal muscle cells. *Amer. J. Physiol.* **250**:c657-660.

Sweeney, H.L. and J. Stull. (1990) Alteration of cross-bridge kinetics by myosin light chain phosphorylation in rabbit skeletal muscle: implications for the regulation of actin-myosin interaction. *Proc. Natl. Acad. Sci. USA* **87**:414-418.

Sweeney, H.L., Z. Yang, G. Zhi, J. Stull and K. Trybus. (1994) Charge replacement near the phosphorylatable serine of the myosin regulatory light chain mimics aspects of phosphorylation. *Proc. Natl. Acad. Sci. USA* **91**:1490-1494.

Taylor, E. W. (1979) Mechanism of actomyosin ATPase and the problem of muscle contraction. *CRC Crit. Rev. Biochem.* **7**:103-164.

Trybus, K. (1994) Role of myosin light chains. J. Mus. Res. Cell Motil. **15**:587-594.

Tohtong, R. H. Yamashita, M. Graham, J. Haeberle, A. Simcox and D. Maughan. (1995) Impairment of muscle function caused by mutations of phosphorylation sites in myosin regulatory light chains. *Nature (London)* **374**:650-653.

Uyeda, T. Q. P., P. D. Abramson, and J. A. Spudich. (1996) The neck region of the myosin motor domain acts as a lever to generate movement. *Proc. Natl. Acad. Sci.* **93**:4459-4464.

Variano-Marston, E., C. Franzini-Armstrong, and J. C. Haselgrove. (1984) the structure and deposition of crossbridges in deep-etched fish muscle. *J. Muscle Res. Cell Motil.* **5**:363-386.

Vaughan, K. T., F. E. Weber, S. Einheber, and D. A. Fischman. (1993b) Molecular cloning of chicken myosin-binding protein H (MyBP) H (96 KDa protein) reveals extensive homology with the MyBR-C (C-protein) with conserved immunoglobulin C2 and fibronectin type III motifs. *J. Biol. Chem.* **268**:3670-3676.

Vaughan, K. T., F. E. Weber, and D. A. Fischman. (1992) cDNA cloning and sequence comparisons of human and chicken muscle C-protein and 86kDA protein. *Symp. Soc. Exp. Biol.* **46**:167-177.

Vaughan, K. T., F. E. Weber, F. C. Reinach, T. Ried, D. Ward, and D. A. Fischman. (1993a) Human myosin-binding protein H (MyBD-H): complete primary sequence, genomic organization, and chromosomal localization. *Genomics* **16**:34-40.

Venema, R.C. and J. Kuo. (1993) Protein kinase C mediated phosphorylation of troponin I and C protein in isolated myocardial cells is associated with inhibition of actomyosin MgATPase. *J. Biol. Chem.* **268**:2705-2711.

Vibert, P. and R. Craig. 1985. Structural changes that occur in scallop myosin filaments upon activation. *J. Cell Biol.* **101**:830-837.

Wakabayashi, K. (1992) Small-angle synchrotron x-ray scattering reveals distinct shape changes of the myosin head during hydrolysis of ATP. *Science.* **258**:443-447.

Wakabayashi, K., T. Akiba, K. Hirose, A. Tomioka, M. Tokunaga, M. Suzuki, C. Toyoshima, K. Sutoh, K.

Yamamoto, T. Matsumoto, K. Saeki, and Y. Amemiya. (1988) Temperature-induced change of thick filament and location of the functional sites of myosin. *Advances in Experimental Medicine and Biology* **226**:39-48.

Walker, M., H. White, B. Belknap, and J. Trinick. (1994) Electron cryomicroscopy of acto-myosin-S1 during steady-state ATP hydrolysis. *Biophys. J.* **66**:1563-1572.

Wang, F., B. Martin and J. Sellers. (1993) Regulation of actomyosin interactions in *Limulus* muscle proteins. *J. Biol. Chem.* **268**:3776-3780.

Wang, K., R. McCarter, J. Wright, J. Beverly, and R. Ramirez-Mitchell. (1993) Viscoelasticity of the sarcomere matrix of skeletal muscles. The titin-myosin composite filament is a dual stage molecular spring. Biophys. J. **64**:1161-1177.

Wang, K., J. McClure, and A. Tu. (1979) Titin: Major myofibrillar components of striated muscle. Proc. Natl. Acad. Sci. USA **76**:3698-3702.

Weber F., K. Vaughan, F. Reinach and D. Fischman. (1993) Complete sequence of human fast-type and slow-type muscle myosin binding protein C (MyBP-C): differential expression, conserved domain structure and chromosome assignment. *Eur. J. Biochem.* **216**:661-669.

Weisberg, A. and S. Winegrad. (1996) Alteration of myosin cross bridges by phosphorylation of myosin-binding protein C in cardiac muscle. *Proc. Natl. Acad. Sci. USA* **93**:8999-9003.

Weisberg, A. and S. Winegrad. (1998) Relation between crossbridge structure and actomyosin ATPase activity in rat heart. *Circ Res.***83**:60-72.

Whittaker, M., E. M. Wilson-Kubalek, J. E. Smith, L. Faust, R. A. Milligan, and H. L. Sweeney. (1995) A 35-A movement of smooth muscle myosin on ADP release. *Nature* **378**:748-753.

Winegrad, S. (1999) Cardiac myosin binding protein C. *Circ. Res.* **84**:1117-1126.

Winegrad, S. and R. Levine. (2000) Effect of extracellular Ca on the structure of thick filaments in cardiac muscle. *Biophys. J.* **78**:119A (abstr.)

Winegrad, S. and A. Weisberg. (1987) Isozyme-specific modifications of myosin ATPase by cyclic AMP in rat heart. *Circ. Res.* **60**:384-392.

Wray, J. S. (1987) Structure of relaxed myosin filaments in relation to nucleotide state in vertebrate skeletal muscle. *J. Muscle Res. Cell Motil.* **8**:62.

Wray, J., R. Goody and K. Holmes. (1986) Toward a molecular mechanism for the crossbridge cycle. *Adv. Exp. Med. Biol.* **266**:49-59.

Xu, S., J. Gu, T. Rhodes, B. Belknap, G. Rosenbaum, G. Offer, H. White and L. Yu. (1999) The M.ADP.Pi state is required for helical order in the thick filaments of skeletal muscle. *Biophys. J.* **77**:2665-2776.

Xu, S., M. Kress, and H. E. Huxley. (1987) X-ray diffraction studies of the structural state of crossbridges in skinned frog sartorius muscle at low ionic strength. *J. Muscle Res. Cell Motil.* **8**:39-54.

Xu, S., S. Malinchik, D. Gilroy, T. Kraft, B. Brenner and L. Yu. (1997) X-ray diffraction studies of cross-bridges weakly bound to actin in relaxed skinned fibers of rabbit psoas muscle. *Biophys. J.* **73**:2292-2303.

Yagi, W., E. J. O'Brien, and I. Matsubara. (1981) Changes of thick filament structure during contraction of frog striated muscle. *Biophys. J.* **33**:121-138.

Yamamoto, K. and C. Moos. (1981) A comparative study of C-proteins from heart and skeletal muscles. *Biophys. J.* **33**:A237.

Yamamoto, K. and C. Moos. (1983) The C-proteins of rabbit red, white, and cardiac muscles. *J. Biol. Chem.* **258**:8395-8401.

Yang, Z., J. Stull, R. Levine and H. L. Sweeney. (1998) Changes in interfilament spacing mimic the effects of myosin regulatory light chain phosphorylation in rabbit psoas fibers. *J. Struct. Biol.***121**:139-148.

Yang, Z. and H. L. Sweeney. (1995) Restoration of phosphorylation-dependent regulation to the skeletal muscle regulatory light chain. *J. Biol. Chem.* **270**:1-4.

Yu, B., J. A. French, L. Carrier, R. W. Jeremy, D. R. McTaggart, M. R. Nicholson, B. Hambly, C. Semsarian, D. R. Richmond, K. Schwartz, and R. J. Trent. (1998) Molecular pathology of familial hypertrophic cardiomyopathy caused by mutations in cardiac myosin binding protein C gene. *J. Med. Genet.* **35**:205-210.

LARRY S. TOBACMAN

STRUCTURE AND REGULATION OF CARDIAC AND SKELETAL MUSCLE THIN FILAMENTS

The biological production of force and movement can be understood only when it is considered as a molecular phenomenon, as has been recognized for many decades (Huxley, 1957). However, detailed molecular-scale investigations of this process had to await atomic resolution structures of the myosin molecular motor (Rayment et al., 1993). In striated muscles such as skeletal muscle and the heart, the primary regulation of myosin is by the reversible binding of Ca^{2+} to the thin filament. As was true for understanding myosin's action, the understanding of myosin's regulation requires detailed information about the structure of the relevant proteins: the thin filament and its components. These data are now increasingly available, thanks to the determined efforts of many investigators. Albeit incomplete, a molecular appreciation for how movement is controlled is now within reach. This review summarizes recent advances in the structures of actin, tropomyosin, troponin, and the assembled thin filament, focusing on how these results help explain the regulation of cardiac and skeletal muscle contraction, and on some of the questions that remain.

1. ACTIN

Unmodified actin tends to polymerize rather than crystallize, and crystallization was required to determine actin's structure at atomic resolution. To obtain actin crystals suitable for X-ray analysis, actin polymerization has been blocked by addition of actin binding proteins (Kabsch et al., 1990; Schutt et al., 1993; McLaughlin et al., 1993) or the marine toxin latrunculin A (Yarmola et al., 2001), or most recently by covalent modification at the actin COOH-terminus (Otterbein et al., 2001). All of the resulting high resolution actin structures have the same overall fold, which is similar to that of hsp70 heat shock proteins, despite the absence of sequence similarity. Actin has two domains and a metal-nucleotide binding pocket lies between them. The nucleotide stabilizes monomeric actin, and the state of the nucleotide strongly influences monomer-monomer binding within the thin filament (Carlier, 1991). Furthermore, either modification of the nucleotide binding site or substitution of Ca^{2+} for Mg^{2+} at the same site has effects on polymerized actin structure as detected by electron microscopy (Orlova et al., 1995; Orlova et al.,

R.J. Solaro and R.L. Moss (eds), Molecular Control Mechanisms in Striated Muscle Contraction, 143-162
©2002 Kluwer Academic Publishers, Printed in the Netherlands

1997). These metal-nucleotide effects exemplify the more general point that the
dynamics of unpolymerized G-actin, and the variations among its crystallo-

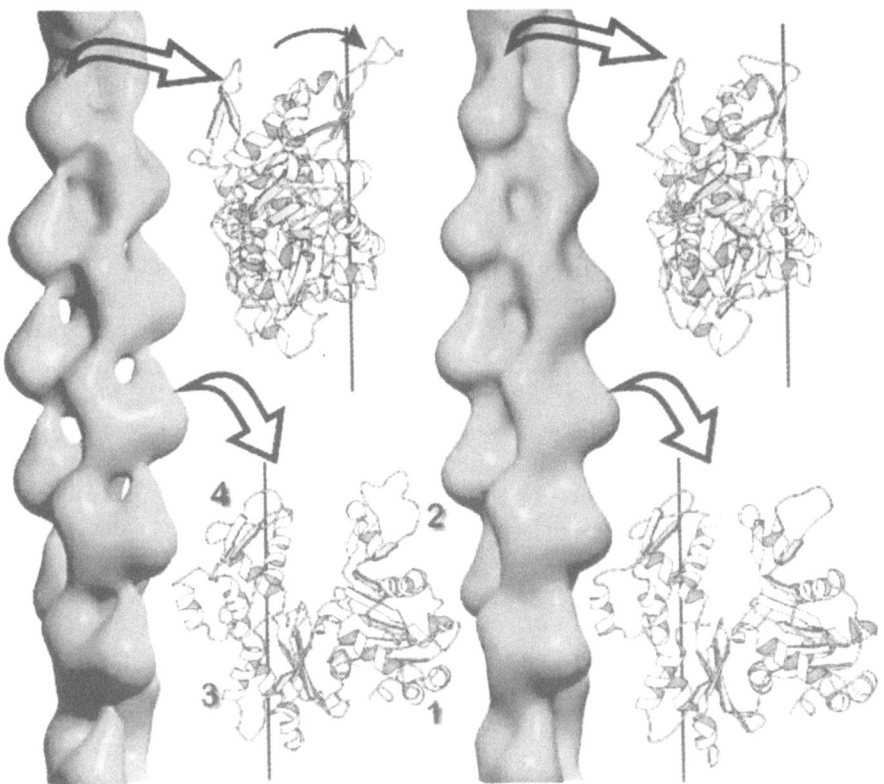

*Figure 1. Two similar structures of F-actin and the actin monomer. Left:
actin as sometimes observed with ADP in the nucleotide pocket. Right:
the more commonly observed filament, corresponding to the monomer
conformation observed in the presence of both ADP and Pi. Adapted from
Belmont, et. al 1999. Color version on page 455.*

graphically determined structures, likely relate to conformational variability within
the polymerized actin filament. For bare actin filaments, in the absence of other
proteins, such changes in polymerized actin can sometimes be detected at the EM
level (Orlova et al., 1997), as exemplified in Fig. 1(Orlova et al., 1995; Steinmetz et
al., 1997; Steinmetz et al., 1997; Egelman and Orlova, 1995).

Simultaneous with the publication of the actin monomer structure, an atomic
model for the actin filament was presented (Holmes et al., 1990). Subsequent studies
have tested and refined this model (Lorenz et al., 1993; Milligan et al., 1990; Tirion
et al., 1995; Mendelson and Morris, 1994; Steinmetz et al., 1997; Egelman and
Orlova, 1995; Vibert et al., 1997), and except for one dissenting viewpoint (Page et

al., 1998) the initial report has been widely supported. Surface views (green in Fig. 1) show the helical, 2-stranded nature of the filament. The larger of actin monomer's two domains, containing subdomains 3 and 4, is closer to the interior of the filament (note numbered monomer). The smaller or outer domain is comprised of subdomain 2 and subdomain 1, the latter of which contains the N- terminus, the COOH-terminus, and the most reactive cysteine, Cys[374]. Consistent with the intermediate resolution level of the experimental filament data, the original Holmes F-actin model proposed that polymerization caused few changes in the monomer. An exception was a hydrophobic loop on the filament interior, which was proposed to extend away from one monomer and form a stabilizing contact with the opposing strand. Subsequently, this mechanism has been supported by results with loop mutants (Chen et al., 1993).

An increasing number of studies indicate small but potentially important variability in actin filament structure. Electron microscopic data imply changes in filament architecture, and spectroscopic results (Prochniewicz and Thomas, 1999) suggest it may be dynamics as well as plasticity that are variable. As exemplified by the two filament states illustrated in Fig. 1, the most variable region in the filament seems to subdomain 2 (Orlova et al., 1995; Belmont et al., 1999; Otterbein et al., 2001). Specifically, the nucleotide binding cleft of filamentous actin is more open in the presence of ADP than in the presence of ADP-Pi, due to a large motion in subdomain 2. A recent crystallographic report defines this movement in more detail. It consists of a 10° subdomain rotation plus formation of an α-helix by the apical loop (Otterbein et al., 2001). Gelsolin has been reported to alter subdomain 2 of an entire filament, when bound only at the filament end (Orlova et al., 1995). Although it is not clear how variability in subdomain 2 may relate to regulation, there are reasons why the variability might be relevant: mutations in this subdomain both affect myosin binding (Razzaq et al., 1999; Korman et al., 2000), and also have large effects on calcium-sensitive regulation ((Korman et al., 2000), refs in same).

Alterations in F-actin structure are not limited to subdomain 2. Subdomain 4 rotates by 4° in the presence of ADP (Otterbein et al., 2001). Another, dramatic example of F-actin plasticity is the effect of cofilin on the filament helix (McGough et al., 1997). Stoichiometric binding of cofilin along actin filaments alters the relationship between successive monomers, changing the twist per subunit without changing the rise. A subsequent report (Galkin et al., 2001) suggests that this is due to stabilization of a twist that can occur even in the absence of cofilin or other proteins, and also suggests that at high pH cofilin can change the tilt of the actin monomers.

In contrast to the above data, it is unknown how troponin, tropomyosin, and calcium binding to troponin affect actin structure. However, small differences between F-actin and F-actin-tropomyosin are implied by x-ray diffraction of ordered filament gels (Lorenz et al., 1995), and calcium influences thin filament flexibility (Isambert et al., 1995). These results indicate that the regulatory proteins alter filament structure in ways that are not yet defined. Helical reconstructions of regulated and unregulated thin filaments indicate that the actin is similar (Lorenz et al., 1995; Milligan et al., 1990; Xu et al., 1999), so changes induced by troponin and

tropomyosin are likely to be subtle. Nevertheless, they could be functionally
significant.

2. TROPOMYOSIN

Tropomyosin is a highly extended alpha-helical dimer that in muscle filaments
stretches along seven actin monomers. The overall stoichiometry of the muscle thin
filament, seven actins : one tropomyosin : one troponin is determined by two
factors: the length

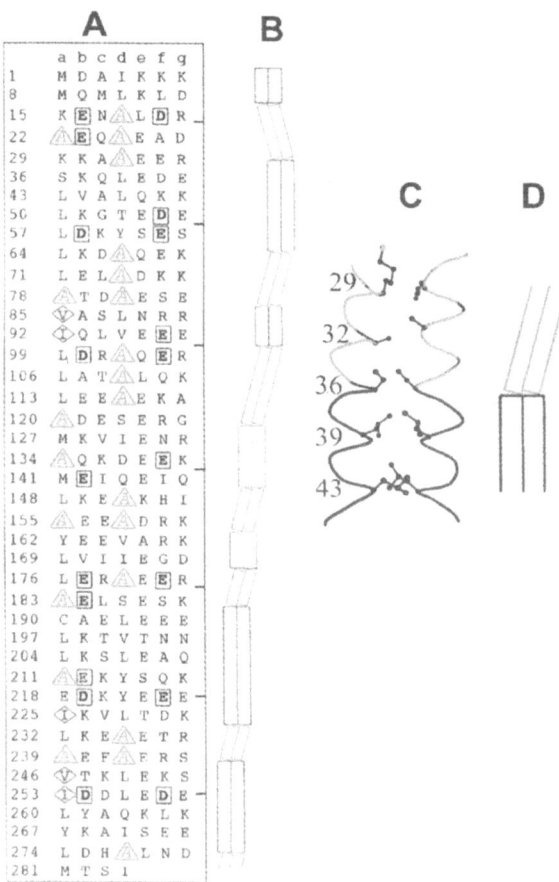

*Figure 2. Segmental bending of α-tropomyosin. Columns a and d of the amino acid
sequence (A) demonstrate periodic hydrophobicity that results in a coiled-coil
dimerization surface or core. Interacting apolar side chains shown in C. Core
regions having many ala residues interact favorably when either strand shifts ahead
(blue regions) producing bends (B and D). Adapted from Brown, et. al, 2001.*
Color version on page 456.

of tropomyosin, and tropomyosin's one to one association with troponin. Tropomyosin's high flexibility and large size have impeded attempts to determine its structure, but this flexibility may be critical to tropomyosin function. Recently, atomic resolution structures of muscle tropomyosin fragments have been solved (Brown et al., 2001; Greenfield et al., 1998). The N-terminus of tropomyosin is a parallel coiled-coil dimer, as expected. When tropomyosin is NH_2-acetylated as it is in muscle cells, the coiled-coil continues to the very NH_2-terminus of the protein. The tropomyosin dimerization surface is formed by the classical heptad repeat motif, in which successive groups of seven residues have hydrophobic amino acids in the first and fourth positions. For each a-b-c-d-e-f-g repeat, amino acids a and d are apolar, and this repeating pattern produces a longitudinal hydrophobic strip that becomes the core in the dimer

Interestingly, at several sites in tropomyosin there is novel deviation from this classical coiled-coil structure (Fig. 2). Muscle tropomyosin has many alanines in the core position (panel A triangles), unlike most coiled-coils. Where these alanines cluster together, which occurs at six or seven regions of the sequence, the dimer core adopts a different structure that depends upon one chain of the dimer shifting about 1 Angstrom ahead of the other (Brown et al., 2001). This results in a bend in the trajectory of the coiled-coil (panel C). Since the bend can occur in either direction and potentially occurs at several sites (panel B), the net effect is a kind of flexibility, which is necessary for tropomyosin to wrap around the actin filament. (Yeast tropomyosins, the only known isoforms that lack these alanine clusters, instead obtain flexibility by a combination of shorter length and interruptions in the phase of the heptad repeat (Strand et al., 2001).) Importantly, this segmental flexibility also may play a role in the shifting of tropomyosin position while bound to the actin surface, as described below for regulated thin filaments.

3. TROPONIN

Muscle contraction is triggered by the reversible binding of calcium ion to the troponin subunit TnC (see reviews (Tobacman, 1996; Gordon et al., 2000)), and TnC is the component of the regulatory complex that is most understood at a structural level. Similar to calmodulin to which it is related , TnC is a dumbbell shaped protein with NH_2- and COOH- domains that are connected by a central helix. (TnC is shown as the space filling component of Fig. 3. The illustration also includes portions of TnI, shown in ribbon format and discussed below). EF-hand metal binding sites are found in each domain, but only the NH_2-domain sites are calcium-specific, and it is the NH_2-domain sites that control muscle contraction. When calcium binds to the two NH_2-domain sites of skeletal muscle TnC (sites I and II), the B-and D-helices adopt a new position relative to the A- and C-helices. This opens the overall conformation of the NH_2-domain, and exposes several hydrophobic residues to solvent. This mechanism was hypothesized based upon the first crystallographic data for TnC, in which the NH_2-domain was in the apo-state (Herzberg et al., 1986). Since then, the mechanism has been abundantly demonstrated by X-ray and by NMR (Gagne et al., 1994; Strynadka et al., 1997;

Houdusse et al., 1997). The Ca^{2+}-saturated form of holo-TnC is the space filling
model in Fig. 3. Importantly, the Ca^{2+}-induced change in skeletal TnC structure
results in substantially increased affinity for the inhibitory subunit of troponin, TnI.

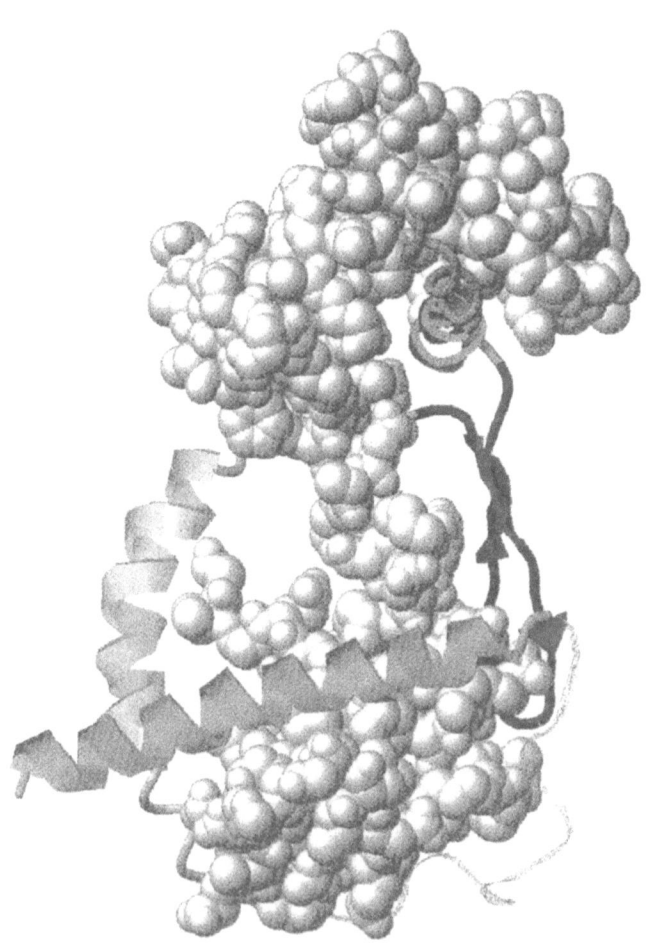

*Figure 3. Proposed structure for the complex of skeletal muscle TnC (space filling)
and TnI (ribbon) in the presence of Ca^{2+}. The NH_2-terminal, regulatory domain of TnC
is at top, with a TnI helix bound to a Ca^{2+}-dependent, hydrophobic patch. From Tung,
et. al, 2000. Color version on page 457.*

This Ca^{2+}-promoted interaction between TnC and TnI is critical for propagation of the Ca^{2+} regulatory signal to the other components of the thin filament, so that myosin binding and cycling can be activated. The opening of the cardiac TnC NH_2-domain can also be promoted by calcium-sensitizing agents that bind to the exposed hydrophobic surface (Li et al., 2000).

Cardiac TnC does not bind calcium at site I, and correspondingly contraction is triggered only by metal binding at site II. Also, calcium binding is not sufficient for opening of the cardiac TnC NH_2-domain for the isolated subunit, unlike skeletal muscle TnC (Sia et al., 1997). Consistent with these observations, calcium has a much smaller effect on TnI affinity for the cardiac TnC NH_2-domain than it does for the skeletal muscle TnC NH_2-domain (Van Eyk et al., 1991; Swenson and Fredrickson, 1992). Nevertheless, despite these qualitative differences, the regulatory mechanisms of the two troponins are similar. To appreciate this fundamental similarity, one needs to consider the structural changes that occur in the TnI-TnC complex.

Although the importance of TnI-TnC interactions has long been recognized, structural information on this complex have been slow to emerge. The first results were from Vassylev et al (Vassylev et al., 1998), who solved the structure of the skeletal muscle TnC COOH-domain in complex with TnI residues 1-33. The TnI peptide forms an extended helix, which binds to the Ca^{2+}-saturated COOH-domain, and a similar binding process occurs in cardiac troponin (Gasmi-Seabrook et al., 1999). Presumably, this complex forms regardless whether Ca^{2+} (active muscle) or Mg^{2+} (relaxed muscle) is bound to TnC COOH-domain sites III and IV. Importantly, the authors proposed that regulation was effected by a similar complex, which formed between the TnC NH_2-domain and a helix comprised of a different portion of TnI (Vassylev et al., 1998). Subsequently, using cardiac troponin, such a complex has been identified. That is, the Ca^{2+}-saturated cardiac TnC NH_2-domain binds to a cardiac TnI helix (Li et al., 1999). This interaction involves an opening of the TnC NH_2-domain to expose a hydrophobic surface, similar to the Ca^{2+}-regulated opening of the skeletal TnC. Thus, in both cardiac and skeletal muscle troponins, the regulatory event involves similar changes in TnC and similar TnI-TnC association, in each case triggered by Ca^{2+} binding to the TnC NH_2-domain.

However mechanistically informative the above results may be, they are less so than what could be learned from the entire, TnI-TnC binary complex, or better yet holo-troponin, which also includes the TnT subunit. In rotary shadowed electron micrographic images troponin is a lollipop shaped molecule, with the elongated stick or tail region comprised of NH_2-terminal portions of TnT (Flicker et al., 1982). More detailed structural models are not available for TnT. Two investigative groups have proposed alternative, lower resolution models of Ca^{2+}-saturated TnI-TnC based upon neutron scattering data (Olah and Trewhella, 1994; Stone et al., 1998). One of these groups has proposed two possible refinements of their model (Tung et al., 2001), based upon a synthesis of scattering data and other experiments, including the partial structure determinations discussed above. "Model L" (Fig. 3) is consistent with X-ray data for the orientation of a TnI helix on the hydrophobic surface of the TnC COOH-domain (across the front of the lower lobe in the figure), and also with NMR data on the orientation of a TnI helix on the regulatory or NH_2-domain (nestled under the upper lobe). Based upon sequence homology, the model also suggests that the TnI inhibitory peptide region (shown in red at right) is a beta hairpin, and is structurally similar to the actin binding region of the G-actin binding protein profilin. The related proposal is that in the absence of Ca^{2+} the troponin-actin interface may, in part, resemble the profilin-actin interface. However, in the presence of Ca^{2+} the inhibitory region is believed to detach from actin, and recent NMR data suggest that it differs from what is shown in Fig. 3 (Blumenschein et al., 2001).

The next step in determination of troponin structure may be a fundamental advance: an atomic structure of a crystallized, complex containing portions of all three subunits (Y. Maeda, personal communication). This work is not finished at the time of this writing, but the investigators have scheduled its presentation at a November 2001 meeting. Preliminary results support the presence of a TnT-TnI coiled-coil, as suggested previously (Stefancsik et al., 1998).

4. THE THIN FILAMENT

Solving the high resolution structures of thin filament proteins is one thing. Determining an atomic model of the assembled thin filament is another. With structures of actin, tropomyosin, and troponin in hand, it will still be challenging to understand how they dock together to form the thin filament, how they shift relative to one another to form distinct quaternary structures, and what other structural changes may occur. At this point, the relative positions of tropomyosin and actin are known, albeit not at atomic resolution, and similar data are now emerging for troponin. Also, it is clear that the thin filament is dynamic, and some of the most prominent movements within the thin filament have been identified. From these results, it is increasingly possible to appreciate the structural basis for the regulation of cardiac and skeletal muscle contraction.

In the absence of calcium, tropomyosin sterically interferes with a large portion of the actin surface where myosin binds to actin, and muscle is correspondingly relaxed. Calcium, which stimulates contraction, alters tropomyosin's position on the actin filament, causing tropomyosin to shift to the edge of the myosin binding site (Lehman et al., 1994; Xu et al., 1999; Milligan et al., 1990). This steric mechanism is critical, because strong myosin binding is a necessary part of the myosin crossbridge cycle, and can not occur unless tropomyosin moves. Evidence for this mechanism, i.e., for shifts in the azimuthal position of tropomyosin, have been determined using helical reconstruction of thin filaments examined in several different settings. Ca^{2+}-induced movement of tropomyosin on actin has been detected in muscle fibers (Holmes, 1995) by x-ray, and by electron microscopy of negatively stained cardiac and skeletal muscle thin filaments (Lehman et al., 1994; Lehman et al., 1995), and by cryo-electron microscopy of unstained, reconstituted thin filaments (Xu et al., 1999). A surface view of actin from this last report is shown in Fig. 4, indicating the positions of tropomyosin in the absence of calcium (EGTA) and in the presence of calcium.

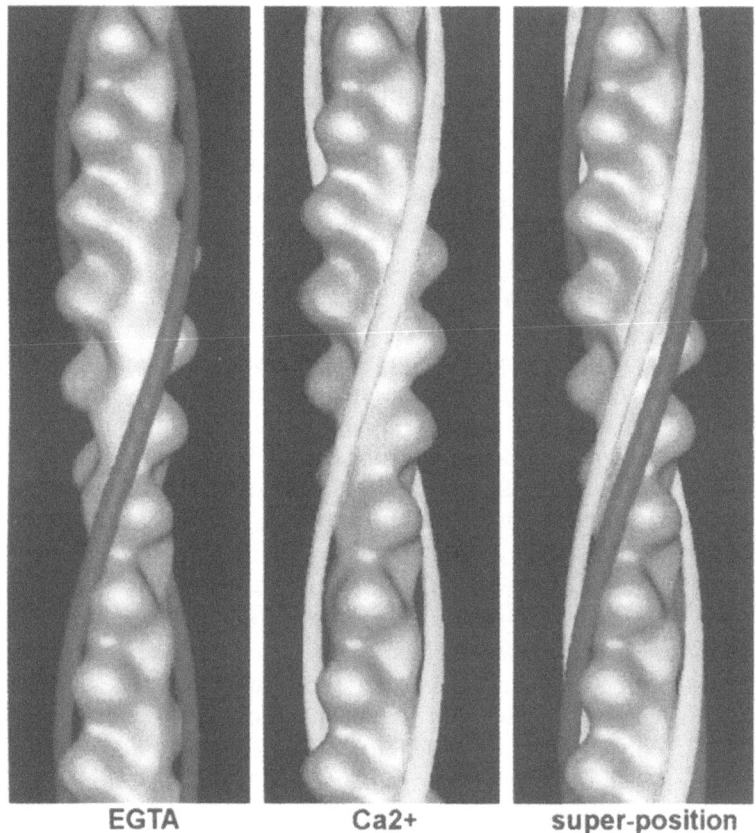

EGTA   Ca2+   super-position

*Figure 4. Ca²⁺ binding to troponin causes tropomyosin to shift position
on the thin filament. In the presence of Ca²⁺ (green), tropomyosin shifts
away from the site on actin where myosin binds, toward the edge of this
binding site. From Xu, et al, 1999. Color version on page 458.*

3-dimensional reconstruction of tropomyosin on F-actin is relatively
straightforward to accomplish, because tropomyosin strands and F-actin have nearly
the same helical symmetry, with tropomyosin making continuous, approximately
equivalent contacts with successive actin monomers.   This experimentally
advantageous symmetry is not true for troponin, which binds once per every seven
actins. Also, this discontinuous spacing of troponin is close enough to the crossover
repeat of actin long-pitch helices (380 *vs.* 360 Å) that successive troponin molecules
have comparable orientations as seen on EM grids, and this is an unfavorable
arrangement for 3-D helical analysis. Moreover, labeling of only every seventh
actin ensures that any troponin signal will be weak. To overcome these difficulties,
and to better visualize troponin, thin filaments were reconstituted with native
troponin and a tropomyosin internal deletion mutant, TmΔ234, in which three of the

seven actin-binding quasi-repeats (quasi-repeats 2, 3 and 4) are missing (Landis et al., 1997). These filament contain one troponin every four actins, almost twice the troponin content of native filaments and additionally offer more varied views of troponin along the helices of individual filaments.

Cross sections of 3-dimensional reconstructions of these filaments (Fig. 5) (Lehman et al., 2001) show an additional, well-defined density (shown in blue) not present in actin-TmΔ234 or in actin-skeletal muscle tropomyosin filaments lacking troponin. This density, which must be due to troponin, was not detected in previous troponin-containing thin filament reconstructions. Helical projections and transverse sections of filaments at low Ca^{2+} (top panel) show clear connectivity between troponin and tropomyosin and between troponin and a broad area of actin (marked by asterisk). In contrast, helical projections of filaments at high Ca^{2+} (bottom panel) show that on average troponin makes less contact with actin and tropomyosin. Transverse sections of Ca^{2+}-treated filaments showed that the contact with tropomyosin remained, while contact with actin falls below statistical significance. Note the Ca^{2+}-induced shift in the site of major contact between actin and tropomyosin, with an arrow placed near the site in the absence of Ca^{2+}.

Three-dimensional views (not shown here) of the same reconstructions show that, in the absence of Ca^{2+}, the troponin density emerges as a narrow stalk from tropomyosin located on the outer domain of actin. The troponin extension widens to form a bulb-like cap that contacts a broad area of the extreme periphery of the actin outer domain. Fitting the atomic model of F-actin within the EM maps indicates that troponin-actin association includes actin outer domain residues thought to be involved in troponin-I – actin interaction. Comparable images demonstrate that, in the presence of Ca^{2+}, both tropomyosin and troponin move away from their low Ca^{2+} positions. The broad contact made by troponin with actin in the absence of Ca^{2+} is no longer seen in the presence of Ca^{2+}. Thus troponin, while still attached to tropomyosin, is released and projects away from actin as tropomyosin moves toward the actin inner domain.

The affinity of actin for troponin-tropomyosin is much stronger than that for tropomyosin alone, and Fig. 5 identifies a major interaction surface between actin and troponin that helps to explain this. Furthermore, the results help to explain how troponin regulates muscle contraction by influencing where tropomyosin binds to actin. Previous studies indicate that tropomyosin, in filaments prepared in the absence of troponin, can alternate freely between the positions on actin that it occupies at high and low Ca^{2+} in the presence of troponin (Lehman et al., 2000). Fig. 5 provides direct structural evidence for models of muscle regulation in which Ca^{2+}-free troponin interacts with actin to constrain tropomyosin in a blocked inhibitory state, and this constraint is released on activation by Ca^{2+} (Potter and Gergely, 1974; Hitchcock et al., 1973; Margossian and Cohen, 1973). The results are also consistent

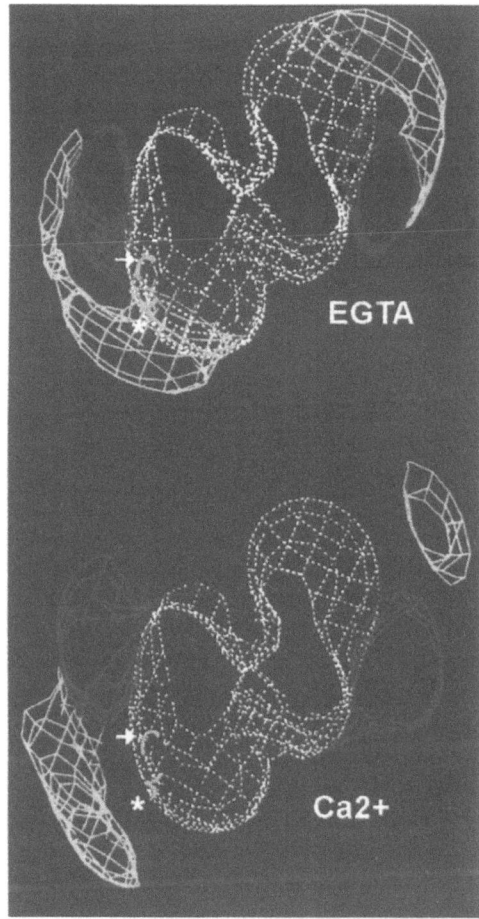

Figure 5. Thin filament transverse sections showing the effects of Ca^{2+} on tropomyosin (red) and troponin (blue). See text for details. Adapted from Lehman, et. al, 2001. Color version on page 459.

with binding studies suggesting that in the absence of Ca^{2+}, but not in its presence, a part of the troponin complex, specifically troponin-I, links the troponin-tropomyosin complex and actin, forming an actin-tropomyosin-troponin-actin loop. These studies suggest that Ca^{2+}-free troponin can act as a latch, which in effect traps tropomyosin on the outer domain of actin in a switched-off state and releases it during Ca^{2+}-activation. Fig. 5 provides structural evidence supporting this regulatory mechanism. However, troponin also binds very tightly to the thin filament in the presence of Ca^{2+}, and structural data to explain this are limited.

5. THE THREE-STATE MODEL OF THE THIN FILAMENT

A consistent finding in the above and other reports is that in the absence of Ca^{2+} tropomyosin is attached to the outer domain of actin, and when Ca^{2+} binds to troponin the tropomyosin density arises instead from the outer edge of the inner domain of actin. Despite this well established change in attachment site, the extent of the azimuthal shift in tropomyosin mass varies somewhat from study to study, and more importantly, in no report does the tropomyosin move sufficiently to fully eliminate steric interference with strong myosin binding to actin. Rather, tropomyosin sterically interferes with strong myosin binding to actin even in the presence of Ca^{2+}, although to a much smaller extent than in the absence of Ca^{2+}. This residual steric hindrance, even in the presence of Ca^{2+} has been noted by at least two different groups (Holmes, 1995; Vibert et al., 1997), and is very significant functionally. Consequently, there is an apparent contradiction between the ability of Ca^{2+} to activate muscle contraction physiologically, and the inability of Ca^{2+} to activate the thin filament structurally. The key to resolving this apparent paradox was the identification of a third state for the thin filament. Specifically, 3-dimensional reconstructions of myosin S1-decorated thin filaments show that the tropomyosin is not in the Ca^{2+} position. Rather, compared to either of the positions shown in Figs. 4 and 5, tropomyosin has undergone a further azimuthal shift across more of the actin inner domain, thereby fully exposing the myosin binding site on actin (Vibert et al., 1997). This is shown in Fig. 6, in which the three different positions for tropomyosin are over layered on the ribbon diagram for one actin monomer. These three positions, from right to left as illustrated in the figure, have been termed the B-state for "blocked" strong binding of myosin, the C-state for "Ca^{2+}-induced" or "closed" but not fully blocked binding state, and the M-state for "myosin-induced" or active state of the thin filament.

With these three structural states identified in 1997, the functional significance of each of them became an obvious and important issue. Long before this work, it was recognized that myosin binding to the thin filament alters its properties in ways that are not produced by Ca^{2+} alone (Bremel and Weber, 1972). However, there was always uncertainty in establishing whether such myosin-induced changes were a necessary part of thin filament activation, or whether they merely modulated activation that was produced by Ca^{2+} alone. This dilemma has now been resolved by the discovery of three structural states, with residual steric interference of myosin binding to actin by C-state tropomyosin, and not only by B-state tropomyosin. These data imply that the Ca^{2+}-induced state of the thin filament (i.e., the C-state) is in fact not an active state. This interpretation from structural data is strongly supported by functional results demonstrating that myosin interacts poorly with thin filaments that are altered so that the C-state forms normally but the M-state does not. Actin-tropomyosin-troponin-Ca^{2+} thin filaments containing specific, inhibitory mutations

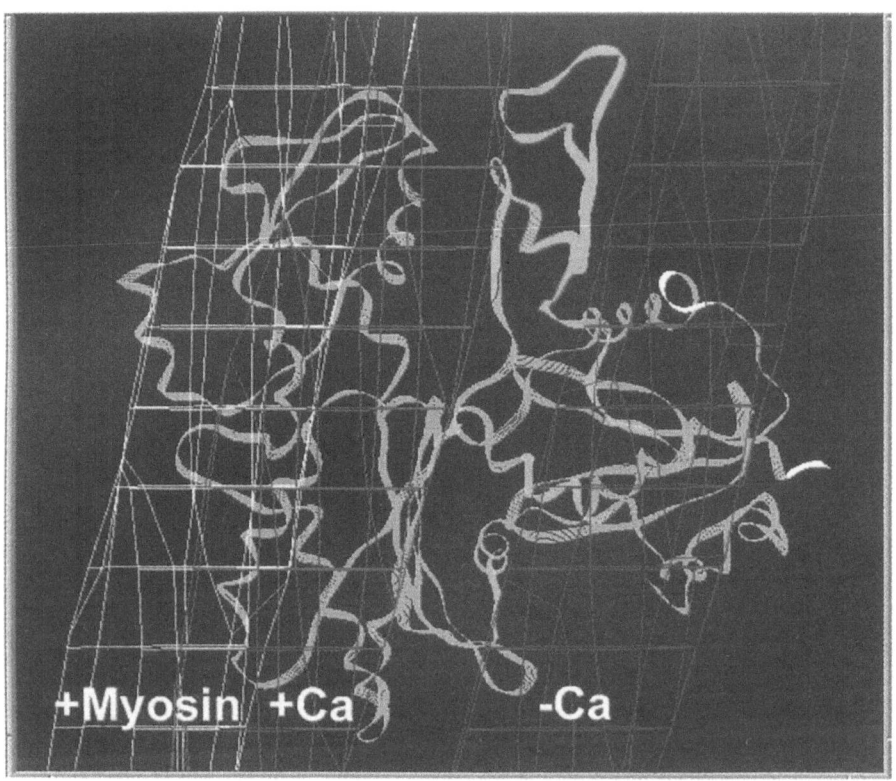

*Figure 6. Three observed positions for tropomyosin on actin describe the
M-state (+Myosin), the C-state (+Ca), and the B-state (-Ca). Blue actin
regions interact with myosin. Courtesy of W. Lehman.
Color version on page 460.*

in either tropomyosin, troponin, or actin adopt the C-state position for tropomyosin
in the presence of Ca^{2+} (Rosol et al., 2000; Korman et al., 2000; Burhop et al.,
2001). Nevertheless, these filaments bind myosin poorly, activate myosin S1
ATPase activity poorly, and do not support *in vitro* motility. These defects are
profound in two of the three cases, essentially indicating thin filaments that are shut
off for myosin interactions. The filaments are inhibitory not because tropomyosin
fails to move to the C-state position, but rather (as solution properties indicate
(Rosol et al., 2000; Landis et al., 1999; Korman et al., 2000)) because of
destabilization of the M-state. These results demonstrate what previously had been
hypothesized: formation of the C-state is not permissive for myosin cycling, and
instead is insufficient to permit myosin function.

6. COMBINED STERIC AND ALLOSTERIC REGULATORY MECHANISM

If the C-state is not an active state, and the M-state has only been observed after myosin is bound, how does the thin filament make the essential switch between these conformations as myosin binds? What are the kinetic steps of this transition, and also, how are conformational changes propagated along the thin filament? For a discussion of the kinetics, the reader is referred to the chapter by Geeves, elsewhere in this volume, and is referred to (Tobacman and Butters, 2000) for discussion of propagation and cooperativity. Here two other, unexpected aspects of this switch are presented: that it is facilitated by tropomyosin and that it involves changes in actin conformation.

Despite steric hindrance between the preferred binding sites for tropomyosin and myosin S1 when they bind to actin separately, there is a paradoxical, positive interaction between tropomyosin and myosin. Myosin S1 binds to actin-tropomyosin filaments, or to actin-troponin-tropomyosin filaments, with an affinity ~4- to 7-fold *higher* than to bare actin (Williams and Greene, 1983; Geeves and Halsall, 1986; Tobacman and Butters, 2000; Geeves and Halsall, 1986). Correspondingly, myosin S1 must increase the actin-affinity of tropomyosin (or of troponin-tropomyosin) by 4^7- to 7^7-fold, i.e., 10,000-fold or more. This very high affinity binding has experimental support: under conditions where muscle tropomyosin binds only weakly to bare actin, tropomyosin binding to actin-S1 is, in contrast, too tight to measure (Eaton, 1976; Cassell and Tobacman, 1996; Maytum et al., 2001). These effects are large, directly opposite to what would be expected based upon steric hindrance, and yet must somehow be consistent with the structural data. The strengthening effects compellingly require the existence of either direct contacts between tropomyosin and myosin, which as yet have not been identified, or else an indirect interaction mediated by a mutually promoted change in actin. High resolution structural data could distinguish between these possibilities. In the absence of such information, there are three lines of evidence to suggest that myosin-tropomyosin interactions are due to indirect, allosteric effects on actin. First, direct myosin-tropomyosin binding would depend upon tropomyosin sequence, yet this sequence differs considerably along the seven quasi-repeating regions of each tropomyosin. A related point is that myosin-actin affinity is similarly increased by vertebrate tropomyosins, regardless of tropomyosin isoform or the presence of troponin.

In addition, it is significant that a set of tropomyosin deletion mutations that greatly suppress myosin-tropomyosin interactions do not simply weaken myosin-thin filament binding as would have been expected if direct interactions were eliminated. Instead, the mutations caused an exaggerated cooperativity in myosin binding, as is characteristic of a ligand-promoted switch in quaternary structure. Therefore, to explain how tropomyosin and myosin promote each other's binding to actin, it was proposed (Tobacman and Butters, 2000) that the changes in actin that accompany myosin binding result in a strengthening of tropomyosin's association with the actin inner domain. Correspondingly, when tropomyosin is bound to the actin inner domain it was proposed (Fig. 7) to promote some of the same changes in actin ($K_T \times L > 1 >> K_T^{\circ}$), thereby increasing the fraction of actin monomers in the M-state. This is because tropomyosin is on the inner domain not only for any actin to which myosin is bound, but also tends to remain in this position for many neighboring actins (Vibert et al., 1997). In this position, tropomyosin modulates myosin's interactions with actin by promoting an actin conformational change that, for example, decreases the work of myosin binding as illustrated. In this view, the M-state is characterized not only by a distinct position for tropomyosin on actin, but also by alterations in actin that are as yet beyond the resolution of thin filament structural experiments. Some effect of myosin on actin structure must occur and

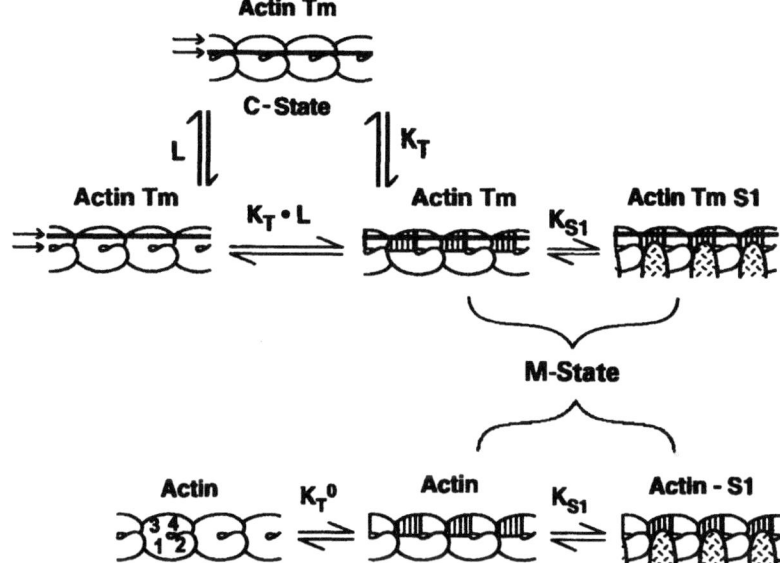

Figure 7. Model of combined steric and allosteric effects of tropomyosin on myosin S1 binding to the thin filament. Tight myosin-actin binding requires tropomyosin to shift from its preferred position (lower arrows of each pair at left) to the actin inner domain (upper arrows). Despite this steric interference in preferred binding sites, myosin and tropomyosin promote each other's binding to actin, explained here by mutually-induced changes in actin (changes schematically indicated by stripes). Note $K_T \times L > 1 >> K_T^{\circ}$.

have been spectroscopically detected (Prochniewicz and Thomas, 1999; Kouyama and Mihashi, 1981; Feng et al., 1997; Feng et al., 1997; Prochniewicz and Thomas, 1999). As the most conservative assumption, the model involves only two states of actin, a resting state and a myosin-induced state, regardless of the presence of the regulatory proteins or the state of the myosin. Since M-state properties are the same, regardless of tropomyosin isoform or tropomyosin presence, this explains the similar effects of different vertebrate tropomyosins and different regions of any single tropomyosin on myosin-actin binding (Lehrer et al., 1997; Strand et al., 2001). It also provides a framework for understanding the many differences between a fully activated thin filament and bare actin.

In a more general sense, this proposal is a solution to the long dispute as to whether regulation of muscle contraction is via steric blocking or allosteric mechanisms. As in hemoglobin, a classical allosteric protein in which steric hindrance prevents 2,3, BPG from binding to one state and not to another, steric and allosteric effects within the thin filament are inseparable. Unlike hemoglobin, the quaternary structures of the thin filament are only beginning to be discovered at atomic resolution. There can be no doubt that many surprises and insights await.

Departments of Internal Medicine and Biochemistry
The University of Iowa,
Iowa City, Iowa

7. REFERENCES

Belmont, L.D., Orlova, A., Drubin, D.G., and Egelman, E.H. (1999). A change in actin conformation associated with filament instability after Pi release. Proc. Natl. Acad. Sci. USA *96*, 29-34.

Blumenschein, T.M.A., Tripet, B.P., Hodges, R.S., and Sykes, B.D. (2001). Mapping the interacting regions between troponins T and C: Binding of TnT and TnI peptides to TnC and NMR mapping of the TnT-binding site on TnC. J. Biol. Chem. *276*, (In Press).

Bremel, R.D. and Weber, A. (1972). Cooperation within actin filament in vertebrate skeletal muscle. Nature New Biology *238*, 97-101.

Brown, J.H., Greenfield, N.J., Dominguez,R., Hitchcock-DeGregori,S.E., and Cohen,C. (2001). Deciphering the design of the tropomyosin molecule. Proc. Natl. Acad. Sci. U. S. A. *98*, 8496-8501.

Burhop, J., Rosol, M., Craig, R., Tobacman, L.S., and Lehman, W. (2001). Effects of a cardiomyopathy-causing troponin T mutation on thin filament function and structure. J. Biol. Chem. *276*, 20788-20794.

Carlier, M.-F. (1991). Actin: protein structure and filament dynamics. J. Biol. Chem. *266*, 1-4.

Cassell, M. and Tobacman, L.S. (1996). Opposite effects of myosin subfragment 1 on binding of cardiac troponin and tropomyosin to actin. J. Biol. Chem. *271*, 12867-12872.

Chen, X., Cook, R.K., and Rubenstein, P. (1993). Yeast actin with a mutation in the "hydrophobic plug" between subdomains 3 and 4 (L266D) displays a cold-sensitive polymerization defect. J. Cell Biol. *123*, 1185-1195.

Eaton, B.L. (1976). Tropomyosin binding to F-actin induced by myosin heads. Science *192*, 1337-1339.

Egelman, E.H. and Orlova, A. (1995). New insights into actin filament dynamics. Curr. Opin. Struc. Biol. *5*, 172-180.

Feng, L., Kim, E., Lee, W.-L., Miller, C.J., Kuang, B., Reisler, E., and Rubenstein, P. (1997). Fluorescence probing of yeast actin subdomain 3/4 loop 262-274. J. Biol. Chem. *272*, 16829-16837.

Flicker, P.F., Phillips, G.N., Jr., and Cohen, C. (1982). Troponin and its interactions with tropomyosin: An electron microscope study. J. Mol. Biol *162*, 495-501.

Gagne, S.M., Tsuda, S., Li, M.X., Chandra, M., Smillie, L.B., and Sykes, B.D. (1994). Quantification of the calcium-induced secondary structural changes in the regulatory domain of troponin-C. Protein Science *3* , 1961-1974.

Galkin, V.E., Orlova, A., Lukoyanova, N., Wriggers, W., and Egelman, E.H. (2001). Actin depolymerizing factor stabilizes an existing state of F-actin and can change the tilt of F-actin subunits. J. Cell Biol. *153*, 75-86.

Gasmi-Seabrook, G.M.C., Howarth ,J.W., Finley, N., Abusamhadneh, E., Gaponenko, V., Brito, R.M.M., Solaro, R.J., and Rosevear, P.R. (1999). Solution structures of the C-terminal domain of cardiac troponin C free and bound to the N-terminal domain of cardiac troponin I. Biochemistry *38*, 8313-8322.

Geeves, M.A. and Halsall, D.J. (1986). The dynamics of the interaction between myosin subfragment 1 and pyrene-labelled thin filaments, from rabbit skeletal muscle. Proc. R. Soc. Lond. *B229*, 85-95.

Gordon, A.M., Homsher, E., and Regnier, M. (2000). Regulation of contraction in striated muscle. Physiol. Rev. *80*, 853-924.

Greenfield, N.J., Montelione, G.T., Farid, R.S., and Hitchcock-DeGregori, S.E. (1998). The structure of the N-terminus of striated muscle alpha-tropomyosin in a chimeric peptide: nuclear magnetic resonance structure and circular dichroism studies. Biochemistry *37*, 7834-7843.

Herzberg, O., Moult, J., and James, M.N.G. (1986). A model for the Ca^{2+}-induced conformational transition of troponin C. A trigger for muscle contraction. J. Biol. Chem. *261*, 2638-2644.

Hitchcock, S.E., Huxley, H.E., and Szent-Gyorgyi, A.G. (1973). Calcium sensitive binding of troponin to actin-tropomyosin: A two-site model for troponin action. J. Mol. Biol. *80*, 825-836.

Holmes, K.C. (1995). The actomyosin interaction and its control by tropomyosin. Biophys. J. *68*, 2s-7s.

Holmes, K.C., Popp, D., Gebhard, W., and Kabsch, W. (1990). Atomic model of the actin filament. Nature *347*, 44-49.

Houdusse, A., Love, M.L., Dominguez, R., Grabarek, Z., and Cohen, C. (1997). Structures of four Ca^{2+}-bound troponin C at 2.0 A resolution: further insights into the Ca^{2+}-switch in the calmodulin superfamily. Structure *5*, 1695-1711.

Huxley, A.F. (1957). Muscle structure and theories of contraction. Prog. Biophys. *7*, 255-318.

Isambert, H., Venier, P., Maggs, A.C., Fattoum, A., Kassab, R., Pantaloni, D., and Carlier, M.-F. (1995). Flexibility of actin filaments derived from thermal fluctuations. Effect of bound nucleotide, phalloidin, and muscle regulatory proteins. J. Biol. Chem. *270*, 11437-11444.

Kabsch, W., Mannerz, H.G., Suck, D., Pai, E.F., and Holmes, K.C. (1990). Atomic structure of the actin:DNase I complex. Nature *347*, 37-44.

Korman, V.L., Hatch, V., Dixon, K., Craig, R., Lehman, W., and Tobacman, L.S. (2000). An actin subdomain 2 mutation that impairs thin filament regulation by troponin and tropomyosin. J. Biol. Chem. *275*, 22470-22478.

Kouyama, T. and Mihashi, K. (1981). Fluorimetry study of N-(1-pyrenyl)iodoacetamide labelled F-actin. Local structural change of actin protomer both on polymerization and on binding of heavy meromyosin. Eur. J. Biochem. *114*, 33-38.

Landis, C.A., Back, N., Homsher, E., and Tobacman, L.S. (1999). Effects of tropomyosin internal deletions on thin filament function. J. Biol. Chem. *274*, 31279-31285.

Landis, C.A., Bobkova, A., Homsher, E., and Tobacman, L.S. (1997). The active state of the thin filament is destabilized by an internal deletion in tropomyosin. J. Biol. Chem. *272*, 14051-14056.

Lehman, W., Craig, R., and Vibert, P. (1994). Ca^{2+}-induced tropomyosin movement in Limulus thin filaments revealed by three-dimensional reconstruction. Nature *368*, 65-67.

Lehman, W., Hatch, V., Korman, V.L., Rosol, M., Thomas, L.T., Maytum, R., Geeves, M.A., Van Eyk, J.E., Tobacman, L.S., and Craig, R. (2000). Tropomyosin and Actin Isoforms Modulate the Localization of Tropomyosin Strands on Actin Filaments. J. Mol. Biol. *302*, 593-606.

Lehman, W., Rosol, M., Tobacman, L.S., and Craig, R. (2001). Troponin Organization on Relaxed and Activated Thin Filaments Revealed by Electron Microscopy and Three-dimensional Reconstruction. J. Mol. Biol *307*, 739-744.

Lehman, W., Vibert, P., Uman, P., and Craig, R. (1995). Steric-blocking by Tropomyosin Visualized in Relaxed Vertebrate Muscle Thin filaments. J. Mol. Biol *251*, 191-196.

Lehrer, S.S., Golitsina, N.L., and Geeves, M.A. (1997). Actin-tropomyosin activation of myosin subfragment 1 ATPase and thin filament cooperativity. The role of tropomyosin flexibility and end-to-end interactions. Biochemistry *36*, 13454.

Li, M.X., Spyracopoulos, L., and Sykes, B.D. (1999). Binding of cardiac troponin-I147-163 induces a structural opening in human cardiac troponin-C. Biochemistry *38*, 8298.

Li, Y., Love, M.L., Putkey, J.A., and Cohen, C. (2000). Bepridil opens the regulatory N-terminal domain lobe of cardiac troponin C. Proc. Natl. Acad. Sci. , USA *97*, 5140-5145.

Lorenz, M., Poole, K.J.V., Popp, D., Rosenbaum, G., and Holmes, K.C. (1995). An atomic model of the unregulated thin filament obtained by x-ray fiber diffraction on oriented actin-tropomyosin gels. J. Mol. Biol. *246*, 108-119.

Lorenz, M., Popp, D., and Holmes, K.C. (1993). Refinement of the F-actin model against X-ray fiber diffraction data by the use of a directed mutation algorithm. J. Mol. Biol. *234*, 826-836.

Margossian, S.S. and Cohen, C. (1973). Troponin subunit interactions. J. Mol. Biol. *81*, 409-413.

Maytum, R., Geeves, M., and Konrad, M. (2001). Actomyosin regulatory properties of yeast tropomyosin are dependent upon N-terminal modification. Biochemistry *39*, 11913-11920.

McGough, A., Pope, B., Chiu, W., and Weeds, A. (1997). Cofilin changes the twist of F-actin: implications for actin filament dynamics and cellular function. J. Cell Biol. *138*, 771-781.

McLaughlin, P.J., Gooch, J.T., Mannerz, H.-G., and Weeds, A.G. (1993). Structure of gelsolin segment 1-actin complex and the mechanism of filament severing. Nature *364*, 685-692.

Mendelson, R.A. and Morris, E.P. (1994). The structure of F-actin. Results of global searches using data from electron microscopy and X-ray crystallography. J. Mol. Biol *240*, 138-154.

Milligan, R.A., Whittaker, M., and Safer, D. (1990). Molecular structure of F-actin and location of surface binding sites. Nature *348*, 217-221.

Olah, G.A. and Trewhella, J. (1994). A model structure of the muscle protein complex $4Ca^{2+}$-troponin C-troponin I derived from small-angle scattering data: implications for regulation. Biochemistry *33*, 12800-12806.

Orlova, A., Chen, X., Rubenstein, P.A., and Egelman, E.H. (1997). Modulation of yeast F-actin structure by a mutation in the nucleotide binding cleft. J. Mol. Biol *271*, 235-243.

Orlova, A., Prochniewicz, E., and Egelman, E.H. (1995). Structural dynamics of F-actin: II. Cooperativity in structural transitions. J. Mol. Biol. *245*, 598-607.

Otterbein, L.R., Graceffa, P., and Dominguez, R. (2001). The crystal structure of uncomplexed actin in the ADP state. Science *293*, 708-711.

Page, R., Lindberg, U., and Schutt, C.E. (1998). Domain motions in actin. J. Mol. Biol *280*, 463-474.

Potter, J.D. and Gergely, J. (1974). Troponin, tropomyosin, and actin interactions in the Ca^{2+} regulation of muscle contraction. Biochemistry *13*, 2697-2703.

Prochniewicz, E. and Thomas, D.D. (1999). Differences in structural dynamics of muscle and yeast actin accompany differences in functional interactions with myosin. Biochemistry *38*, 14860-14867.

Rayment, I., Holden, H.M., Whittaker, M., Yohn, C.B., Lorenz, M., Holmes, K.C., and Milligan, R.A. (1993). Structure of the actin-myosin complex and its implications for muscle contraction. Science *261*, 58-65.

Razzaq, A., Schmitz, S., Veigel, C., Molloy, J.E., and Geeves, M.A. (1999). Actin residue Glu93 is identified as an amino acid affecting myosin binding. J. Biol. Chem. *274*, 28321-28328.

Rosol, M., Lehman, W., Craig, R., Landis, C., Butters, C., and Tobacman, L.S. (2000). Three-dimensional reconstruction of thin filaments containing mutant tropomyosin. Biophys. J. *78*, 908-917.

Schutt, C.E., Myslik, J.C., Rozycki, M.D., Goonesekere, N.C.W., and Lindberg, U. (1993). The structure of crystalline profilin:β-actin. Nature *365*, 810-816.

Sia, S.K., Li, M.X., Spyracopoulos, L., Gagne, S.M., Liu, W., Putkey, J.A., and Sykes, B.D. (1997). Structure of cardiac muscle troponin C unexpectedly reveals a closed regulatory domain. J. Biol. Chem. *272*, 18216-18221.

Stefancsik, R., Jha, P.K., and Sarkar, S. (1998). Identification and mutagenesis of a highly conserved domain in troponin T responsible for troponin I binding: Potential role for coiled coil interaction. Proc. Natl. Acad. Sci. USA *95*, 957-962.

Steinmetz, M.O., Stoffler, D., Hoenger, A., Bremer, A., and Aebi, U. (1997). Actin: from cell biology to atomic detail. J. Struct. Biol. *119*, 295-320.

Stone, D.B., Timmins, P.A., Schneider, D.K., Krylova, I., Ramos, C.H., Reinach, F.C., and Mendelson, R.A. (1998). The effect of regulatory Ca^{2+} on the in situ structures of troponin C and troponin I: a neutron scattering study. J. Mol. Biol *281*, 689-704.

Strand, J., Nili, M., Homsher, E., and Tobacman, L.S. (2001). Strand, J., Nili, M. Homsher, E. and Tobacman, L.S.: Modulation of Myosin Function by Isoform-Specific Properties of *S. Cerevisiae* and Muscle Tropomyosins. J. Biol. Chem. *276*, (In Press).

Strynadka, N.C., Cherney, M., Sielecki, A.R., Li, M.X., Smillie, L.B., and James, M.N. (1997). Structural details of a calcium-induced molecular switch: X-ray crystallographic analysis of the calcium-saturated N-terminal domain of troponin C at 1.75 A resolution. J. Mol. Biol *273*, 238-255.

Swenson, C.A. and Fredrickson, R.S. (1992). Interaction of troponin C and troponin C fragments with troponin I and the troponin I inhibitory peptide. Biochemistry *31*, 3420-3429.

Tirion, M.M., ben-Avraham, D., Lorenz, M., and Holmes, K.C. (1995). Normal modes as refinement parameters for the F-actin model. Biophys. J. *68*, 5-12.

Tobacman, L.S. (1996). Thin filament-mediated regulation of cardiac contraction. Annu. Rev. Physiol. *58*, 447-481.

Tobacman, L.S. and Butters, C.A. (2000). A new model of cooperative myosin-thin filament binding. J. Biol. Chem. *275*, 27587-27593.

Tung, C.-S., Wall, M.E., Gallagher, S.C., and Trewhella, J. (2000). A model of troponin-I in complex with troponin-C using hybrid experimental data: The inhibitory region is a β-hairpin. Protein Science *9*, 1312-1326.

Van Eyk, J.E., Kay, C.M., and Hodges, R.S. (1991). A comparative study of the interactions of synthetic peptides of the skeletal and cardiac troponin I inhibitory region with skeletal and cardiac troponin C. Biochemistry *30*, 9974-9981.

Vassylev, D., Takeda, S., Wakatsuki, S., Maeda, K., and Maeda, Y. (1998). Crystal structure of troponin C in complex with troponin I fragment at 2.3-A resolution. Proc. Natl. Acad. Sci. USA *95*, 4847-4852.

Vibert, P., Craig, R., and Lehman, W. (1997). Steric-model for activation of muscle thin filaments. J. Mol. Biol. *266*, 8-14.

Williams, D.L. and Greene, L.E. (1983). Comparison of the effects of tropomyosin and troponin-tropomyosin on the binding of myosin subfragment 1 to actin. Biochemistry *22*, 2770-2774.

Xu, C., Craig, R., Tobacman, L.S., Horowitz, R., and Lehman, W. (1999). Tropomyosin postitions in regulated thin filaments revealed by cryoelectron microscopy. Biophys. J. *77*, 985-992.

Yarmola, E.G., Sumasundaram, T., Boring, T.A., Spector, I., and Bubb, M.R. (2001). Actin-latrunculin A structure and function. Differential modulation of actin-binding protein function by latrunculin A. J. Biol. Chem. *275*, 28120-28127.

ALDRIN V. GOMES, KEITA HARADA & JAMES D. POTTER

CATION SIGNALING IN STRIATED MUSCLE CONTRACTION

1. INTRODUCTION

Vertebrate striated (skeletal and cardiac) muscle contraction is regulated in a Ca^{2+}-dependent manner by the troponin (Tn) complex through interactions with tropomyosin (Tm) and the actin filament. The Tn complex consists of three proteins: troponin I (TnI) which inhibits the actomyosin Mg^{2+}-ATPase activity, troponin C (TnC) which binds Ca^{2+} ions and removes TnI inhibition, and troponin T (TnT) which makes primary interaction with Tm. These three subunits interact in a cooperative manner with each other and with Tm and actin. Binding of Ca^{2+} to TnC initiates structural changes in TnC which lead to structural changes in the other Tn subunits (TnI and TnT) as well as Tm. The movement of Tm on the actin filament facilitates the interaction between myosin and actin leading to muscle contraction. The cation, Ca^{2+}, is the main regulatory and signaling molecule in striated muscle. Another cation, Mg^{2+}, is also important in the regulation of striated muscle contraction mainly via its ability to inhibit the Ca^{2+} release channel, especially in skeletal muscle.

The main proteins involved in Ca^{2+} signaling include TnC, the ryanodine receptor, the sarcoplasmic reticulum (SR) Ca^{2+} pump, myosin light chains, calsequestrin, parvalbumin (in skeletal tissue), calmodulin (CaM), annexins, sorcin, calcineurin, S100, and calpain, all influence the contractile properties of striated muscle (see Table 1). In addition to Ca^{2+}, some of these proteins are also modulated by Mg^{2+}, such as the ryanodine receptor, parvalbumin and TnC. In this section, the major biochemical aspects of the mechanism of muscle contraction with particular reference to the role of cation signaling, will be discussed.

2. INTRACELLULAR CALCIUM TRANSIENTS

2.1. Senders and Transducers of Ca^{2+} Signals

Calcium (Ca^{2+}) forms about 2% of the human body, 99% of which is found in bones and teeth and is one of the most important second messengers found in cells. The concentration of free Ca^{2+} is extremely low in the cytosol ($\sim 10^{-7}$M) compared to the extracellular fluid ($\sim 10^{-3}$ M) or to the endoplasmic reticulum (ER). Dynamic

R.J. Solaro and R.L. Moss (eds), Molecular Control Mechanisms in Striated Muscle Contraction, 163-197
©2002 Kluwer Academic Publishers, Printed in the Netherlands

changes in intracellular Ca^{2+} concentrations are essential for a great variety of cellular processes, including intra- and extra-cellular signaling processes, such as neuronal signal transduction, hormone secretion, cell division, cell adhesion, cell motility and muscle contraction. A variety of reactions, such as receptor-ligand interactions, mediate changes in the concentration of free intracellular Ca^{2+} (Carafoli, 1987).

3. INITIATION OF "ON-STATE" OR SYSTOLE BY CA^{2+}: MUSCLE EXCITATION

The sequence of events leading to contraction is initiated in the central nervous system, either as voluntary activity from the brain or as reflex activity from the spinal cord. A motor neuron in the ventral horn of the spinal cord is activated, and an action potential propagates down the ventral root of the spinal cord. During an action potential the membrane potential becomes reversed so that for a few milliseconds the inside of the muscle cell becomes positively charged with respect to the outside of the cell. The inside of the cell then returns to its initial negative charge and after a latency period of a few milliseconds, the muscle contracts. The muscle membrane (sarcolemma) extends into the interior of the muscle fiber. These extensions of membrane are called transverse tubules (T-tubules). At the opening of each T tubule onto the muscle fiber surface, the action potential spreads inside the muscle fiber. The T-tubule voltage sensor, called the dihydropyridine (DHP) receptor is a multi-subunit protein that is integral in the T-tubule membrane. The ends of the T-tubule comes in close contact with the sarcoplasmic reticulum (SR), which is a form of the endoplasmic reticulum (ER), that has evolved into a specialized release and uptake organ in muscle tissue. In skeletal muscle, depolarization of the T-tubule causes a conformational change in the DHP receptor resulting in the SR Ca^{2+} release channel, called the ryanodine receptor type 1 (RyR1), opening due to a direct coupling between the RyR1 and the T-tubule DHP receptor in skeletal muscle (Table 1) (Schneider and Chandler, 1973). In cardiac muscle depolarization of the T-tubule triggers the opening of voltage-gated Ca^{2+} channels (also a DHP receptor but a different isoform from that found in skeletal muscle) permitting a small influx of extracellular Ca^{2+} ions. The Ca^{2+} entering the cell binds to and opens the Ca^{2+} release channel type 2 (RyR2) in nearby SR and the Ca^{2+} released from the SR activates other Ca^{2+} release channels, amplifying the initial Ca^{2+} release. This process is unique to cardiac muscle and is known as "Ca^{2+} induced Ca^{2+} release" (CICR). In both skeletal and cardiac muscle the RyR is a large ion channel found in the membrane of the terminal cisternae of the SR which conducts stored Ca^{2+} in the SR lumen into the sarcoplasm (Meissner, 1994), resulting in Ca^{2+} availability for TnC binding. Differences between the cardiac and skeletal T-tubule systems are explained in the variations in the heart and skeletal muscle systems section.

Table 1. Cation Binding Proteins Potentially involved in Striated Muscle Contraction

Protein	Isoform	Molecular Weight	Localization	Ca²⁺ Binding (affinity, in M^{-1}, shown in brackets)	Mg²⁺ Binding (affinity, in M^{-1}, shown in brackets)	Proposed Function	References
Troponin C*	Fast skeletal	18,000	Fast skeletal muscle	Yes, 4 sites Two high affinity sites III and IV (5×10^8) Two low affinity sites I and II (5×10^6)	Yes, 2 sites # (5×10^4)	Ca²⁺ sensor protein in myofibril	(Potter and Gergely, 1975; Robertson et al., 1981)
	Slow skeletal/ cardiac	18,000	Slow skeletal/cardiac muscle	Yes, 3 sites Two high affinity sites III and IV (3×10^8) One low affinity site II (2×10^6)	Yes, 2 sites # (7×10^4)		
Myosin Regulatory Light chain*	Fast skeletal	19,000	Fast skeletal muscle	Yes, 1 high affinity site (3×10^7)	Yes, 1 site ($2.5\text{-}3 \times 10^5$)	Modulator of muscle contraction	(Holroyde et al., 1979)
	cardiac	19,000	cardiac muscle	Yes, 1 high affinity site (3×10^7)	Yes, 1 site ($2.5\text{-}3 \times 10^5$)		
Calmodulin		17,000	Ubiquitous	Yes, 4 sites (4.2×10^5)	Yes, 4 sites (2×10^2)	Multifunctional regulator	(Potter et al., 1983)
Parvalbumin		12,000	Fast skeletal muscle (does not occur in cardiac muscle)	Yes, 2 sites (2.5×10^8)	Yes, 2 sites (1.1×10^4)	Ca²⁺ transport from myofibrils to SR	(Johnson et al., 1981)
Annexin	VI	67,000	Muscle sarcolemma	Yes, 2 sites	Yes	Possible SR Ca²⁺ release channel modulator	(Matteo and Moravec, 2000)
Ca²⁺ pump	Fast skeletal (SERCA1a)	110,000	Fast skeletal muscle	Yes, binds and transports 2 Ca²⁺ ions per ATP molecule		Ca²⁺ transport into the SR	(Guerini et al., 1998)
	Slow skeletal/ cardiac (SERCA2a)	110,000	Slow skeletal/cardiac muscle	Yes, binds and transports 2 Ca²⁺ ions per ATP molecule			

Table 1 Continued. Cation Binding Proteins Potentially involved in Striated Muscle Contraction

Protein	Isoform	Molecular Weight	Localization	Ca²⁺ Binding (affinity, in M^{-1}, shown in brackets)	Mg²⁺ Binding (affinity, in M^{-1}, shown in brackets)	Proposed Function	References
Calpain	μ-calpain m-calpain	Large subunit 80,000 Small subunit 28,000	All muscle types	Yes, 3 sites which bind Ca²⁺ at physiological Ca²⁺ concentration	Yes	Ca²⁺-dependent protease	(Blanchard et al., 1997; Dutt et al., 2000)
Calsequestrin	Fast skeletal	45,000	Fast skeletal muscle	Yes, many sites (> 40)	Yes, many sites	Ca²⁺-storage protein of the SR	(Cozens and Reithmeier, 1984; Yano and Zarain-Herzberg, 1994; Krause et al., 1991)
	Slow skeletal/cardiac	45,000	Slow skeletal/cardiac muscle	Yes, many sites (> 40)	Yes, many sites but different sites from Ca²⁺-binding sites		
Calcineurin		Subunit A 59,000 Subunit B 18,000	All muscle types	No / Yes, 4 sites. 1 high affinity site (2.4x10⁸) 3 low affinity sites (1.5x10⁵)	No / Yes	Ca²⁺-dependent phosphatase	(Rusnak and Mertz, 2000)
Ryanodine Receptor	Fast skeletal (RyR1)	550,000	Fast skeletal muscle	Yes, at least two classes of Ca²⁺ binding sites	Yes	Ca²⁺ release channel of the SR	(Meissner, 1994; Xiong et al., 1998)
	cardiac (RyR2)	550,000	cardiac muscle	Yes, at least two classes of Ca²⁺ binding sites	Yes		
S100	α	10,000	cardiac and skeletal muscle	Yes, 2 sites	Yes	Activation of twitching	(Engelkamp et al., 1992; Schafer and Heizmann, 1996)
Sorcin		12,000	Several tissues including cardiac and skeletal muscle	Yes, two sites		Inhibits ryanodine receptor	(Valdivia, 1998)

* These values are for Ca²⁺ binding to TnC in the whole Troponin complex and for Ca²⁺ binding to RLC in Myosin. Values for Ca²⁺ binding to isolated TnC are different from Ca²⁺ binding to Troponin. # physiologically relevant sites. Two other sites also bind Mg²⁺ but with very weak affinity.

4. CONTROL MECHANISMS TO KEEP MYOPLASMIC CA^{2+} CONCENTRATIONS IN THE "OFF-STATE" OR DIASTOLE

The concentration of free Ca^{2+} ions in the cytoplasm is critical for the control of cardiac contraction. The normal cardiac free Ca^{2+} ion concentration rises from ~ 160 nM to value of ~ 2700 nM at systole (Kirschenlohr et al., 2000). These values are the result of Ca^{2+} entering and leaving the cytosol during each cardiac cycle and the Ca^{2+}-binding and Ca^{2+} release properties of the cytoplasmic Ca^{2+}-binding proteins or "buffers". Endogenous Ca^{2+} buffers in the heart can be divided into different groups based on their affinities. Most Ca^{2+}-binding sites on high-affinity "buffers" (proteins with association constant > 3 x 10^7 M^{-1}) will have Ca^{2+} bound in resting cells and will not contribute significantly to buffering of the cardiac Ca^{2+} transient. Ca^{2+}-binding proteins with low affinities (association constant < 10^4 M^{-1}) will not bind an appreciable amount of Ca^{2+} when Ca^{2+} is released from the SR. The major contributors to cytosolic Ca^{2+} buffering during the Ca^{2+} release and uptake by the SR (Ca^{2+} transient) will be from Ca^{2+}-binding proteins with affinities in the range 10^4-10^7 M^{-1}. *In vitro* studies suggest that cardiac TnC (cTnC) and intact cardiac myofibrils have only one Ca^{2+}-binding site per TnC with an affinity in the physiologically relevant range (Pan and Solaro, 1987; Holroyde et al., 1980; Robertson et al., 1981) and as such the TnC component of the myofibrils is considered the major protein thought to buffer Ca^{2+} during the cardiac Ca^{2+} transient. Recent biochemical studies have shown that in the presence of ATP, cardiac myofibrils bind Ca^{2+} co-operatively, but still only one Ca^{2+} ion per TnC, with an apparent K$_d$ in the μM range (Morimoto and Ohtsuki, 1994a). Skeletal TnC (sTnC) has two Ca^{2+}-specific sites with an affinity (K$_d$ 5x10^{-6}M) in the physiologically relevant range (Melzer et al., 1995; Potter and Gergely, 1975; Robertson et al., 1981) and is the major protein in skeletal muscle that would buffer Ca^{2+} during muscle excitation.

While the resting myocyte contains an intracellular free Ca^{2+} concentration of ~ 0.1-0.2μM, the extracellular fluid contains ~1mM Ca^{2+}, thereby resulting in a large concentration gradient favoring the entry of Ca^{2+} into the myoplasm. Ca^{2+} must be continuously removed from the cardiomyocytes as the cell membrane is somewhat leaky to Ca^{2+}, by the Na-Ca exchanger or by the ATP-dependent Ca^{2+} pump of the sarcolemma (Egger and Niggli, 1999).

The contractile unit of striated muscle, the sarcomere, consists of the thick and thin filaments, composed mainly of the two most abundant proteins, myosin (60-70%) and actin (20-25%). The Ca^{2+}-binding proteins found in the sarcomere as well as other Ca^{2+}-binding proteins potentially involved in muscle contraction are shown in Table 1. Calsequestrin is a major muscle SR Ca^{2+}-binding protein that has the ability to bind at least 40 Ca^{2+} ions (at salt concentrations close to physiological levels) for each molecule of protein, thereby increasing the Ca^{2+} storage capacity of the SR (Jones et al., 1998). The affinity of calsequestrin for Ca^{2+} is relatively low and this protein is described as a high capacity, low affinity Ca^{2+}-binding protein. Some studies suggest that it may also have an active role in regulating the release of Ca^{2+} from the SR (Ikemoto et al., 1991). Transgenic mice with a 10 fold overexpression of cardiac calsequestrin developed severe cardiac hypertrophy with increases in overall heart mass and size of the myocytes (Jones et al., 1998). In recent years, several reports show evidence for calsequestrin influencing RyR activity. A direct protein-protein interaction between calsequestrin and RyR has also been documented (Murray and Ohlendieck, 1998). Moreover, phosphorylation of calsequestrin was shown to selectively control the RyR channel activity (Herzog et al., 2000).

Other possible modulatory proteins have been found in myofibrils but their physiological roles are not yet fully established. The S100α protein is a cytoplasmic Ca^{2+}-binding protein that is capable of stimulating CICR *in vitro* at a protein concentration of 5μM (Fano et al., 1989). Sorcin is a 22kDa Ca^{2+}-binding protein that has recently been shown to regulate the activity of the cardiac ryanodine receptor (Lokuta et al., 1997) and has also been shown to be present in skeletal muscle (Valdivia, 1998). Sorcin inhibits ryanodine receptor when added to the cytoplasmic side of the channel and this inhibition could be relieved by phosphorylation of sorcin with protein kinase A (Lokuta et al., 1997). Each sorcin molecule contains two EF-hand Ca^{2+}-binding domains that bind Ca^{2+} with high affinity (~1μM) (Meyers et al., 1995). Annexin VI is a 67kDa protein that has been shown to modulate the SR Ca^{2+} release channel activity *in vitro* at protein concentrations of 5-40nM (Diaz-Munoz et al., 1990). However, these effects may be an artifact since the localization studies suggest that annexin VI is associated with the SR of skeletal but not heart muscle (Luckcuck et al., 1998). Calcineurin is a widely distributed protein phosphatase that has been implicated in transducing signals responsible for cardiac and skeletal hypertrophy (Dunn et al., 1999; Force et al., 1999). Transgenic mice overexpressing calcineurin developed cardiac hypertrophy and heart failure (Molkentin et al., 1998). Calpain is an endogenous Ca^{2+} dependent protease that selectively degrades the RyR resulting in impaired activation *in vitro* (Seiler et al., 1984; Gilchrist et al., 1992).

5. INTRACELLULAR MAGNESIUM

Magnesium (Mg^{2+}) is an important intracellular cation that contributes to many enzymatic and metabolic processes including protein synthesis. In fact Mg^{2+} is the second most abundant intracellular divalent cation and hypomagnesemia is known to cause hyperirritability, arrhythmias, and vasomotor changes while hypermagnesemia causes muscle weakness, bradycardia, weak pulse, hypotension, heart block, and cardiac arrest (Laurant and Touyz, 2000; Romani and Scarpa, 2000). Some of these symptoms and diseases could be attributed to changes in MgATP concentrations. MgATP is the substrate for both the Ca^{2+} pump and myosin. Relatively little is known about the regulation of cellular Mg^{2+} homeostasis, in contrast to the detailed knowledge of the Ca^{2+} homeostasis. This lack of knowledge is due to difficulties in measuring cytoplasmic free Mg^{2+}. The Mg^{2+} concentration in mammalian cells has been estimated by several different labs and ranges from 0.4-0.9mM (Murphy et al., 1989a; Murphy et al., 1989b; Handy et al., 1996). Although the Mg^{2+} concentration remains roughly constant during contractile activity, it is displaced by Ca^{2+} from binding sites on parvalbumin and the C-terminal of TnC and simultaneously moves into the SR where it acts as a counter ion to the released Ca^{2+} (Somlyo et al., 1985).

6. REGULATION OF THE RYANODINE RECEPTOR BY MG^{2+}

The physiologically occurring divalent cations, Ca^{2+} and Mg^{2+}, act in an antagonistic way in the regulation of the ryanodine receptor (RyR). While micromolar Ca^{2+} concentrations up to 100 μM (in the absence of ATP and Mg^{2+}) activate the skeletal Ca^{2+} release channel *in vitro*, higher concentrations inhibit channel activity (Ma et al., 1988). No sequence with the exact characteristics of an EF hand Ca^{2+}-binding region has been found in the human ryanodine receptor. The EF-hand consists of α-helix E (residues 1-4), a loop that binds the Ca^{2+} ion (residues 10-21) and a second α-helix F (residues 19-29) (Kawaski and Kretsinger, 1994). Recently, a modulator binding region was localized which appears to bind Ca^{2+} and modulate Ca^{2+} release from the RyR (Meissner, 1994). Less information is available for the binding site involved in the Mg^{2+} regulation of the RyR. In contrast to Ca^{2+}-binding sites, no consensus sequences have yet been defined for Mg^{2+}-binding sites and as such they cannot be predicted from the primary structure of proteins. Both in skinned fiber studies and in Ca^{2+}-flux measurements, Mg^{2+} has been found to inhibit SR Ca^{2+}-release in millimolar concentrations (Meissner, 1994).

A key difference between the Ca^{2+}-release channel of skeletal (RyR1) and cardiac muscle (RyR2) is the affinity of an inhibition site on these proteins for Mg^{2+}. The inhibitory site in the cardiac release channel (RyR2) has a much greater affinity for Ca^{2+} and Mg^{2+} than the inhibition site in the skeletal Ca^{2+}-release channel (RyR1). The affinity of the inhibition site for both Ca^{2+} and Mg^{2+} are nearly identical and the binding of Mg^{2+} is as effective as Ca^{2+} in inducing inhibition. Since the physiological concentration of intracellular Mg^{2+} is relatively high ($\sim$1mM) and both Mg^{2+} and Ca^{2+} are able to induce inhibition at $\sim$0.1mM concentrations (K_i $\sim$0.1mM in skeletal muscle for both cations), the inhibitory site would be occupied mainly by Mg^{2+} since Ca^{2+} levels are unlikely to reach mM levels even during peak Ca^{2+} release. The RyR inhibitory site could therefore be regarded as an "Mg^{2+} inhibition site" (Lamb, 1993). As such, in skeletal muscle, Mg^{2+} exerts a powerful and important inhibitory effect on the Ca^{2+}-release channels (RyR1) under resting conditions, resulting in the Ca^{2+} release channel being closed at rest (Lamb and Stephenson, 1994). When the intracellular Mg^{2+} concentration is lowered significantly below its physiological level, the Ca^{2+} release channels open spontaneously and Ca^{2+} is released from the SR. The voltage sensor activation of skeletal muscle overcomes this inhibitory effect of Mg^{2+}, via an unknown mechanism, possibly by lowering the affinity of the release channels for Mg^{2+}.

7. STRUCTURE AND FUNCTION OF SKELETAL MUSCLE TNC

7.1. Size, Structural Properties and Location of sTnC

Troponin C (TnC), the highly acidic Ca^{2+}-binding protein, is the most extensively investigated subunit of the troponin complex. Two isoforms of TnC exist in striated muscle, fast skeletal TnC (sTnC) and cardiac TnC (cTnC), which is identical to the slow skeletal isoform of TnC. These isoforms are highly homologous (see Table 1 and Figs. 1 and 2). Skeletal TnC (sTnC) is a single 159 amino acid polypeptide which contains four Ca^{2+}-binding sites (termed sites I to IV), beginning from the N-terminus. Each Ca^{2+}-binding site is composed of a Ca^{2+}-coordinating loop rich in aspartic (Asp) and glutamic acid (Glu) residues and two α-helical segments on either side of the loop (called E–F-hand or the helix–loop–helix structure). Each EF-hand motif consists of ~ 30 amino acid residues including 12 amino acids of the loop flanked by α-helices. Residues 1, 3, 5, 7, 9 and 12 (termed positions X, Y, Z, -Y, -X, and -Z) of the Ca^{2+}-binding loop have carboxyl or hydroxyl group where the oxygen ligands of those groups coordinate the divalent cation ion (Ca^{2+} or Mg^{2+}). Residue 12 is always a conserved Glu residue which donates two oxygen atoms from its carboxyl group to bind Ca^{2+} (Kawaski and Kretsinger, 1994).

The specificity and affinity of Ca^{2+}-binding sites are determined by several factors in addition to the primary structure of the loop, including the properties of the helices flanking the Ca^{2+}-binding loop, the helices located in the vicinity of the Ca^{2+}-binding loop and the interaction between neighboring loops.

Analysis of turkey Ca^{2+}-sTnC at high resolution revealed a dumbbell shaped molecule with its N and C-terminal domains well separated by a single central nine-turn α helix without direct interdomain interactions (Herzberg and James, 1985). The two Ca^{2+}-binding regions in both the N and C-terminal domain of sTnC are able to coordinate divalent cations. The numerous crystal structures of sTnC have all been carried out at acidic pH in the absence or presence of divalent cations such as Ca^{2+} (10 mM) (Mercola et al., 1975; Herzberg and James, 1985). The structure of TnC at physiologically neutral pH, would likely be somewhat different from that of TnC crystals prepared under acidic conditions since the affinity for Ca^{2+} of both high- and low-affinity sites of TnC is known to be greatly lowered (Hincke et al., 1978).

8. STNC-CATION COUPLING: BINDING AND ACTIVATION

Ca^{2+}-induced structural changes in troponin C have been investigated extensively (see Leavis et al., 1978). The conformational change between apo (the absence of Ca^{2+}) and the four Ca^{2+} ions bound state of sTnC is essentially the same as seen in CaM, and consists mainly of a reorganization of existing secondary structural elements in each domain. Ca^{2+}-binding to the high-affinity sites of sTnC increased the helical content by ~ 50%, attributed almost entirely to the C-terminal domain (Leavis et al., 1978). Mg^{2+} ions produced similar structural changes in the C-

A

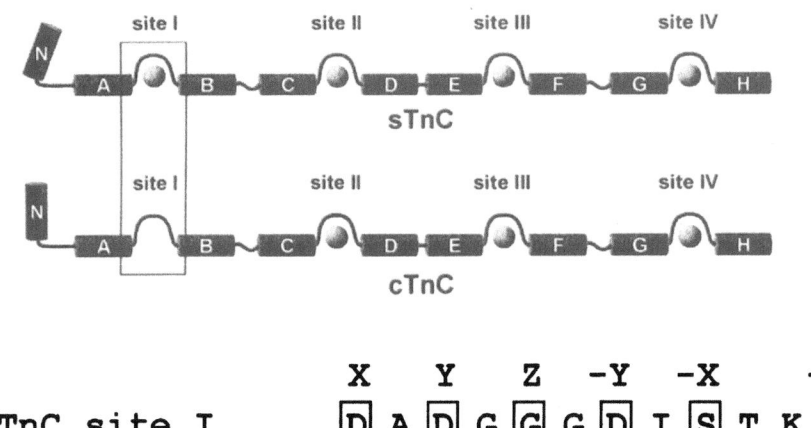

B

	X	Y	Z	-Y	-X	-Z
csTnC site I	D A D	G	G G D	I	S	T K E
bcTnC site I	V L G A	E	D G C	I	S	T K E

Figure 1. **Diagram showing the helices and Ca²⁺-binding sites of fast skeletal and cardiac** *TnC. A, schematic representation of the secondary structure of sTnC and cTnC. Helical regions are indicated by columns in gray and are termed N and A-H, beginning from the N-terminus. The Ca²⁺-binding sites, I-IV, are marked with balls within the corresponding loops. Note that site I of the cTnC isoform is unable to bind Ca²⁺. B, primary structures of chicken sTnC (csTnC) and bovine cTnC (bcTnC) site I. X, Y, Z, -Y, -X, and –Z indicate the position of specific residues which contribute to Ca²⁺ coordination of site I. There are differences in site I between sTnC and cTnC with an insertion of Val in the beginning of the loop and substitutions of key ligands relative to sTnC (Leu and Ala at the positions of X and Y in cTnC instead of Asp and Asp at the corresponding positions in sTnC).*

terminal domain of TnC. Binding of Ca^{2+} to the low-affinity sites causes a small change in the structure corresponding to the elongation of the α helix by only 1 or 2 residues (Nagy and Gergely, 1979). Mg^{2+} ions do not cause any significant change in the structure of low-affinity sites. The presence of Asp in position 3 of the third and fourth Ca^{2+}–binding loops accommodates the binding of Mg^{2+} in the C-terminal domain (Houdusse et al., 1997).

The four Ca^{2+}-binding sites of sTnC can be classified into two high affinity sites and two low affinity sites (Potter and Gergely, 1975). Investigation of the Ca^{2+}-binding properties of various proteolytic fragments of sTnC indicate that the C-terminal sites III and IV are the high affinity sites of sTnC. These high affinity sites bind Ca^{2+} with an affinity of $K_a=2.1\times10^7$ M^{-1}, and also bind Mg^{2+} but ~ 4200-fold weaker ($K_a=5\times10^3$ M^{-1}) compared with that of Ca^{2+}, and are therefore referred to as Ca^{2+}-Mg^{2+} sites (Potter and Gergely, 1975). Note that the Ca^{2+}-binding affinity values in Table 1 are different from those above because those in the table are for Ca^{2+}-binding to the Tn complex and not the isolated TnC. sTnC within the Tn complex binds Ca^{2+} with moderate affinity ($5\times10^6M^{-1}$) and fast binding rate in the N-terminal region while the two C-terminal binding sites shows slow kinetics and high Ca^{2+} affinity ($5x10^8M^{-1}$) (Robertson et al., 1981). These sites I and II are called Ca^{2+}-specific sites because at relatively low Mg^{2+} concentrations, Ca^{2+}-binding to the TnC low-affinity sites is not affected by Mg^{2+}. In contrast to its C-terminal sites, III and IV, the Ca^{2+}-binding affinity of these sites I and II are weak ($K_a=3.2x10^5$ M^{-1}) (Potter and Gergely, 1975) and also the Mg^{2+}-binding affinity of these sites are significantly lower than that of Ca^{2+}-Mg^{2+} sites ($K_a=520$ M^{-1}) (Ogawa, 1985). In the complexes of TnI-TnC and TnC-TnI-TnT (Tn), all four TnC sites have the same affinity for Ca^{2+} in the presence of 2 mM Mg^{2+} due to the competitive binding of Mg^{2+} which reduces the binding constant of the high affinity Ca^{2+}-binding sites (Potter and Gergely, 1975). Since a change in free Mg^{2+} concentration by about 2 mM (which is higher than the physiological Mg^{2+} concentration) did not affect the Ca^{2+} sensitivity of myofibrillar ATPase activity, it was concluded that the low-affinity sites are related to the regulation of contraction (Potter and Gergely, 1975).

The release of Ca^{2+} from the low-affinity sites of TnC is much faster than that from the high-affinity sites (Johnson et al., 1979; Andersson et al., 1981). Johnson et al. (Johnson et al., 1979; Johnson et al., 1980) previously showed that Ca^{2+}-binding constants of both high- and low-affinity sites of TnC–TnI and Tn are about 10 times higher than that of isolated TnC in the absence of Mg^{2+}. Other labs have also shown that the affinity of TnC for Ca^{2+} is modified by the interaction with other components (Berchtold et al., 2000). These findings suggest that the Ca^{2+}-binding properties of TnC are modified by other endogenous proteins in the thin filament.

While intracellular Ca^{2+} changes in concentration from about 200nM to between 2-10μM during muscle contraction, the concentration of Mg^{2+} is constant in the mM range. Therefore, the C-terminal sites of TnC are probably occupied by divalent metal ions, either Ca^{2+} or Mg^{2+}, and remain saturated with these divalent cations throughout the contraction cycle. Structural differences between the Ca^{2+}-bound and Mg^{2+}-bound N-and C-terminal domains of sTnC were determined by multi-dimension nuclear magnetic resonance (NMR) (Tsuda et al., 1990). They found that Mg^{2+} binding to the N-terminal domain of sTnC did not induce a conformational change in the hydrophobic region of this domain.

Johnson and co-workers determined the Ca^{2+} off-rate of the high and low affinity sites of sTnC by stopped flow fluorometry (Johnson et al., 1979), and demonstrated that the rate of structural change of high affinity sites of sTnC following Ca^{2+} release from these sites was ~1 s^{-1}, so that the structural change occurring with Ca^{2+} exchange of the Ca^{2+}-Mg^{2+} sites would be too slow to be directly involved in the regulation of contraction. Furthermore, according to computer simulations of the time course of Ca^{2+}-binding to Tn, saturation of the Ca^{2+}-Mg^{2+} sites with Ca^{2+} changed very little in response to the Ca^{2+} transient in the presence of physiological Mg^{2+} concentrations (Robertson et al., 1981). Hence, the Ca^{2+}-Mg^{2+} exchange accompanied with the Ca^{2+} transition does not seem to influence the Ca^{2+} regulation of muscle contraction, although the physiological importance of this Ca^{2+}-Mg^{2+} exchange (including structural changes) is still unclear. Several studies have demonstrated that these sites are responsible for maintaining the proper protein conformation for activation by Ca^{2+}. For example, TnC molecules can be extracted from myofibrils by removal of Ca^{2+} or Mg^{2+} bound to these sites by using a Ca^{2+} chelator such as EDTA (ethylenediamine-tetraacetic acid) (Cox et al., 1981; Zot and Potter, 1982) or CDTA (*trans*-1, 2-cyclohexanediamine-N,N,N',N'-tetraacetic acid) (Morimoto and Ohtsuki, 1987). Moreover, a TnC mutant that had sites III and IV inactivated showed a decreased affinity for the thin filament without a major change in regulatory properties (Negele et al., 1992; Szczesna et al., 1996a). These results strongly suggest that the binding of divalent cations to the C-terminal sites is necessary to stabilize the Tn complex (especially the TnI-TnC binary complex). Since Ca^{2+} binding to sites III and IV causes large secondary structural changes (Nagy and Gergely, 1979), these conformational changes may be critical for interaction with TnI.

9. DIVALENT CATION-BINDING TO N-TERMINAL SITES

sTnC has two N-terminal metal-binding sites called low affinity or Ca^{2+}- specific sites (Potter and Gergely, 1975). Ca^{2+}-binding to and release from the N-terminal Ca^{2+}-specific sites on both sTnC and cTnC can take place on a time scale that is consistent with muscle activation and relaxation *in vivo* (Lowey et al., 1993b; Robertson et al., 1981). The crystal structure of sTnC (from turkey skeletal muscle) was first solved in 1985 (Herzberg and James, 1985; Sundaralingam et al., 1985). In this structure only the C-terminal metal-binding sites were occupied by Ca^{2+} or Mn^{2+} due to the crystallization conditions (low pH) utilized. Subsequently, the structure of Ca^{2+}-saturated regulatory N-domain was solved by nuclear magnetic resonance (NMR) and reported in 1995 (Gagne et al., 1995; Slupsky and Sykes, 1995), and as the model structure previously proposed by Herzberg et al., it was shown that helices B and C had moved away from helices A and D compared to the structure of the apo N-domain when Ca^{2+} was bound to these N-terminal metal-binding sites. This movement may cause exposure of a hydrophobic patch around their helices where the N-terminal of sTnC can interact with the inhibitory region of TnI, neutralizing the inhibitory action of TnI. However, until the structure of the Tn complex in the presence or absence of Ca^{2+} is resolved the role of these different helices in TnC will not be fully understood.

Studies on chicken sTnC (CsTnC) mutations in which site I (VG1) or site II (VG2) have been inactivated by amino acids substitution, have shown that Ca^{2+}-binding to both of the Ca^{2+} specific sites of sTnC is necessary for full activation of force development in skinned fiber studies (Sheng et al., 1990). Since the steepness of pCa-force relationship curves of skinned fibers reconstituted with these sTnC mutants were much less than that of CsTnC or bovine cTnC (BcTnC)-incorporated fibers, it is likely that there are cooperative interactions between site I and site II in sTnC. Interestingly, site II-inactivated mutant (VG2) with Ca^{2+} bound to site I was more efficient in restoring force to muscle fibers than site I-inactivated mutant (VG1) with Ca^{2+} bound to site II.

10. REGULATION OF SKELETAL MUSCLE CONTRACTION AT THE LEVEL OF THIN FILAMENT VIA TROPONIN AND TROPOMYOSIN BY CA^{2+}

Contraction of skeletal muscle is regulated by Ca^{2+} at the level of the thin filament via Tn and Tm. Strong cross-bridge binding is also involved in the activation of the thin filament. The movement of myosin and actin filaments is controlled by several regulatory proteins and their position is supported by cytoskeletal proteins (Zot and Potter, 1987).

The interaction between actin and myosin is suppressed by Tm and Tn in the absence of Ca^{2+}. When Ca^{2+} concentration increases, and TnC binds Ca^{2+}, this suppression is removed and the contractile actin-myosin interaction is activated. Interactions of both the N and C terminal domains of TnC are necessary for the full regulatory activity of this protein. Both sites I and II in sTnC are known to be important for the regulatory activity of sTnC since inactivation of either of these sites results in significant decreases in the regulatory activity of the sTnC (Sheng et al., 1990). The action of Ca^{2+} on Tn is transmitted to actin molecules through Tm. TnI alone is capable of inhibiting the contractile interaction between actin and myosin in the presence or absence of Tm. This inhibition by TnI is removed by TnC independent of the Ca^{2+} ion concentration. However, for Ca^{2+} regulation of contraction, the presence of TnT is required. TnT is a highly charged 30kDa protein which has the Tm-binding component necessary for the Ca^{2+} regulation of contractile interaction. TnT interacts with Tm, TnI, TnC and possibly actin. TnT binds to the N-terminal half of TnC (Grabarek et al., 1981). TnI interacts with TnC at resting myoplasmic Ca^{2+} concentrations ($<0.2\mu M$) (Ca^{2+}-Mg^{2+} site dependent interaction), but this interaction is greatly strengthened in the presence of elevated Ca^{2+} concentrations ($>2\mu M$) (Ca^{2+} specific site dependent interaction). TnI is able to interact with TnC in the myofibril even at low Ca^{2+} concentrations, and may contribute to the stability of the Tn complex. TnC interacts with both TnI and TnT via multiple Ca^{2+}-dependent and Ca^{2+}-independent interactions (Potter et al., 1981). A model of sTnI in complex with sTnC consistent with available cross-linking and fluorescence data was recently presented by Tung et al., (Tung et al., 2000). In this model, the inhibitory region of TnI (residues 95-114) was modeled as a flexible β-hairpin which may be an important structural motif for transmitting the Ca^{2+} activated Tn regulatory signal to actin as the inhibitory region is similar to a β-hairpin region of the actin-binding protein profilin.

The major contractile protein myosin together with the major thin filament protein actin, transform the chemical energy of ATP in the myofibrillar system to the mechanical work in striated muscle. The globular head region of myosin, called subfragment 1 (S1), contains both the ATPase activity that provides the energy for contraction, and the actin binding site which is the site of cross-bridge formation. The arrangement and flexibility of the many myosin heads allow each thick filament to interact with several thin filaments. During ATP binding, ATP hydrolysis to ADP + P_i and the release of ADP and P_i, the myosin head undergoes different conformational changes. The myosin-ATP complex has a very low affinity for actin. As such, ATP binding to myosin promotes its dissociation from the actin-myosin complex. Release of ADP from myosin results in tight binding to actin (~10,000 fold increase in affinity) and the movement of the upper portion of the myosin head and the thin filament away from the Z line (called the power stroke). The Z lines are the perpendicular bands that separate the longitudinal arrays of thick and thin filaments in the sarcomere. The release of inorganic phosphate (P_i) from the quaternary actin-myosin.ADP.P_i complex plays a pivotal role in force generation. ATP cleavage to ADP + P_i during contraction is readily resynthesized during the metabolism of the energy from the energy store glycogen. Glycogen is broken down into glucose which stimulates glycolysis, the citric acid cycle and oxidative phosphorylation pathways. The relative contributions of the latter processes to the generation of ATP depend on the muscle type. Vertebrate muscle contains a reservoir of another energy store called phosphocreatine, which is able to transfer its phosphoryl group to ADP thereby replenishing ATP concentrations by creatine kinase. Phosphocreatine is the major source of phosphoryl groups during short periods of muscular exertion.

Ca^{2+}-binding to Tn and the CaM subunit of phosphorylase kinase triggers contraction and glycogen breakdown simultaneously. Therefore, besides contraction, the intracellular Ca^{2+} concentration also coordinates metabolism.

Actin-myosin interaction is mediated by intermolecular forces exerted at the actin-myosin interface. The relatively loose interaction of Tm with actin allows it to readily alter its conformation in response to changes in the Ca^{2+} concentration. Each Tm molecule interacts with one Tn molecule and seven actin monomers helping to stabilize the thin filament (Zot and Potter, 1987). At low Ca^{2+} levels the regulatory proteins bound to the actin filaments (Tn and Tm) inhibit the ATP-driven actomyosin cycle during relaxation by either blocking the attachment of myosin cross-bridges to actin or subsequently reactions of the bound cross-bridges. X-ray diffraction analysis showed that in the relaxed state some crossbridges remain attached, possibly important in maintaining resting tone.

11. CARDIAC MUSCLE

11.1. Size, Structural Properties and Location of cTnC

Although the primary structure of cTnC is 70 % identical to that of sTnC, there are significant differences in the first 40 amino acid residues. The regulatory N-terminal domain of Ca^{2+} saturated cTnC is more compact than the N-terminal domain of Ca^{2+} saturated sTnC. cTnC also forms a dumbbell-shaped molecule with separate N- and C-terminal domains connected by a flexible central linker (Sia et al., 1997). The C-terminal of cTnC is structurally similar to the C-terminal domain of sTnC.

These structural differences between cTnC and sTnC results from inactivation of site I where insertion of Val in the beginning of the loop of site I and substitutions of key ligands, Leu and Ala at the positions of X and Y in cTnC (see Fig. 1) instead of Asp and Asp at the corresponding positions in sTnC (van Eerd and Takahashi, 1975).

12. CTNC-CATION COUPLING: BINDING AND ACTIVATION

The amino acid sequence of bovine cardiac TnC was first reported by Van Eerd & Takahashi in 1975, and they proposed that cTnC would bind only three mol of Ca^{2+}/ mol since Ca^{2+}-binding site I was thought to be eliminated due to amino acid substitution which are located in the important region for Ca^{2+} coordination (van Eerd and Takahashi, 1975). This was later confirmed and the three Ca^{2+}-binding sites were characterized as two high affinity Ca^{2+}-Mg^{2+} sites and one low affinity Ca^{2+}-binding site (Holroyde et al., 1980). The first cation-binding site of cTnC (site I) contains an insertion and nonconservative replacements disrupting its ability to bind Ca^{2+}. The second Ca^{2+}-binding site of cTnC, the functional N-terminal metal-binding site II, has an affinity and specificity similar to the Ca^{2+}-binding site II of sTnC and binds Ca^{2+} weakly but specifically with a K_a of ~5x10^5 M^{-1} (Johnson et al., 1980). Recently, the NMR structures of the Ca^{2+}-saturated N-terminal cTnC fragment and the whole cTnC have been reported (Spyracopoulos et al., 1997; Sia et al., 1997). Surprisingly, the regulatory domain of these cTnCs exist in a "closed" conformation even in the Ca^{2+}-bound state, in contrast to the Ca^{2+}-saturated N-terminal of sTnC which exists in an "open" conformation. This "open" conformation had been thought to be important for interaction with TnI.

13. DIVALENT CATION-BINDING TO N-TERMINAL DOMAIN METAL-BINDING SITES

Like sTnC, cTnC has two classes of divalent cation-binding sites (described above). The NMR structure of a site I-inactivated mutant of N-terminal sTnC fragment with Glu41 mutated to Ala (Glu41Ala) with Ca^{2+} bound at site II was solved, resulting in a structure in which site I remained in a closed conformation upon Ca^{2+}-binding, similar to that observed for the Ca^{2+}-saturated structure of cTnC (Li et al., 1997). However, this mutant has less ability to activate muscle contraction (Pearlstone et al., 2000). On the other hand, a chimeric sTnC mutant which has sites II-IV from sTnC and the N-terminal portion (including site I) from cTnC produces an activated Ca^{2+}-sensitive force generation in skinned fibers similar to cTnC or sTnC. However, other chimeric sTnC mutants containing site I (A, B helices and intermediate loop) from cTnC, or the N-terminal side of site I from cTnC (not including site I) plus the inactivated-site I of sTnC, are not capable of initiating Ca^{2+}-sensitive force generation in skinned fibers (Gulati et al., 1992). These results suggest that the N-terminal portion including site I of cTnC, may play an important role for regulation of cardiac muscle contraction, although site I of cTnC lacks the ability to bind Ca^{2+} and an "open" conformation has not been observed in isolated cTnC even upon binding Ca^{2+}.

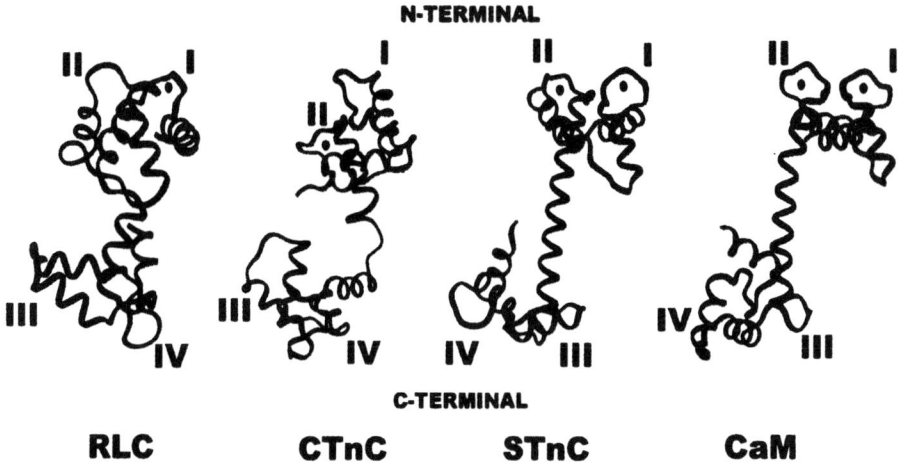

N-TERMINAL

C-TERMINAL

RLC **CTnC** **STnC** **CaM**

Figure 2. Tertiary structures of RLC, cTnC, sTnC and CaM. These structures were prepared using Insight II from the PDB database; 1ALM (RLC), 1AJ4 (cTnC), 2TN4 (sTnC) and 1CLM (CaM). The EF-hand sites I-IV in RLC, cTnC, sTnC and CaM are labeled. For simplicity Ca^{2+} (represented by a black circle within the Ca^{2+}-binding loops) is only shown bound to the N-terminal sites of these Ca^{2+}-binding proteins. Note that in cTnC site I is unable to bind Ca^{2+}. The differences between the central helix of sTnC and cTnC shown above could be due to the different methods utilized for structure determination. sTnC and CaM were determined by X-ray crystallography while cTnC was determined by NMR. The structure of the human RLC shown above is a theoretical model based on X-ray structures of RLC from other species.

One very important parameter determining muscle contractility is the number of Ca^{2+}-binding sites on the TnC molecule that become occupied by Ca^{2+}. Contractility is affected by changing the Ca^{2+} sensitivity of TnC. As Ca^{2+} is released from the SR it delivers enough Ca^{2+} to the contractile apparatus to cause less than 25% of the full activation possible (Fabiato, 1981; Solaro et al., 1974). As such, it is possible for subtle variations in the Ca^{2+} concentration to affect the rate of crossbridge formation (which does not seem to be the case in skeletal muscle). A non-diseased heart is never maximally stimulated allowing for increases as well as decreases in contractility according to the variable hemodynamic demands of circulating blood.

14. DIVALENT CATION-BINDING TO C-DOMAIN METAL-BINDING SITES

The C-terminal domain of cTnC is highly homologues to that of sTnC, with the high affinity Ca^{2+}-Mg^{2+} binding sites of cTnC being located in the C-terminal metal-binding sites III and IV. The Ca^{2+} binding constants of these sites of bovine cTnC were determined by using equilibrium dialysis method ($K_a=1 \times 10^7$ M^{-1}). These sites also bind Mg^{2+} competitively with a binding constant of 0.7×10^3 M^{-1} (Holroyde et al., 1980). While it seems that cation binding to one site of the C-terminal domain of cTnC is sufficient for TnC to bind to the thin filament (Negele et al., 1992), both C-terminal sites are needed for binding of sTnC to the thin filament (Sorenson et al., 1995). Like the C-terminal Ca^{2+}-Mg^{2+} site of sTnC, kinetic evidence suggests that the C-terminal sites III and IV of cTnC might be responsible for maintaining the proper protein conformation for activation by Ca^{2+} (Robertson et al., 1981) and for interaction of TnC with TnI.

15. CA^{2+}-BINDING PROPERTIES OF CTNC WITHIN THE TN COMPLEX AND MYOFILAMENTS

Both the Ca^{2+}-binding affinity of the Ca^{2+} specific and Ca^{2+}-Mg^{2+} sites in cTnC are increased about 10-fold when the Tn complex is formed (K_a=2x10^6 M^{-1} for Ca^{2+} specific site, K_a=3x10^8 M^{-1} for Ca^{2+}-Mg^{2+} sites) (Holroyde et al., 1980). In the myofilament, the cooperativity of Ca^{2+}-binding to sTnC is enhanced by cycling cross-bridges, reflecting cooperative Ca^{2+}-sensitive ATPase activity of fast skeletal myofibrils (Grabarek et al., 1983; Morimoto, 1991). On the other hand, the Ca^{2+} sensitive ATPase activity of cardiac myofibrils has shown less cooperativity than that of fast skeletal myofibrils. While sTnC-incorporated fast skeletal and cardiac myofilaments showed high cooperativity on both Ca^{2+} sensitive ATPase activity of myofibrils and force generation of skinned fibers, cTnC-incorporated fast skeletal and cardiac myofilaments showed low cooperativity compared to that of sTnC-incorporated myofilaments (Morimoto and Ohtsuki, 1994b). These results suggest that the difference in cooperativity between fast skeletal and cardiac muscle contraction is caused by differences within sTnC and cTnC. Since one of the regulatory sites, site I in cTnC lacks the ability to bind Ca^{2+}, cooperative Ca^{2+}-binding to the regulatory site in cTnC would not be induced by cycling cross-bridge as in the case of sTnC where cooperative interactions on Ca^{2+}-binding may occur between site I and site II (Sheng et al., 1990).

16. REGULATION OF CARDIAC MUSCLE CONTRACTION AT THE LEVEL OF THIN FILAMENT VIA TROPONIN AND TROPOMYOSIN BY CA^{2+}

In cardiac muscle, the regulation of the intracellular Ca^{2+} concentration consumes ~25% of the total energy output of the myocyte (Langer, 1992) and the signal of Ca^{2+}-binding to cTnC is conducted basically via the same interaction between myosin and actin as described for fast skeletal muscle (see sTnC section). However, there are significant isoform differences in the regulatory proteins between fast skeletal and cardiac muscle and these isoform differences would cause varied regulation in different muscle types. cTnC, which shows significant differences from sTnC (Fig. 1) in its N-terminal site I region is still able to regulate the interaction between actin and myosin. This means that site II plays an especially important role in the regulation of actomyosin interaction in cardiac muscle. The isoform differences for the other sarcomeric regulatory proteins (TnI, TnT and Tm) probably help to compensate for some the differences between sTnC and cTnC. The interaction between cTnI and cTnC is weaker than the corresponding interaction between sTnI and sTnC (Liao et al., 1994). This could be due to differences between the sites of interaction for the Tn subunits in skeletal and cardiac muscle. It could also be due to different conformational changes in cTnI and cTnC to allow the opening of site I in the N-terminal region of cTnI.

In striated muscles, there are two major Tm isoforms called α and β-Tm. In small animals, cardiac Tm is predominantly α -Tm, while an equal mixture of α and β-Tm has been found in cardiac muscles from large animals. Cardiac Tn components are also different from fast skeletal Tn components. cTnT is longer than sTnT due to an elongated amino acid sequence in the hypervariable region near the N-terminus (Leszyk et al., 1987).

The Ca^{2+} sensitivity of myofilaments changes as a function of sarcomere length (Endo, 1972). Changes in the sarcomere length of cardiac muscle myofilaments results in dynamically different contraction, therefore the length dependence of Ca^{2+} sensitivity of cardiac muscle is one of the most important physiological properties of cardiac muscle contractility (Wannenburg et al., 2000). This sarcomere length-Ca^{2+} sensitivity relationship is a critical factor in regulating force on a regular beat-to-beat basis. This sarcomere length dependence of Ca^{2+} sensitivity may be caused by changes in the interfilament lattice spacing rather than changing in the sarcomere length per se (Fuchs and Wang, 1996).

On a longer term basis, β-adrenergic stimulation leads to positive inotropic and chronotropic effects in the heart, which are mediated through the cAMP-dependent protein kinase (PKA) system. The cTnI isoform (which is longer than the sTnI isoform), has two serine residues in its extended N-terminal region which are phosphorylated by PKA after β-adrenergic stimulation (Robertson et al., 1982; Zhang et al., 1995). This phosphorylation of cTnI molecule by PKA modulates cardiac muscle contractility.

As discussed above, cTnC is different in its structural, and functional properties from those of sTnC. Although all of these cardiac specific regulatory protein isoforms could be necessary for efficient physiological heart functions, detail molecular mechanism of Ca^{2+} regulation of cardiac muscle contraction by cardiac regulatory proteins is still unclear compared to skeletal muscle.

17. VARIATIONS IN HEART AND SKELETAL MUSCLE SYSTEMS

The basic sequence of events that leads to the activation of contraction in cardiac muscle is similar to that of skeletal muscle. In both muscles, the depolarization of the sarcolemma induces the release of Ca^{2+} from the SR into the myoplasm. Binding of the released Ca^{2+} to troponin activates the actomyosin interaction resulting in muscle contraction. In skeletal muscle this mechanism is essentially an "all or none" response, that is, the fiber is either turned on (active) or turned off (relaxed). In contrast, the twitch of a cardiac muscle fiber can be strongly influenced by a number of external factors and by the recent history of the muscle fiber itself. This variability is due to additional control mechanisms that operate in the cardiac fiber and to variation in the amount of Ca^{2+} released into the myoplasm during a twitch. Some of these additional control mechanisms in cardiac muscle are discussed below.

The contractile machinery in cardiac muscle is switched on and off in a regular manner by the rise and fall of intracellular free Ca^{2+} allowing the heart to contract and relax rhythmically without fatigue throughout life. The force or "contractility" increases as the Ca^{2+} released into the myoplasm increases. The SR in the cardiac muscle is much less abundant than in skeletal muscle (Lamb, 2000). As a result, the amount of Ca^{2+} that can be released or taken up by the cardiac SR is small (0.35nmol/ms/g tissue, (Levitsky et al., 1981)). This amount of Ca^{2+}, which is much smaller than the total cellular Ca^{2+}, is only sufficient to induce one twitch contraction after which most of the Ca^{2+} is re-accumulated by the SR at a rate corresponding to the speed of relaxation. In cardiac muscle CaM is a receptor for Ca^{2+}, which is able to bind 4 Ca^{2+} ions and activate several enzymes including a Ca^{2+}/CaM-dependent protein kinase found in the SR which then phosphorylates the small regulatory protein phospholamban (Pifl et al., 1984). Recent evidence suggests a critical role for SR bound CaM in controlling the velocity of Ca^{2+} pumping in native cardiac SR (Xu and Narayanan, 2000). Phospholamban is a 52 residue membrane (SR) protein that controls the SR Ca^{2+} pump by direct physical interaction. Dephosphorylated phospholamban suppresses the Ca^{2+} pump activity while phosphorylated phospholamban increases the activity of this pump. As a result phospholamban controls Ca^{2+} flux to the SR (Pan et al., 1999).

The Ca^{2+} transients that occur during the activation of cardiac muscle are also more complex than those in skeletal muscle. In skeletal muscle the Ca^{2+} that activates the contraction is released primarily from the SR and the SR releases a fairly constant amount of Ca^{2+} that is sufficient to fully bind the sites of sTnC. Although the release of Ca^{2+} from the SR in cardiac muscle plays a role in the activation of the contraction a significant portion of the Ca^{2+} that activates this muscle comes in directly from the extracellular space through the sarcolemma. Due to the small size of myocytes Ca^{2+} can diffuse through the interior of the cell with sufficient speed to activate the myocyte. In contrast to cardiac muscle, the contraction of skeletal muscle *in vitro* is not influenced by the level of Ca^{2+} in the external medium, and skeletal muscles can remain active in Ca^{2+} free solutions for hours.

Other variations between heart and skeletal contractile systems include differences in the thick and thin filament components. Actin, exists as a variety of highly conserved isoforms whose distribution is tissue-specific in vertebrates (Khaitlina, 2001). Actin isoforms cannot substitute for each other, and synthesis of exogenous actins leads to alterations in cell morphology. Differences between other cardiac and skeletal sarcomeric protein isoforms are discussed in their relevant sections.

18. EXCITATION-CONTRACTION COUPLING (E-C COUPLING)

In cardiac muscle, when the membrane depolarization reaches the end of the transverse tubule, ionized Ca^{2+} is released from the T-tubule. Ca^{2+} inflow through

DHP Ca^{2+} channels (L-type) in the T-tubule act as signal molecules, stimulating the RyR2 to release the Ca^{2+} stored in the cisternae of the SR into the sarcoplasm enabling muscle contraction. This Ca^{2+} induced Ca^{2+} release (CICR) only occurs in cardiac muscle. This is in contrast to what happens in skeletal muscle where the very little Ca^{2+} influx through these Ca^{2+} channels is not involved in activating SR Ca^{2+} release (Rios and Brum, 1987; Schneider, 1994). The main difference between the mechanism of excitation-contraction coupling in cardiac and skeletal muscle is due to distinct types and amounts of DHP receptors and ryanodine receptors found in these muscles. In cardiac muscle the DHP receptors are also Ca^{2+} channels while in skeletal muscle the DHP receptors are voltage sensors. A direct coupling occurs between the skeletal DHP receptor and the Ca^{2+} channel (RyR1) of the SR occurs which allows changes in the skeletal DHP receptor to be translated to the SR. No such coupling occurs between the cardiac DHP receptors and RyR2. The number of ryanodine receptors significantly outnumbers the number of DHP receptors in cardiac ventricular cells (Wibo et al., 1991) in contrast to skeletal muscle where the DHP receptors have a role as voltage-sensors. The SR ATP-driven Ca^{2+} pump in skeletal muscle is present at such high concentrations that only 2-3 pump cycles are needed to re-sequester all the Ca^{2+} released during muscle excitation (lowering myoplasmic Ca^{2+} concentration to <1μM). Each Ca^{2+} ATPase pump transports two Ca^{2+} ions per molecule of ATP hydrolyzed resulting in the build up of high Ca^{2+} concentration gradients.

19. STRUCTURE AND FUNCTION OF THE REGULATORY LIGHT CHAIN

Cardiac myosin is a 480kDa hexameric protein with two heavy chains and four light chains. The heavy chains of myosin (MHC) are coiled-coil at ~4/5 of the length of the molecule forming a filamentous rod and are globular at their N-termini forming so called heads (S1). Each head (which is the site of biological activity in myosin) contains one regulatory light chain (RLC) and one essential light chain (ELC). In the normal heart α-MHC is preferentially expressed in the atrium while the β-MHC is almost exclusively expressed in the ventricle (Bouvagnet et al., 1984). In the hypertrophied atrium considerable amounts of β-MHC are expressed resulting in a decrease in the maximal shortening velocity of this tissue (Morano, 1999). α-MHC confers a higher ATPase activity and higher shortening velocity to the heart than β-MHC (Morano, 1999).

RLC is also called myosin light chain 2 (MLC2) or 5,5′-dithio-2-nitrobenzoic acid (DTNB) light chain. ELC is also called myosin light chain 1 (MLC1) or alkaline light chain. The regulatory and essential light chains are collectively called myosin light chains (MLCs). X-ray crystallographic analyses demonstrated that both RLC and ELC are associated with the 8.5nm long α-helical region of the myosin head (Rayment et al., 1993a; Rayment et al., 1993b) and together these light chains are responsible for stabilizing this neck region. ELC and RLC both contain 4 putative EF-hand domains and belong to the CaM superfamily (Kawaski and Kretsinger, 1994). While ELC from myosin in invertebrates such as scallop is able to bind one Ca^{2+}, the 4 EF-hands of the vertebrate ELC have all lost their ability to bind Ca^{2+} during the course of evolution. About 1/3 of the known EF-hand domains of Ca^{2+}-binding proteins do not bind Ca^{2+} while most of the EF-hands that bind Ca^{2+} also binds Mg^{2+} but with significantly lower affinity (Kawaski and Kretsinger, 1994). Unlike other EF-hand Ca^{2+}-binding proteins, RLC contains only one Ca^{2+}-binding site, located between amino acids 37-48 in human cardiac RLC (HcRLC) and between amino acids 39-50 in rabbit skeletal RLC. The tertiary structure of RLC is shown in Fig. 2. The Ca^{2+} binding site of RLC occurs at site I and also binds Mg^{2+}.

20. KINETICS OF CATION BINDING TO RLC

Little is known about the kinetics of Ca^{2+}-binding to LCs and even less is known about the kinetics of Mg^{2+} binding to LCs. Isolated HcRLC binds Ca^{2+} with relatively low affinity, $K_{Ca} = 6.7 \times 10^5$ M^{-1} (Szczesna et al., 2001). Phosphorylation of this protein with Ca^{2+}-calmodulin activated myosin light chain kinase (MLCK) decreased its Ca^{2+}-binding affinity 7.4-fold ($K_{Ca} = 0.9 \times 10^5$ M^{-1}) (Szczesna et al., 2001). Rabbit skeletal RLC (RsRLC) also bind Ca^{2+} with relatively low affinity, $K_{Ca} = 2.5 \times 10^5$ M^{-1} (Alexis and Gratzer, 1978). Mg^{2+} also binds to this protein competively, but ~200 fold weaker than Ca^{2+} ($K_{Mg} = 1.2 \times 10^3$ M^{-1}). Phosphorylation of SRLC resulted in an ~5 fold reduction in the Ca^{2+}-binding affinity ($K_{Ca} \sim 5 \times 10^4$ M^{-1}) while the Mg^{2+} binding affinity remained the same (Alexis and Gratzer, 1978).

In myosin, the Ca^{2+}-binding affinity of sRLC and cRLC are increased when compared to the RLCs alone. Both types of striated muscle myosins (skeletal and cardiac) bound Ca^{2+} with similar affinities of 3×10^7 M^{-1} (Holroyde et al., 1979). In the presence of 3×10^{-4} M Mg^{2+} the myosins bound Ca^{2+} with a reduced affinity of 3.5×10^5 M^{-1} (Holroyde et al., 1979). Assuming competition between Mg^{2+} and Ca^{2+} for this Ca^{2+}- binding site on myosin, the Mg^{2+} affinity for sRLC and cRLC were calculated to be $\sim 2.75 \times 10^5$ M^{-1} (Holroyde et al., 1979).

21. PHYSIOLOGICAL FUNCTION OF MYOSIN LIGHT CHAINS

In rabbit skeletal myosin, the site I of RLC bind Ca^{2+} with an affinity of $\sim 10^5 M^{-1}$ in the presence of mM Mg^{2+} (Watterson et al., 1979). Since the rate constant for Mg^{2+} dissociation (off-rate) from RLC is low (0.05-$0.06s^{-1}$), the Ca^{2+} to Mg^{2+} exchange at this RLC site would take tens of seconds to complete and would be too slow to contribute to the rapid initiation of contraction in striated muscle and is thought to have bound Mg^{2+} during contraction (Bagshaw and Reed, 1977). Therefore, it was suggested that myosin light chain phosphorylation in skeletal muscle is not an obligatory event in the production of force, and any role it does play must involve a modulation of the contractile interaction.

In 1978, Lehman (Lehman, 1978) observed that the ATPase activity of myosin in solution was substantially increased by the addition of Ca^{2+} and suggested that there was a thick filament-linked Ca^{2+} regulation in vertebrate striated muscle. Several more recent biochemical and physiological studies have supported Lehman's suggestion (Diffee et al., 1995; Lowey et al., 1993b; Hofmann et al., 1990). Diffee et al., (Diffee et al., 1995) found that skeletal muscle containing RsRLC that had significantly less affinity for divalent cations (due to an Asp to Ala mutation, D47A, in the Ca^{2+}/Mg^{2+} binding loop) had significantly reduced maximum tension (at pCa 4.5) compared with pre-exchange tension, and the amount of decrease in tension was directly related to the extent of D47A exchange. Lowey et al. (Lowey et al., 1993b) showed that the sliding velocity of actin in an *in vitro* motility assay was reduced in the presence of RLC-deficient myosin. Removal of both ELC and RLC from the chicken skeletal muscle myosin reduced the velocity of actin filament movement by 90% without significant loss of the myosin ATPase activity in an *in vitro* motility assay (Lowey et al., 1993a). In skinned muscle fibers, extraction of up to 50% of the endogenous RLC has been shown to decrease the maximal shortening velocity (V_{max}) (Hofmann et al., 1990) while the partial extraction of RLC from skinned skeletal muscle fibers modulated the Ca^{2+} sensitivity of the rate constant of force redevelopment (Metzger and Moss, 1992). Extraction of RLC from skinned fibers has also been shown to result in a decrease in the Ca^{2+} sensitivity of force development (Szczesna et al., 1996b).

In the mammalian heart, atrial (a)- and ventricular (v)-specific isoforms exist for both ELC and RLC. Unique functional roles for the cardiac muscle specific expression of MLC isoforms are not well understood, but in hyperthyroid rats, the cross-bridge kinetics of atrial fibers are faster than that of ventricular fibers, despite having the same MYHC composition (Bottinelli et al., 1995). These results suggest that the unique MLC isoform compositions of the different cardiac compartments play a role in modulating the cross-bridge kinetics of the ventricles and atria.

Mutations in either light chain (ELC or RLC) can lead to cardiovascular disease (Epstein, 1998; Poetter et al., 1996). A mutation in human cRLC (Glu22Lys) resulted in an unusual pattern of hypertrophy in transgenic mice, as well as changes at the myofilament and cellular levels, with the myofibrils showing increased Ca^{2+} sensitivity and significant deficits in relaxation in a transgene dose-dependent manner (Sanbe et al., 2000). These results all suggest that the RLC plays an important role in normal muscle function and in the manner in which force is generated during the actin-myosin cross-bridge cycle. *In vitro* studies have been able to reveal the structural basis for the lever-arm action of the light chain domain of the myosin motor during force development (Corrie et al., 1999). During contraction, the RLC binding domain of the myosin head undergoes repetitive conformational changes and may therefore play an active role during force generation in muscle (Corrie et al., 1999). The myosin light-chain domain is able to tilt and twist with respect to the actin filaments. Other results also suggest that RLC can control aspects of cross-bridge cycling and alter force development (Szczesna et al., 1996b).

22. ROLE OF REGULATORY LIGHT CHAIN (RLC) PHOSPHORYLATION

Smooth muscle contraction is initiated by the phosphorylation of RLC by the Ca^{2+}/CaM dependent myosin light chain kinase (MLCK). However, in striated muscle RLC phosphorylation does not activate muscle contraction but seems to play a modulatory role (Sweeney et al., 1993).

Early experiments to determine a physiological role for RLC relied on myosin ATPase activity measurements which do not change significantly after RLC phosphorylation. A recent report by our group showed that several RLC familial hypertrophic cardiomyopathy (FHC) mutants had significantly altered properties with respect to phosphorylation and Ca^{2+}-binding properties (Szczesna et al., 2001) and their effects varied depending on the location of the FHC mutation. An RLC mutant containing Arg58Gln FHC mutation had Ca^{2+}-binding completely eliminated. Phosphorylation of this RLC Arg58Gln mutant by MLCK restored its ability to bind Ca^{2+}. This is quite interesting since other EF hand Ca^{2+}-binding proteins, e.g. TnC and CaM contain the Gln residue (and not Arg) in the equivalent position in the helix C-terminal of the Ca^{2+}-binding site. The arginine residue of HcRLC however, is highly conserved across different species and a wide spectrum of other RLCs contains the R residue in this position. Another RLC FHC mutant, Glu22Lys could not be phosphorylated. Patients with such RLC FHC mutations show cardiac hypertrophy and in some cases abnormal ECGs (Flavigny et al., 1998).

Results from other labs also suggest that phosphorylation of the RLCs may have an important physiological role in the regulation of muscle contraction (Morano et al., 1985). The Ca^{2+} sensitivity of force development in skinned skeletal and cardiac muscle fibers increases after RLC phosphorylation (Morano et al., 1985; Szczesna et al., 1996b). Transgenic mice expressing nonphosphorylatable cRLC showed a spectrum of cardiovascular changes including atrial hypertrophy and dilatation (Sanbe et al., 1999). Structural effects of RLC phosphorylation have also been recently shown. In relaxed muscle the myosin heads are arranged close to the backbone (myosin tails) and upon activation of the muscle the heads move away from the backbone and binds actin. Electron microscopy revealed that phosphorylation of RLC caused the movement of myosin heads away from the backbone of isolated skeletal muscle thick filaments (Levine et al., 1998). This is most likely due to a change in charge in the N-terminus of RLC which allows increased myosin head mobility and increased calcium sensitivity to tension development (Levine et al., 1998). Hence, RLC phosphorylation may represent a thick filament regulatory system capable of downward modulation of actomyosin ATPase *in vivo* during maintained contraction in fast twitch muscle (Crow and Kushmerick, 1982). Podlubnaya et al, 1999 (Podlubnaya et al., 1999) demonstrated Ca^{2+}-induced reversible structural changes in synthetic cardiac and skeletal myosin filaments (with dephosphorylated RLCs) using different electron microscopic techniques. As such, the relationship between phosphorylation and Ca^{2+}-binding to RLC seems to play a key role in striated muscle contraction (Morano, 1999; Sanbe et al., 2000).

23. MODULATORS OF Ca^{2+} SENSITIVITY IN STRIATED MUSCLE

23.1. Direct Modulators of Ca^{2+} Sensitivity

23.1.1. PH Dependent Ca^{2+} Sensitivity of Contraction

In vertebrate striated muscle, contraction is inhibited under acidic conditions depending on muscle type in the order of cardiac > fast skeletal > slow skeletal, and the inhibition of muscle contraction due to acidosis is thought to occur during myocardial ischemia (Donaldson et al., 1978; Fabiato and Fabiato, 1978; Lynch and Williams, 1994). Although the contractile properties of striated muscles are affected by changes in several sarcomeric proteins interactions, such as myosin and actin interaction (Metzger and Moss, 1987), the intermolecular interactions of Tn components play an important role in determining the sensitivity of Ca^{2+}-activated muscle contraction to acidic pH (Metzger et al., 1993). It is known that developmental isoform switching of TnI (i.e. slow type to cardiac type) occurs in cardiac muscle and this isoform switching may be responsible for the difference reported for the Ca^{2+} sensitivity of cardiac muscle contraction between neonatal and adult cardiac muscles (Solaro et al., 1986; Solaro et al., 1988). The pH dependence of the Ca^{2+} binding affinity of isolated sTnC and cTnC has also been examined (Ogawa, 1985; Iida, 1988; Palmer and Kentish, 1994). The Ca^{2+} binding affinity of cTnC is decreased by reducing the pH from 7.0 to 6.2. This acidic pH-induced depression of Ca^{2+} sensitivity is greater than that of sTnC (Palmer and Kentish, 1994) where the Ca^{2+} binding affinity is not affected (Ogawa, 1985) or slightly decreased by acidic pH (Iida, 1988), although it has also been reported that the effect of acidic pH on the Ca^{2+} affinity of sTnC is enhanced when sTnC is complexed with TnI (el-Saleh and Solaro, 1988). Many studies have showed that isoform specific interactions between TnI and TnC appear to be critical for the pH sensitivity of muscle contraction (Ball et al., 1994; Ding et al., 1995; Parsons et al., 1997). When skinned cardiac muscle fibers are exchanged with different combinations of skeletal and cardiac TnI and TnC, it was found that cTnC conferred a higher sensitivity to acidic pH than sTnC on the Ca^{2+} dependent force generation independent of the TnI isoform used (Morimoto et al., 1999). Fibers which contained the slow skeletal TnI showed the greatest resistance to acidic pH when either TnC isoform was present. Parsons et al, 1997 utilizing fluorescently labeled cTnC and sTnC showed that the changes in Ca^{2+} sensitivity when the pH was lowered or raised are caused by a decrease or increase, respectively, in the Ca^{2+} affinity of the Ca^{2+}-specific site(s) of TnC. These results suggest that the difference in pH sensitivity between fast skeletal and cardiac muscle isoforms depend solely on the TnC isoform present while the pH sensitivity of slow skeletal muscle is determined more by the TnI isoform present.

23.1.2. Ca^{2+} Sensitizers

Both the Ca^{2+} sensitivity of muscle contraction and the Ca^{2+}-binding affinity of TnC are increased in the presence of certain chemicals, called Ca^{2+} sensitizers (Kurebayashi and Ogawa, 1988; Ogawa, 1985). It has been reported that two Ca^{2+} sensitizers, trifluoperazine (TFP) a known tranquilizer, and the calmodulin antagonist bepridil, bind to both of the N-terminal and C-terminal domains of isolated cTnC molecule noncovalently only in the presence of Ca^{2+} (Kleerekoper et al., 1998). This suggests that bound Ca^{2+} sensitizers alter the conformation of cTnC within the Tn complex and that conformational changes affect TnC-TnI interaction and increases both the Ca^{2+}-binding affinity of TnC and the Ca^{2+} sensitivity of muscle contraction.

23.2. Indirect Modulators – Tropomyosin and Troponin Isoform Composition, Protein Phosphorylation Level

Tn and Tm are the key regulatory proteins controlling striated muscle contraction. Tissue and species specific Tm differences occur with two well established Tm isoforms, α and β. The Tm molecule is made up of two identical (e.g. αα) or different polypeptides (e.g. αβ), called homo- or heterodimer respectively. Expression of β-Tm in the heart of transgenic mice resulted in an increase in Ca^{2+} sensitivity (Palmiter et al., 1996). Phosphorylation of α-Tm on Ser283 *in vitro* enhances actin activated ATPase activity without changing the Ca^{2+}-sensitivity (Heeley, 1994). At least 32 potential isoforms of TnT occur in chicken skeletal muscle of which at least five isoforms are expressed at the protein level (Perry, 1998). In vertebrates, several isoforms of cTnT have been found. In rat skeletal muscle as many as 128 potential isoforms of sTnT (differing in amino acid content) could exist due to several sTnT exons undergoing alternative splicing (Perry, 1998). In humans four isoforms have so far been reported (TnT1 to TnT4), with the TnT3 isoform predominating in adult human heart. Reexpression of the fetal cardiac isoform TnT4 occurs in some cases of end-stage heart failure (Anderson et al., 1995). In permeabilized rabbit heart muscle fibers a relationship between the force-pCa characteristics of the rabbit myocardium and the cTnT isoforms that it expresses (Nassar et al., 1991).

Sarcomeric protein phosphorylation is a significant modulator of Ca^{2+} sensitivity. The unique 32 residue N-terminal extension of cTnI, a region not found in sTnI, is phosphorylated mainly at serine residues 23 and 24 upon β-adrenergic stimulation and when both of these residues are phosphorylated the Ca^{2+} affinity of TnC decreases (Zhang et al., 1995). cTnI and cTnT are phosphorylated by several kinases *in vitro* (Solaro and Pan, 1989). Phosphorylation of cTnI by protein kinase C (PKC) decreases the maximal Ca^{2+} stimulated MgATPase activity in reconstituted actomyosin without significantly affecting the sensitivity of the reconstituted contractile system (Noland and Kuo, 1991). Like TnI, TnT has been shown to be phosphorylated at multiple sites by PKC leading to reduced Ca^{2+}-dependent actomyosin MgATPase activity (Jideama et al., 1996). Jideama et al. (Jideama et al., 1996) found that different PKC isoforms had distinct specificities with respect to BcTnT phosphorylation resulting in different functional consequences ranging from decreases in both Ca^{2+} sensitivity and maximal activity to slight increases in Ca^{2+} sensitivity without any effect on the maximal activity of MgATPase.

Differences in the level of cTnI phosphorylation between the normal (where cTnI is phosphorylated to a greater extent) and failing heart (Bodor et al., 1997) suggest that factors which affect protein phosphorylation would have functional consequences in the heart. In the last ten years several different pathologies besides heart failure have been found to affect the extent of cTnI phosphorylation. As such, there is no doubt that phosphorylation of cTnI plays an important role in the regulation of the contractile system in the heart. sTnI and sTnT are also known to be phosphorylated by PKC (Mazzei and Kuo, 1984) but the physiological significance of this is not yet well understood.

24. ALTERATIONS IN CA^{2+}-BINDING IN DISEASE STATES SUCH AS FHC

As discussed above, intermolecular interactions within the Tn complex are responsible for the Ca^{2+}-binding property of TnC. Therefore, impairment of any of these interactions would cause alterations in Ca^{2+}-binding to Tn or in the Ca^{2+} signal profile of muscle contraction. Recently, several mutations in genes encoding muscle regulatory proteins, HcTnT, HcTnI and α-Tm have been found in familial hypertrophic cardiomyopathy (FHC) (Kimura et al., 1997; Watkins' et al., 1995), which is an autosomal dominant disease associated with a high incidence of sudden cardiac death in young adults. FHC has also been shown to be caused by mutations in other major sarcomeric proteins including α-myosin heavy chain, myosin essential and regulatory light chains, myosin-binding protein C (Redwood et al., 1999), α-cardiac actin (Mogensen et al., 1999), and titin (Satoh et al., 1999). Several studies have been carried out to explore the functional consequences of these FHC mutations in the cTnT, cTnI, and Tm genes. Tobacman et al. (Tobacman et al., 1999) reported that one FHC-causing mutant, ΔGlu160 HcTnT (residue Glu160 is deleted) increased Ca^{2+} sensitivity of actomyosin S1 ATPase activity and that thin filament reconstituted with this ΔGlu160 mutation in BcTnT showed higher Ca^{2+}-binding affinity than that of wild-type BcTnT-included in the thin filament. Furthermore, it has been reported that this mutant potentiated the neutralizing action (i.e. releasing action from inhibition of TnI) of HcTnC, while the inhibitory action of HcTnI was not affected by this HcTnT mutant in the absence of HcTnC, implying that the apparent affinity of TnC for TnI is increased by this mutant HCTnT (Harada et al., 2000). These results suggest that this HCTnT mutant has altered intermolecular interaction within Tn complex or thin filament, and that leads to increased Ca^{2+}-binding affinity of TnC and Ca^{2+} sensitivity of muscle contraction. Several other studies have also reported that many FHC-causing TnT mutations (i.e. Ile79Asn, Arg92Gln, Phe110Ile, Glu163Lys, Glu244Asp, Arg278Cys, and splice donor site mutants) showed Ca^{2+} sensitizing effect on contractile interaction of muscle *in vitro* and *in vivo* (Elliott et al., 2000; Redwood et al., 1999; Szczesna et al., 2000; Takahashi-Yanaga et al., 2000; Knollmann et al., 2001; Miller et al., 2001). The Ca^{2+} sensitivity of muscle contraction has also been modulated (increased) by the FHC HcTm mutants, Asp175Asn and Glu180Gly (Bing et al., 2000). Since the inhibition of the stoichiometric amount of TnI is incomplete in the absence of TnT (which assists TnI in its inhibitory action through it's interaction with Tm) or Tm altered functions of TnT and Tm by FHC mutations would perturb inhibitory interaction between TnI and actin-Tm through TnT-TnI interaction resulting in abnormal (especially increase) Ca^{2+} sensitivity of muscle contraction. For the HcTnI gene, all FHC mutations identified to date are located in the C-terminal region of the cTnI molecule, including the inhibitory region that interacts with actin-Tm and inhibits cross-bridge interaction in the absence of Ca^{2+}. One FHC HcTnI mutation, Arg145Gly, is located in the inhibitory region of TnI and it has been reported that this mutant has less inhibitory ability compared to normal HcTnI molecule and also increased Ca^{2+} sensitivity of contractile properties of muscle contraction *in vitro* (Elliott et al., 2000; Lang et al., 1999; Takahashi-Yanaga et al., 2000). Elliott et al. (Elliott et al., 2000) have also reported another FHC HcTnI mutant Arg162Trp, which is located outside of inhibitory region but also may interact with N-terminal region of HcTnC, showed increased Ca^{2+} sensitivity of actomyosin ATPase activity and potentiated the binding affinity for HcTnC. Other FHC-causing TnI mutations (Arg145Gln, ΔLys183, Ser199Asn, Gly203Ser and Lys206Gln) have also been reported but little is known about the biophysical effects of these mutations (Hernandez et al., 2001). These findings indicate that these FHC HcTnI mutations, Arg145Gly and Arg162Trp, reduced inhibitory interaction between HcTnI and actin-Tm or increased regulatory interaction between HcTnI and HcTnC, and these alterations of intermolecular interaction within the regulatory proteins of the thin filament resulted in increased Ca^{2+} sensitivity of muscle contraction *in vitro*. The functional consequences of all the mutations in these regulatory protein genes, identified to date, are not yet reported and further studies are required in order to understand the detailed molecular properties of these mutations. These findings, however, suggest that an abnormal Ca^{2+} sensitivity, especially an increase in Ca^{2+} sensitivity, of *in vitro* muscle contraction caused by the mutations in these regulatory protein genes is responsible for the pathogenesis of FHC.

25. CONCLUDING REMARKS

The cations Mg^{2+} and Ca^{2+} are the major ions responsible for the regulation of striated muscle contraction. Ca^{2+} in particular plays a crucial role in the regulation of striated muscle contraction. This may be due to Ca^{2+} ions having a smaller charge density and hydration energy than Mg^{2+}, which results in faster binding and dissociation rates. The critical mechanism of Ca^{2+} regulation is the change in the steric relationship of Tm and the actin filament. An action potential leads to the release of Ca^{2+} from the SR resulting in the myoplasmic Ca^{2+} concentration being increased to $\sim 10 \mu M$. Ca^{2+}-binding to the N-terminal low affinity TnC sites activates the actomyosin-ATPase activity and this results in muscle contraction. The binding of TnT to Tm seems to be the point of integration of the physiological function of the Tn complex. The extent of activation depends on the free Ca^{2+} concentration and on other factors that affect the Ca^{2+} sensitivity of the thin filament. The Ca^{2+} sensitivity of the thin filament is affected by the troponin isoform composition of the thin filament, phosphorylation of certain sarcomeric proteins and Ca^{2+} sensitizing drugs that act on the sarcomeric proteins. In the absence of Ca^{2+}, the role of TnT is mainly to adjust the positioning of TnI relative to Tm-actin preventing the interaction of actin with myosin. When the free Ca^{2+} level falls, Ca^{2+} dissociates from TnC preventing the recruitment of new cross-bridges and also increases the rate of decay of the force generating cross-bridges (Hoskins et al., 1999). The myosin light chains lie at the base of the myosin head where they wrap around the α-helical neck region in a position where they are able to modulate the contractile activity of the head domain.

Both cardiac and skeletal muscle has evolved mechanisms for controlling intracellular Ca^{2+} levels, mainly by controlling the release and uptake by the SR. The SR plays a central role in the contraction-relaxation sequence in muscle by regulating the cytosolic Ca^{2+} concentration. It is the Ca^{2+} released from the SR that ultimately regulates force development by the contractile apparatus. The SR is able to rapidly fine-tune cytoplasmic Ca^{2+} back down to resting levels (100-200nM). Mitochondria seem to play only a minor role in the control of cytosolic Ca^{2+}. Modulation of Ca^{2+} activation by protein phosphorylation is more prominent in cardiac muscle than in skeletal muscle. Also the recruitment of crossbridges in cardiac muscle seems to occur by means of changes in the amount of Ca^{2+} delivered to the contractile apparatus. Although Mg^{2+} seems to be an important regulator of SR Ca^{2+} release *in vivo,* especially in skeletal muscle, the precise mechanism of its interaction with the channel protein remains to be explained.

It seems likely that the vertebrate striated muscle is dually regulated with the primary mode of regulation occurring at the thin filament level via Ca^{2+}-interacting with Tn. A secondary mode of regulation seems to occur at the thick filament level via phosphorylation and/or Ca^{2+}-binding of RLC.

Department of Molecular and Cellular Pharmacology
University of Miami School of Medicine
1600 N.W. 10th Ave.
Miami, FL 33136
305-243-5874
305-243-4555 (FAX)
e-mail: jdpotter@miami.edu

26. REFERENCES

Alexis, M. N., and Gratzer, W. B. (1978). Interaction of skeletal myosin light chains with calcium ions, Biochemistry *17*, 2319-25.

Anderson, P. A., Greig, A., Mark, T. M., Malouf, N. N., Oakeley, A. E., Ungerleider, R. M., Allen, P. D., and Kay, B. K. (1995). Molecular basis of human cardiac troponin T isoforms expressed in the developing, adult, and failing heart, Circ Res *76*, 681-6.

Andersson, T., Drakenberg, T., Forsen, S., and Thulin, E. (1981). A 43Ca NMR and 25Mg NMR study of rabbit skeletal muscle troponin C: exchange rates and binding constants, FEBS Lett *125*, 39-43.

Bagshaw, C. R., and Reed, G. H. (1977). The significance of the slow dissociation of divalent metal ions from myosin 'regulatory' light chains, FEBS Lett *81*, 386-90.

Ball, K. L., Johnson, M. D., and Solaro, R. J. (1994). Isoform specific interactions of troponin I and troponin C determine pH sensitivity of myofibrillar Ca^{2+} activation, Biochemistry *33*, 8464-71.

Berchtold, M. W., Brinkmeier, H., and Muntener, M. (2000). Calcium ion in skeletal muscle: its crucial role for muscle function, plasticity, and disease, Physiol Rev *80*, 1215-65.

Bing, W., Knott, A., Redwood, C., Esposito, G., Purcell, I., Watkins, H., and Marston, S. (2000). Effect of hypertrophic cardiomyopathy mutations in human cardiac muscle alpha -tropomyosin (Asp175Asn and Glu180Gly) on the regulatory properties of human cardiac troponin determined by in vitro motility assay, J Mol Cell Cardiol *32*, 1489-98.

Blanchard, H., Grochulski, P., Li, Y., Arthur, J. S., Davies, P. L., Elce, J. S., and Cygler, M. (1997). Structure of a calpain Ca^{2+}-binding domain reveals a novel EF-hand and Ca^{2+}-induced conformational changes [see comments], Nat Struct Biol *4*, 532-8.

Bodor, G. S., Oakeley, A. E., Allen, P. D., Crimmins, D. L., Ladenson, J. H., and Anderson, P. A. (1997). Troponin I phosphorylation in the normal and failing adult human heart, Circulation *96*, 1495-500.

Bottinelli, R., Canepari, M., Cappelli, V., and Reggiani, C. (1995). Maximum speed of shortening and ATPase activity in atrial and ventricular myocardia of hyperthyroid rats, Am J Physiol *269*, C785-90.

Bouvagnet, P., Leger, J., Pons, F., Dechesne, C., and Leger, J. J. (1984). Fiber types and myosin types in human atrial and ventricular myocardium. An anatomical description, Circ Res *55*, 794-804.

Carafoli, E. (1987). Intracellular calcium homeostasis, Annu Rev Biochem *56*, 395-433.

Corrie, J. E., Brandmeier, B. D., Ferguson, R. E., Trentham, D. R., Kendrick-Jones, J., Hopkins, S. C., van der Heide, U. A., Goldman, Y. E., Sabido-David, C., Dale, R. E., et al. (1999). Dynamic measurement of myosin light-chain-domain tilt and twist in muscle contraction, Nature *400*, 425-30.

Cox, J. A., Comte, M., and Stein, E. A. (1981). Calmodulin-free skeletal-muscle troponin C prepared in the absence of urea, Biochem J *195*, 205-11.

Cozens, B., and Reithmeier, R. A. (1984). Size and shape of rabbit skeletal muscle calsequestrin, J Biol Chem *259*, 6248-52.

Crow, M. T., and Kushmerick, M. J. (1982). Myosin light chain phosphorylation is associated with a decrease in the energy cost for contraction in fast twitch mouse muscle, J Biol Chem *257*, 2121-4.

Diaz-Munoz, M., Hamilton, S. L., Kaetzel, M. A., Hazarika, P., and Dedman, J. R. (1990). Modulation of Ca^{2+} release channel activity from sarcoplasmic reticulum by annexin VI (67-kDa calcimedin), J Biol Chem *265*, 15894-9.

Diffee, G. M., Greaser, M. L., Reinach, F. C., and Moss, R. L. (1995). Effects of a non-divalent cationbinding mutant of myosin regulatory light chain on tension generation in skinned skeletal muscle fibers, Biophys J *68*, 1443-52.

Ding, X. L., Akella, A. B., and Gulati, J. (1995). Contributions of troponin I and troponin C to the acidic pH-induced depression of contractile Ca^{2+} sensitivity in cardiotrabeculae, Biochemistry *34*, 2309-16.

Donaldson, S. K., Hermansen, L., and Bolles, L. (1978). Differential, direct effects of H+ on Ca^{2+}-activated force of skinned fibers from the soleus, cardiac and adductor magnus muscles of rabbits, Pflugers Arch *376*, 55-65.

Dunn, S. E., Burns, J. L., and Michel, R. N. (1999). Calcineurin is required for skeletal muscle hypertrophy, J Biol Chem *274*, 21908-12.

Dutt, P., Arthur, J. S., Grochulski, P., Cygler, M., and Elce, J. S. (2000). Roles of individual EF-hands in the activation of m-calpain by calcium, Biochem J *348*, 37-43.

Egger, M., and Niggli, E. (1999). Regulatory function of Na-Ca exchange in the heart: milestones and outlook, J Membr Biol *168*, 107-30.

Elliott, K., Watkins, H., and Redwood, C. S. (2000). Altered regulatory properties of human cardiac troponin I mutants that cause hypertrophic cardiomyopathy, J Biol Chem 275, 22069-74.

el-Saleh, S. C., and Solaro, R. J. (1988). Troponin I enhances acidic pH-induced depression of Ca^{2+} binding to the regulatory sites in skeletal troponin C, J Biol Chem 263, 3274-8.

Endo, M. (1972). Stretch-induced increase in activation of skinned muscle fibres by calcium, Nat New Biol. 237, 211-3.

Engelkamp, D., Schafer, B. W., Erne, P., and Heizmann, C. W. (1992). S100 alpha, CAPL, and CACY:molecular cloning and expression analysis of three calcium-binding proteins from human heart, Biochemistry 31, 10258-64.

Epstein, N. D. (1998). The molecular biology and pathophysiology of hypertrophic cardiomyopathy dueto mutations in the beta myosin heavy chains and the essential and regulatory light chains, Adv Exp Med Biol 453, 105-14; discussion 114-5.

Fabiato, A. (1981). Myoplasmic free calcium concentration reached during the twitch of an intact isolatedcardiac cell and during calcium-induced release of calcium from the sarcoplasmic reticulum of a skinned cardiac cell from the adult rat or rabbit ventricle, J Gen Physiol 78, 457-97.

Fabiato, A., and Fabiato, F. (1978). Effects of pH on the myofilaments and the sarcoplasmic reticulum of skinned cells from cardiace and skeletal muscles, J Physiol (Lond) 276, 233-55.

Fano, G., Marsili, V., Angelella, P., Aisa, M. C., Giambanco, I., and Donato, R. (1989). S-100a0 protein stimulates Ca^{2+}-induced Ca^{2+} release from isolated sarcoplasmic reticulum vesicles, FEBS Lett 255, 381-4.

Flavigny, J., Richard, P., Isnard, R., Carrier, L., Charron, P., Bonne, G., Forissier, J. F., Desnos, M., Dubourg, O., Komajda, M., et al. (1998). Identification of two novel mutations in the ventricular regulatory myosin light chain gene (MYL2) associated with familial and classical forms of hypertrophic cardiomyopathy, J Mol Med 76, 208-14.

Force, T., Rosenzweig, A., Choukroun, G., and Hajjar, R. (1999). Calcineurin inhibitors and cardiachypertrophy, Lancet 353, 1290-2.

Fuchs, F., and Wang, Y. P. (1996). Sarcomere length versus interfilament spacing as determinants of cardiac myofilament Ca^{2+} sensitivity and Ca^{2+} binding, J Mol Cell Cardiol 28, 1375-83.

Gagne, S. M., Tsuda, S., Li, M. X., Smillie, L. B., and Sykes, B. D. (1995). Structures of the troponin C regulatory domains in the apo and calcium-saturated states, Nat Struct Biol 2, 784-9.

Gilchrist, J. S., Wang, K. K., Katz, S., and Belcastro, A. N. (1992). Calcium-activated neutral protease effects upon skeletal muscle sarcoplasmic reticulum protein structure and calcium release, J Biol Chem 267, 20857-65.

Grabarek, Z., Drabikowski, W., Leavis, P. C., Rosenfeld, S. S., and Gergely, J. (1981). Proteolyticfragments of troponin C. Interactions with the other troponin subunits and biological activity, J Biol Chem 256, 13121-7.

Grabarek, Z., Grabarek, J., Leavis, P. C., and Gergely, J. (1983). Cooperative binding to the Ca^{2+}-specific sites of troponin C in regulated actin and actomyosin, J Biol Chem 258, 14098-102.

Guerini, D., Garcia-Martin, E., Zecca, A., Guidi, F., and Carafoli, E. (1998). The calcium pump of the plasma membrane: membrane targeting, calcium binding sites, tissue-specific isoform expression, Acta Physiol Scand Suppl 643, 265-73.

Gulati, J., Babu, A., and Su, H. (1992). Functional delineation of the $Ca^{(2+)}$-deficient EF-hand in cardiac muscle, with genetically engineered cardiac-skeletal chimeric troponin C, J Biol Chem 267, 25073-7.

Handy, R. D., Gow, I. F., Ellis, D., and Flatman, P. W. (1996). Na-dependent regulation of intracellular free magnesium concentration in isolated rat ventricular myocytes, J Mol Cell Cardiol 28, 1641-51.

Harada, K., Takahashi-Yanaga, F., Minakami, R., Morimoto, S., and Ohtsuki, I. (2000). Functional consequences of the deletion mutation deltaGlu160 in human cardiac troponin T, J Biochem (Tokyo) 127, 263-8.

Heeley, D. H. (1994). Investigation of the effects of phosphorylation of rabbit striated muscle alpha alpha-tropomyosin and rabbit skeletal muscle troponin-T, Eur J Biochem 221, 129-37.

Hernandez, O. M., Housmans, P. R., and Potter, J. D. (2001). Invited Review: Pathophysiology of cardiac muscle contraction and relaxation as a result of alterations in thin filament regulation, J. Appl. Physiol. 90, 1125-1136.

Herzberg, O., and James, M. N. (1985). Structure of the calcium regulatory muscle protein troponin-C at 2.8 A resolution, Nature 313, 653-9.

Herzog, A., Szegedi, C., Jona, I., Herberg, F. W., and Varsanyi, M. (2000). Surface plasmon resonance studies prove the interaction of skeletal muscle sarcoplasmic reticular Ca($2+$) release channel/ryanodine receptor with calsequestrin, FEBS Lett 472, 73-7.

Hincke, M. T., McCubbin, W. D., and Kay, C. M. (1978). Calcium-binding properties of cardiac and skeletal troponin C as determined by circular dichroism and ultraviolet difference spectroscopy, Can J Biochem 56, 384-95.

Hofmann, P. A., Metzger, J. M., Greaser, M. L., and Moss, R. L. (1990). Effects of partial extraction of light chain 2 on the Ca^{2+} sensitivities of isometric tension, stiffness, and velocity of shortening in skinned skeletal muscle fibers, J Gen Physiol 95, 477-98.

Holroyde, M. J., Potter, J. D., and Solaro, R. J. (1979). The calcium binding properties of phosphorylated and unphosphorylated cardiac and skeletal myosins, J Biol Chem 254, 6478-82.

Holroyde, M. J., Robertson, S. P., Johnson, J. D., Solaro, R. J., and Potter, J. D. (1980). The calcium and magnesium binding sites on cardiac troponin and their role in the regulation of myofibrillar adenosine triphosphatase, J Biol Chem 255, 11688-93.

Hoskins, B. K., Lipscomb, S., Mulligan, I. P., and Ashley, C. C. (1999). How do skinned skeletal muscle fibers relax?, Biochem Biophys Res Commun 254, 330-3.

Houdusse, A., Love, M. L., Dominguez, R., Grabarek, Z., and Cohen, C. (1997). Structures of four Ca^{2+}-bound troponin C at 2.0 A resolution: further insights into the Ca^{2+}-switch in the calmodulin superfamily, Structure 5, 1695-711.

Iida, S. (1988). Calcium binding to troponin C. II. A Ca^{2+} ion titration study with a Ca^{2+} ion sensitive electrode, J Biochem (Tokyo) 103, 482-6.

Ikemoto, N., Antoniu, B., Kang, J. J., Meszaros, L. G., and Ronjat, M. (1991). Intravesicular calcium transient during calcium release from sarcoplasmic reticulum, Biochemistry 30, 5230-7.

Jideama, N. M., Noland, T. A., Jr., Raynor, R. L., Blobe, G. C., Fabbro, D., Kazanietz, M. G., Blumberg, P.M., Hannun, Y. A., and Kuo, J. F. (1996). Phosphorylation specificities of protein kinase C isozymes for bovine cardiac troponin I and troponin T and sites within these proteins and regulation of myofilament properties, J Biol Chem 271, 23277-83.

Johnson, J. D., Charlton, S. C., and Potter, J. D. (1979). A fluorescence stopped flow analysis of Ca^{2+} exchange with troponin C, J Biol Chem 254, 3497-502.

Johnson, J. D., Collins, J. H., Robertson, S. P., and Potter, J. D. (1980). A fluorescent probe study of Ca^{2+} binding to the Ca^{2+}-specific sites of cardiac troponin and troponin C, J Biol Chem 255, 9635-40.

Johnson, J. D., Robinson, D. E., Robertson, S. P., Schwartz, A., and Potter, J. D. (1981). Ca^{2+} exchange with troponin and regulation of muscle contraction. In Regulation of Muscle Contraction: Excitation-contraction Coupling, A. Grinnel, ed. (Academic Press), pp. 241-259.

Jones, L. R., Suzuki, Y. J., Wang, W., Kobayashi, Y. M., Ramesh, V., Franzini-Armstrong, C., Cleemann, L., and Morad, M. (1998). Regulation of Ca^{2+} signaling in transgenic mouse cardiac myocytes overexpressing calsequestrin, J Clin Invest 101, 1385-93.

Kawaski, H., and Kretsinger, R. (1994). Calcium-binding Proteins 1: EF-hands, Vol 1, Academic Press.

Khaitlina, S. Y. (2001). Functional specificity of actin isoforms, Int Rev Cytol 202, 35-98.

Kimura, A., Harada, H., Park, J. E., Nishi, H., Satoh, M., Takahashi, M., Hiroi, S., Sasaoka, T., Ohbuchi, N., Nakamura, T., et al. (1997). Mutations in the cardiac troponin I gene associated with hypertrophiccardiomyopathy, Nat Genet 16, 379-82.

Kirschenlohr, H. L., Grace, A. A., Vandenberg, J. I., Metcalfe, J. C., and Smith, G. A. (2000). Estimation of systolic and diastolic free intracellular Ca^{2+} by titration of Ca^{2+} buffering in the ferret heart, Biochem J 346, 385-91.

Kleerekoper, Q., Liu, W., Choi, D., and Putkey, J. A. (1998). Identification of binding sites for bepridil and trifluoperazine on cardiac troponin C, J Biol Chem 273, 8153-60.

Knollmann, B. C., Blatt, S. A., Horton, K., Freitas, F., Miller, T. E., Bell, M., Housmans, P. R., Weissman, N. J., Morad, M., and Potter, J. D. (2001). Inotropic Stimulation Induces Cardiac Dysfunction in Transgenic Mice Expressing a Troponin T (I79n) Mutation Linked to Familial Hypertrophic Cardiomyopathy, J Biol Chem 276, 10039-10048.

Krause, K. H., Milos, M., Luan-Rilliet, Y., Lew, D. P., and Cox, J. A. (1991). Thermodynamics of cationbinding to rabbit skeletal muscle calsequestrin. Evidence for distinct Ca($2+$)- and Mg($2+$)-binding sites, J Biol Chem 266, 9453-9.

Kurebayashi, N., and Ogawa, Y. (1988). Increase by trifluoperazine in calcium sensitivity of myofibrils in a skinned fibre from frog skeletal muscle, J Physiol (Lond) 403, 407-24.

Lamb, G. D. (1993). Ca^{2+} inactivation, Mg^{2+} inhibition and malignant hyperthermia [news], J Muscle Res Cell Motil *14*, 554-6.

Lamb, G. D. (2000). Excitation-contraction coupling in skeletal muscle: comparisons with cardiac muscle, Clin Exp Pharmacol Physiol *27*, 216-24.

Lamb, G. D., and Stephenson, D. G. (1994). Effects of intracellular pH and [Mg^{2+}] on excitation-contraction coupling in skeletal muscle fibres of the rat, J Physiol (Lond) *478*, 331-9.

Lang, R., Zhao, J., and Potter, J. D. (1999). Impaired Inhibition and activation observed with the arginine 145 to glycine mutation of human cardiac troponin I, Biophys J *76*, A274.

Langer, G. A. (1992). Calcium and the heart: exchange at the tissue, cell, and organelle levels, Faseb J *6*, 893-902.

Laurant, P., and Touyz, R. M. (2000). Physiological and pathophysiological role of magnesium in thecardiovascular system: implications in hypertension, J Hypertens *18*, 1177-91.

Leavis, P. C., Rosenfeld, S. S., Gergely, J., Grabarek, Z., and Drabikowski, W. (1978). Proteolytic fragments of troponin C. Localization of high and low affinity Ca^{2+} binding sites and interactions with troponin I and troponin T, J Biol Chem *253*, 5452-9.

Lehman, W. (1978). Thick-filament-linked calcium regulation in vertebrate striated muscle, Nature *274*, 80-1.

Leszyk, J., Dumaswala, R., Potter, J. D., Gusev, N. B., Verin, A. D., Tobacman, L. S., and Collins, J. H. (1987). Bovine cardiac troponin T: amino acid sequences of the two isoforms, Biochemistry *26*, 7035-42.

Levine, R. J., Yang, Z., Epstein, N. D., Fananapazir, L., Stull, J. T., and Sweeney, H. L. (1998). Structural and functional responses of mammalian thick filaments to alterations in myosin regulatory light chains, J Struct Biol *122*, 149-61.

Levitsky, D. O., Benevolensky, D. S., Levchenko, T. S., Smirnov, V. N., and Chazov, E. I. (1981). Calcium-binding rate and capacity of cardiac sarcoplasmic reticulum, J Mol Cell Cardiol *13*, 785-96.

Li, M. X., Gagne, S. M., Spyracopoulos, L., Kloks, C. P., Audette, G., Chandra, M., Solaro, R. J., Smillie, L. B., and Sykes, B. D. (1997). NMR studies of Ca^{2+} binding to the regulatory domains of cardiac and E41A skeletal muscle troponin C reveal the importance of site I to energetics of the induced structural changes, Biochemistry *36*, 12519-25.

Liao, R., Wang, C. K., and Cheung, H. C. (1994). Coupling of calcium to the interaction of troponin I with troponin C from cardiac muscle, Biochemistry *33*, 12729-34.

Lokuta, A. J., Meyers, M. B., Sander, P. R., Fishman, G. I., and Valdivia, H. H. (1997). Modulation of cardiac ryanodine receptors by sorcin, J Biol Chem *272*, 25333-8.

Lowey, S., Waller, G. S., and Trybus, K. M. (1993a). Function of skeletal muscle myosin heavy and light chain isoforms by an in vitro motility assay, J Biol Chem *268*, 20414-8.

Lowey, S., Waller, G. S., and Trybus, K. M. (1993b). Skeletal muscle myosin light chains are essential for physiological speeds of shortening, Nature *365*, 454-6.

Luckcuck, T., Trotter, P. J., and Walker, J. H. (1998). Localization of annexin VI in the adult and neonatal heart, Cell Biol Int *22*, 199-205.

Lynch, G. S., and Williams, D. A. (1994). The effect of lowered pH on the Ca($^{2+}$)-activated contractile characteristics of skeletal muscle fibres from endurance-trained rats, Exp Physiol *79*, 47-57.

Ma, J., Fill, M., Knudson, C. M., Campbell, K. P., and Coronado, R. (1988). Ryanodine receptor of skeletal muscle is a gap junction-type channel, Science *242*, 99-102.

Matteo, R. G., and Moravec, C. S. (2000). Immunolocalization of annexins IV, V and VI in the failing and non-failing human heart, Cardiovasc Res *45*, 961-70.

Mazzei, G. J., and Kuo, J. F. (1984). Phosphorylation of skeletal-muscle troponin I and troponin T by phospholipid-sensitive Ca^{2+}-dependent protein kinase and its inhibition by troponin C and tropomyosin, Biochem J *218*, 361-9.

Meissner, G. (1994). Ryanodine receptor/Ca^{2+} release channels and their regulation by endogenous effectors, Annu Rev Physiol *56*, 485-508.

Melzer, W., Herrmann-Frank, A., and Luttgau, H. C. (1995). The role of Ca^{2+} ions in excitation-contraction coupling of skeletal muscle fibres, Biochim Biophys Acta *1241*, 59-116.

Mercola, D., Bullard, B., and Priest, J. (1975). Crystallisation of tropinin-C, Nature *254*, 634-5.

Metzger, J. M., and Moss, R. L. (1987). Greater hydrogen ion-induced depression of tension and velocity in skinned single fibres of rat fast than slow muscles, J Physiol (Lond) *393*, 727-42.

Metzger, J. M., and Moss, R. L. (1992). Myosin light chain 2 modulates calcium-sensitive cross-bridge transitions in vertebrate skeletal muscle, Biophys J 63, 460-8.

Metzger, J. M., Parmacek, M. S., Barr, E., Pasyk, K., Lin, W. I., Cochrane, K. L., Field, L. J., and Leiden, J.M. (1993). Skeletal troponin C reduces contractile sensitivity to acidosis in cardiac myocytes from transgenic mice, Proc Natl Acad Sci U S A 90, 9036-40.

Meyers, M. B., Zamparelli, C., Verzili, D., Dicker, A. P., Blanck, T. J., and Chiancone, E. (1995). Calcium-dependent translocation of sorcin to membranes: functional relevance in contractile tissue, FEBS Lett 357, 230-4.

Miller, T., Szczesna, D., Housmans, P. R., Zhao, J., deFreitas, F., Gomes, A. V., Culbreath, L., McCue, J., Wang, Y., Xu, Y., et al. (2001). Abnormal Contractile Function in Transgenic Mice Expressing an FHC-Linked Troponin T (179N) Mutation, J Biol Chem 276, 2743-3755.

Mogensen, J., Klausen, I. C., Pedersen, A. K., Egeblad, H., Bross, P., Kruse, T. A., Gregersen, N., Hansen, P. S., Baandrup, U., and Borglum, A. D. (1999). Alpha-cardiac actin is a novel disease gene in familial hypertrophic cardiomyopathy, J Clin Invest 103, R39-43.

Molkentin, J. D., Lu, J. R., Antos, C. L., Markham, B., Richardson, J., Robbins, J., Grant, S. R., and Olson, E. N. (1998). A calcineurin-dependent transcriptional pathway for cardiac hypertrophy, Cell 93, 215-28.

Morano, I. (1999). Tuning the human heart molecular motors by myosin light chains, J Mol Med 77, 544-55.

Morano, I., Hofmann, F., Zimmer, M., and Ruegg, J. C. (1985). The influence of P-light chain phosphorylation by myosin light chain kinase on the calcium sensitivity of chemically skinned heart fibres, FEBS Lett 189, 221-4.

Morimoto, S. (1991). Effect of myosin cross-bridge interaction with actin on the $Ca^{(2+)}$-binding properties of troponin C in fast skeletal myofibrils, J Biochem (Tokyo) 109, 120-6.

Morimoto, S., Harada, K., and Ohtsuki, I. (1999). Roles of troponin isoforms in pH dependence of contraction in rabbit fast and slow skeletal and cardiac muscles, J Biochem (Tokyo) 126, 121-9.

Morimoto, S., and Ohtsuki, I. (1987). Ca^{2+}- and Sr^{2+}-sensitivity of the ATPase activity of rabbit skeletal myofibrils: effect of the complete substitution of troponin C with cardiac troponin C, calmodulin, and parvalbumins, J Biochem (Tokyo) 101, 291-301.

Morimoto, S., and Ohtsuki, I. (1994a). Ca2+ binding to cardiac troponin C in the myofilament lattice and its relation to the myofibrillar ATPase activity, Eur J Biochem 226, 597-602.

Morimoto, S., and Ohtsuki, I. (1994b). Role of troponin C in determining the $Ca^{(2+)}$-sensitivity and cooperativity of the tension development in rabbit skeletal and cardiac muscles, J Biochem (Tokyo) 115, 144-6.

Murphy, E., Freudenrich, C. C., Levy, L. A., London, R. E., and Lieberman, M. (1989a). Monitoring cytosolic free magnesium in cultured chicken heart cells by use of the fluorescent indicator Furaptra, Proc Natl Acad Sci U S A 86, 2981-4.

Murphy, E., Steenbergen, C., Levy, L. A., Raju, B., and London, R. E. (1989b). Cytosolic free magnesium levels in ischemic rat heart, J Biol Chem 264, 5622-7.

Murray, B. E., and Ohlendieck, K. (1998). Complex formation between calsequestrin and the ryanodine receptor in fast- and slow-twitch rabbit skeletal muscle, FEBS Lett 429, 317-22.

Nagy, B., and Gergely, J. (1979). Extent and localization of conformational changes in troponin C caused by calcium binding. Spectral studies in the presence and absence of 6 M urea, J Biol Chem 254, 12732-7.

Nassar, R., Malouf, N. N., Kelly, M. B., Oakeley, A. E., and Anderson, P. A. (1991). Force-pCa relation and troponin T isoforms of rabbit myocardium, Circ Res 69, 1470-5.

Negele, J. C., Dotson, D. G., Liu, W., Sweeney, H. L., and Putkey, J. A. (1992). Mutation of the high affinity calcium binding sites in cardiac troponin C, J Biol Chem 267, 825-31.

Noland, T. A., Jr., and Kuo, J. F. (1991). Protein kinase C phosphorylation of cardiac troponin I or troponin T inhibits $Ca^{(2+)}$-stimulated actomyosin MgATPase activity, J Biol Chem 266, 4974-8.

Ogawa, Y. (1985). Calcium binding to troponin C and troponin: effects of Mg^{2+}, ionic strength and pH, J Biochem (Tokyo) 97, 1011-23.

Palmer, S., and Kentish, J. C. (1994). The role of troponin C in modulating the Ca^{2+} sensitivity of mammalian skinned cardiac and skeletal muscle fibres, J Physiol 480, 45-60.

Palmiter, K. A., Kitada, Y., Muthuchamy, M., Wieczorek, D. F., and Solaro, R. J. (1996). Exchange of beta-for alpha-tropomyosin in hearts of transgenic mice induces changes in thin filament response to Ca^{2+}, strong cross-bridge binding, and protein phosphorylation, J Biol Chem 271, 11611-4.

Pan, B. S., Hannon, J. D., Wiedmann, R., Potter, J. D., Kranias, E. G., Shen, Y. T., Johnson, R. G., and Housmans, P. R. (1999). Effects of isoproterenol on twitch contraction of wild type and phospholamban-deficient murine ventricular myocardium, J Mol Cell Cardiol 31, 159-66.

Pan, B. S., and Solaro, R. J. (1987). Calcium-binding properties of troponin C in detergent-skinned heart muscle fibers, J Biol Chem 262, 7839-49.

Parsons, B., Szczesna, D., Zhao, J., Van Slooten, G., Kerrick, W. G., Putkey, J. A., and Potter, J. D. (1997). The effect of pH on the Ca^{2+} affinity of the Ca^{2+} regulatory sites of skeletal and cardiac troponin C in skinned muscle fibres, J Muscle Res Cell Motil 18, 599-609.

Pearlstone, J. R., Chandra, M., Sorenson, M. M., and Smillie, L. B. (2000). Biological function and site II Ca^{2+}-induced opening of the regulatory domain of skeletal troponin C are impaired by invariant site I or II glu mutations, J Biol Chem 275, 35106-15.

Perry, S. V. (1998). Troponin T: genetics, properties and function, J Muscle Res Cell Motil 19, 575-602.

Pifl, C., Plank, B., Wyskovsky, W., Bertel, O., Hellmann, G., and Suko, J. (1984). Calmodulin X $(Ca^{2+})4$ is the active calmodulin-calcium species activating the calcium-, calmodulin-dependent protein kinase of cardiac sarcoplasmic reticulum in the regulation of the calcium pump, Biochim Biophys Acta 773, 197-206.

Podlubnaya, Z., Kakol, I., Moczarska, A., Stepkowski, D., and Udaltsov, S. (1999). Calcium-induced structural changes in synthetic myosin filaments of vertebrate striated muscles, J Struct Biol 127, 1-15.

Poetter, K., Jiang, H., Hassanzadeh, S., Master, S. R., Chang, A., Dalakas, M. C., Rayment, I., Sellers, J. R., Fananapazir, L., and Epstein, N. D. (1996). Mutations in either the essential or regulatory light chains of myosin are associated with a rare myopathy in human heart and skeletal muscle, Nat Genet 13, 63-9.

Potter, J. D., and Gergely, J. (1975). The calcium and magnesium binding sites on troponin and their role in the regulation of myofibrillar adenosine triphosphatase, J Biol Chem 250, 4628-33.

Potter, J. D., Robertson, S. P., and Johnson, J. D. (1981). Magnesium and the regulation of muscle contraction, Fed Proc 40, 2653-6.

Potter, J. D., Strang-Brown, P., Walker, P. L., and Iida, S. (1983). Ca^{2+} binding to calmodulin, Methods Enzymol 102, 135-43.

Rayment, I., Holden, H. M., Whittaker, M., Yohn, C. B., Lorenz, M., Holmes, K. C., and Milligan, R. A. (1993a). Structure of the actin-myosin complex and its implications for muscle contraction [see comments], Science 261, 58-65.

Rayment, I., Rypniewski, W. R., Schmidt-Base, K., Smith, R., Tomchick, D. R., Benning, M. M., Winkelmann, D. A., Wesenberg, G., and Holden, H. M. (1993b). Three-dimensional structure of myosin subfragment-1: a molecular motor [see comments], Science 261, 50-8.

Redwood, C. S., Moolman-Smook, J. C., and Watkins, H. (1999). Properties of mutant contractile proteins that cause hypertrophic cardiomyopathy, Cardiovasc Res 44, 20-36.

Rios, E., and Brum, G. (1987). Involvement of dihydropyridine receptors in excitation-contraction coupling in skeletal muscle, Nature 325, 717-20.

Robertson, S. P., Johnson, J. D., Holroyde, M. J., Kranias, E. G., Potter, J. D., and Solaro, R. J. (1982). The effect of troponin I phosphorylation on the Ca^{2+}-binding properties of the Ca^{2+}-regulatory site of bovine cardiac troponin, J Biol Chem 257, 260-3.

Robertson, S. P., Johnson, J. D., and Potter, J. D. (1981). The time-course of Ca^{2+} exchange with calmodulin, troponin, parvalbumin, and myosin in response to transient increases in Ca^{2+}, Biophys J 34, 559-69.

Romani, A. M., and Scarpa, A. (2000). Regulation of cellular magnesium, Front Biosci 5, D720-34.

Rusnak, F., and Mertz, P. (2000). Calcineurin: form and function, Physiol Rev 80, 1483-521.

Sanbe, A., Fewell, J. G., Gulick, J., Osinska, H., Lorenz, J., Hall, D. G., Murray, L. A., Kimball, T. R., Witt, S. A., and Robbins, J. (1999). Abnormal cardiac structure and function in mice expressing nonphosphorylatable cardiac regulatory myosin light chain 2, J Biol Chem 274, 21085-94.

Sanbe, A., Nelson, D., Gulick, J., Setser, E., Osinska, H., Wang, X., Hewett, T. E., Klevitsky, R., Hayes, E., Warshaw, D. M., and Robbins, J. (2000). In vivo analysis of an essential myosin light chain mutation linked to familial hypertrophic cardiomyopathy, Circ Res 87, 296-302.

Satoh, M., Takahashi, M., Sakamoto, T., Hiroe, M., Marumo, F., and Kimura, A. (1999). Structural analysis of the titin gene in hypertrophic cardiomyopathy: identification of a novel disease gene, Biochem Biophys Res Commun 262, 411-7.

Schafer, B. W., and Heizmann, C. W. (1996). The S100 family of EF-hand calcium-binding proteins: functions and pathology, Trends Biochem Sci 21, 134-40.

Schneider, M. F. (1994). Control of calcium release in functioning skeletal muscle fibers, Annu Rev Physiol 56, 463-84.

Schneider, M. F., and Chandler, W. K. (1973). Voltage dependent charge movement of skeletal muscle: a possible step in excitation-contraction coupling, Nature 242, 244-6.

Seiler, S., Wegener, A. D., Whang, D. D., Hathaway, D. R., and Jones, L. R. (1984). High molecular weight proteins in cardiac and skeletal muscle junctional sarcoplasmic reticulum vesicles bind calmodulin, are phosphorylated, and are degraded by Ca^{2+}-activated protease, J Biol Chem 259, 8550-7.

Sheng, Z., Strauss, W. L., Francois, J. M., and Potter, J. D. (1990). Evidence that both $Ca(^{2+})$-specific sites of skeletal muscle TnC are required for full activity [published erratum appears in J Biol Chem 1993 May 5;268(13):9936], J Biol Chem 265, 21554-60.

Sia, S. K., Li, M. X., Spyracopoulos, L., Gagne, S. M., Liu, W., Putkey, J. A., and Sykes, B. D. (1997). Structure of cardiac muscle troponin C unexpectedly reveals a closed regulatory domain, J Biol Chem 272, 18216-21.

Slupsky, C. M., and Sykes, B. D. (1995). NMR solution structure of calcium-saturated skeletal muscle troponin C, Biochemistry 34, 15953-64.

Solaro, R. J., Kumar, P., Blanchard, E. M., and Martin, A. F. (1986). Differential effects of pH on calcium activation of myofilaments of adult and perinatal dog hearts. Evidence for developmental differences in thin filament regulation, Circ Res 58, 721-9.

Solaro, R. J., Lee, J. A., Kentish, J. C., and Allen, D. G. (1988). Effects of acidosis on ventricular muscle from adult and neonatal rats, Circ Res 63, 779-87.

Solaro, R. J., and Pan, B.-S. (1989). Control and modulation of contractile activity of cardaic myofilaments. In Physiology and Pathophysiology of the Heart., N. Spereliakis, ed. (Boston, KlumerAcademic), pp. 291-303.

Solaro, R. J., Wise, R. M., Shiner, J. S., and Briggs, F. N. (1974). Calcium requirements for cardiac myofibrillar activation, Circ Res 34, 525-30.

Somlyo, A. V., McClellan, G., Gonzalez-Serratos, H., and Somlyo, A. P. (1985). Electron probe X-ray microanalysis of post-tetanic Ca^{2+} and Mg^{2+} movements across the sarcoplasmic reticulum in situ, J Biol Chem 260, 6801-7.

Sorenson, M. M., da Silva, A. C., Gouveia, C. S., Sousa, V. P., Oshima, W., Ferro, J. A., and Reinach, F. C. (1995). Concerted action of the high affinity calcium binding sites in skeletal muscle troponin C, J. Biol Chem 270, 9770-7.

Spyracopoulos, L., Li, M. X., Sia, S. K., Gagne, S. M., Chandra, M., Solaro, R. J., and Sykes, B. D. (1997). Calcium-induced structural transition in the regulatory domain of human cardiac troponin C, Biochemistry 36, 12138-46.

Sundaralingam, M., Bergstrom, R., Strasburg, G., Rao, S. T., Roychowdhury, P., Greaser, M., and Wang, B.C. (1985). Molecular structure of troponin C from chicken skeletal muscle at 3-angstrom resolution, Science 227, 945-8.

Sweeney, H. L., Bowman, B. F., and Stull, J. T. (1993). Myosin light chain phosphorylation in vertebrate striated muscle: regulation and function, Am J Physiol 264, C1085-95.

Szczesna, D., Ghosh, D., Li, Q., Gomes, A., Guzman, G., Arana, C., Zhi, G., Stull, J. T., and Potter, J. D. (2001). Familial hypertrophic cardiomyopathy mutations in the regulatory light chains of myosin affect their structure, Ca^{2+} binding and phosphorylation, J Biol Chem 276, 7086-7092.

Szczesna, D., Guzman, G., Miller, T., Zhao, J., Farokhi, K., Ellemberger, H., and Potter, J. D. (1996a). The role of the four Ca^{2+} binding sites of troponin C in the regulation of skeletal muscle contraction, J Biol Chem 271, 8381-6.

Szczesna, D., Zhang, R., Zhao, J., Jones, M., Guzman, G., and Potter, J. D. (2000). Altered regulation of cardiac muscle contraction by troponin T mutations that cause familial hypertrophic cardiomyopathy, J Biol Chem 275, 624-30.

Szczesna, D., Zhao, J., and Potter, J. D. (1996b). The regulatory light chains of myosin modulate cross-bridge cycling in skeletal muscle, J Biol Chem 271, 5246-50.

Takahashi-Yanaga, F., Morimoto, S., and Ohtsuki, I. (2000). Effect of Arg145Gly mutation in human cardiac troponin I on the ATPase activity of cardiac myofibrils, J Biochem (Tokyo) *127*, 355-7.

Tobacman, L. S., Lin, D., Butters, C., Landis, C., Back, N., Pavlov, D., and Homsher, E. (1999). Functional consequences of troponin T mutations found in hypertrophic cardiomyopathy, J Biol Chem *274*, 28363-70.

Tsuda, S., Ogura, K., Hasegawa, Y., Yagi, K., and Hikichi, K. (1990). 1H NMR study of rabbit skeletal muscle troponin C: $Mg^2(+)$-induced conformational change, Biochemistry *29*, 4951-8.

Tung, C. S., Wall, M. E., Gallagher, S. C., and Trewhella, J. (2000). A model of troponin-I in complex with troponin-C using hybrid experimental data: the inhibitory region is a beta-hairpin, Protein Sci *9*, 1312-26.

Valdivia, H. H. (1998). Modulation of intracellular Ca^{2+} levels in the heart by sorcin and FKBP12, two accessory proteins of ryanodine receptors, Trends Pharmacol Sci *19*, 479-82.

van Eerd, J. P., and Takahashi, K. (1975). The amino acid sequence of bovine cardiac troponin-C. Comparison with rabbit skeletal troponin-C, Biochem Biophys Res Commun *64*, 122-7.

Wannenburg, T., Heijne, G. H., Geerdink, J. H., Van Den Dool, H. W., Janssen, P. M., and De Tombe, P. P. (2000). Cross-bridge kinetics in rat myocardium: effect of sarcomere length and calcium activation, Am J Physiol Heart Circ Physiol *279*, H779-90.

Watkins, H., McKenna, W. J., Thierfelder, L., Suk, H. J., Anan, R., O'Donoghue, A., Spirito, P., Matsumori, A., Moravec, C. S., Seidman, J. G., and et al. (1995). Mutations in the genes for cardiac troponin T and alpha-tropomyosin in hypertrophic cardiomyopathy, N Engl J Med *332*, 1058-64.

Watterson, J. G., Kohler, L., and Schaub, M. C. (1979). Evidence for two distinct affinities in the binding of divalent metal ions to myosin, J Biol Chem *254*, 6470-7.

Wibo, M., Bravo, G., and Godfraind, T. (1991). Postnatal maturation of excitation-contraction coupling in rat ventricle in relation to the subcellular localization and surface density of 1,4-dihydropyridine and ryanodine receptors, Circ Res *68*, 662-73.

Xiong, H., Feng, X., Gao, L., Xu, L., Pasek, D. A., Seok, J. H., and Meissner, G. (1998). Identification of a two EF-hand Ca^{2+} binding domain in lobster skeletal muscle ryanodine receptor/Ca^{2+} release channel, Biochemistry *37*, 4804-14.

Xu, A., and Narayanan, N. (2000). Reversible inhibition of the calcium-pumping ATPase in native cardiac sarcoplasmic reticulum by a calmodulin-binding peptide. Evidence for calmodulin-dependent regulation of the V(max) of calcium transport, J Biol Chem *275*, 4407-16.

Yano, K., and Zarain-Herzberg, A. (1994). Sarcoplasmic reticulum calsequestrins: structural and functional properties, Mol Cell Biochem *135*, 61-70.

Zhang, R., Zhao, J., and Potter, J. D. (1995). Phosphorylation of both serine residues in cardiac troponin I is required to decrease the Ca^{2+} affinity of cardiac troponin C, J Biol Chem *270*, 30773-80.

Zot, A. S., and Potter, J. D. (1987). Structural aspects of troponin-tropomyosin regulation of skeletal muscle contraction, Annu Rev Biophys Biophys Chem *16*, 535-59.

Zot, H. G., and Potter, J. D. (1982). A structural role for the Ca^{2+}-Mg^{2+} sites on troponin C in the regulation of muscle contraction. Preparation and properties of troponin C depleted myofibrils, J Biol Chem *257*, 7678-83.

HERBERT C. CHEUNG

CALCIUM-INDUCED MOLECULAR AND STRUCTURAL SIGNALING IN STRIATED MUSCLE CONTRACTION

1. INTRODUCTION

Regulation of contraction in vertebrate striated muscle is linked to the thin filament. The initial and critical step in activation is the binding of regulatory Ca^{2+} to specific sites in troponin. This binding triggers a series of protein-protein interactions which in turn induce structural changes in the proteins and lead to a cyclic interaction between actin and force-generating myosin crossbridge. These molecular interactions are dependent upon transduction of the initial Ca^{2+} signal along the thin filament. The transduction mechanism is viewed as a component of the overall contractile mechanism.

A full understanding of the transduction mechanism requires knowledge of structural information on individual proteins and protein complexes and changes in their structures during the course of activation, and energetic aspects of relevant molecular interactions. Recent advances in structural studies of small proteins and the advent of molecular biologic techniques have made possible during the past decade dissection of the structure/function relationships of some of these proteins and their interactions in detail. A general picture has emerged on many molecular aspects of the troponin subunits in response to the binding of regulatory Ca^{2+} to its receptor. These recent findings have provided a basis for construction of models to probe transduction of the Ca^{2+} signal and Ca^{2+} activation of muscle functions.

This chapter gives a glimpse of our current knowledge on several structural aspects of troponin C and troponin I from both skeletal and cardiac muscle, and their interactions with each other. Also will be discussed are kinetic mechanisms of Ca^{2+} binding to troponin C and troponin. Although experimental details are kept to a minimum, some details are given for studies that have been carried out using fluorescence spectroscopy in our laboratory in the past five years. Whenever practical, topics are introduced with a historical perspective so as to provide a sense of how the present knowledge has evolved.

R.J. Solaro and R.L. Moss (eds), Molecular Control Mechanisms in Striated Muscle Contraction, 199-245

2. STRUCTURAL COMPONENTS OF THE THIN FILAMENT

The thin filament is a pseudodouble helical filament of polymerized actin decorated with the dimeric coiled-coil helices of tropomyosin (Tm) and the heterotrimeric troponin complex (Tn). The Tm molecule may be homodimeric or heterodimeric dependent upon muscle type, and the two strands of the coiled-coil are parallel in register. Each strand of the Tm dimer has 284 amino acids, and each dimer is considered to be a cable-like structure that winds along the surface of the actin filament in contact with seven consecutive actin monomers (A) of the same strand of the two-stranded actin helix. Contiguous Tm molecules are non-covalently bonded to each other in a head-to-tail fashion on the actin surface. Because of a small head-to-tail overlap in the bonding region, the effective length of Tm in muscle is about 405 to 410 Å (Phillips et al., 1986). Each Tm is associated with one Tn. This composition gives rise to the regulatory unit with stoichiometry A_7-Tm-Tn. Tm appears to be neither extensible nor compressible in its axial direction, although it has considerable flexibility in the transverse direction. The Tn subunit troponin C (TnC) is the Ca^{2+} receptor, the subunit troponin I (TnI) binds to both TnC and actin in relaxed muscle and inhibits actomyosin ATPase, and the subunit troponin T (TnT) binds to Tm.

The atomic structure of monomeric globular actin (G-actin) has been determined from its complex with DNase I (Kabsch et al., 1990). A model of the actin filament has been constructed on the basis of the G-actin structure and X-ray diffraction patterns of oriented actin filaments (Holmes et al., 1990). Single crystals of Tm prepared from native tissues have a mesh-like packing due to the long head-to-tail linked chains and high water content. These crystals give poor X-ray diffraction patterns and are not suitable for crystallographic studies at atomic resolution. To circumvent some of these problems, a new form of crystal from lobster skeletal muscle was obtained from non-polymerizable Tm mutant which was expressed in a baculovirus-based system (Miegel et al., 1996). A model at 7 Å resolution was recently reported for cardiac Tm with crystals containing spermine (Whitby and Phillips, 2000), but it still lacks sufficient detail to define the three-dimensional structure of Tm.

Calcium activation of striated muscle is triggered by the binding of activator Ca^{2+} to specific sites in TnC within the troponin complex. Elucidation of the structural mechanisms by which this initiation is achieved requires detailed structural information on the heterotrimeric Tn complex. No structure at atomic resolution is available for this complex from vertebrate muscle that could contribute to the understanding of how the three subunits regulate the actin-myosin interaction. In an early study, skeletal Tn was shown by low-resolution rotatory shadowing to have a bipartite structure with a length of 265 Å (Flicker et al., 1982) in which TnC and TnI form the globular domain which binds to the C-terminal end of the long rod-like TnT. The length of skeletal TnT is about 180 Å and that of cardiac TnT is about 200 Å because of an amino terminal extension in the cardiac isoform. A unique property of TnT is its ability to form a complex with Tm. In the myofilament, TnT and Tm run anti-parallel to each other (Fig. 1). Co-crystals of the skeletal Tn-Tm complex

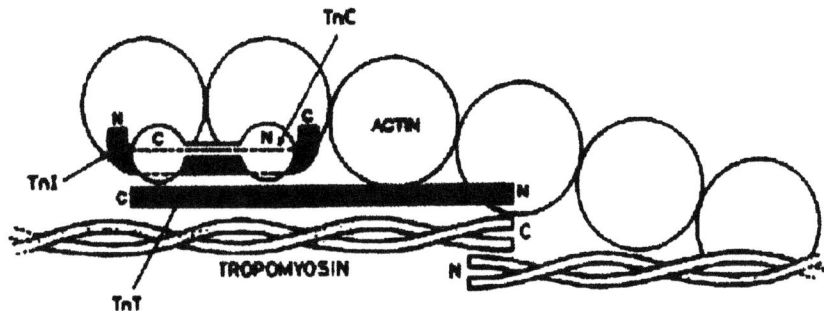

Figure 1. A schematic diagram showing the approximate interrelationships of the proteins of the striated muscle thin filament. (From Perry, 1998).

revealed that the N-terminal end of TnT binds to the head-to-tail junction of the Tm chain and the remainder of the TnT extends about 180 Å along the Tm molecule (White et al., 1987). A recent single particle analysis of electron micrographs of the Tm-Tn complex prepared from insect flight muscle provided a three-dimensional reconstruction of the Tn complex at 26 Å resolution. The model at this resolution provides little information on the topography of individual subunits within the whole complex and gives no clue on potential changes in the overall topography induced by Ca^{2+} under conditions that would lead to activation of muscle functions.

At the molecular level, the binding of Ca^{2+} to Tn elicits a cascade of changes in protein conformations and protein-protein interactions within the Tn trimer. These changes ultimately result in a lateral movement of the Tm chain on the surface of the double helical actin helix and activation of actomyosin ATPase, and attachment of force generating myosin crossbridges of the thick filament to actin sites of the thin filament. Thus, the molecular events that are believed to be central to regulation in striated muscle are Ca^{2+}-induced structural changes involving the regulatory unit (A_7-Tm-Tn) and myosin crossbridge binding to actin in the regulatory unit. These changes will be described with emphasis on TnC and TnI, and the differences in these changes between skeletal and cardiac subunits will be discussed

3. SKELETAL MUSCLE TROPONIN SUBUNITS

3.1. Troponin C

As will be apparent, there are substantial differences in the structures and properties between skeletal Tn subunits and cardiac Tn subunits. Theses differences can be better understood if the properties of the skeletal isoform are discussed first. TnC is the only troponin subunit whose crystal structure has been solved and whose solution NMR structure has been determined. Early biochemical studies established that TnC from fast skeletal muscle (fsTnC) has four cation-binding sites. Two sites (sites I and II) located in the N-terminal half of TnC bind Ca^{2+} with an affinity of about 10^5 M^{-1} and has no known affinity for Mg^{2+} (Ca^{2+}-specific, regulatory sites); the other two sites (sites III and IV) located in the C-terminal half of TnC bind Ca^{2+} with a higher affinity (10^7 M^{-1}) and have a weak affinity for Mg^{2+} (10^3 M^{-1}). In relaxed muscle, sites III and IV are likely saturated with Mg^{2+} because of the relatively high intracellular Mg^{2+} concentration (millimolar), but sites I and II are unoccupied because the intracellular Ca^{2+} level is 0.1 μM or less. When intracellular [Ca^{2+}] is elevated and reaches a critical level of about 10 μM, bound Mg^{2+} at sites III and IV is exchanged with bound Ca^{2+} and the two Ca^{2+}-specific regulatory sites I and II become saturated. The two high-affinity Ca^{2+} sites have no known functional role and likely play a structural role. It is widely accepted that Ca^{2+} binding to the two regulatory sites elicits a conformational change that constitutes the major intracellular Ca^{2+} signal for triggering activation of muscle functions such as elevation of actomyosin ATPase activity and generation of force. In a solution containing [Ca^{2+}] < 10^{-7} M and [Mg^{2+}] ~ 1-2 mM (ionic conditions that mimic intracellular environment of muscle cells in relaxed muscle), TnI binds strongly to actin and also to TnC with an affinity of about 6 x 10^7 M^{-1}. Under these conditions, the regulatory sties of TnC are unoccupied, but the sites III and IV are saturated with Mg^{2+} and actomyosin ATPase is inhibited. This TnI-TnC affinity is enhanced 100-fold if all four TnC sites are saturated with Ca^{2+}. This enhanced binding of TnC with TnI weakens the interaction of TnI with actin, which is responsible for relieving the inhibitory effect of TnI. In the presence of bound TnI, the affinity of TnC sites I and II for Ca^{2+} is 10-fold enhanced (10^6 M^{-1}) and that of sites III and IV is similarly elevated (10^8 M^{-1}). These reciprocal affinity enhancements arise from free energy coupling for the binding of Ca^{2+} and TnI to TnC (Wang and Cheung, 1985; Cheung et al., 1988). Although TnT binds to both TnC and TnI, it has no apparent effect on the TnC affinity for Ca^{2+}, at both sets of sites within the ternary troponin complex. The enhanced affinity of TnI for TnC elicited by bound Ca^{2+} at the two regulatory sites is due to a second interaction between TnI and TnC, and the TnC sites for this second interaction are likely hydrophobic residues that are exposed by the Ca^{2+} binding (Drabikowski et al., 1985; Levine et al., 1977).This Ca^{2+}-dependent TnC-TnI interaction is the critical step leading to alterations of other protein-protein interactions within the regulatory unit of the thin filament. These alterations provide a mechanism by which force-generating interactions are made between myosin crossbridges and actin sites. Thus, the Ca^{2+} switch in regulation of muscle activation is in the TnI-TnC linkage.

3.1.1. Crystallographic Structures

The initial crystallographic structures of avian fsTnC (162 residues) revealed a dumbbell-shaped molecule with both the N-terminal and C-terminal segments folded into two globular domains, which are linked by a 22-residue 45 Å central helix (Sundalingam et al., 1985; James and Herzberg, 1985). These structures, which were derived from crystals containing bound Ca^{2+} at sites III and IV (holo C-domain) and no bound Ca^{2+} at sites I and II (2Ca-state, apo N-domain), were refined to 2.0 Å resolution for both the chicken protein (Satyshur et al., 1988) and the turkey protein (Herzberg and James, 1988). More recently, the crystal structure of rabbit fsTnC (159 residues) fully saturated with Ca^{2+} ($4Ca^{2+}$-state) was reported at 2.0 Å resolution (Houdusse et al., 1997). The regulatory N-terminal domain consists of five helices, which are designated as helices N, A, B, C, and D starting from the N-terminus. The D helix continues into an interdomain helical linker, followed by four helices designated as helices E, F, G, and H, which make up the C-terminal domain. The long central helix is composed of helix D, the interdomain linker, and helix E. Each domain consists of a pair of helix-loop-helix EF-hand motif (Kretsinger and Nockolds, 1973) in which a Ca^{2+} ion is coordinated to the 12-residue loop. In the N-domain, the N helix is connected to helix A by a short loop. Ca^{2+}-binding site I (first EF- hand) begins with helix A, followed by the 12-residue loop (loop I), then helix B. A flexible linker (B-C linker) connects helices B and C, which is followed by a 12-residue loop linked to helix D. The helix C-loop-helix D unit (second EF-hand) makes up the site II. In the C-terminal domain, the four helices are similarly connected giving rise to site III (helix E-loop-helix F) and site IV (helix G-loop-helix H), followed by a short segment to the C-terminus. In the crystal structures, the flanking helices of the EF-hands at sites III and IV are oriented at angles close to 90 degrees, whereas the angles of the flanking helices at sites I and II are considerably larger than 90 degrees. These angles define the global conformation of the domains: interhelical angles near 90 degrees giving rise to an "open" domain conformation, and angles >> 90 degrees giving rise to a "closed" or compact domain conformation.

3.1.2. An Early Model of the Ca^{2+}-induced Structure of the Regulatory Domain

The early crystallographic structure of the 2Ca-bound TnC provides a view of the structural features of the apo-state of the regulatory N-domain. If the two domains are structurally homologous, the helices in the N-domain are expected to experience reorientations upon Ca^{2+} binding to sites I and II. Using the atomic coordinates of the Ca^{2+}-saturated C-domain, a model was constructed for the conformational changes in the regulatory N-domain induced by bound Ca^{2+} at sites I and II (Herzberg et al., 1986; Strynadka and James, 1989). This model (the HMJ model, Fig. 2) predicts that helices B and C and the linker between them move away as a unit from their dispositions in the apo structure (Fig. 2B). The relative dispositions of helices N, A, and D remain unchanged. These helix reorientations result in a less compact or more open conformation of the N-domain and expose a patch of hydrophobic residues in helix B and helix D. These exposed hydrophobic residues are believed to be the sites for the Ca^{2+}-dependent interaction with TnI. The proposed conformational changes are in general agreement with the recently determined crystallographic structure of holo N-domain in the 4Ca-state of rabbit fsTnC (159 residues) at 2 Å resolution (Houdusse et al., 1997), and the structure at 1.75 Å resolution of an isolated recombinant N-terminal domain (N-fsTnC, residues 1-90) derived from chicken fsTnC (Strynadka et al., 1997). The crystallographic data, however, show a more open Ca^{2+}-induced conformation in the N-domain when compared with the HMJ model.

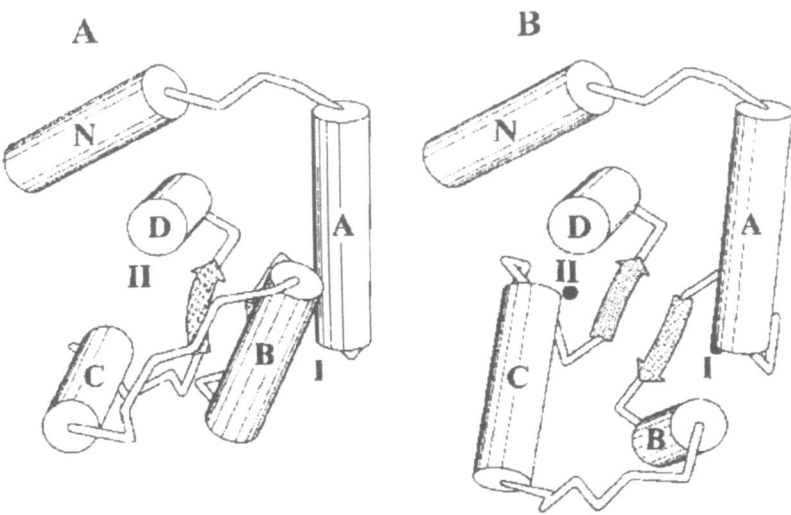

Figure 2. A proposed model for Ca^{2+}-induced conformational changes in the regulatory N-domain of fsTnC. The five helices are labeled helix N and helices A-D starting from the N-terminus. Helix D is linked by the central helix to the C-domain which are not shown. (A) Conformation of apo N-domain showing the locations of the two unoccupied Ca^{2+} sites (I and II), based on X-ray crystallographic data. (B) Proposed conformation of the holo state of the N-domain with two bound Ca^{2+} ions (closed circles) at sites I and II. In this proposed model, the relative dispositions of helices N, A, and D are unchanged as shown in (A), whereas helices B and C and the peptide linker (B-C linker) between the two helices move as a unit from their locations in the apo conformation. The Ca^{2+}-induced reorientations of the helices give rise to a more open conformation and expose a hydrophobic pocket. This model is referred to as the Herzberg-Moult-James (HMJ) model. (From Strynadka and James, 1989).

3.1.3. Solution NMR Structures

It is generally accepted that proteins in solution have multiple conformations. A given crystallographic structure is a static picture of a conformation that is more favorable for crystallization than the other conformations. A complementary structural method that can yield a solution three-dimensional structure with atomic resolution is heteronuclear multidimensional NMR spectroscopy. Briefly, this method requires production of recombinant protein with uniformly labeled carbon (^{13}C) and nitrogen (^{15}N) atoms and involves triple-resonance NMR experiments. The spectroscopic data are used for assignment of resonances and determination of distance and dihedral angle restrains. These restraints are used to calculate a set of structures of the protein using a simulated annealing protocol. NMR structures give a view of the dynamics of the structure and can provide structural insights that are not revealed in crystallographic structure.

The first three-dimensional NMR solution structures of the N-terminal domain of chicken fsTnC (residues 1-90) was reported in 1995 (Gagné et al., 1995), followed by the structures of full-length fsTnC (Slupsky and Sykes, 1995). Although the two-domain feature of fsTnC is evident in the NMR structure, the segment of residues 85-94 in the central helix appears highly flexible and conformationally heterogeneous such that the relative orientations of the two domains cannot be defined. In the 4Ca-state, the holo regulatory N-domain has an open conformation (small interhelical angles near 90 degrees) when compared to the apo domain (large interhelical angles). This opening exposes a large hydrophobic pocket consistent with the HMJ model. The area of this exposed hydrophobic pocket is larger than that observed in the C-domain. This larger hydrophobic pocket may facilitate its presentation to the target protein TnI to turn on the Ca^{2+}switch. Accumulating high-resolution structural information (from X-ray crystallography and NMR) indicates somewhat different Ca^{2+}-saturated structures of the two domains. Earlier circular dichroism studies of fsTnC reported significant increases in negative ellipticity induced by Ca^{2+} binding (Hincke et al., 1978; Johnson and Potter, 1978). For many years, these changes were interpreted as increased α-helicity. A recent NMR study showed that the observed changes in CD signal only marginally relate to changes in secondary structure (Gagné et al., 1994), but more likely to changes in tertiary structure involving Ca^{2+}-induced reorientation of flanking helices in the EF-hand motif (Manning, 1989; Pearlstone et al., 2000). This Ca^{2+}-induced change in CD signal provides a convenient qualitative spectroscopic method to follow the transition between the closed and open domain conformations (Pearlstone et al., 2000).

3.1.4. Global Conformations of the Regulatory Domain

While X-ray crystallographic and NMR analyses reveal high-resolution structural changes in TnC that accompany Ca^{2+} binding, a third approach has been used to investigate Ca^{2+}-induced global conformational changes in the regulatory N-domain of TnC. The separation between two chromophores attached to specific sites in a macromolecule can be determined by the Förster type of fluorescence resonance energy transfer (FRET), if they meet certain spectroscopic requirements (Cheung, 1991). Briefly, if the fluorescence emission spectrum of one chromophore (energy donor) overlaps the absorption spectrum of the other chromophore (energy acceptor), a dipolar interaction between the emission dipole and the absorption dipole can occur if the dipoles are favorably oriented with respect to each other. This interaction results in the transfer of excitation energy from the donor to the acceptor. The transfer efficiency (E) falls off rapidly according to an inverse sixth power of the separation (R) between the donor and acceptor sites: $E \propto (1/R^6)$. The useful range of R that can be determined by FRET is from about 10 to 60 Å when both donor and acceptor are aromatic molecules. This range is well within the long dimension of TnC. Discussed below is a strategy for determination of intersite distances in fsTnC by FRET and investigation of molecular changes in TnC as related to Ca^{2+} activation.

Tryptophanyl residues in proteins are good energy donor for FRET study. Appropriate energy acceptors include a number of naphthalene derivatives that have sulfhydryl reactivity. In a recent FRET study (She et al., 1988a), we used three chicken fsTnC mutants each containing a single Trp and a single Cys produced via site-directed mutagenesis and expressed in *E. coli* : (1) F22W/N52C, (2) A90W/N52C, and (3) F22W. In native protein, F22 is located in helix A, N52 in the B-C linker, and A90 in the N-terminal end of the interdomain D/E helix linker. The first two residues are in the regulatory N-domain, and N52 is the only one of the three residues that was predicted by the HMJ model to move away upon Ca^{2+} binding to the N-domain sites. Mutants (1) and (2) each also contained a third mutation (C101L) and were designed to study the distances between residues 22 and 52 (distance 22-52), and between residues 90 and 52 (distance 90-22). The endogenous single cysteine (Cys101) was retained in the single mutant F22W, and this mutant was used to determine the distance between residues 22 and 101 (distance 22-101). The single cysteine in the mutants was modified with the probe IAEDANS (5-(iodoacetamidoethyl)aminonaphthalene-1-sulfonic acid) and the attached probe served as energy acceptor from the tryptophan donor. The locations of the four residues are shown in Fig. 3. For determination of transfer efficiency, the intensity decay (fluorescence lifetime) of donor tryptophan was measured in the time domain. In the absence of the acceptor, the decay of F22 was monoexponential (lifetime 5.75 ns), but in the presence of the acceptor attached to Cys52, it became biexponential (Fig. 4A). The fast decay component (0.76 ns) with a 92% fractional amplitude is a clear demonstration of energy transfer from donor to acceptor. Fig. 4B shows the two decay curves obtained for the 4Ca-state. The tryptophan fluorescence of F22W responded to Ca^{2+} binding to sites I and II, but not to binding to sites III and IV (She et al., 1998b). Thus, changes in the intensity decay pattern shown in Fig. 4B were due entirely to the effect of bound Ca^{2+} at the regulatory N-domain. The data in Fig. 4 can be quantified and are displayed as distributions of the 22-52 distances (Fig. 5). Also shown in Fig. 5 are the distributions for the 90-52 distances (She et al., 1998a). The distance parameters recovered from the distributions are listed in Table 1. The transition of the regulatory N-domain from the apo state to the holo state ($2Mg^{2+}$-state to $4Ca^{2+}$-state) results in an increase in the mean distance between donor and acceptor sites by 8.4 and 10.7 Å for the two distances. These increases are accompanied by a large narrowing of the distribution with half-widths decreasing from > 10 Å to less than 4 Å. The magnitude of the Ca^{2+}-induced increase of the distance Trp22-Cys52 is remarkably similar to the increase between the α-carbon atoms of the two residues as calculated from crystallographic structures. These results also are consistent with NMR structures derived from the N-terminal half molecule N-fsTnC (Gagné et al., 1995) and whole fsTnC (Slupsky and Sykes, 1995).

The tryptophan fluorescence properties of F22W provide additional information regarding Ca^{2+}-induced opening of the regulatory N-domain (She et al., 1997). The very high quantum yield (0.33) and a single lifetime (5.65 ns) observed for the apo state of the N-domain are unusual for single-tryptophan proteins. In the holo state, the quantum yield is reduced to 0.25, the steady-state spectrum is slightly red-shifted, and the intensity becomes biexponential with two lifetimes. These and other spectroscopic properties suggest that, in the apo state of the N-domain, the Trp22 environment is highly non-polar and inaccessible to solvent, and in the holo state it becomes more polar and the residue is more accessible. These changes are compatible with an opening of the N-domain in which Trp22 (and helix A) becomes more exposed to solvent.

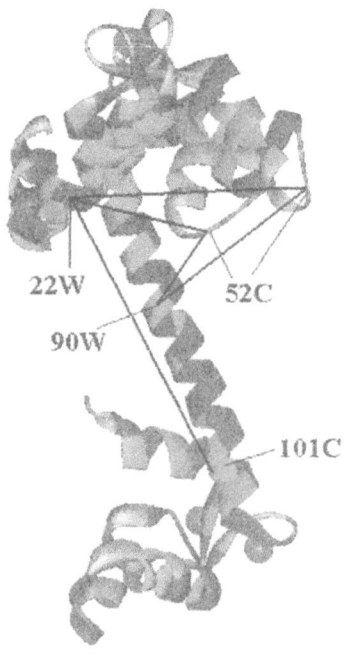

Figure 3. A ribbon representation of the crystal structure of fsTnC in the 2Ca-state. The two circles in the C-terminal domain (bottom of structure) are bound Ca^{2+} ions at sites III and IV. Sites I and II in the N-terminal domain are unoccupied. For FRET studies of intersite distances between residues 22 and 52, residues 90 and 52, and residues 22 and 101, several mutants were used. These involved the following mutations: 22F → 22W, 52N → 52C, 90A → 90W. Residue 101C is in the native sequence. The two helices in the upper right-hand corner are helices B and C. Residue 52C is located in the B-C linker (see Fig. 2). In the presence of bound Ca^{2+} at sites I and II in the N-domain, helices B and C and the B-C linker are proposed to move as a unit away from the long central helix (HMJ model, Fig. 2), and the new locations of this unit are clearly indicated by the two positions of residue 52C. The crystal structure is the PDB 5tnc structure, and the 4Ca-state model is based on the HMJ model (Herzberg et al., 1986). Color version on page 461.

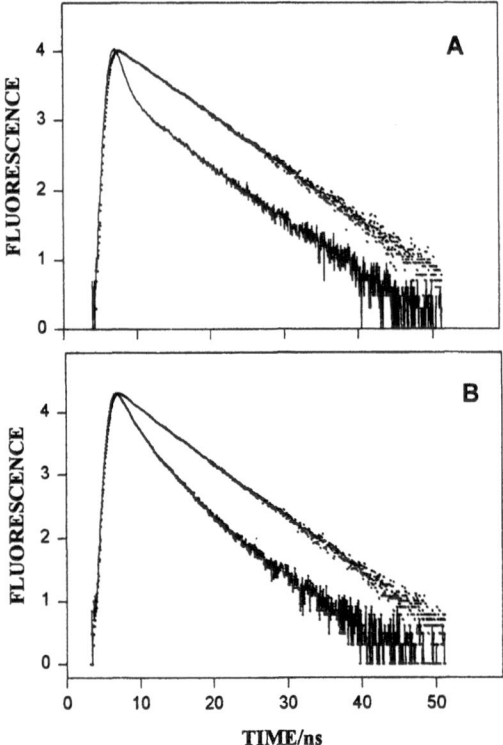

Figure 4. Fluorescence intensity decay curves of Trp22 in chicken fsTnC mutant F22W/N52C/C101L used to determine FRET efficiency and calculate intersite distance. (A) Decays determined in the presence of Mg^{2+}; the top curve is the decay of the Trp22 donor in the absence of energy transfer, and the lower curve is the decay in the presence of the acceptor IAEDANS attached to Cys52. (B) Decays determined in the presence of Mg^{2+} and Ca^{2+} ($4Ca^2$-state); top curve is the donor-alone decay, and lower curve is the decay in the presence of the acceptor. (From She et al., 1998).

The results obtained from three very different types of structural studies discussed above provide unequivocal evidence for an opening of the regulatory N-domain as the major conformational change induced by Ca^{2+} binding to this domain. The fsTnC mutants used for the FRET study were shown to bind Ca^{2+} at the sites I and II with apparent affinities (pK_d ~5.3-5.7) that fall within the narrow range expected of native fsTnC, and support Ca^{2+} activation of myofibrillar ATPase using fresh myofibrils in which endogenous fsTnC was exchanged with fsTnC mutants (She et al., 1998b). The Ca^{2+}-activated ATPase activities were within 10% of controls, and the acceptor label had little or only negligible effect on the activity. These results indicate that Ca^{2+} activation of ATPase is accompanied by an opening of the regulatory N-domain.

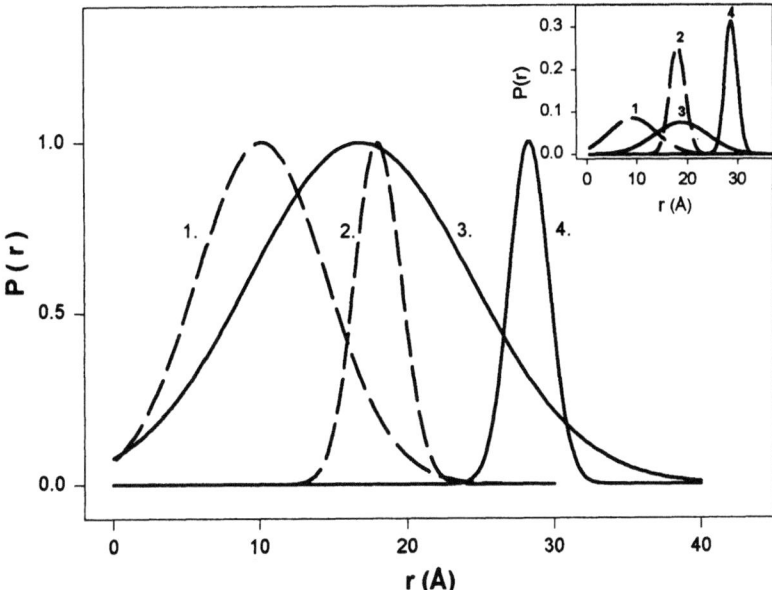

Figure 5. Distributions of distances for chicken fsTnC in which Trp is the FRET energy donor and IAEDANS attached to the single cysteine is the acceptor. Broken curves, distance between Trp22 and Cys52 (distance 22-52); solid curves, distance between Trp90 and Cys52 (distance 90-52). Curves 1 and 3, $2Mg^{2+}$-state; curves 2 and 4, $4Ca^{2+}$-state. These P(r) curves are peak-normalized to facilitate visual comparison. The inset shows area-normalized distributions for the same two sets of distances; note overlaps between curves 1 and 2 ($2Mg^{2+}$-state and $4Ca^2$-state, respectively) for distance 22-52, and between curves 3 and 4 for distance 90-52. (From She et al., 1998).

The FRET study revealed an additional feature of the global conformation of the Ca^{2+}-activated regulatory domain. Solution FRET results are frequently analyzed on the assumption that both donor and acceptor are rigidly held in space with a static protein conformation. This assumption is questionable since many proteins or segments of a given protein may exist in multiple conformational states or experience dynamic fluctuation. These conformational properties are contained in the distance distribution function (Lakowicz et al., 1988; Cheung, 1991), such as those shown in Fig. 5. The transition of the N-domain from the apo state to the holo state is accompanied by large decreases in the half-width of the distance distributions. These changes suggest a flexible N-domain in the apo state and a highly constrained N-domain in the holo state. An open conformation is needed to expose critical hydrophobic residues for interaction with TnI as the molecular trigger of the contractile cycle. Whether or not an open conformation is both necessary and sufficient for interaction with TnI may be dependent upon the bimolecular rate of interaction between the two protein sites and the rate at which

Table 1. Distribution of FRET intersite distances in fsTnC

Distance	State	$\langle r \rangle$ (Å)	hw (Å)	Distance change (Å) FRET	Crystallography
22-52	$2Mg^{2+}$	9.7	10.8		
	$4Ca^{2+}$	18.1	3.6	8.4	8.8
90-52	$2Mg^{2+}$	17.9	15.1		
	$4Ca^{2+}$	28.6	3.1	10.7	

Distance refers to donor-acceptor separation between the residues indicated. The donor is Trp22 or Trp90, and the acceptor is IAEDANS attached to the common residue Cys52. $\langle r \rangle$ is the mean distance of the Gaussian distribution, and hw is the half-width of the distribution. The distance change from FRET is the change in the observed mean distance between the $4Ca^{2+}$-state and the $2Mg^{2+}$-state. The distance change determined from crystallography between the two alpha carbon atoms of the two residues was obtained from two sets of atomic coordinates: (1) the crystal structure of the $2Ca^{2+}$-state of turkey fsTnC (Strynadka and James, 1988 and PDF 5tnc) for the $2Mg^{2+}$-state (apo N-domain), and (2) the crystal structure of the Ca^{2+}-saturated chicken N-fsTnC (residues 1-90) (Strynadka et al., 1997 and PDF 1AVS) for the $4Ca^{2+}$-state (Ca^{2+}-saturated N-domain). No comparison is included between FRET and crystallography results for distance 90-52 because the N-fsTnC crystal structure contains no information on the C-terminal residue 90.

the open conformation fluctuates. If the two rates are not compatible, this interaction may be difficult. The constrained conformation may be necessary to ensure the bimolecular reaction to take place with rates competent to support physiological demand.

The area-normalized distributions (inset, Fig. 5) show a 10% overlap between the conformations of the apo and holo states. The overlap suggests that a fraction of the TnC molecules with apo N-domain may be in the open/partially open conformation. There are two potential pathways by which activator Ca^{2+} confers an open and constrained conformation. One possibility is that the binding of Ca^{2+} to the closed/partially open conformation forces a domain opening and imposes a rigid open structure of the domain. The other possibility is that Ca^{2+} prefers binding to those apo N-domain with an open or partially open conformation and this binding shifts the closed = open equilibrium and stabilizes the open conformation. This initial Ca^{2+} complex may undergo further conformational rearrangement to reach the final open conformation. At this time, there is no strong evidence for either of these possibilities.

3.1.5. Equilibrium Ca^{2+} Binding to the Regulatory Domain

Bound Ca^{2+} is seven-coordinated to the 12-residue loop via oxygens in a pentagonal bipyramidal arrangement in all known EF-hands. The coordinating ligands are oxygen atoms from the side chains of amino acid residues or peptide carbonyl groups in the loop sequence and water molecules. The first residue in the loop sequence is always Asp and the 12th residue is always Glu. Mutation of either residue usually results in a total loss or a large reduction of Ca^{2+} affinity. The sequences of the two loops in the regulatory N-domain of chicken fsTnC are shown below:

	1	3	5	7	9		12
Loop I	[30]Asp-Ala-Asp-Gly-Gly-Gly-Asp-Ile-Ser-Thr-Lys-Glu[41]						
Loop II	[66]Asp-Ala-Asp-Gly-Ser-Gly-Thr-Ile-Asp-Phe-Glu-Glu[77]						

The positions of the six residues in the loop sequence contributing to liganding groups are indicated. These positions are also designated as X, Y, Z, -Y, -X, and -Z, respectively. In the site I of fsTnC, the residues in positions 1 and 3 each provides one direct oxygen ligand from the side chains, the Asp residue in position 7 provides its carbonyl oxygen atom, and the bidentate Glu residue in position 12 contributes both of its side chain oxygen atoms to Ca^{2+} coordination. The other two ligands in positions 5 and 9 are water molecules. In site II, the residue in position 5 is Ser, which contributes its side chain oxygen to coordination; a water molecule contributes to the 7th ligand at position 9. The contribution of the residue at position 7 is always from the carbonyl group in all EF-hands. The differences in the coordination ligands between sites I and II would suggest different Ca^{2+} affinities for the two sites, although it has been generally assumed that both sites have the same Ca^{2+} affinity. Resolution of the two affinities has been difficult and was not achieved until recently. A Ca^{2+} titration monitored by ($^1H,^{15}N$)-HMQC (heteronuclear multiple quantum coherence) NMR spectral changes yielded results indicating a stepwise Ca^{2+} binding to N-fsTnC with two Ca^{2+} dissociation constants in the range 0.8-3 µM and 5-23 µM (Li et al., 1995). Although there is no direct evidence for this, Ca^{2+} likely binds to site II first with the higher affinity because the cation at the site II is coordinated by six protein ligands and only one water molecule, whereas two water molecules contribute to the coordination at site I. This assignment is consistent with the general observation that the more water ligands a Ca^{2+}-binding site has, the weaker is the apparent Ca^{2+} affinity (Strynadka and James, 1988).

The structures of the apo N-domain of avian fsTnC derived from both crystallographic and NMR data show a kink in the B helix arising from irregular backbone dihedral angles at Glu41 (Herzberg and James, 1988; Satyshur et al., 1988; Gagné et al., 1994). This Glu residue is in position 12 within the binding loop of site I. In the absence of bound Ca^{2+} at site II, the kink would prevent Ca^{2+} binding to site I because of geometric constraint in which the two coordinating carboxylate oxygens of Glu41 side chain are too far from the Ca^{2+} position (10.8 Å) (Gagné et al., 1997). This kink is straightened upon Ca^{2+} binding to site II (Slupsky et al., 1995), resulting in reorientations of the helix B-helix C unit. This change in tertiary structure brings the coordination oxygens of Glu41 to within 3 Å of the Ca^{2+} position, thus allowing Ca^{2+} coordination to Glu41 and ensuring opening of the N-domain. This structural change appears critical for the opening of the regulatory domain and strongly coupled to Ca^{2+} affinity. In mutant E41A, the Ca^{2+} affinity is reduced by two orders of magnitude for site I and one order of magnitude for site II (Li et al., 1997). These reductions result from replacement of a bidentate coordination Glu residue to a nonligand Ala residue and consequently a reduction of two Ca^{2+}-coordinating ligands. The large suppressed Ca^{2+} affinities in the mutant disrupt the strong coupling between Ca^{2+} binding and induced structural changes. An earlier study of rabbit fsTnC mutants E38D and E38N in which the equivalent Glu was mutated had impaired Ca^{2+} binding and greatly suppressed force regulation in skinned fiber (Babu et al., 1992). The inability of the rabbit mutants to support Ca^{2+} activation of force has been corroborated by the results of a very recent study of mutant E41A from chicken which showed a recovery of only 19% of steady state tension in skinned fiber after reconstituted with the mutant (Pearlstone et al., 2000). If an open N-domain of TnC is critical for Ca^{2+} activation of force, these results would suggest that the regulatory domain of these TnC mutants remain closed even in the presence of the other two troponin subunits. Solutions NMR studies have recently shown that the binding of the fsTnI peptide P(115-131) to N-fsTnC mutant E41A in the presence of Ca^{2+} induced an open conformation, but this structure is not as fully open as Ca^{2+}-loaded N-fsTnC (McKay et al., 2000). This is an important result because the TnI peptide has been shown to interact with the hydrophobic patch of native fsTnC. Taken together, these different results strongly suggest that Ca^{2+} activation of force requires a transition between a closed and an open conformation of the regulatory domain of fsTnC and provide a general frame work to understand the molecular basis of Ca^{2+}-induced opening of the regulatory domain and Ca^{2+} activation of muscle function.

3.1.6. Kinetics of Ca^{2+} Binding to the Regulatory Domain of fsTnC.
Equilibrium studies such as those described in preceding sections have suggested that activation of actomyosin ATPase and force may be correlated with a Ca^{2+}-induced opening of the regulatory N-domain. The time course of the opening has not been directly established, but is expected to be correlated with the kinetics of Ca^{2+}-induced activation. Maximum tension in fsTnC is observed in about 12 -13 ms after excitation and decays after an additional 40-50 ms (Close, 1965). Thus, during a cycle of contraction and relaxation, Ca^{2+} must bind to the regulatory sites and induce conformational changes (such as an opening of the N-domain) in about 12 ms after excitation. Ca^{2+} dissociation from the sites and reversal of the Ca^{2+}-induced conformational change must occur within a range of time constants coincident with force decay. These physiologic constraints set a time window within which reversible Ca^{2+} binding and molecular changes must occur if these events are involved in Ca^{2+} activation.

Stopped-flow spectrometry has been the major technique to investigate the kinetic problem. In these experiments, a sample of apo TnC is rapidly mixed with a solution containing a well-defined concentration of Ca^{2+} ion which is controlled by the presence of a chelator such as EDTA/EGTA. The time course of the reaction between Ca^{2+} and TnC sites can be monitored by the change in an optical signal originating from TnC. Under optimal conditions, rapid mixing usually can be achieved within 1-2 ms, and fast kinetic events with time constants in the millisecond regime can be resolved. Several preparations of fsTnC have been used for this purpose. These include, but not limited to, native rabbit TnC labeled with extrinsic fluorescent probes such as dansylaziridine (Johnson et al., 1979) and I A N B D   (4 - (N-iodoacetoxyethyl-N-methyl)-7-nitrobenz-2-oxa-1,3-diazole) (Rosenfeld and Taylor, 1985a). Native TnC containing tyrosine (Iio and Knodo, 1980) and chicken mutants containing a single tryptophan (Johnson et. al., 1994) also have been used in stopped-flow studies. The extrinsic or intrinsic fluorophores are sensitive to environmental changes and are well suited for monitoring Ca^{2+}-binding kinetics. Ca^{2+} dissociation kinetics can be similarly monitored by mixing a sample of Ca^{2+}-saturated TnC with a solution containing a chelator which has a higher affinity for Ca^{2+} than TnC sites. The rate at which the chelator removes bound Ca^{2+} is the same as the rate at which bound Ca^{2+} dissociates. The dissociation kinetics can also be tracked by using fluorescent Ca^{2+} chelators (e.g., Quin-2) whose fluorescence increases upon binding Ca^{2+}. The important experimental design in these experiments must be capable of delineating the kinetics of Ca^{2+} interaction with the sites I and II in the N-domain from that involving the sites III and IV in the C-domain. The rates of Ca^{2+} binding and release are two orders of magnitude faster for the sites I and II than for the sites III and IV in the C-domain (Rosenfeld and Taylor, 1985b; Johnson et al., 1979; Johnson et al., 1994). Thus, the rates of binding and release from sites I and II are easily resolved from those slow rates at the other two sites, and are more commensurate with the speeds of contraction and relaxation, respectively, in fast skeletal muscle. These kinetic results provide additional support that the sites I and II in the N-domain, and not sites III and IV, are involved in Ca^{2+} activation.

Early kinetic studies interpreted the binding of Ca^{2+} to the high-affinity sites III and IV as a simple diffusion controlled process (Johnson et al., 1979). Other studies with a labeled fsTnC preparation from rabbit (Rosenfeld and Taylor, 1985a) established that the observed fluorescence signal for Ca^{2+} binding could be fitted to a single exponential function over a wide range of Ca^{2+} concentration. The apparent rate constant increased linearly with $[Ca^{2+}]$ in the low concentration range and approached a limiting value at high $[Ca^{2+}]$, and this plot could be fitted to a rectangular hyperbola. These results are consistent with a two-step binding mechanism in which the apparent bimolecular binding rate constant was 1.5×10^6 M^{-1} s^{-1}, a value more than two orders of magnitude lower than for a diffusion-controlled process.

More recently, the Ca^{2+} binding kinetics at the regulatory sites was studied with a single-tryptophan mutant (F29W) and a dansylaziridine-labeled preparation, both derived from chicken fsTnC (Johnson et al., 1994). The binding kinetic tracings were reported to be single-exponential over a very narrow Ca^{2+} concentration range (0.1 - 5.1 μM), and the apparent rate varied linearly with $[Ca^{2+}]$ within this narrow concentration range. The apparent bimolecular binding rate constant calculated from the kinetic data was in the range $(1-2) \times 10^8$ M^{-1} s^{-1}. At 5.1 μM $[Ca^{2+}]$, the observed rate was in excess of 700 s^{-1}, which likely approached the upper end of resolution by stopped-flow measurement. It is not known whether the observed rate would have saturated at high [Ca^{2+}] since no information was available on this. Although the calculated binding rate constant is valid, in the absence of additional data at higher [Ca^{2+}] it is not possible to establish the kinetic mechanism for the binding. Of importance here is whether there are first order transitions following formation of the initial encounter complex between Ca^{2+} the regulatory sites. In our laboratory, we also observed very high apparent Ca^{2+} binding rates with a chicken fsTnC mutant (F22W) and were unable to measure the rate beyond a few micromolar Ca^{2+}. The Ca^{2+} binding rate at the regulatory sites of chicken fsTnC appears to be too fast to be measured by stopped-flow at high [Ca^{2+}]. The rate of Ca^{2+} dissociation from the regulatory sites as determined by Quin-2 and EGTA with several different preparations of fsTnC was about 460 s^{-1} at 4°C. Thus, both the on-rate and off-rate for the regulatory sites are very fast. It must be noted that these rates are likely slowed down when TnC is incorporated into the ternary troponin complex or the thin filament.

Using a reconstituted rabbit fast skeletal troponin in which TnI was labeled with IANBD, Rosenfeld and Taylor (1985b) measured the Ca^{2+} concentration dependence of the pseudo-first-order rate for Ca^{2+} binding to the regulatory sites. The rate was saturated with a limiting value of 300 s^{-1}, and the apparent bimolecular binding rate constant was 1 x 10^7 M^{-1} s^{-1}. These results are consistent with a two-step binding model: Tn + 2Ca = Tn(Ca_2) = Tn*(Ca_2), where the first step is a rapid equilibrium for formation of an initial complex, followed by a first-order conformational change with an apparent rate constant about 300 s^{-1} at 4°C. The dissociation rate was 1.0 - 1.5 s^{-1}. These rates are considerably slower than the corresponding rates reported for isolated fsTnC as summarized above. Since Ca^{2+} activation in muscle occurs with the binding of the cation to TnC within the troponin complex, kinetic results obtained with troponin provide more meaningful insights on the activation mechanism than isolated TnC. It is noted that the two-step mechanism for the binding of Ca^{2+} to Tn holds also for the binding to the regulatory complex, Tn-Tm, with essentially the same binding and dissociation rates.

3.2. Models of the fsTnC-fsTnI Complex

There is general agreement that in the binary complex TnC -TnI from both skeletal (Farah et al., 1994) and cardiac muscle (Krudy et al., 1994) the two proteins are arranged in an anti-parallel manner. This polarity is preserved in the ternary complex with the extended TnT molecule (Fig. 1). Numerous biochemical and structural studies have been reported to identify interactions sites between the two proteins. The main approaches to these investigations include (1) chemical crosslinking, (2) multinuclear multidimensional NMR, and (3) FRET. Short peptides derived from TnI have provided insights not only on potential TnI segments that interact with the two domains and the central helix of TnC, but also on potential changes of specific interactions which occur during the transition of the regulatory N-domain of TnC from the apo state to the Ca^{2+}-loaded state. The latter information provides a molecular basis to propose mechanisms of Ca^{2+} activation.

As will be seen in Sec. 4, there are substantial differences in the interactions between TnC and TnI for the skeletal and cardiac isoforms, and possibly there are subtle differences in the activation mechanism for the two types of striated muscle. It should be noted, however, any mechanism based on biochemical and structural information derived from the simple binary TnC-TnT complex is a minimum mechanism that needs to be refined when additional/new information is available from more complex systems such as the regulated thin filament/actomyosin.

3.2.1. An Overview of the Ca^{2+}-dependent Activation and Regulation

Synthetic actomyosin can be prepared by mixing actin and myosin or actin and myosin subfragment 1 (acto-S1). The actomyosin Mg^{2+}-ATPase activity is relatively high and insensitive to Ca^{2+} concentration. Upon addition of TnI, the ATPase activity is inhibited and this TnI inhibitory effect is potentiated in the presence of Tm. The addition of TnC in the presence of Ca^{2+} (4Ca-state TnC) neutralizes the inhibition. Full activation of the ATPase to a level above that of acto-S1-Tm ATPase requires the presence of TnT. Since the affinities of TnC for TnI in the presence of Ca^{2+} ($\sim 10^9$ M^{-1}) and Mg^{2+} ($\sim 10^8$ M^{-1}) are very high (Ingraham and Swenson, 1984; Wang and Cheung, 1985; Cheung et al., 1987), the two proteins certainly remain associated under *in vivo* conditions favorable for both relaxation and contraction. The regulation by TnI must involve Ca^{2+}-induced altered interactions between TnC and TnI and between TnI and actin without complete dissociation of TnC from TnI.

Earlier biochemical studies focused on identification of regions of TnI that could be directly responsible for inhibition of actomyosin ATPase. The first such inhibitory region was identified from proteolytic fragments to be residues 96 - 116 (inhibitory region) in rabbit fsTnI (Syska et al., 1976), and the minimal sequence that caused inhibition was later localized by using synthetic peptides to residues 104-115 (Talbot and Hodges, 1981). This fsTnI peptide (104-115) was shown to interact with actin-Tm and inhibit acto-S1-Tm ATPase. This peptide inhibition was reversed by TnC in a Ca^{2+}-dependent manner, and the peptide exhibited properties very similar to full-length TnI including its ability to substitute for intact TnI in skinned muscle fibers to regulate Ca^{2+}-dependent contraction and relaxation (Van Eyk et al., 1993). These results strongly suggest that the Ca^{2+}-dependent switching between relaxation and contraction requires a molecular gating involving the inhibitory region of TnI which transduces the Ca^{2+} signal between TnC and actin-Tm. Recent NMR and other studies have provided additional information regarding other regions of fsTnI that bind specific regions of fsTnC. For example, peptides derived from the inhibitory region (residues 96-115) have been shown to bind Ca^{2+}-saturated C-domain of TnC (McKay et al., 1999), and peptides derived from TnI(116-131) (Tripet et al., 1997) or TnI(116-135) (McKay et al., 1999) with many positively charged residues have been found to bind to the exposed hydrophobic residues in the open Ca^{2+}-loaded regulatory N-domain. A short segment in TnI (residues 140 -148) is known as the second interaction site for actin-Tm. The binding of this second inhibitory region and TnI (140-148) to actin-Tm confers inhibition on ATPase activity. The N-terminal segment (residues 1- 40) of TnI has been shown to bind the C-domain of TnC, either in the presence of Ca^{2+} or in the presence of Mg^{2+}. Tripet et al. (1997) have summarized these results as a schematic model for Ca^{2+}-dependent regulation of muscle contraction (Fig. 6).

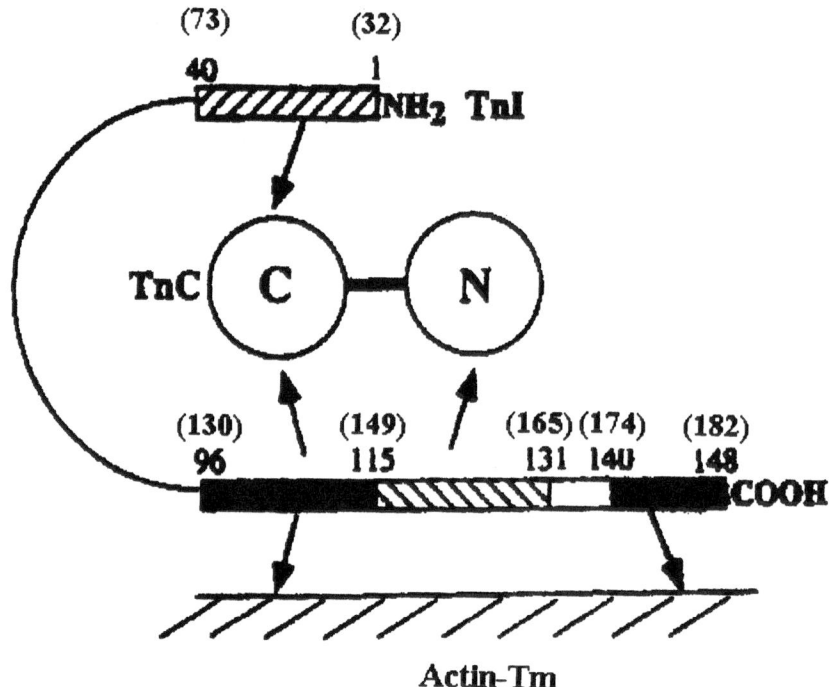

Actin-Tm

Figure 6. A TnI switching model for Ca^{2+} regulation showing the proposed interactions of TnI segments with TnC. Numbers indicate residues for rabbit fsTnI, and the numbers in parentheses correspond to the residues in rat cTnI. The N-terminal 31 residues of cTnI are not shown. Arrows indicate various regions of TnI involved in binding TnC and the actin-Tm. The two proteins are in an anti-parallel arrangement. In the absence of bound regulatory Ca^{2+} in the N-domain of TnC, the N-terminal segment of fsTnI (residues 1-40) interacts with the C-domain of TnC, and the inhibitory region (residues 96-115 in fsTnI) and the C-terminal region (residues 140-148) are bound to the actin-Tm filament, thus inhibiting force generation and the ATPase activity. In the presence of bound regulatory Ca^{2+} in the N-domain of TnC, the inhibitory region (residues 96-115) switches from the actin-Tm filament to the C-domain of TnC and displaces the bound N-terminal segment (residues 1-40 in fsTnI). The region with residues 116-131 (the Ca^{2+}-dependent interaction site with TnC) moves to the Ca^{2+}-loaded open regulatory N-domain of TnC and binds the exposed hydrophobic pocket. (Adapted from Tripet et al., 1997).

Figure 6 is based on the idea of a TnI switching between the N-terminal segment (residues 1- 40) and the inhibitory region (residues 96-115) of fsTnI in the interaction with the C-domain of fsTnC (Nagi and Hodges, 1992; Olah et al., 1994;

Ngai et al., 1994). In the 2Ca-state of TnC (relaxed muscle), TnI residues 1- 40 are bound to the C-domain and anchors the TnC to the actin-Tm filament via an interaction between TnT and TnI in the region of residues 41-95. The inhibitory region (residues 96-115) is tightly bound to actin-Tm, as is the second actin-Tm binding site region (residues 140-148). These tight interactions shut off interactions between actin-Tm and strong force-generating myosin crossbridge (S1) in the thick filament, thus inhibiting actomyosin ATPase and development of force on muscle fibers. When the regulatory N-domain of TnC is saturated with Ca^{2+} (4Ca-state), the TnI segment 116-131 moves toward the open N-domain and interacts with the exposed hydrophobic pocket in that domain. It is thought that the latter interaction pulls the inhibitory region and the 140-148 region away from actin and moves residues 96-148 toward TnC. This movement relieves the inhibitory effect of TnI and allows the inhibitory region (residues 96-115) to switch from the actin-Tm site to the C-domain of TnC. In this switching, the inhibitory region is postulated to displace the N-terminal TnI segment (residues 1- 40) from the C-domain. The effects of these changes in conformations and protein-protein interactions would cause movement of TnT and Tm, and initiation of contraction. Some of the studies that yielded information contributing to this switching scheme were done with peptides.

A recent NMR study detected no interaction between cardiac TnI inhibitory peptide (residues129-149) and the C-domain of Ca^{2+}- loaded cardiac TnC in the presence of a second cTnI peptide (residues 1-80) bound to the C-domain of cTnC (Abbott et al., 2000). This result shows that cardiac inhibitory peptide cannot displace bound N-terminal peptide of cTnI (residues 1-80) from the C-domain. A similar results was reported in another NMR study in which fsTnI inhibitory peptide (residues 96-115) and fsTnI N-terminal peptide corresponding to residues1-40 were used (Mercier et al., 2000). Thus, results from studies with both skeletal and cardiac proteins/peptides question the TnI switching model (Tripet et al., 1997; see Fig. 6) as a component of the mechanism of Ca^{2+} regulation in striated muscle

Biochemical results showed that isolated inhibitory peptides and the inhibitory region of full-length TnI were crosslinked to both the N- and C-domain of TnC. The binding of small inhibitory peptides was found by NMR to perturb both TnC domains (Dalgarno et al., 1982; Tsuda et al.,1992), and fluorescence titration studies showed binding of the inhibitory peptide to both TnC domains (Swenson and Fredricksen, 1992). There is evidence that the interactions of the C-domain of TnC with isolated peptide and with the peptide region in whole TnI may have subtle differences (Ngai and Hodges, 1982). In an earlier NMR study of the TnI peptide (104-115) bound to TnC, Campbell and Sykes (1991) reported that the peptide had two short helical regions interrupted by two central prolines, Pro109 and Pro110, and the bound peptide formed a hairpin conformation. The inhibitory peptide (residues 96-115) bound to a deuterated TnC was subsequently shown by NMR spectroscopy to have an extended conformation with a kink at Gly104 (Hernández et al., 1999). It is not clear where the TnC site is for the bound peptide, although it is likely in the C-domain. The properties previously observed for the shorter peptide (104-115) were not observed in the Hernández et al. study. These observed conformational differences likely are due to the different lengths of the   peptides and point to potential problems in extending results from short peptides to whole proteins. There is no direct evidence for the inhibitory region displacing bound TnI N-terminal segment (residues 1-40) in the presence of bound regulatory Ca^{2+}.

3.2.2. Low Resolution Models from Small-Angle Scattering Data

A low-resolution model for fsTnC-fsTnI (Olah and Trewhella, 1994) was constructed from small-angle X-ray and neutron scattering profiles determined for the binary complex formed between the 4Ca-state of native rabbit fsTnC and fsTnI in 2-3 M urea (Olah et al., 1994). In the scattering study, the Ca^{2+}-saturated TnC was deuterated and the TnI was not. Because of the different neutron scattering properties of hydrogen and deuterium, the scattering profiles of the two proteins in the complex can be distinguished. The model, which was constructed using a Monte Carlo modeling method, shows a TnI structure with two similar toroidal caps connected by a central helix of ~70 Å which spirals around the entire length of the dumbbell-shaped TnC structure. The centers- of-mass of the two structures and their long axes are approximately coincident (with a resolution of 10 Å) and the TnI mass is almost evenly distributed on either sides of the TnC center of mass. This feature would place the region of TnI with residues 50-130 in the central spiral, and this placement is consistent with previously published studies suggesting that the inhibitory region (residues 96-115) binds to a segment of the E helix in the C-domain in the 4Ca-state of TnC (Chong and Hodges, 1981; Leszyk et al., 1987). For an anti-parallel arrangement, the N-terminal residues 1- 40 of TnI would be in the cap region and associated with the TnC C-domain. This model would not allow extensive contacts between the TnI inhibitory region and the N- and C-domain of TnC, in contrast to reports that the inhibitory sequence interacts with both TnC domains (Leszyk et al., 1987; Kobayashi et al., 1991). Together with previously published biochemical data on the interaction between the two proteins, the model suggests that the Ca^{2+} switch may involve a signal transduction from one end of the complex to the other end via the spiral region of TnI. This possibility will be further discussed in Sec. 5.2.

A second model of the structures of fsTnC and fsTnI within the ternary complex reconstituted with fsTnT was reported in 1998 with chicken recombinant proteins, using neutron scattering data (Stone et al., 1998). Specific attention was paid to identify global changes in both TnC and TnI structures induced by bound Ca^{2+} at the two regulatory sites in TnC. The striking feature of this model is that bound TnI is represented by two domains: (1) a highly prolate ellipsoid of revolution consisting of about 35% of the mass and (2) a highly oblate ellipsoid of revolution containing about 65% of the mass. These two ellipsoids are not coaxial, but have arbitrary orientations (Fig 7). The binding of regulatory Ca^{2+} results in a structure in which

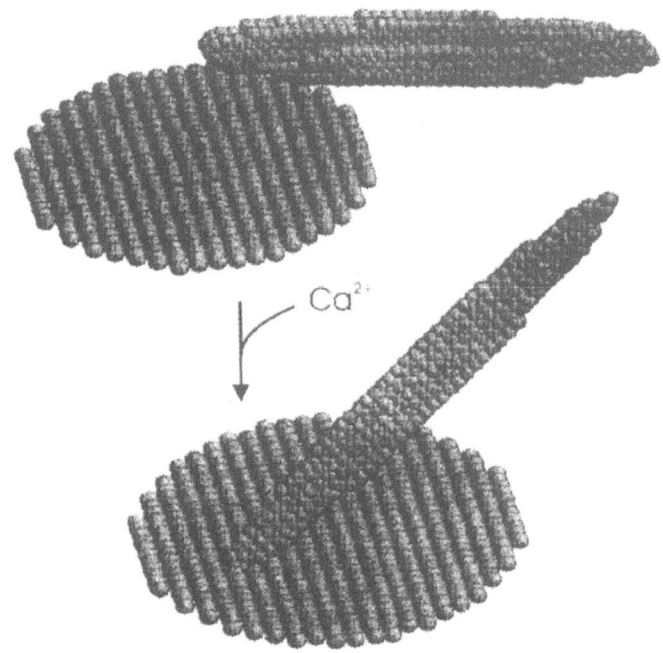

Figure 7. Space-filling models of recombinant chicken fsTnI derived from neutron scattering data obtained with the complex fsTnC-fsTnT-(deuterated fsTnI). Upper figure, without bound regulatory Ca^{2+} in the TnC; lower figure, in the presence of bound regulatory Ca^{2+}. (From Stone et al., 1988). Color version on page 462.

the small tail portion (prolate ellipsoid) moves about 15 Å closer to the large disk-like (oblate ellipsoid) component. It is not possible to identify the residues of the two ellipsoidal domains. This model differs significantly from the spiral TnI model proposed by Olah and Trewheller (1994) and summarized in the preceding paragraph. Other hydrodynamic data also indicate differences between the two models. The ellipsoidal model of TnI in the 4Ca-state of the ternary troponin complex has a substantially smaller radius of gyration than the spiral model of TnI in the binary TnC-TnI complex. The differences between the two models for bound TnC are small. It is not apparent how the two-domain TnI binds to an extended TnC with extensive contacts within the complex.

3.2.3. Two Recent Models of the fsTnC-fsTnI Complex

The two models summarized above provide low-resolution information on the overall topography of the TnC-TnI complex. High-resolution structural data are needed to define the secondary and tertiary structures of the complex. At this time, the binary complex is too large for detailed NMR studies with full-length proteins and has not been successfully crystalized. A model was constructed for the fsTnC-fsTnI complex based on crosslinking and FRET data (Luo et al., 2000a; Luo et al., 2000b). This model (Fig. 8) is a map showing proximity relationship of TnI with

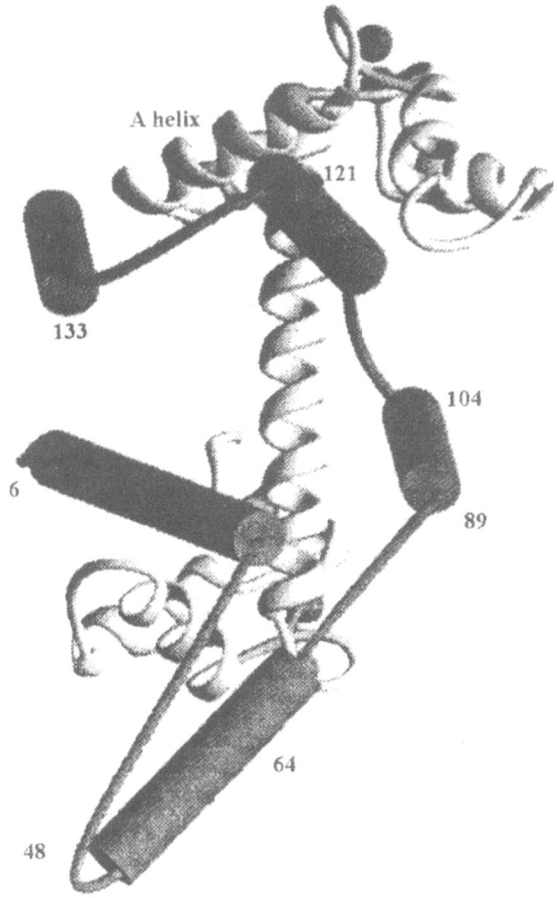

Figure 8. A model for the structure of rabbit fsTnC-fsTnI in the ternary fsTnC-fsTnI-fsTnT complex in the presence of bound Ca²⁺ at all four sites of TnC. The extended dumbbell-shaped helical molecule is TnC with the N-domain at top. Bound TnI is shown from residue 6 to 150. The C-terminal residues beyond 151 are not included. TnI segments shown as cylinders are α-helices predicted by sequence analysis. (From Luo et al., 2000a)
Color version on page 463.

TnC in their complexes. These workers used a photosensitive heterobifunctional crosslinker (4-maleimidobenzophenone) to label several specific single-cysteine mutants of rabbit fsTnI which was then reconstituted into the fsTnI-fsTnC complex, the reconstituted troponin, and the synthetic thin filament (Tn-Tm-actin). Upon photolysis, the reagent attached to different fsTnI sites was found to crosslink to different partners in the complexes. These crosslinking results and the observed FRET distances from fsTnI residue 6 to several sites in the N-domain of fsTnC indicate that the N-terminal region of fsTnI interacts with the C-terminal domain of fsTnC in both the absence and presence of Ca^{2+}.

To obtain a more detailed view of the dispositions of one protein in relation to the other, it is necessary to combine available NMR and crystallographic data derived from studies with whole proteins with model-building tools to construct TnC-TnI models. Such constructed models can be validated using available distance data. Tung et al. (2000) started with the atomic model of fsTnC as a scaffold for five contiguous segments of fsTnI between residues 3 and 134. The sixth segment comprising residues 135-181 is not included in the model because no structural constraints are available for this C-terminal segment, which is unstructured in the complex with the N-fsTnC. The TnI sequence was fitted to each of two previously published alternate neutron scattering profiles of TnI bound to the 4Ca-state of TnC (Olah and Trewhellar, 1994), using all-atom models for TnI. Two alternate models were obtained which differ in that TnI winds spirally about the TnC central helix in a left-handed sense in one model (L model) and in a right-handed sense in the other model R model). Figure 9 shows the L model. Two prominent features of the models stand out. The first feature is the structure of the inhibitory segment (segment IV, residues 95-114) which is modeled as a flexible β-hairpin in the Ca^{2+}-saturated fsTnC-fsTnI complex localized to the same region on the central helix of TnC in both L and R models. The choice of a β-hairpin as a model for the inhibitory segment in the complex is based, in part, on the observation that the sequence of this segment is similar to that of a β-hairpin in profilin. The long β-hairpin in profilin is a site for interaction with actin. In the proposed model, the flexible β-hairpin interacts alternately with TnC and actin during a cycle of contraction and relaxation.

The second feature is that the N-terminal segment (segment I, residues 3-33) is located in two distinctly different sites of TnC in the two models. These two features are inconsistent with the TnI switching model summarized by Tripet et al. (1997) in which the TnI inhibitory region and residues 1-40 compete for binding to the C-domain of $4Ca^{2+}$-TnC. The proposed β-hairpin structure for the inhibitory segment is different from the proposed α-helical structure for this segment in the model by Luo et al. (Fig. 8). Aside from these two main differences, there is apparent general agreement in the secondary structure of TnI between the model of Luo et al. and the models of Tung et al. With respect to the interaction between TnC and TnI in their complex, the model of Luo et al. and the L model of Tung et al. are generally consistent with each other, but with significant differences in the placement of TnI segments II and III. The R model is very different from the model of Luo et al. in the interaction along the entire length of TnI. These initial models serve as starting models for future studies of the fsTnC-fsTnI complex when additional structural data become available.

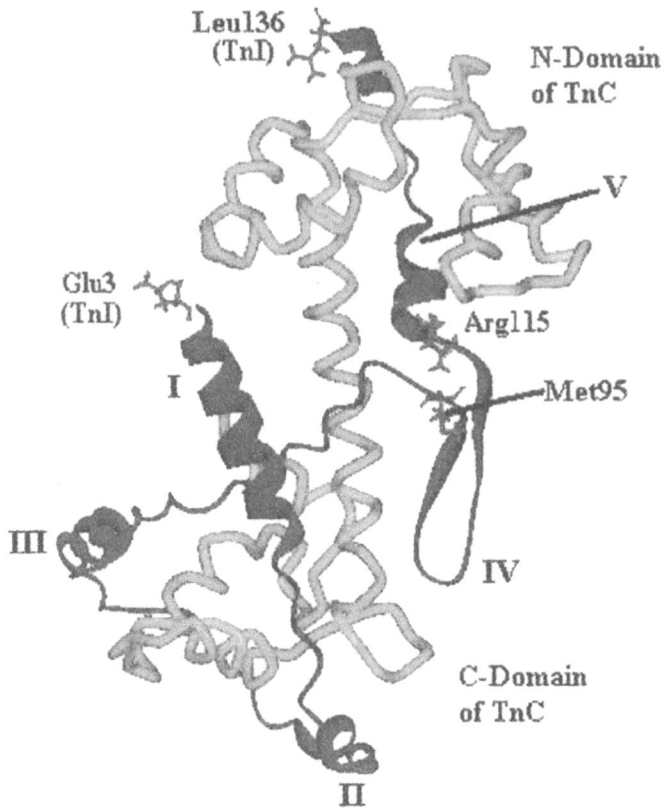

Figure 9. A ribbon representation of rabbit fsTnI in complex with rabbit fsTnC. TnI is shown in five segments, starting with Glu3 ending with Leu136. These segments are indicated by I-V, where segments I, II, III, and V are shown as helices, and segment IV is modeled as a flexible β-hairpin. Segments III and V wrap around TnC in a right-handed sense. TnC is shown with the N-domain at the top. (Adapted from Tung et al., 2000). Color version on page 464.

4. CARDIAC MUSCLE TROPONIN SUBUNITS

4.1. Structures of cTnC and cTnI

In comparison with the other troponin subunits, TnC is unique in that only two isoforms are known. The cardiac isoform is identical to TnC from slow skeletal muscle and has161 amino acid residues. Cardiac TnCs derived from both types of muscle are collectively designated as cTnC. From species to species, the cTnC sequence differs at only two or three positions. Chicken fsTnC (162 residues) has two additional residues at the N-terminus, and cTnC has an insertion (Val) at position 28. These differences account for the difference of one residue between the two isoforms of TnC. The sequences of the two binding loops in Ca^{2+}-binding sites I and II of cTnC are shown below:

$$\begin{array}{cccccc}
1 & 3 & 5 & 7 & 9 & 12
\end{array}$$

Loop I: ^{29}Leu-Gly-Ala- Glu-Asp-Gly-Cys-Ile- Ser- Thr-Lys- Glu^{40}

Loop II: ^{65}Asp-Glu-Asp-Gly-Ser- Gly-Thr-Val-Asp-Phe-Asp-Glu^{76}

When compared with the loop I sequence of fsTnC (Sec. 3.1.5), cTnC loop I has two substitutions at positions 1 and 3. These substitutions abolish the ability of the site I loop to coordinate Ca^{2+}. Thus, site II is the only regulatory Ca^{2+} site in cTnC. Native cTnC has two cysteine residues, one within the inactive loop I (Cys35) and the other (Cys84) located in the C-terminal end of helix D in the N- domain. These endogenous cysteines have been extensively used as sites for attachment of extrinsic probes.

High-resolution NMR structures of whole cTnC (Sia et al., 1997), isolated recombinant N-domains with residues 1-89 (Spyracopoulos et al., 1997) and residues 1-91 (Pääkkönen et al., 1998), and isolated C-domain with residues 81-161 (Gasmi-Seabrook et al., 1999) have been solved. Although the crystal structure of the $3Ca^{2+}$-state of cTnC is not available, the structure of the $3Ca^{2+}$- state of cTnC complexed with the Ca^{2+}-sensitizer bepridil has been determined (Li et al., 2000) . The NMR structures have provided insights into molecular changes in cTnC upon binding regulatory Ca^{2+}. These changes will be discussed in Sec. 4.3.

Cardiac TnI is substantially longer than the skeletal isoform due to a unique 32-residue extension at the N-terminus. This cardiac muscle-specific extension contains two consecutive serine residues (Ser23 and Ser24) that are sites of phosphorylation by PKA (protein kinase A). Early studies demonstrated that phosphorylation of cTnI resulted in an increase in the Ca^{2+} concentration needed to achieve half-maximal ATPase activity (Holroyde, et al., 1979; Bailin, 1979) or steady-state force (Mope, et al., 1980) determined with isolated myofibrils or skinned fiber preparations, respectively. This requirement of a higher Ca^{2+} concentration is usually expressed as a decrease in pCa which corresponds to a loss in Ca^{2+} sensitivity. In addition, cTnI phosphorylation also reduced the amount of myofibrillar bound Ca^{2+} (Holroyde et al, 1979; Holroyde, et al., 1980). The phosphorylation was later shown to increase the rate of Ca^{2+} dissociation from cardiac troponin reconstituted with phosphorylated cTnI by a factor of 1.45 (Robertson et al., 1982). These results suggest that phosphorylation of cTnI by PKA is potentially an important signaling pathway in cardiac myofilaments.

TnI has a single tryptophan residue that is highly conserved among many isoforms and this is Trp192 in cTnI. The apparent rotational correlation time of cTnI derived from anisotropy decay of the tryptophan residue is 23.5 ns, indicative of a very asymmetric shape with an axial ratio of 4-5 (Liao and Cheung, 1992), in general agreement with the conclusion of an early hydrodynamic study (Byers and Kay,1982). Upon phosphorylation, the correlation time decreased to 14.6 ns, suggesting that phosphorylated cTnI is considerably more symmetric than nonphosphorylated cTnI with an axial ratio of about 2. This reduction in cTnI correlation time is carried over to the complex cTnI-cTnC, and the phosphorylation-induced global conformational change of cTnI persists in the binary complex. In Sec. 5.1, this conformational change will be examined in some detail.

4.2. Kinetics of Ca^{2+} Binding to Cardiac Troponin

Of special interest in the kinetics of Ca^{2+} binding to cTnC is the kinetic mechanism for binding the cation to the single regulatory site in the N-domain. Prior to a complete kinetic study, it is necessary to first obtain equilibrium binding data to establish the binding constant of cTnC for Ca^{2+}. Although radioactive ^{45}Ca was initially used to determine the binding constant, this method is not likely very sensitive to detect small changes in the affinity. Since native cTnC has two cysteines (Cys35 and Cys84), the fluorescent sulfhydryl probe IAANS (2-[(4'-iodoacetamido)anilino]-naphthalene-6-sulfonic acid) was covalently attached to the two residues and the probe fluorescence intensity was used to monitor Ca^{2+} binding to cTnC (Johnson et al., 1980). The IAANS signal can be used in stopped flow kinetic experiments, thus allowing both equilibrium and kinetic experiments to be performed with the same fluorescent probe.

We labeled Cys84 with IAANS, using a single-cysteine cTnC mutant cTnC(C35S) derived from rat heart, and performed stopped-flow Ca^{2+} binding experiments by mixing the labeled protein with Ca^{2+} over the range 0.1- 480 µM (Dong et al., 1996). In the presence of Mg^{2+} (to prevent Ca^{2+} from binding to the sites III and IV), the kinetic tracings were resolved into two phases with positive amplitude which were completed in less than 100 ms. At 4°C, the apparent rate of the fast phase increased rapidly with increasing Ca^{2+} concentration, reaching a maximum of ~ 590 s^{-1}, and that of the slow phase was approximately 100 s^{-1} and did not depend on Ca^{2+} concentration. The binding rates are attributed to the binding of Ca^{2+} to the regulatory site II. Dissociation of bound Ca^{2+} from the regulatory site occurred with a rate of 102 s^{-1}, whereas dissociation from the sites III and IV was about two orders of magnitude slower. Shown in Fig 10 is a plot of the two observed binding rates vs. free $[Ca^{2+}]$. The upper curve is a rectangular hyperbola. These results are consistent with a three-step scheme for the reversible binding of Ca^{2+} to the regulatory site of isolated cTnC.

$$Ca + TnC \overset{K_o}{=} CaTnC \underset{k_{-1}}{\overset{k_1}{=}} (CaTnC)* \underset{k_{-2}}{\overset{k_2}{=}} (CaTnC)**$$

where the asterisks denote states with enhanced fluorescence. The kinetic parameters for this scheme are summarized in Table 2. The apparent bimolecular rate constant for Ca^{2+} binding is $K_o k_1 = 1.4 \times 10^8$ M^{-1} s^{-1}. The sum of k_1 and k_{-1} is

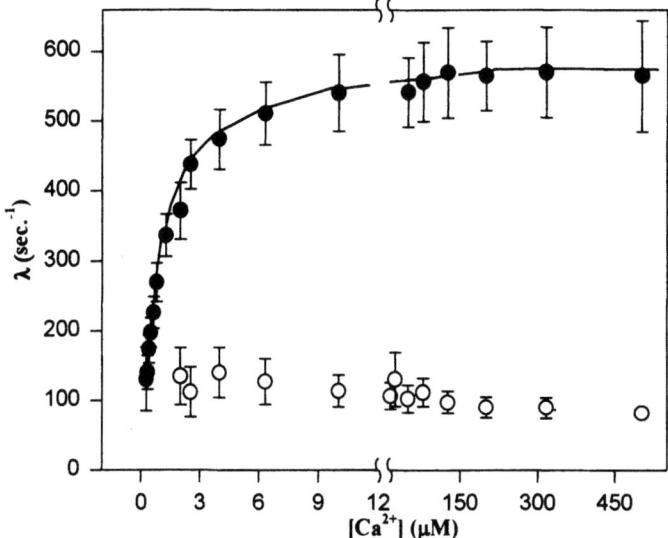

Figure 10. Ca^{2+} *dependence of the two observed binding rate constants (λ) for* Ca^{2+} *binding to the single regulatory site (site II) of cTnC labeled with IAANS at Cys84 at 4°C , using mutant cTnC (C35S). The fast rate increases rapidly over a narrow range of* $[Ca^{2+}]$ *and approaches a limiting value of ca. 590* s^{-1} *at high* $[Ca^{2+}]$*. The slow rate is insensitive to* $[Ca^{2+}]$ *and is approximately 100* s^{-1}*. The initial slope of the upper hyperbolic curve is equal to the bimolecular binding rate constant for the model given in the text. (From Dong et al., 1996).*

approximately 100 s $^{-1}$. Although the bimolecular rate constant is on the order of 10^8 M $^{-1}$ s $^{-1}$, the binding reaction is not a simple diffusion-controlled reaction. In this model, the first step is a rapid equilibrium for formation of an initial complex, followed by two first-order transitions which can be attributed to Ca^{2+}-induced structural changes. These changes are considerably slower than the binding of Ca^{2+}.

The rate of the decrease of the IAANS fluorescence (102 s^{-1}) determined in dissociation experiment was obtained using EGTA to displace bound Ca^{2+} and reflected dissociation of bound Ca^{2+}, rather than protein conformational changes following Ca^{2+} removal. This interpretation was supported in an experiment by mixing unlabeled cTnC saturated with Ca^{2+} at all three sites with the fluorescent chelator Quin-2. The observed kinetic tracings were biphasic with positive amplitude (increasing fluorescence intensity). The two observed rates were 133 s $^{-1}$ (35% amplitude) and 7 s $^{-1}$ (65% amplitude). Since the Quin-2 signal arose not from cTnC, these amplitudes suggested that the fast phase reflected removal of bound Ca^{2+} from the single regulatory site, and the slow phase was associated with Ca^{2+}

Table 2. Kinetic parameters for the binding of Ca^{2+} to cardiac troponin

Protein	Dissociate rate		K_o	K_ok_1	k_1	k_{-1}	k_2	k_{-2}	K_{eq}
	fast	slow							
	s^{-1}		M^{-1}	$M^{-1}s^{-1}$	s^{-1}				$10^5\ M^{-1}$
cTnC	101.0	—	0.30×10^6	14×10^7	470	117	20	80	3.2
cTnC–cTnI	22.6	4.3	2.6×10^6	5.9×10^7	22.7	14.0	2.4	2.0	49.0
cTn	20.6	4.7	2.2×10^6	5.1×10^7	23.2	13.3	2.7	2.1	47.3
p-(cTnC–cTnI)	58.0	10.1	0.37×10^6	2.2×10^7	58.3	30.3	20.1	14.9	9.2
p-(cTn)	53.0	9.3	0.33×10^6	1.7×10^7	52.0	28.2	20.0	14.0	8.7

Mutant cTnC(C35S) was labeled at Cys84 with IAANS and used as unbound cTnC in stopped-flow experiments at 4°C. In cTnC-cTnI and cTn (cTnC-cTnI-cTnT), mutant cTnC(C85S) was labeled at Cys35 and used in reconstitution with native cTnI and cTnT into the binary and ternary complexes. Dissociation of Ca^{2+} from unbound cTnC was monophasic with a single dissociation rate constant, and dissociation from the binary and ternary complexes was biphasic with two rate constants. The other parameters are derived from observed rate data for the three-step sequential reaction. p-(cTnC-cTnI) and p-(cTn) were reconstituted with PKA-phosphorylated cTnI. The results for phosphorylated proteins are discussed in Sec. 5.1. K_{eq} is the overall equilibrium binding constant separately determined from Ca^{2+} titration experiments.

dissociation from the sites III and IV. These results are in agreement with previous Ca^{2+} dissociation data (dissociation rates162 s^{-1} and 0.6-1.6 s^{-1} from site II and from sites III and IV, respectively, at 15°C) obtained from native cTnC using Quin-2 as the chelator (Smith and England, 1990; MacLachlan et al.,1990), but in conflict with another study which reported a dissociation rate > 700 s^{-1} as monitored by Quin-2 (Hazard et al., 1998).

Dong et al. (1997a) reported a similar kinetic study of Ca^{2+} binding to the binary complex cTnC-cTnI and the ternary complex cTnC-cTnI-cTnT (cTn), using cTnC mutant (C84S) labeled with IAANS at Cys35. In the complexes, the fluorescence of IAANS attached to Cys84 was not sensitive to Ca^{2+}, but when the probe was attached to Cys35 the fluorescence in both complexes decreased by a factor of 3 (Dong et al., 1997b), thus providing a convenient signal to monitor the Ca^{2+} binding and dissociation kinetics. Very similar Ca^{2+} binding and dissociation kinetics was observed for both the binary and ternary complexes. For the binding kinetics, the kinetic tracings obtained from reconstituted cTn were resolved into two phases as with uncomplexed cTnC. In the presence of Mg^{2+}, the rate of the fast phase increased hyperbolically with increasing $[Ca^{2+}]$, reaching a maximum of ~ 35 s^{-1} at 4 °C, and the rate of the slow phase was about 5 s^{-1} and did not depend on Ca^{2+} concentration. These rates are at least an order of magnitude smaller than those observed with free cTnC. Dissociation of bound Ca^{2+} from the regulatory site in the complexes as sensed by IAANS occurred in two phases, with rates in the range 21-23 s^{-1} and 4.3-4.7 s^{-1}, rather than in a single phase as was observed with isolated cTnC (rate 101 s^{-1}) . Results from dissociation experiments performed with native subunits and Quin-2 supported the conclusion that the biphasic IAANS transient reflected removal of bound Ca^{2+} from the regulatory site, and this removal was not limited by slow conformational changes. These results are adequately described by a three-step sequential model similarly to the previously reported mechanism for the binding of Ca^{2+} to free cTnC:

$$Ca + cTn \overset{K_o}{=} Ca\text{-}cTn \underset{k_{-1}}{\overset{k_1}{=}} (Ca\text{-}cTn)^* \underset{k_{-2}}{\overset{k_2}{=}} (Ca\text{-}cTn)^{**}$$

The kinetic parameters for this model are given in Table 2. The apparent bimolecular binding rate constant (K_ok_1) is 5.1 x 10^7 M^{-1} s^{-1}, a factor of 3 smaller than that for cTnC. Unpublished results from the same group showed that the kinetics of Ca^{2+} binding to cTn-Tm and regulated thin filament (actin-cTn-Tm) followed closely the kinetics of Ca^{2+} binding to free cTn and was also consistent with a three-step sequential mechanism. The bimolecular binding rate constants determined with cTn-Tm and actin-cTn-Tm are ~ 4.8 x 10^7 M^{-1} s^{-1} and ~2.3 x 10^7 M^{-1} s^{-1}, respectively. In dissociation experiments of Ca^{2+} from native cTn-Tm, the Quin-2 transients were biphasic with rates 33 s^{-1} and 7.2 s^{-1}, very similar to the previously reported Ca^{2+} dissociation rates (23 s^{-1} and 1.2 s^{-1}) for cTn-Tm using Quin-2 as the chelator (Smith and England, 1990). These results indicate that the binding of regulatory Ca^{2+} to either free cTn or regulated thin filament is not diffusion-controlled, and that the binding step is followed by two slow first-order transitions which are attributed to conformational changes with time constants much slower than the binding step.

For free cTnC, step 1 (the first first-order transition) is thermodynamically favorable with an equilibrium constant of 4 (k_1/k_{-1} = 470/117), but step 2 is unfavorable (k_2/k_{-2} = 20/80). The dominant equilibrium species is (Ca-cTnC)*, and the single-exponential IAANS transient observed in Ca^{2+} dissociation experiment likely arises from this dominant species. With cTn (and the binary complex), both steps 1 and 2 are thermodynamically favorable with comparable equilibrium constants (1.2 and 1.3). The IAANS transient observed in dissociation experiments is expected to arise from both Ca^{2+}- bound species, (Ca-cTn)* and (Ca-cTn)**. Thus, the biphasic dissociation kinetics with cTn (and cTnC-cTnI) likely reflects dissociation from both species. The overall equilibrium binding constant of the Ca^{2+}-bound complex is determined by the product of the individual equilibrium constants of the three steps. When compared to free cTnC, the presence of bound cTnI in cTnC-cTnI and cTn increases the equilibrium constant of the initial bimolecular kinetic step (K_o) by a factor of 9. This stabilization and the stabilization of the second first-order step are expected to result in a large increase in the overall equilibrium Ca^{2+} binding constant for both cTn and cTnC-cTnI relative to the Ca^{2+} affinity of free cTnC. The experimental equilibrium binding constant (K_{eq}) listed in Table 2 were obtained independently of kinetic measurements and confirm this expectation. The enhanced Ca^{2+} affinity of troponin is well known for both skeletal and cardiac isoforms, and has been attributed to a free energy coupling by TnI. The kinetic results show for the first time that this free energy coupling is related to alteration of the rate constants of the elementary kinetic steps of Ca^{2+} binding to cTnC. It is also known that TnT does not contribute to the free energy coupling in the stabilization of the Ca^{2+}-troponin complex. This is also reflected in the kinetic results that both cTnC-cTnI and cTn have essentially the same rates. It should not be concluded, however, that cTnT and fsTnT play no role in Ca^{2+} activation. It has a potentiating effect via an interaction with Tm in fully regulated thin filament. This aspect of Ca^{2+} regulation is not discussed in this chapter.

The kinetic results suggest that at a saturating level of Ca^{2+} at 4°C, the conformational transitions in cardiac troponin would take about 40 ms to reach a 95% completion, and the dissociation of bound Ca^{2+} would take about 50 ms. The time-to-peak tension after excitation and the relaxation time are species-dependent and can vary widely with temperature. Under physiological conditions, the observed Ca^{2+} transients could be faster. If tension transients follow Ca^{2+} transients, the anticipated conformational transition rates may be kinetically competent to support contraction. In contrast, the time constants for the binding of Ca^{2+} to free cTnC are one order of magnitude faster than for cTn/cTn-Tm and may be too fast to be compatible with physiological events.

A recent stopped-flow kinetic study of Ca^{2+} binding to free cTnC with both native protein and mutant C35S labeled with IAANS at Cys84 proposed a two-step mechanism in which the binding step is followed by one conformational step (Hazard et al.1998). These authors did not observe biphasic IAANS transients in binding experiments, and reported the rate of the IAANS transient as Ca^{2+}-independent with values ($\sim$210 s^{-1}) which are within a factor of two of the Ca^{2+}-independent rate of the slow phase reported by Dong et al. (1996). The rate from the single phase of IAANS transient was attributed to Ca^{2+}-induced protein conformational change. It is not obvious why the previously observed fast Ca^{2+}-dependent phase of IAANS transient was not detected in this study. The macroscopic Ca^{2+} on-rate ($k_{on} = K_{eq} \times k_{off}$), which was calculated from the experimentally determined equilibrium Ca^{2+} binding constant and an observed Ca^{2+} dissociation rate (> 700 s^{-1}), was in good agreement with the bimolecular binding rate constant determined by Dong et al. (1996). As already indicated in a preceding paragraph, the very large dissociation rate was not observed by other investigators. More recently, Kohout and Falke (1999) reported detection of at least two conformational steps following Ca^{2+} binding to the regulatory domain of cTnC. No detail, however, was available in this report.

Previous kinetic studies of Ca^{2+} binding to cTnC and cardiac troponin focused largely on dissociation kinetics; only a few studies have reported both binding and dissociation kinetics over a wide range of Ca^{2+} concentration required to establish kinetic models. Our recent studies have established that the binding of Ca^{2+} to the regulatory site of cTnC is rapid, followed by two slow structural transitions. With reconstituted cTn and cTn-Tm/actin-cTn-Tm, the kinetic events are considerably slower with time constants that may be physiologically competent to support activation and relaxation of cardiac muscle. Of interest here is the origin of the structural transitions that occur with observed time constants. Kinetic studies alone cannot provide an answer, but it can be inferred from structural information. The transitions may involve movements of helices B and C and the B-C linker, and these changes will be discussed in the next section.

4.3. Structural Signaling of Ca^{2+} Activation

4.3.1. High-Resolution Ca^{2+}-induced Structural Changes in cTnC

Experimental studies have firmly established that the binding of regulatory Ca^{2+} induces movements of helices B and C and the linker between the two helices in skeletal TnC. These are exciting findings that have provided a basis for understanding the molecular aspects of Ca^{2+} activation of skeletal muscle functions, and verified the HMJ model for a Ca^{2+}-induced open conformation of the regulatory domain in the 4Ca-state of fsTnC. A computer model for the $3Ca^{2+}$-state of cTnC also proposed an open regulatory domain (Ovaska and Taskinen, 1991), and it was anticipated that regulatory Ca^{2+} binding to cTnC would elicit a similar open conformation in the regulatory N-domain as in fsTnC. In the absence of experimental results, much of the thinking about the molecular aspects of Ca^{2+} activation in cardiac muscle was largely based on what was known about fsTnC, in spite of the knowledge that cTnC has only one regulatory Ca^{2+} binding site. Many binding studies of potential therapeutic agents to increase the Ca^{2+} sensitivity of cardiac myofilaments were interpreted on the basis of an open conformation of the Ca^{2+}-loaded regulatory domain of cTnC.

The first unequivocal experimental evidence that the conformation of the Ca^{2-} loaded regulatory domain of cTnC does not resemble that previously observed for 4Ca-fsTnC came from an analysis of triple-resonance NMR experiments. In the 2Ca-state of fsTnC the two interhelical angles between helices A and B and between helices C and D (in apo regulatory N-domain) are 138 and 145 degrees, respectively. As has been pointed out in Sec. 3.1.1, these large angles (>> 90 degrees) define a closed domain conformation. In the 4Ca-state of fsTnC, the corresponding interhelical angles are reduced to 81 ± 5 and 78 ± 7 degrees, respectively. These smaller angles are structural features that define an open domain conformation. In the 3Ca-state of chicken cTnC, the A-B and C-D interhelical angles are 138 ± 3 and 108 ± 4 degrees, respectively (Sia et al., 1997), indicating that the Ca^{2+}-loaded regulatory N-domain of cTnC remains in an essentially closed conformation and is significantly more compact than the corresponding regulatory domain of fsTnC. In addition, the total exposed surface area observed in the cTnC regulatory domain is approximately 800 $Å^2$ less than in the fsTnC regulatory domain. Subsequent NMR studies of isolated regulatory N-domain of human cTnC residues 1-89 (Spyracopoulos et al., 1997) and residues 1-91 (Pääkkönen et al., 1998) have confirmed and extended the initial observations. Results from ^{15}N NMR resonances revealed the existence of conformational exchange involving several residues in the Ca^{2+}-saturated regulatory domain (Pääkkönen et al., 1998; Gaponeko et al., 1999). These findings are consistent with an equilibrium between a closed and open conformation of the regulatory domain.

It is of interest to compare the structures of both the sites I and II of fsTnC with those of cTnC for clues as to why the regulatory domain of cTnC remains closed in the 3Ca-state. As pointed out in Sec. 3.1.5, Glu41 in fsTnC is the 12th residue of the Ca^{2+} binding loop I and contributes both oxygens of its carboxylate group to Ca^{2+} coordination. The presence of a kink at this residue prevents Ca^{2+} binding to site I, unless site II is already filled. The structure of the apo regulatory domain of cTnC reveals a kink at Glu40 (corresponding to Glu41 in fsTnC), and this kink is not straightened upon Ca^{2+}binding to site II. It was suggested that this structural feature may be responsible for site I in cTnC to be defective in coordinating Ca^{2+}. By inference, bound Ca^{2+} at site I in cTnC would be needed to achieve an opening of the regulatory domain. The question is whether Ca^{2+} activation of cardiac muscle functions follows the same mechanisms that have been put forward for skeletal muscle.

4.3.2. Conformational Changes in cTnC N-domain Induced by Ca^{2+} and Bound cTnI
The apparent inability of bound regulatory Ca^{2+} to induce an opening of the regulatory domain of cTnC was corroborated by FRET studies which were designed for comparison with previous FRET studies of fsTnC as described in Sec. 3.1.4. We determined three intersite distances of cTnC between pairs of residues: (1) Trp20-Cys51, (2) Trp12-Cys51, and (3) Trp20-Cys89. The first distance is equivalent to the distance Trp22-Cys52 in fsTnC, with the tryptophan as the energy donor and the probe IAEDANS attached to the cysteine as energy acceptor. The second distance is

between a site in the N helix and the common acceptor site Cys51, and the third distance is between the same donor site as in (1) and the acceptor site Cys89 located at the C-terminus of helix D. Initial measurements of distance (1) showed that a substantial increase in the distance could be elicited by Ca^{2+} binding to the site II only in the presence of bound cTnI (Gong et al., 1998). This was the first demonstration of bound cTnI as a prerequisite to achieve a Ca^{2+}-induced domain opening in cTnC.

Figure 11 shows the distributions of two of the distances which provide additional insights on the properties of the cTnC regulatory domain (Dong et al., 1999). The distance parameters from the distributions are given in Table 3. In the apo state of the regulatory domain, the mean Trp20-Cys51 distance (15.7 Å) is considerably longer than the corresponding distance Trp22-Cys52 in fsTnC (9.2 Å, Table 1), with a substantially smaller half-width (2.8 Å) for the distribution. When the single regulatory site is saturated with Ca^{2+}, the mean distance increases by ~1 Å with little or no change in the half-width. A careful analysis of the χ_R^2 surface of the mean distances indicated that this small   increase is not statistically significant. When compared with the apo regulatory domain of fsTnC, these distributions parameters suggest that apo regulatory domain of cTnC is already partially open and highly constrained.   In the absence of Ca^{2+}, bound cTnI does not change the distribution (curve 1 vs. curve 3, Fig. 11A), but in the Ca^{2+} saturated cTnC-cTnI complex, the distribution (curve 4) shifts toward longer distances by 6.5 Å, leaving the half-width unchanged. Similar changes were observed for the distance Trp12-Cys51. These results are unequivocal evidence that bound cTnI is a prerequisite to achieve an open Ca^{2+}-saturated regulatory domain in cTnC. The properties of this open domain elicited by both bound Ca^{2+} and cTnI are similar to those previously observed for the regulatory domain of fsTnC in the presence of bound Ca^{2+} and absence of bound fsTnI.

The requirement of bound cTnI for domain opening leads to the question of which segment of cTnI would be responsible for the observed global conformational change. Several peptides derived from cTnI were used to investigate this problem: (1) P(1-128), (2) P(128-211), (3) (129-149), (4) P(150-166), and (5) P(129-166). The N-terminal half peptide P(1-128) had little or no effect on the cTnC distance 12-51, while the C-terminal half cTnI peptide P(128-211) induced an increase in distance approaching that of cTnI itself. P(129-149) was no more effective than P(1-128), whereas P(150-166) was substantially more effective than P(129-149) in inducing an increase in the 12-51 distance. The effectiveness was enhanced when P(150-166) was extended in the N-terminal end to include the sequence 129-149. The enhanced effectiveness observed with P(129-166) (peptide 5) was not detected when individual peptides P(129-149) and P(150-166) were added together to cTnC, suggesting an important role of the sequence 150-166 in opening the regulatory domain upon Ca^{2+} binding to site II of cTnC. This conclusion is corroborated by an NMR study of the structure of cTnC N-domain complexed with a cTnI peptide consisting of residues 147-163 (Li et al., 1999). Results from two complementary

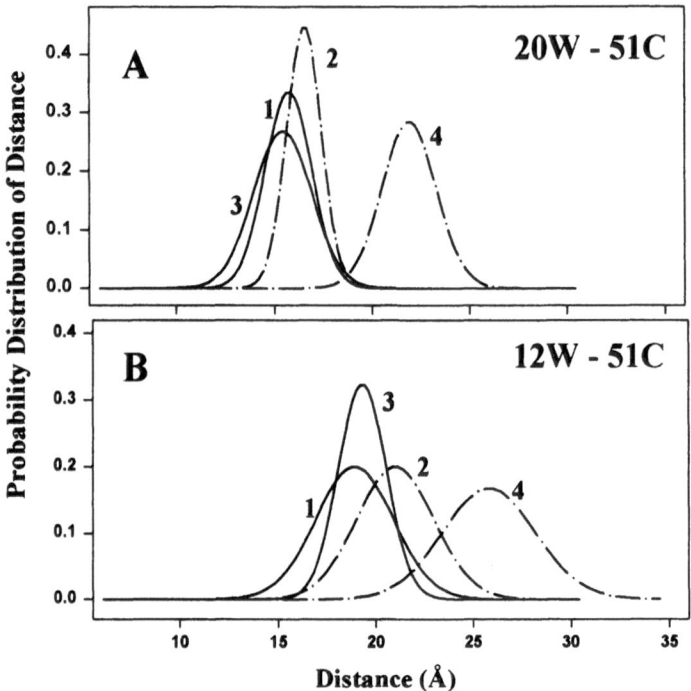

Figure 11. Area-normalized distributions of intersite distances determined from cTnC mutants and their complexes with cTnI. (A) Distributions for the distance between Trp20 and Cys51, (B) distributions for the distance between Trp12 and Cys51. Curves 1 (apo) and 2 (holo) are the distributions for free cTnC; curves 3 (apo) and 4 (holo) are the distributions for cTnC bound to cTnI. (From Dong et al., 1999)

spectroscopic methods have yielded evidence for the role of a segment of cTnI adjacent to the inhibitory region in modulating the Ca^{2+}-dependent opening of the regulatory domain in cTnC.

Table 3. Intersite distances in the N-domain of cTnC

Distance	Sample	$\langle r \rangle$ Å	hw Å	Distance change Å	NMR distance Å
20-51	apo cTnC	15.7	2.8		
	holo cTnC	16.5	2.1	0.8	16.0
	apo cTnC-cTnI	15.4	3.5		
	holo cTnC-cTnI	21.9	3.3	6.5	21.4
12-51	apo cTnC	18.9	4.7		
	holo cTnC	21.0	4.7	2.1	21.6
	apo cTnC-cTnI	19.3	2.9		
	holo cTnC-cTnI	25.8	5.1	6.5	29.1

The two distances 20-51 and 12-51 in the N-domain of cTnC were determined with free cTnC and cTnC bound to cTnI in both the apo and holo states of the N-domain. The intrinsic FRET donor was Trp in positions 20 and 12, and the acceptor IAEDANS was attached to the common residue Cys51. The magnitude of distance change refers to the observed change in the distance between the holo and apo states. The NMR distance of holo cTnC is the distance between the two alpha carbon atoms of the residues from the structure of chicken N-fsTnC (residues 1-89) (PDF 1AJ4), and the NMR distance of holo cTnC-cTnI is from the structure of chicken N-fsTnC (residues 1-89) in its complex with cTnI peptide (residues 147-163), taken from the best representative conformers in an ensemble of 40 conformers (Li et al., 1999 and RCSB protein data bank, accession code 1MXL).

4.3.3. Comparison of Cardiac and Skeletal TnC Regulatory Domains

Failure of bound Ca^{2+} at the single regulatory site of cTnC to elicit an open conformation of the regulatory domain raises some interesting questions. When compared with fsTnC, is this failure due to differences in energetics or structure? It is recalled that fsTnC has two regulatory Ca^{2+} sites, whereas cTnC has only one such site. At 20°C, the binding of Ca^{2+} to sites I and II in full-length rabbit fsTnC and chicken mutant N-fsTnC stabilizes the regulatory domains by -14.9 (Wang and Cheung, 1985) and -15.0 kcal (Li et al., 1997), respectively. The energetics required for domain opening and exposure of a hydrophobic pocket presumably is balanced by this free energy drop. On the other hand, the Ca^{2+}-saturated N-domain of cTnC is stabilized by only -7.7 kcal (Dong et al., 1996; Li et al., 1997) and is less stable than that of fsTnC by 7.3 kcal. This destabilization of cTnC N-domain relative to Ca^{2+}-loaded fsTnC can be compensated for by the formation of an apo cTnC-cTnI, which releases -8.5 kcal free energy (Dong et al., 1997a). Thus, holo cTnC-cTnI would be energetically favorable for Ca^{2+}-induced domain opening to occur (Dong et al., 1999). The opening of the N-cTnC-Ca complex in the presence of a cTnI peptide (residues 147-163) bound to cTnC has been similarly interpreted (McKay et al., 2000). It is noted that this cTnI peptide stabilizes the N-cTnC(Ca) complex by -5.5 kcal (Li et al., 1999), and this smaller free energy drop may account for the FRET results showing that the binding of cTnI peptide P(150-166) to cTnC is not as effective as whole cTnI in eliciting a domain opening (Dong et al., 1999). Since the binding of one Ca^{2+} to site II in cTnC is accompanied by little structural change, the binding of cTnI to cTnC is a major contribution to domain opening and a major pathway in the initial step in Ca^{2+} activation of cardiac muscle. This pathway is not an energetic prerequisite for skeletal muscle.

FRET results suggest that holo N-domain of cTnC is already partially open and constrained (Table 2) in comparison with apo fsTnC (Table 1). These features may provide a favorable, although not optimal, structural basis for Ca^{2+}-dependent interaction between cTnC and cTnI. The holo N-domain, however, is not stable enough to overcome the barrier of full domain opening and exposure of a hydrophobic patch. This deficit is removed by bound cTnI. The initial pathways by which an open regulatory domain is achieved are different for cardiac and skeletal muscles, but the Ca^{2+} switch for activation is the TnC-TnI linkage and is similar in both types of striated muscle.

5. REGULATION OF CARDIAC MYOFILAMENT ACTIVATION BY TNI PHOSPHORYLATION

5.1. Alteration of cTnI Conformation and Ca^{2+} Binding by Phosphorylation

The unique amino extension of cTnI contains two sites (Ser23 and Ser24 in rat heart) that are the only known targets of PKA among thin filament proteins (Solaro and Rarick, 1998). This phosphorylation occurs *in vivo* and by β-adrenergic stimulation of the heart *in situ*. An early stopped-flow study showed that PKA-dependent phosphorylation of cTnI resulted in a small loss of myofilament sensitivity to Ca^{2+} and a small increased rate of Ca^{2+} dissociation (Robertson et al., 1982). The phosphorylation was later shown to be both necessary and sufficient to elicit the reduction in myofilament Ca^{2+} sensitivity (Solaro and Van Eyk, 1996). As part of the study of the kinetic mechanism of Ca^{2+} binding to reconstituted cTn, we also investigated the effect of cTnI phosphorylation on the Ca^{2+} binding kinetic mechanism (Sec. 4.2 and Table 2).

Although the phosphorylation itself does not alter the three-step sequential kinetic mechanism, it affects not only the rate of Ca^{2+} dissociation as previously reported, but also the rates and equilibrium constants of the kinetic steps. The bimolecular binding rate constant is decreased by a factor of 2-3 and the first-order rate constants (k_1 and k_{-1}) associated with step 1 are increased by a factor of 2-3. The rate constants of step 2 (k_2 and k_{-2}) are enhanced by a factor of 7-8, and the two observed Ca^{2+} dissociation rates are also enhanced. The net effect of these changes is a substantial decrease in K_0, the equilibrium constant for formation of the initial cTn-Ca complex, and reduction of the overall equilibrium binding constant (K_{eq}). The latter result is in agreement with a similar decrease induced by cTnI phosphorylation in the equilibrium constant for formation of cTnI-cTnC (Liao et al., 1994) and cTn (Robertson et al., 1982). These changes suggest that phosphorylation induces a new cTnI conformation.

The amino acid sequence of the segment containing Ser23 and Ser23 is ^{19}VRRRSSANYRA^{29}. This segment may have an extended conformation due to electrostatic repulsion of the four positively charged arginine side chains. When the two serines are phosphorylated, the incorporated phosphate groups could be involved in electrostatic interaction with the adjacent arginine side chains, leading to a collapse of the extended conformation and a folding of the N-terminal segment towards the C-terminus. We investigated this possibility by determination of the distribution of the distances between Cys5 and the single Trp192, using IAANS attached to the cysteine residue as the energy acceptor from the tryptophan residue (Dong et al.,1997c). The mean intersite distance was 45.3 Å in unphosphorylated cTnI and 35.8 Å in phosphorylated cTnI. This large distance decrease, which was carried over to the cTnC-(phosphorylated cTnI) complex, supports the notion of a phosphorylation-induced folding of the N-terminal segment. The half-width of the distribution was reduced from 9.5 Å for unphosphorylated cTnI to 3.7 Å for phosphorylated cTnI. This decrease suggests that the segment of cTnI between donor and acceptor sites is held more rigidly in the phosphorylated state. Such a folded conformation of phosphorylated cTnI must be more compact or less asymmetric than non-phosphorylated cTnI. This agrees with our earlier observation of a decrease of the rotational correlation time of cTnI from 23.5 to 14.6 ns upon PKA phosphorylation (Liao and Cheung, 1992).

Contacts between residues of cTnC and a cTnI peptide (residues 1-80) in their complex were recently determined in NMR experiments (Finley et al., 1999). Non-phosphorylated cTnI peptide was found to make contacts with both the C-domain and the regulatory N-domain of cTnC as evidenced from relaxation data showing that, in the complex, the two cTnC domains do not tumble independently. When bound to an analog of phosphorylated cTnI peptide in which Ser23 and Ser24 were mutated to Asp, the two cTnC domains tumble independently, suggesting disruption of the interaction of peptide with the N-terminal domain of cTnC. Although results obtained with a peptide rather than full-length cTnI should be interpreted with some caution, the loss of interaction with the regulatory domain of cTnC could partially explain the reduction in the affinity of cTnC for phosphorylated cTnI previously observed by us (Liao et al., 1994).

Since the cTnC-cTnI complex is antiparallel, the question arises as to how the phosphorylation signal at one end of the complex is transduced to the Ca^{2+} binding site at the other end. The interaction of cTnI with two fragments of cTnC, cTnC(1-89) and cTnC(90-162), was studied with the same cTnI mutant fluorescently labeled at Cys5 that was used in FRET measurement to demonstrate a phosphorylation induced folding of cTnI. An important finding is that phosphorylation of cTnI at the N-terminal segment results in a reduced affinity of cTnI for cTnC(1-89), a change in affinity similar to that previously observed with intact cTnI. With this complex, the fluorescence spectral properties of the label at Cys5 was not significantly affected, suggesting no direct contact between the N-terminal segment of cTnI and the regulatory N-domain of cTnC. Taken together, these results indicate that transduction of the cTnI phosphorylation signal from cTnI to the regulatory N-domain of cTnC does not require the presence of the cTnC central helix, but involves propagation of a global conformational change along the cTnI polypeptide (Chandra et al., 1997).

5.2. Structural Mapping of cTnI and cTnI-cTnC

Several models, both low-resolution and high-resolution, have been proposed for the structure of the fsTnC-fsTnI complex and have been discussed in this chapter. No similar attempt has been reported for the binary complex from cardiac proteins. Differences between the two complexes can be expected because cardiac TnI is bigger and it has a unique function that is not found in fsTnI. We recently investigated the global conformation of free cTnI and cTnI complexed with cTnC (Dong et al., 2000a). In this work, we used a series of nine single-cysteine mutants of cTnI and a combination of sulfhydryl reactivity and FRET to determine cysteine accessibility and intersite distances from Trp192 to different cysteine residues singly located from positions 167 to 5. Solvent accessibility of a cysteine residue in cTnI was determined by the reaction of the fluorescent sulfhydryl reagent CPM [7-(diethylamino)-3-(4.-maleimidylphenyl)-4-methylcourmarin]. The pseudo-first-order time constant of this reaction was taken as a measure of solvent accessibility. A major finding is that all nine cysteines are similarly accessible and highly exposed, with a standard deviation of the accessibility only 14% from the mean value. If any of the cysteines was buried or significantly protected from collision with solvent, its accessibility would be expected to be much lower than the observed values.

The FRET distances from Trp192 at the C-terminal end to eight upstream cysteines increase in a systematic manner, reaching a maximum value of 45 Å at Cys5. This pattern of intersite separations is expected if the long polypeptide is not globular, but more or less an extended linear chain that does not cross itself in multiple regions or fold back with sharp turns, and has a large curvature. Such a global open conformation is consistent with the large exposure of cysteine residues distributed along the linear sequence. These conformational features are largely retained in the sequence between residues 40 and 192 upon phosphorylation at Ser23 and Ser24. In cTnC-cTnI and cTn, every intramolecular distance in cTnI is lengthened when compared to free cTnI and the overall conformation of the bound cTnI remains elongated with reduced exposure for the cysteines. The highly flexible nature of the N-terminal extension of cTnI is preserved in the complexes, suggesting that this segment is either not bound or only loosely bound to the C-domain of cTnC. In a related study in which energy transfer between sensitized Tb^{3+} bound to site III in the C-domain and Cys34 in the N-domain, we established that the separation between the two domains of cTnC remains unchanged upon complex formation with either cTnI or cTnI plus cTnT (Dong et al., 2000b). Thus, cTnC in the complexes remains extended. The global conformation of cTnC-cTnI appears to be extended with constrained flexibility. The sulfhydryl groups of Cys133 and Cys150 of cTnI incorporated into cTnC-cTnI and cTn experience large reduced exposure resulting from the binding of Ca^{2+} to the single regulatory site of cTnC. Key regions of cTnI involved in Ca^{2+} activation become highly shielded upon activation.

The crystal structure of fsTnC complexed with an N-terminal fsTnI peptide (residues 1-47) shows a partially unwound and bent central α-helix at the junction of helices D and E. This bound peptide-induced structural perturbation brings the N- and C-domain to close proximity where the two domains are in direct contact, and the long dimension of the fsTnC molecule is reduced from about 65 to 50 Å (Vassylyev et al., 1998). Interestingly, in the crystal structure the N-terminus of the bound peptide appears to be in contact with helix A in the N-domain of fsTnC and the C-terminus extends beyond the C-domain of the TnC molecule. In contrast, our FRET results show that cTnC in the binary and ternary complexes with the other subunits is an extended molecule (Dong et al., 2000b). The different results are not likely due to differences between fsTnC and cTnC, but rather due to the properties of the short bound peptide. In the presence of bound fsTnI inhibitory peptide (residues 96-116), the rotational and translational motions of fsTnC were interpreted to suggest that, within the troponin complex, the binding of regulatory Ca^{2+} induces a transition of the conformation of fsTnC from a near spherical shape to an elongated shape (Moncrieffe et al., 1999). FRET results obtained with full-length cTnI do not support such a large scale conformational transition of cTnC within cTnC-cTnI or cTn. These different results point out potential difficulties in relating results from studies of some peptides to those obtained with full-length proteins.

The β-hairpin motif proposed for the inhibitory region in holo fsTnC-fsTnI (Tung et al., 2000) has not been experimentally explored using full-length fsTnI. This region (segment IV) spans residues 95-114 (rabbit fsTnI). In the hairpin model, Met95 and Ala118 are in close proximity and their $C^{\alpha-}$ atoms are 13.1 Å apart (Fig. 9). The corresponding residues in cTnI are 129 and 152. Since the sequence of the cardiac inhibitory region (residues 128-148) is almost identical to that of the skeletal inhibitory region, a similar β-hairpin would be expected for the cardiac inhibitory region in the $3Ca^{2+}$-state of cTnC-cTnI, and the separation between cTnI residues 129-152 in the complex would be about 13 Å. Our recent FRET results showed that the distance between Trp129 and Cys52 in cTnI is 19.4 Å when cTnI is free and essentially unaltered when cTnI is in binary complexes or fully reconstituted troponin. In the $3Ca^{2+}$-state of cTnC-cTnI or cTn, this FRET distance increases to about 29 Å, a factor of 2 longer than would be expected for a β-hairpin of the inhibitory region (Dong et al., 2001). The large intersite distance observed with holo complexes suggests an extended conformation of the segment and rules out a hairpin conformation for the cTnI inhibitory region in holo cardiac troponin. It is plausible that the inhibitory segment is positioned along the long central helix of TnC in TnC-TnI, as suggested in previous models for fsTnC-fsTnI (Luo et al., 2000; Tung et al., 2000), but FRET results suggest that the cardiac inhibitory region in this position in holo cTn is consistent with an extended conformation. A recent multidimensional heteronuclear NMR study showed that the cTnI inhibitory segment (residues 129-147) may not specifically interact with regions of the central helix of cTnC in the presence of bound regulatory Ca^{2+} (Abbott et al., 2001). This is an important finding because the NMR results were obtained with a long peptide containing both the inhibitory and regulatory regions (residues 129-166). In this system, the inhibitory region is constrained by the contiguous regulatory region which binds to the exposed hydrophobic pocket of the open regulatory N-domain of cTnC. This strategy precludes the possibility that an isolated short inhibitory peptide may bind to other regions of cTnC. FRET results indicated that the separations between residue 89 of the central helix of cTnC and several residues within the inhibitory region of cTnI in cTnC-cTnI were greater than 30 Å, and these distances decreased by only 2-3 Å in the $3Ca^{2+}$ state (Xing et al., 1999). These relatively long separations suggest weak or no specific interactions between the central helix of cTnC and the cTnI inhibitory region, consistent with NMR results (Abbott et al., 2001). Taken together, currently available structural information suggests that the extended inhibitory region of cTnI appears not to have specific interaction with the central helix of cTnI and does not displace the bound N-terminal segment of cTnI from the C-domain of cTnC.

On the basis of FRET and NMR results, we have proposed a conceptual model for the Ca^{2+}-induced conformational transition of the cTnI inhibitory region in cardiac troponin (Fig. 12). In the absence of bound regulatory Ca^{2+}, the cTnI segment between residues 129 and 152 has a short intersite distance of 19.9 Å and is depicted with a loop in the middle of the segment. This helix-loop-helix feature is consistent with secondary structure prediction. When Ca^{2+} binds to the regulatory site in cTnC, movement of the adjacent regulatory region (residues 150-165) to the exposed hydrophobic pocket of the open regulatory N-domain of cTnC is facilitated by a global conformational change of the inhibitory region in which the loop curvature is reduced leading to a "stretching" of the "end-to-end" distance of the inhibitory segment and a more extended conformation of the segment. The movement of the regulatory region toward cTnC pulls open the helix-loop-helix motif . If the loop shown in Fig. 12A is the binding site for actin in relaxed muscle, then the stretching of the loop induced by regulatory Ca^{2+} would allow the inhibitory region to pull away from actin and would serve as a structural switch between actin and cTnC to facilitate binding of the regulatory region to the open N-domain of cTnC.

6. SUMMARY AND PROSPECTS

Calcium activation and regulation in striated muscle differ from smooth muscle in that the Ca^{2+} receptor protein, troponin C, is part of the thin filament. There has been a rapid advance during the past few years in our understanding of the structure of TnC in different biochemical states and structural changes in this protein as related to muscle functions, such as development of tension and activation of myofibrillar ATPase. Although less is known about the structural aspects of the other two troponin subunits, either as unbound proteins or in complexes with TnC, our present knowledge is at a level where the interaction of TnC with TnI is being investigated with increasing detail. Biochemical studies of the TnC-TnI interaction and interactions involving the other subunits have paved the way for spectroscopic studies of these interactions. Ultimately, we need to interpret biochemical and structural information, derived from both low-resolution and high-resolution studies, in terms of transduction of the initial Ca^{2+} signal to distant sites within the regulatory unit. We have not yet reached this stage at this time.

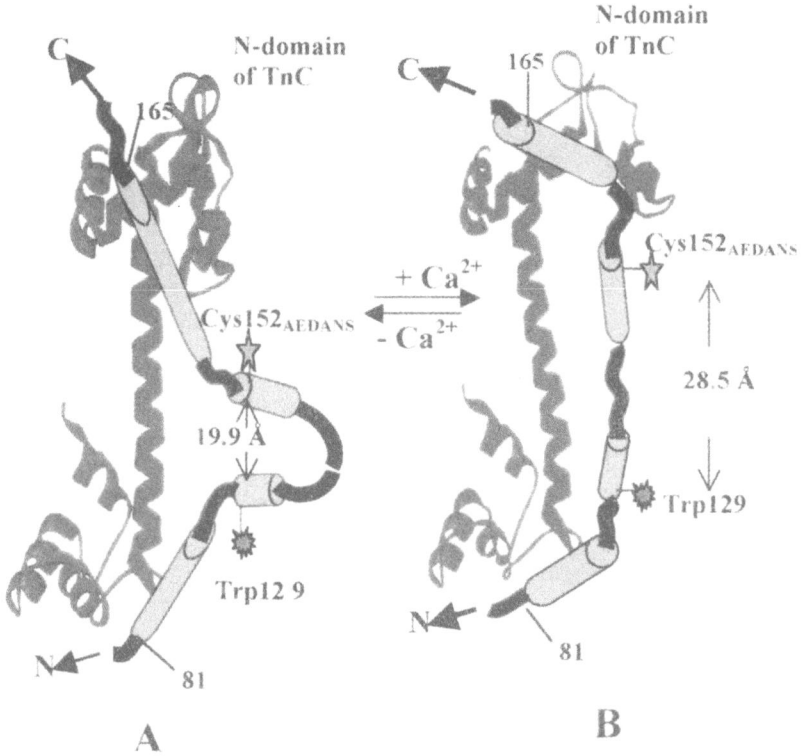

Figure 12. A model for Ca²⁺-induced conformational transition of cTnI inhibitory region in cardiac troponin. Shown are schematic proximity relationships between cTnC (black ribbon) and cTnI (grey) arranged in an antiparallel fashion. cTnT is omitted. cTnI is shown for the segment between residues 81 and 165. Grey cylinders denote helices. The segment Trp129-Cys152 brackets the inhibitory region (residues 130-149) and corresponds to rabbit fsTnI segment between Met95 and Ala118 (see Fig. 9). cTnI residues 150-165 represent the regulatory region. Color version on page 465.

In this chapter, the discussion is focused on structural, molecular, and kinetic aspects of TnC from both skeletal and cardiac muscle, with an emphasis on our recent studies of the interaction between cTnC and cTnI using intact recombinant proteins. The first step in activation is creation of an open regulatory N-domain in TnC to allow a critical Ca²⁺-dependent interaction of an exposed hydrophobic patch with its target, a segment of TnI. We have established that the initial pathway of Ca²⁺ activation is different between the two isoforms of TnC. Bound cTnI is a prerequisite to achieve a Ca²⁺-induced opening of the regulatory N-domain of cTnC, whereas domain opening in fsTnC can be achieved in the absence of fsTnI. This difference is related to the fact that the regulatory domain of cTnC has only one bound Ca²⁺ ion and is less stable than that of fsTnC with two bound regulatory Ca²⁺ ions, although other factors including differences in the structures of key amino acid residues in the Ca^{2+}-binding loops may also contribute to the requirement of bound cTnI.

We have a reasonable, although still incomplete, working frame work for Ca^{2+} regulation in striated muscle derived from solution studies. Structural changes induced by regulatory Ca^{2+} in TnC and in TnI and TnC complexed with each other are known with some detail, at high resolution derived from NMR/crystallography and at low resolution derived from fluorescence spectroscopy. These changes are believed to be responsible for activation and regulation. To what extent these structural changes take place in intact muscle and are relevant in activation of ATPase and force cannot be easily established. High-resolution structural information is difficult, if not impossible at this time, to obtain from a contracting fiber. On the other hand, global conformational changes involving individual regulatory proteins that can be determined by FRET are amenable for detection in skinned fibers via exchange of endogenous proteins with appropriately labeled mutants. Simultaneous recording of steady-state force and emission spectra of both energy donor and acceptor can be implemented. Success in this new direction will take us one step further in identification of Ca^{2+}-modulated molecular events in triggering and regulation of muscle contraction.

7. ACKNOWLEDGEMENT

I thank Wen-Ji Dong for helpful discussions and contributions to many of the fluorescence studies described in this chapter. His assistance in preparing the figures is also appreciated. The work carried out in my laboratory and described herein was supported, in part, by the National Institute of Heart, Lung and Blood (HL52508).

Department of Biochemistry and Molecular Genetics
University of Alabama at Birmingham
Birmingham, AL 35294-2041
(205) 934-2485
FAX: (205) 975-4621
hccheung@uab.edu

8. REFERENCES

Abbott, M.B., Dvoretsky, A., Gaponenko, V., and Rosevear, P.R. (2000) Cardiac troponin I inhibitory peptide: location of interaction sites on troponin C, *FEBS Lett.* **469**, 168-172.

Abbott, M.B., Dong, W-J., Dvoretsky. A., DaGue, B., Caprioli, R.M., Cheung, H.C., and Rosevear, P.R. (2001) Modulation of cardiac troponin C-cardiac troponin I regulatory interactions by the amino-terminus of cardiac troponin I. *Biochemistry* **40**, 5992-6001.

Babu, A., Su, H., Ryu, Y., and Gulati, J. (1992) Determination of residues specificity in the EF-hand of troponin C for Ca^{2+} coordination, by genetic engineering, *J. Biol. Chem.* **267**, 15469-15474.

Bailin, G. (1979) Phosphorylation of a bovine cardiac actin complex, *Am. J. Physiol.* **236**, C41-C46.

Byers, D.M. and Kay, C.M. (1982) Hydrodynamic properties of bovine cardiac troponin C, *Biochemistry*, **21**, 229-233.

Campbell, A.P. and Sykes, B.D. (1991) Interaction of troponin I and troponin C. Use of the two-dimensional nuclear magnetic resonance transferred nuclear Overhauser effect to determine the structure of the inhibitory troponin I peptide when bound to skeletal troponin C, *J. Mol. Biol.* **222**, 405-421.

Chandra, M., Dong, W.-J., Pan, B.-S., Cheung, H.C., and Solaro, R.J. (1997) Effects of protein kinase A phosphorylation on signaling between cardiac troponin I and the N-terminal domain of cardiac troponin C, *Biochemistry* **36**, 13305-13311.

Cheung, H.C. (1991) Resonance Energy Transfer, in Lakowciz, J.R. (ed.), *Topics in Fluorescence Spectroscopy, Vol. 2: Principles*, Plenum Press, New York, pp. 127-178.

Cheung, H.C., Wang, C.-K., and Malik, N.A. (1987) Interactions of troponin subunits: free energy of binary and ternary complexes, *Biochemistry* **26**, 5904-5907.

Cheung, H.C., Wang, C.-K., Gryczynski, I., Wiczk, W., Laczko, G., Johnson, M.K., and Lakowicz, J.R. (1991) Distance distributions and anisotropy decays of troponin C and its complex with troponin I, *Biochemistry* **30**, 5238-5247.

Chong, P.C. and Hodges, R.S. (1981) A new bifunctional cross-linking reagent for the study of biological interactions between proteins, *J. Biol. Chem.* **256**, 5071-5076.

Close, R. (1965) The relation between intrinsic speed of shortening and duration of the active state of muscle, *J. Physiol.* (London) **180**, 542-559.

Dalgarno, D.C., Grand, R.J.A., Levine, B.A., Moir, A.J.G., Scott, G.M., and Perry, S.V. (1982) Interaction between troponin I and troponin C, *FEBS Lett.* **150**, 54-59.

Dong, W-J., Rosenfeld, S.S., Wang, C.-K., Gordon, A.M., and Cheung, H.C. (1996) Kinetic studies of calcium binding to the regulatory site of troponin C from cardiac muscle, *J. Biol. Chem*, **271**, 688-694.

Dong, W.-J., Wang, C.-K., Gordon, A.M., Rosenfeld, S.S., and Cheung, H.C. (1997a) A kinetic model for the binding of Ca^{2+} to the regulatory site of troponin from cardiac muscle, *J. Biol. Chem.* **272**, 19229-19235.

Dong, W.-J., Wang, C.-K., Gordon, A.M., and Cheung, H.C. (1997b) Disparate fluorescence properties of 2-[4.-(iodoacetamido)anilino]-naphthalene-6-sulfonic acid attached to Cys-84 and Cys-35 of troponin C in cardiac muscle troponin, *Biophys. J.* **72**, 850-857.

Dong, W.-J., Chandra, M., She, M., Solaro, R.J., and Cheung, H.C. (1997c) Phosphorylation-induced distance change in a cardiac muscle troponin I mutant, *Biochemistry* **36**, 6754-6761.

Dong, W.-J., Xing, J., Villain, M., Hellinger, M., Robinson, J.M., Chandra, M., Solaro, R.J., Umeda, P.K., and Cheung, H.C. (1999) Conformation of the regulatory domain of cardiac muscle troponin C in its complex with cardiac troponin I, *J. Biol. Chem.* **274**, 31382-31390.

Dong, W.-J., Xing, J., Chandra, M., Solaro, R.J., and Cheung, H.C. (2000a) Structural mapping of single cysteine mutants of cardiac troponin I, *Proteins* **41**, 438-447.

Dong, W.-J., Robinson, J.M., Xing, J., Umeda, P.K., and Cheung, H.C. (2000b) An interdomain distance in cardiac troponin C determined by fluorescence spectroscopy, *Protein Sci.* **9**, 280-289.

Dong, W.-J., Xing, J., Robinson, J.M., and Cheung, H.C. (2001) Ca^{2+} induces an extended conformation of the inhibitory region of troponin I in cardiac muscle troponin, *J. Mol. Biol.* **315**, xxx.

Drabikowski, W., Dalgarno D.C., Levine, B.A., Gergely, J., Grabarek, Z., and Leavis, P.C. (1985) Solution conformation of the C-terminal domain of skeletal troponin C, *Eur. J. Biochem.* **151**, 17-28.

Farah, C.S., Miyamoto, C.A., Ramos, C.H.I., da Silva, A.C.R., Quaggio, R.B., Fujimori, K., Smillie, L.B., and Reinach, F.C. (1994) Structural and regulatory functions of the NH$_2$- and COOH-terminal regions of skeletal muscle troponin I, *J. Biol. Chem.* **269**, 5230-5240.

Finley, N., Abbott, M.B., Abusamhadneh, E., Gaponenko, V., Dong, W.-J., Gasmi-Seabrood, G., Howarth, J.W., Rance, M., Solaro, R.J., Cheung, H.C., and Rosevear, P.R. (1999) NMR analysis of cardiac troponin C-troponin I complexes: effects of phosphorylation, *FEBS Lett.* **453**, 107-112.

Flicker, P.F., Phillips, G.N., and Cohen, C. (1982) Troponin and its interactions with tropomyosin: an electron microscope study, *J. Mol. Biol.* **162**, 495-501.

Gagné, S.M., Tsuda, S., Li, M.X., Chandra, M., Smillie, L.B., and Sykes, B.D. (1994) Quantification of the calcium-induced secondary structural changes in the regulatory domain of troponin-C, *Protein Sci.* **3**, 1961-1974.

Gagné, S.M., Tsuda, S., Li, M.X., Smillie, L.B., and Sykes, B.D. (1995) Structure of the troponin C regulatory domains in the apo and calcium-saturated states, *Nat. Struc. Biol.* **2**, 784-789.

Gagné, S.M., Li, M.X., and Sykes, B.D. (1997) Mechanism of direct coupling between binding and induced structural change in regulatory calcium binding proteins, *Biochemistry* **36**, 4386-4392.

Gaponeko, V., Abusamhadneh, E., Abbott, M.B., Finley, N., Gasmi-Seabrook, G., Solaro, R.J., Rance, M. and Rosevear, P.R. (1999) Effects of troponin I phosphorylation on conformational exchange in the regulatory domain of cardiac troponin C, *J. Biol. Chem.* **274**, 16681-16684.

Gasmi-Seabrook, G.M.C., Howarth J. W., Finley, N., Abusamhadneh, E., Gaponenko, V., Brito, R.M.M., Solaro, R.J., and Rosevear, P.R. (1999) Solution structures of the C-terminal domain of cardiac troponin C free and bound to the N-terminal domain of cardiac troponin I, *Biochemistry* **38**, 8313-8322.

Gong, Z., Xing, J., Chandra, M., Dong, W.-J., Solaro, R.J., Umeda. P.K., and Cheung, H.C. (1998) Comparison of the regulatory domain conformation of troponin C from cardiac and skeletal muscle, *Biophys. J.* **74**, A51 (abs).

Hazard, A.L., Kohout, S.C., Stricker, N.L., Putkey, J.A., and Falke, J.J. (1998) The kinetic cycle of cardiac troponin C: calcium binding and dissociation at site II trigger slow conformational rearrangements, *Proteins Sci.* **7**, 2451-2459.

Hernández, G., Blumenthal, D.K., Kennedy, M.A., Unkerfer, C.J., and Trewhella, J. (1999) Troponin I inhibitory peptide (96-115) has an extended conformation when bound to skeletal muscle troponin C, *Biochemistry* **38**, 6911-6917.

Herzberg, O. and James, M.N.G. (1985) Structure of the calcium regulatory muscle protein troponin C at 2.8 Å resolution, *Nature* **313**, 653-659.

Herzberg, O. and James, M.N.G. (1988) Refined crystal structure of troponin C from turkey skeletal muscle at 2.0 Å resolution, *J. Mol. Biol.* **203**, 761-779.

Herzberg, O., Moult, J., and James, M.M.G. (1986) A model for the Ca^{2+}-induced conformational transition of troponin C. A trigger for muscle contraction, *J. Biol. Chem.* **261**, 2638-2644.

Hincke, M.T., McCubbin, W.D., Kay, C.M.. (1978) Calcium-binding properties of cardiac and skeletal troponin C as determined by circular dichroism and ultraviolet difference spectroscopy, *Can. J. Biochem.* **56**, 384-395.

Holmes, K.C., Popp, D., Gebhard, W., Kabsch, W. (1990) Atomic model of the actin filament, *Nature* **374**, 44-49.

Holroyde, M.J., Howe, E., and Solaro, R.J. (1979) Modification of calcium requirements for activation of cardiac myofibrillar ATPase by cAMP dependent phosphorylation. *Biochim. Biophys. Acta,* **586**, 63-69.

Holroyde, M.J., Robertson, S.P., Johnson, J.D., Solaro, R.J., and Potter, J.D. (1980) The calcium and magnesium binding sites on cardiac troponin and their role in the regulation of myofibrillar adenosine triphosphatase, *J. Biol. Chem.* **255**, 11688-11693.

Houdusse, A., Love, M.L., Dominguez, R., Grabarek, Z., and Cohen, C. (1997) Structures of four Ca^{2+}-bound troponin C at 2.0 Å resolution: further insights into the Ca^{2+}-switch in the calmodulin superfamily, *Structure* **5**, 1695-1711.

Iio, T. and Knodo, H. (1980) Comparison of the kinetic properties of troponin-C and dansylazididine-labeled troponin-C1, *J. Biochem. (Tokyo)* **88**, 547-556.

Ingrahm, R.H. and Swenson, C.A. (1984) Binary interactions of troponin subunits, *J. Biol. Chem.* **259**, 9544-9548.

Johnson, J.D. and Potter, J.D. (1978) Detection of two classes of Ca^{2+} sites in troponin C with circular dichroism and tyrosine fluorescence, *J. Biol. Chem.* **253**, 3775-3777.

Johnson, J.D., Charlton, S.C., and Potter, J.D. (1979) A fluorescence stopped flow analysis of Ca^{2+} exchange with troponin C, *J. Biol. Chem.* **254**, 3497-3502.

Johnson, J.D., Collins, J.H., Robertson, S.P., and Potter, J.D. (1980) A fluorescent probe study of Ca^{2+} binding to the Ca^{2+}-specific sites of troponin and troponin C, *J. Biol. Chem.* **255**, 9635-9640.

Johnson, J.D., Nakkula, R.J., Vasulka, C., and Smillie, L.B. (1994) Modulation of Ca^{2+} exchange with the Ca^{2+}-specific regulatory sites of troponin C, *J. Biol. Chem.* **269**, 8919-8923.

Kabsch, W., Mannherz, H.G., Suck, D., Pai, E.F., and Holems, K.C. (1990) Atomic structure of the actin:DNase I complex, *Nature* **374**, 37-44.

Kobayashi, T., Tao, T., Grabarek, Z., Gergely, J., and Collins, J.H. (1991) Cross-linking of residue 57 in the regulatory domain of a mutant rabbit skeletal muscle troponin C to the inhibitory region of troponin I, *J. Biol. Chem.* **266**, 13746-13751,

Kohout, S.C. and Falke, J.J. (1999) Comparison of cardiac and skeletal troponin C, *Biophys. J.* **76**, A158 *(abstract).*

Kretsinger, R.H. and Nockolds, C.E. (1973) Carp muscle calcium-binding protein II. Structure determination and general description, *J. Biol. Chem.* **248**, 3313-3326.

Krudy, G.A., Kleerekoper, Q., Guo, X., Howarth, J. W., Solaro, R.J., and Rosevear, P.R.. (1994), NMR studies delineating spatial relationships within the cardiac troponin I-troponin C complex, *J. Biol. Chem.* **269**, 23731-23735.

Lakowicz, J.R., Gryczynski, I., Cheung, H.C., Wang, C.-K., Johnson, M.L., and Joshi, N. Distance distributions in proteins recovered using frequency-domain fluorometry: application to troponin I and its complex with troponin C, *Biochemistry* **27**, 9149-9160.

Leszyk, J., Dumaswala, R., Potter, J.D., Gusev, N.B., Verin, A.D., Tobacman, L.S., Collins, J.H. (1987) Bovine cardiac troponin T: amino acid sequence of the two isoforms, *Biochemistry*, **26**, 7035-7042.

Levine, B.A., Coffman, D.M., and Thornton, J.M. (1977) Calcium binding by troponin-C. A proton magnetic resonance study, *J. Mol. Biol.* **115**, 743-760.

Li, M.X., Gagné, S.M., Tsuda, S., Kay, C.M., Smillie, L.B., and Sykes, B.D. (1995) Calcium binding to the regulatory N-domain of skeletal muscle troponin C occurs in a stepwise manner, *Biochemistry* **34**, 8330-8340.

Li, M.X., Gagné, S.M., Spyracopoulos, L., Kloks, C., Audette, G., Chandra, M., Solaro, R.J., Smllie, L.B., and Sykes, B.D. (1997) NMR studies of Ca^{2+} binding to the regulatory domains of cardiac and E41A skeletal muscle troponin C reveal the importance of site I to energetics of the induced structural changes, *Biochemistry* **36**, 12519-12525.

Li, M.X., Spyracopoulos, L., and Sykes, B.D. (1999) Binding of troponin-$I_{147-163}$ induces a structural opening in human cardiac troponin-C, *Biochemistry* **38**, 8289-8298.

Li, Y., Move, M.L., Putkey, J.A., and Cohen (2000) Bepridil opens the regulatory N-terminal lobe of cardiac troponin C, *Proc. Natl. Acad. Sci. U.S.A.* **97**, 5140-5145.

Liao, R. and Cheung, H.C. (1992) Time-resolved tryptophan emission study of cardiac troponin I, *Biophys. J.* **63**, 986-995.

Liao, R., Wang, C.-K., and Cheung, H.C. (1994) Coupling of calcium to the interaction of troponin I with troponin C from cardiac muscle, *Biochemistry* **33**, 12729-12734.

Luo, Y., Wu, J.-L., Li, B., Langsetmo, K., Gergely, J., and Tao, T. (2000a) Photocrosslinkg of benzophenone-labeled single cysteine troponin I mutants to other thin filament proteins, *J. Mol. Biol.* **296**, 899-910.

Luo, Y. Leszyk, J., Li, B., Gergely, J., and Tao, T. (2000b) Proximity relationships between residue 6 of troponin I and residues in troponin C: further evidence for extended conformation of troponin C in the troponin complex, *Biochemistry* **39**, 15306-15315.

Manning, M.C. (1989) Underlying assumption in the estimation of secondary structure content in proteins by circular dichroism spectroscopy - a critical review, *J. Pharm. Biomed. Analysis* **7**, 1103-1119.

MacLachlan, L.K., Reid, D.G., Mitchell, R. C., Salter, C.J., and Smith, S.J. (1990) Binding of a calcium sensitizer, Bepridil, to cardiac troponin C, *J. Biol. Chem.* **265**, 9764-9770.

McKay, R.T., Triplet, B.P., Pearlstone, J.R., Smillie, L.B., and Sykes, B.D. (1999) Defining the region of troponin-I that binds troponin-C, *Biochemistry*, **38**, 5478-5489.

McKay, R.T., Salibus, L.F., Li, M.X., and Sykes, B.D. (2000) Energetics of the induced structural change in a Ca^{2+} regulatory protein: Ca^{2+} and troponin I peptide binding to E41A mutant of the N-domain of skeletal troponin C, *Biochemistry* **39**, 12731-12738.

Mercier, P., Li, M.X., and Sykes, B.D. (2000) Role of the structural domain of troponin C in muscle regulation: NMR studies of Ca^{2+} binding and subsequent interactions with regions 1-40 and 96-115 of troponin I, *Biochemistry* **39**, 2902-2911.

Miegel, A., Sano, K.I., Yamamoto, K., Maeda, K., Tanigiuchi, H., Yao, M., and Wakatsuki, S. (1996) Production and crystallization of lobster muscle tropomyosin expressed in Sf9 cells, *FEBS Lett.* **394**, 201-205.

Moncrieffe, M.C., Eaton, S., Bajzer, _, Haydock, C., Potter, J.D., Laue, T.M., Prendergast, F.G. (1999) Rotational and translational motions of troponin C, *J. Biol. Chem.*, **274**, 17464-17470.

Mope, L., McClellan, G.B., and Winegrad, S. (1980) Calcium sensitivity of the contractile system and phosphorylation of troponin in hyperpermeable cardiac cells, *J. Gen. Physiol.* **75**, 271-282.

Nagi, S.M. and Hodges, R.S. (1992) Biologically important interactions between synthetic peptides of the N-terminal region of troponin I and troponin C, *J. Biol. Chem.* **267**, 15715-15720.

Nagi, S.M., Sonnichsen, F.D., and Hodges, R.S. (1994) Photochemical cross-linking between native rabbit skeletal troponin C and benzoylbenzoyl-troponin I inhibitory residues 104-115, *J. Biol. Chem.* **269**, 2165-2172.

Olah, G.A. and Trewhella, J. (1994) A model structure of the muscle protein complex $4Ca^{2+}$/troponin C/troponin I derived from small-angle scattering data: implication for regulation, *Biochemistry* **33**, 12800-12806.

Olah, G.A., Rokop, S.E., Wang, C.L., Blechner, S.L., and Trewhella, J. (1994) Troponin I encompasses an extended troponin C in the Ca^{2+}-bound complex, a small-angle X-ray and neutron scattering study, *Biochemistry*, **33**, 8233-8239.

Ovaska, M. and Taskinen, J. (1991) A model for human cardiac troponin C and for modulation of its Ca^{2+} affinity by drugs, *Proteins* **11**, 79-94.

Pääkkönen, K.A., Annila, A., Sorsa, T., Pollesello, P., Tilgmann, C., Kilpeläinen, P., Ulmamed, I., and Drakenbert, T. (1998) Solution structure and main chain dynamics of the regulatory domain (residues 1-91) of human cardiac troponin C, *J. Biol. Chem.* **273**, 15633-15638.

Pearlstone, J.R., Chandra, M., Sorenson, M.M., and Smillie, L.B. (2000) Biological function and site II Ca^{2+}-induced opening of the regulatory domain of skeletal troponin C are impaired by invariant site I or II Glu mutation, *J. Biol. Chem.* **275**, 35106-35115.

Perry, S.V. (1998) Troponin T:genetics, properties and function, *J. Muscle Res. Cell Motil.* **19**, 575-602.

Phillips, G.N., Jr., Fillers, J.P., and Cohen, C. (1986) Tropomyosin crystal structure and muscle regulation, *J. Mol. Biol.* **192**, 111-127.

Robertson, S.P., Johnson, J.D., Holroyde, M.J., Kranias, E.G., Potter, J.D., and Solaro, R.J. (1982) The effect of troponin I phosphorylation on the Ca^{2+}-regulatory site of bovine cardiac troponin, *J. Biol. Chem.* **257**, 260-263.

Rosenfeld, S.S. and Taylor, E.W. (1985a) Kinetics studies of calcium and magnesium binding to troponin C, *J. Biol. Chem.* **260**, 242-251.

Rosenfeld, S.S. and Taylor, E.W. (1985b) Kinetic studies of calcium binding to regulatory complexes from skeletal muscle, *J. Biol. Chem.* **260**, 252-261.

Satyshur, K.A., Rao, S.T., Pyzalska, D., Drendel, W., Greaser, M., and Sundaralingam, M. (1988) Refined structure of chicken skeletal muscle troponin C in the two-calcium state at 2 Å resolution, *J. Mol. Biol.* **263**, 1628-1647.

She, M., Dong, W.-J., Umeda, P.K., and Cheung, H.C. (1997) Time-resolved fluorescence study of the single tryptophan of engineered skeletal muscle troponin C, *Biophys. J.* **73**, 1042-1055

She, D., Xing, J., Dong, W.-J., Umeda, P.K., and Cheung, H.C. (1998a) Calcium binding to the regulatory domain of skeletal muscle troponin C induces a highly constrained open conformation, *J. Mol. Biol.* **281**, 445-452.

She, M., Dong, W.-J., Umeda, P.K., and Cheung, H.C. (1998b) Tryptophan mutants of troponin C from skeletal muscle. An optical probe of the regulatory domain, *Eur. J. Biochem.* **251**, 600-607.

Sia, S.K., Li, M.X., Spyracopoulos, L., Gagné, S.M., Liu, W., Putkey, J.A., and Sykes, B.D. (1997) Structure of cardiac muscle troponin C unexpectedly reveals a closed regulatory domain, *J. Biol. Chem.* **272**, 18216-18227.

Slupsky, C.M. and Sykes, B.D. (1995) NMR solution structure of calcium-saturated skeletal muscle troponin C, *Biochemistry* **34**, 15953-15964.

Slupsky, C.M., Reinach, F.C., Smillie, L.B., and Sykes, B.D. (1995) Solution secondary structure of calcium-saturated troponin C monomer determined by multidimensional heteronuclear NMR spectroscopy, *Protein Sci.* **4**, 1279-1290.

Smith, S.J. and England, P.J. (1990) The effects of reported Ca^{2+} sensitisers on the rates of Ca^{2+} release from cardiac troponin C and troponin-tropomyosin complex, *Br. J. Pharmacol.* **100**, 779-784.

Spyracopoulos, L., Li, M.X., Sia, S. K., Gagné, S. M., Chandra, M., Solaro, R. J., and Sykes, B.D. (1997) Calcium-induced structural transition in the regulatory domain of human cardiac troponin C, *Biochemistry* **36**, 12138-12146.

Stone, D.B., Timmins, P.A., Schneider, D.K., Krylova, I., Ramos, C.H.I., Reinach, F.C., and Mendelson, R.A. (1998) The effect of regulatory Ca^{2+} on the *in situ* structures of troponin C and troponin I: a neutron scattering study, *J. Mol. Biol.* **281**, 689-704.

Solaro, R.J. and Van Eyk, J.(996) Altered interactions among thin filament proteins modulate cardiac function, *J. Mol. Cell. Cardiology* **28**, 217-230.

Solaro, R.J. and Rarick, H.M. (1998) Troponin and tropomyosin, proteins that switch on and tune in the activity of cardiac myofilaments, *Circ. Res.* **83**, 471-480.

Strynadka, N.C. and James, M.N.G. (1989) Crystal structures of the helix-loop-helix calcium-binding proteins, *Annu. Rev. Biochem.* **58**, 951-988.

Strynadka, N.C., Cherney M., Sielecki, A.R., Li, M.X., Smillie, L.B., and James, M.N.G. (1997) Structural details of a calcium-induced molecular switch: X-ray crystallographic analysis of the calcium-saturated N-terminal domain of troponin C at 1.75 Å resolution, *J. Mol. Biol.* **273**, 238-255.

Sundaralingam, M., Bergstrom, M., Strasburg, G., Rowchowdhury, P., Greaser, M., and Wang, B.-C. (1985) Molecular structure of troponin C from chicken skeletal muscle at 3 Å resolution, *Science* **227**, 945-948.

Syska, H., Wilkinson, J.M., Grand, R.J., and Perry, S.V. (1976) The relationship between biological activity and primary structure of troponin I from white skeletal muscle of the rabbit, *Biochem. J.* **153**, 375-387.

Swenson, C.A.. and Fredricksen, R.S. (1992) Interaction of troponin C and troponin C fragments with troponin I and the troponin I inhibitory peptide, *Biochemistry* **31**, 3420-3429.

Talbot, J. and Hodges, R.S. (1981) Synthetic studies on the inhibitory region of rabbit skeletal troponin I, *J. Biol. Chem.* **256**, 2798-2802.

Tung, C.-S., Wall, M. E., Gallagher, S.C., and Trewhella, J. (2000) A model of troponin-I in complex with troponin-C using hybrid experimental data: the inhibitory region is a β-hairpin, *Protein Sci.* **9**, 1313-1326.

Tsuda, S., Aimoto, S., and Hikichim K. (1992) ^{1}H-NMR study of Ca^{2+}- and Mg^{2+}-dependent interaction between troponin C and troponin I inhibitory peptide (96-116), *J. Biochem.* **112**, 665-670.

Tripet, B., Van Eyke, J., and Hodges, R.S. (1997) Mapping of a second actin-tropomyosin and a second troponin C binding site within the C terminus of troponin I, and their importance in the Ca^{2+}-dependent regulation of muscle contraction, *J. Mol. Biol.* **271**, 728-750.

Van Eyk, J.E., Strauss, J.D., Hodges, R.S., and Ruegg, J.C. (1993) A synthetic peptide mimics troponin I function in the calcium-dependent regulation of muscle contraction, *FEBS Lett.* **323**, 223-228.

Vassylyev, D.G., Takeda, S., Wakatsuki, S., Maeda, K., and Maeda Y. (1998) Crystal structure of troponin C in complex with troponin fragment at 2.3- Å resolution, *Proc. Natl. Acad. Sci. U.S.A.* **95**, 4847-4852.

Wang, C.-K. and Cheung, H.C. (1985) Energetics of the binding of calcium and troponin I to troponin C from rabbit skeletal muscle, *Biophys. J.* **48**, 727-739.

Whitby, F.G. and Phillips, G.N. (2000) Crystal structure of tropomyosin at 7 Angstroms resolution, *Proteins* **38**, 49-59.

White, S.P., Cohen, C., and Phillips, G.N. (1987) Structure of co-crystals of tropomyosin and troponin, *Nature* **325**, 826-828.

Xing, J., Dong, W-J., Chandra, R., Solaro, R.J., Umeda, P.K., and Cheung, H.C. (1999) Proximity mapping of the cardiac cTnC-cTnI complex, *Biophys. J.* **76**, A279.

M. A. GEEVES & S. S. LEHRER

COOPERATIVITY IN THE Ca^{2+}-REGULATION OF SKELETAL MUSCLE CONTRACTION

1. INTRODUCTION

An increase in intracellular Ca^{2+} initiates muscle contraction for all types of muscle. In striated muscles the Ca^{2+} binds to the muscle thin filaments and relieves the inhibition of myosin binding to actin allowing the ATP driven actomyosin cross bridge cycle which results in contraction. Although we know all of the essential components and the signal transduction pathway which leads from the change in free Ca^{2+} concentration to the contraction of a muscle fiber, the molecular details of the turning-on process are poorly understood. The change in Ca^{2+} concentration does not simply turn the contraction on and off but the exact concentration and the rate of change of concentration produces a graded response in the cross bridge cycle. The system is complex and involves at least 6 distinct protein subunits assembled into μm long filaments that are themselves assembled into a precise 3-dimensional array of interdigitating filaments. Each subunit interacts with more than one partner in a series of steric and allosteric interactions which may propagate along the thin filament. The potential for long range cooperative interactions involving hundreds or thousands of molecules is clearly present and part of the object of this chapter is to define the extent to which such long range interactions occur. The level of cooperativity is illustrated in Fig. 1, which shows the Ca^{2+} dependence of the force developed by an isometrically contracting skinned muscle fiber.

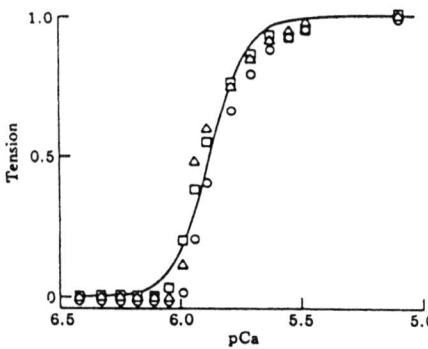

Figure 1. Normalized isometric tension vs. pCa. The line is a fit to a Hill plot with log K=5.87 and Hill coefficient = 5.8 (from Brandt, Cox, & Kawai, 1980).

247

R.J. Solaro and R.L. Moss (eds), Molecular Control Mechanisms in Striated Muscle Contraction, 247-269
©2002 Kluwer Academic Publishers, Printed in the Netherlands

The large Hill coefficient for this Ca^{2+} induced turning-on of force (>5 for the example shown here but can range from 2.5 to >5 depending upon conditions) indicates a high degree of cooperativity yet each troponin C binds only 2 Ca^{2+} ions. Since any model of cooperativity requires the Hill coefficient to be less than or equal to the number of binding sites this large Hill coefficient establishes that cooperative binding of Ca^{2+} to each individual TnC cannot on its own account for the Ca^{2+} switch. Furthermore, the data in Fig.1 (as well as from other laboratories; Moss, 1992) shows a non-symmetrical force/pCa curve indicating a greater Hill coefficient at high pCa compared to low pCa. The large Hill coefficient and the asymmetry in the force pCa curve indicate that the switching on process involves complex cooperative interactions beyond the binding of Ca^{2+} to a single troponin. To begin to understand this cooperative behavior we must first review the fundamental nature of cooperative processes.

2. ESSENTIALS OF COOPERATIVITY

Consider a ligand binding to a single site on a protein

$$E + L \leftrightarrow E'L$$ Scheme I

where the binding of a ligand, L, induces a change in the properties of the protein e.g. E is inactive E' is active. The affinity for the ligand is defined by the association constant K_L ($= k_{+1}/k_{-1}$) $= [E'L]/[E][L]$, (the more familiar dissociation constant, $K_d = 1/K_L$, can also be used but K_L is simpler to use here). Rearranging the equation: $[L]K_L = [L]/K_d = [E'L]/[E]$. This equation tells us that to change the fraction of enzyme in the active bound form from 10% ($[E'L]/[E] = 0.1/0.9$) to 90% ($[E'L]/[E] = 0.9/0.1$), L must increase from $1/9^{th}$ of K_d to $9K_d$. i.e. an 80 fold increase in ligand concentration. Thus if K_d is 1 μM and the enzyme activity is proportional to the concentration of E'L then Ca^{2+} concentration must increase from 0.1 μM to 9 μM to switch the system from 10% on to 90% on. To go from 5% to 95% the concentration must increase 360 fold from 0.052 to 19 μM.

If the protein has more than a single binding site (on a single polypeptide as in TnC or in a multisubunit protein such as hemoglobin) then the change in the whole protein structure (E to E' in the example above) can be induced by the binding of one or more ligands. This is a case of the MWC model of cooperative binding (Monod, Wyman, & Changeux 1965) as shown in Fig 2. Initially the ligand binds with low affinity (K_L, association constant are simpler to use here) to the protein in the E state but with a c fold higher affinity ($K_L.c$) to the protein in the E' form. If the initial equilibrium between the two forms of E is defined by K_T ($= [E']/[E]$) then detailed balance requires that each ligand that binds will alter the value of K_L by the factor c as shown in Fig 2 for a 4 site protein.

A

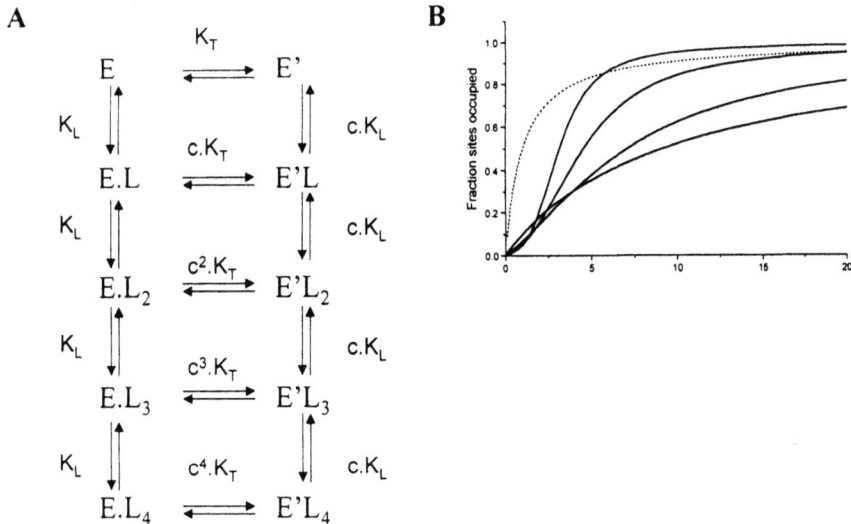

B

Figure 2. The MWC Cooperative binding model. A) The generalized scheme for a ligand (L) binding to an enzyme (E) with 4 identical binding sites. The enzyme can exist in two forms E and E' (K_T = [E']/[E]) which differ in their affinity for ligand (K_L = [E.L]/[E][L]) by a factor c. B) The fraction of binding sites occupied by ligand for the MWC cooperative model for a protein with 1,2,3 or 4 identical binding sites with K_T = 0.1, c = 10 and $1/K_L$ = 10 μM. The dotted line represents the plot for a single binding site with $1/K_L$ = 1 μM.

If the affinity of the ligand for E' is 10 times that of E (i.e. c =10) and K_T = 0.01, it follows that the equilibrium between the two forms of the protein shifts towards the E' form as the ligand binding sites become occupied. For 1, 2, 3 or 4 sites occupied the equilibrium constant switches from a value of 0.01 to 0.1, 1, 10 and 100 respectively (see Fig 2b). An increase in the fraction of protein sites occupied from 10% to 90% requires only a 4-5 fold increase in ligand concentration (1.5 to 6.9 μM if K_d ($1/K_L$) is 10 μM) compared to 90 fold for independent sites. For 2 or 3 ligand binding sites the change in ligand concentration required is 24 or 8.8 fold respectively. Thus the cooperativity allows for a more efficient switching of the system from off to on and back again. In the case of the muscle system the Ca^{2+} concentration needs to change over a much narrower range allowing both a more rapid response and better fine-tuning of the control of contraction.

The above outline describes the MWC model of cooperativity. There are other versions that differ in details (Koshland et al, 1966; Hill, 1960) but the essential nature of cooperativity is that there is a concerted change of state of a multi protein complex coupled to ligand binding.

The introduction section established that the cooperativity within the Ca^{2+} switch in a muscle must involve interactions in addition to the Ca^{2+} binding to sites on a single TnC. Once there is a requirement for additional cooperative interactions then we have to consider the whole of the muscle structure. The possible sources of additional cooperative interactions are difficult to establish since there are many potential interactions, for example, between adjacent TmTn components along the thin filament and with other components of either the thick or thin filament. The complexity involved is illustrated in Fig 3a that shows a longitudinal view of a section of a muscle sarcomere in which each half sarcomere consists of a set of interdigitating thick and filaments in a hexagonal array. The detailed structures of the thick and thin filaments are dealt with elsewhere. We are primarily concerned with the thin filament (Fig 3b) which consists of a polymer of globular actin subunits and forms the characteristic arrangement of a twisted double string of beads. Tropomyosin (Tm), a coiled-coil alpha helix, interacts with 7 actin monomers and with itself head-to-tail. Each Tm is associated with one Tn consisting of three subunits, troponin I (TnI) and troponin C (TnC) and troponin T, (TnT). TnC is the Ca^{2+} binding subunit; TnI is the inhibitory subunit that interacts with actin and TnC in a Ca^{2+}-dependent manner, and with TnT. TnT, an elongated molecule, interacts with both TnI and TnC through its C-terminal domain and with the head-to-tail region of Tm through its N-terminal domain. Other thin filament components include actin capping proteins and nebulin which do not appear to be involved in the regulatory mechanism. Thus the repeating structural unit of the regulated actin filament is A7TmTn.

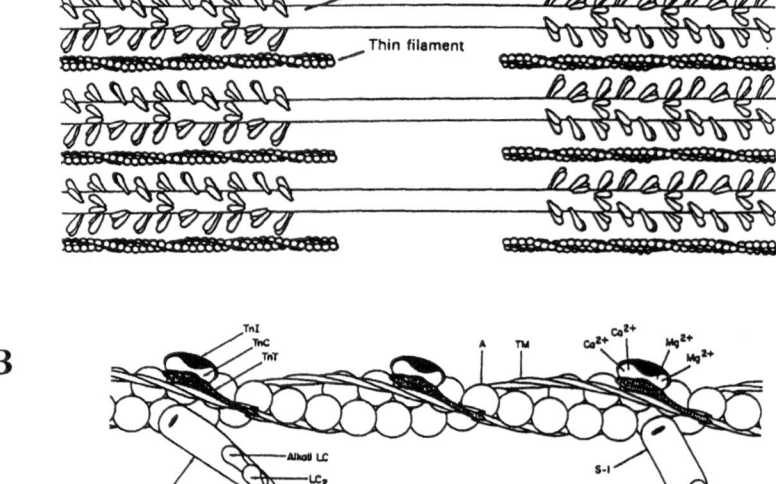

Figure 3. A, Schematic diagram of the components of the filament arrangement in the central region of the striated muscle sarcomere (Offer & Knight 1998). B, Schematic diagram of the components of the thin and thick filaments of striated muscle (Hoffman, Greaser, & Moss, 1991).

The complexity of the structure means that unlike the classical, cooperative globular protein systems, it is not possible to define limits of cooperativity in either thick or thin filaments. Potential sources of cooperativity are along the thin filament, within and between A_7TmTn structural units, along the thick filament, between groups of myosins heads and between myosin heads and the A_7TmTn units. However, the regular geometry of the fiber does define some limits to the cooperativity. For example the stoichiometry of the protein complexes within each filament is fixed, as is the ratio of actin to myosin sites. In rigor all of the myosin heads can bind to actin with a maximum of 67% of actin occupied by myosin in the overlap zone. Under contracting conditions (in the presence of ATP and Ca^{2+}) the maximum number of myosin heads attached to actin at any one time remains undefined (see section 7).

Cooperativity of the complete muscle is beyond the scope of the current overview and we will therefore concentrate on the thin filament and its interaction with isolated myosin heads. We will attempt to reduce the problem to the smallest cooperative unit and then consider to what extent the complete system can be rebuilt from these simple units.

3. COOPERATIVITY OF CA^{2+} BINDING TO TROPONIN

Skeletal TnC has four divalent cation binding sites. Two high affinity sites which bind either Ca^{2+} or Mg^{2+} are thought to play a structural role and exchange the tightly bound ions relatively slowly and will not be considered further. The low affinity, specific Ca^{2+} binding sites trigger the muscle contraction. Data on Ca^{2+} binding to isolated TnC or to TnC complexed with TnI, whole Tn, TmTn or thin filaments suggest that Ca^{2+} binds with a Hill coefficient of between 1.5 and 2 (Potter & Gergely, 1975; Grabarek et al, 1983). A smaller Hill coefficient is observed for cardiac Tn and other Tn's which have a single regulatory Ca^{2+} binding site. Since the Hill coefficient is <2 these observations are compatible with cooperative Ca^{2+} binding to the TnC without longer-range cooperativity between TnC's along the filament. An interesting observation is that the affinity of Ca^{2+} for TnC does vary for the different TnC complexes. For isolated TnC and isolated thin filaments (A_7TmTn), the affinity is relatively low with K_d = 2-5 µM but is almost 10 fold tighter for the TnI-TnC complex, for whole Tn, for TnTm and for isolated thin filaments decorated with S1 (Bremel, et al., 1972; Grabarek et al, 1983; Potter & Gergely, 1975; Murray, et al., 1975). It is possible therefore that the two values of K_d reflect two different conformations of TnC in these complexes. A high affinity for Ca^{2+} is observed in any complex in which there is a strong TnI-TnC interaction. In the absence of TnI or when the TnI is strongly complexed to actin (in whole thin filaments but not in S1 decorated thin filaments) TnC has a lower affinity for Ca^{2+}. Thus TnI can be viewed as an allosteric effector of Ca^{2+} binding to TnC and reciprocally Ca^{2+} is an allosteric effector of the TnI- thin filament interaction and thereby an effector of the conformation of the thin filament.

Equilibrium measurements estimate the cooperativity of Ca^{2+} binding to TnC specifically. They do not estimate how many S1 binding sites on actin have been affected by the Ca^{2+}-binding to each TnC in the thin filament. What the measurements do show is that for isolated TnC or complexes of TnC in thin filaments with and without S1 bound the Hill coefficient is similar and has a value of less than 2. Thus the data can be explained without requiring any interaction between adjacent A_7TmTn structural units. Such longer range interactions may exist along the thin filament but the fact that the Hill coefficient < 2 for Ca^{2+} binding to isolated TnC and to TnC in thin filaments, suggests that any such longer range effect on Ca^{2+} is relatively small.

4. COOPERATIVITY IN MYOSIN HEADS (S1) BINDING TO THIN FILAMENTS

4.1. Kinetics of Myosin Binding and the B or blocked-state of the thin filament

Information on the influence of Ca^{2+}/TnC on myosin binding to actin within the A_7TmTn structural unit comes from studies of the kinetics and equilibrium binding of S1 to thin filaments. Kinetic studies of the rate of S1 binding to actin allowed the presence of a blocked or B-state of actin filaments to be identified (see Fig. 6). This conformation of the thin filament which is unable to bind S1 and is most simply identified with the original steric blocking position of Tm over the S1 binding site on actin (Trybus & Taylor, 1980, McKillop & Geeves, 1993).

If the rate of an excess of S1 binding to pure actin filaments is followed in a stopped flow spectrophotometer a simple exponential binding reaction is observed. For S1 (M) in excess of actin (A) i.e [M]>>[A] and an irreversible process:

$$A + M \xrightarrow{k} A.M \qquad \qquad \text{Scheme II}$$

then the rate of change in [A] is given by

$$-d[A]/dt = [A][M].k \qquad \qquad (1)$$

where k is the rate constant of the process.

On integration this predicts the observed exponential decrease in [A].

$$[A]_t = [A]_o exp(-k_{obs}t) \text{ where } k_{obs} = [M]_{total} k \qquad (2)$$

For [A]>>[M] a similar equation gives:

$$[M]_t = [M]_o exp(-k_{obs}t) \text{ where } k_{obs} = [A]_{total}.k \qquad (3)$$

The equations accurately described the kinetics and predict a linear dependence of k_{obs} on the concentration of the protein in excess, with the same slope (k). In fact, measurements made using pure actin filaments, ATm or ATmTn in the presence of

Ca^{2+} give identical results. Thus, the kinetics indicate that k (the second order rate constant of association) is identical in each case and that all of the actin sites are available to the S1 in the presence of Ca^{2+}.

When the reaction is followed with actin in excess of S1 in the absence of Ca^{2+}, an exponential is also observed but k_{obs} is a factor of 3 lower than in the presence of Ca^{2+}. In this measurement only very few actin sites become occupied by S1, ideally ≤ one S1 per cooperative unit and the rate of the first S1 binding to the "*off-state*" is observed. Since k_{obs} is 1/3 of the value in the presence of Ca^{2+} then either k is lower or [A] is lower because a fraction of the actin sites is unavailable. If there are two states of the filament in equilibrium one of which is unable to bind S1, called the B or blocked state, as shown below:

$$A_{off} \overset{K_B}{\leftrightarrow} A_{on} + M \overset{k}{\rightarrow} A.M \quad where \ K_B = [A_{on}]/[A_{off}] \quad \text{Scheme III}$$

$$d[M]/dt = -[A_{on}][M].k \tag{4}$$

then assuming the two forms of actin are in rapid equilibrium, the fraction of the total actin in the form A_{on} is $K_B/(1 + K_B)$. Hence

$$d[M]/dt = -[A_{total}][M]k.K_B/(1+K_B) \tag{5}$$

$$and \ k_{obs} = [A]_{total.}k.K_B/(1+K_B) \tag{6}$$

As k_{obs} $(-Ca^{2+})$ is 1/3 of k_{obs} $(+Ca^{2+})$ the data can be explained by 30% of the actin monomers occupying the on-state i.e. K_B is 0.5 and $K_B/1+K_B = 0.3$. The alternative is that k is 3 fold smaller in the absence of Ca^{2+}. By varying the nucleotide bound to S1 (which is known to change k without affecting [A]) it was found that the value of k_{obs} in the absence of Ca^{2+} was always reduced to 30% of the value in the presence of Ca^{2+}. Thus, the more likely explanation is that removal of Ca^{2+} affects K_B and not k. The off state is termed the blocked of B-state of the filament by analogy with the steric blocking model of thin filament regulation.

4.2. Cooperativity of B TO C Transition

The kinetic experiments of S1 binding to excess actin and which reported the fraction of available actin sites was repeated over a range of Ca^{2+} concentrations. The resulting data (Fig. 4a & b), shows that the fraction of actin sites in the B-state switches from 0 to 30% over the pCa range 4 - 8 (Head et al, 1995). A fit to the Hill equation gave a mid point at pCa = 5.6 and a Hill coefficient of 1.8 similar to the values observed for Ca^{2+} binding to the thin filament (Potter, & Gergely, 1975). Similar results were obtained for cardiac Tn with a smaller Hill coefficient (Reiffert et al., 1996). Thus an equilibrium exists between the B & C conformations of the actin filament with $K_B = 0.3$ in the absence of Ca^{2+}, increasing to a value >10 in the presence of Ca^{2+}. We still do not know what the size of the unit is (the number of actin sites) which switches between the B & C-states. Assuming an A_7TmTn unit and a two site Ca^{2+} switch in a simple MWC system of TnC we can predict the relationships in Scheme IV to define the ability of Ca^{2+} to affect the B/C transition:

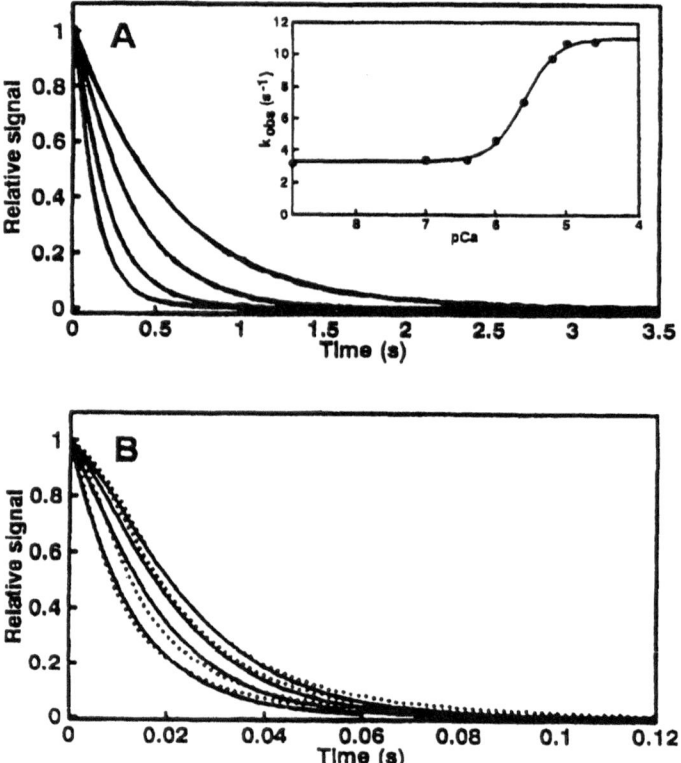

Figure 4. The Ca^{2+}- dependent binding kinetics of S1 to ATmTn. A. Binding of 1 µM S1 to 5 µM actin, 1.4 µM TmTn at different pCa values, with fitted exponentials as full lines. insert. pCa dependence of observed exponential rate constant fitted to the Hill equation. B. Binding of 40 µM S1 to 1 µM actin, 0.28 µM TmTn at the above pCa values. The solid lines represent the modeled transients for a reaction scheme with A_7TmTn^B in equilibrium with A_7TmTn^C , the first S1 to bind traps the cooperative unit in the C-state subsequent S1 bind to this state (Head, et al., 1995).

If the affinity of Ca^{2+} to the blocked state is that measured for TnC and ATmTn ($1/K_{Ca}$ = 2-5 µM) and that for the closed state is that measured for a TnIC complex or S1.ATmTn ($1/K'_{Ca}$ = 0.2-0.5 µM) (see section 3), then from the relationship $K_B.K'_{Ca} = K'_B.K_{Ca}$ we can calculate that K'_B is approximately 10x0.3= 3 and K''_B = 30. Thus the binding of two Ca^{2+} would switch the actin sites from 77% blocked to 3%. For a smaller difference between the two affinities the switching is less complete (i.e. for a factor of 5 the fraction blocked would be 77, and 11%).

B-States   C-States

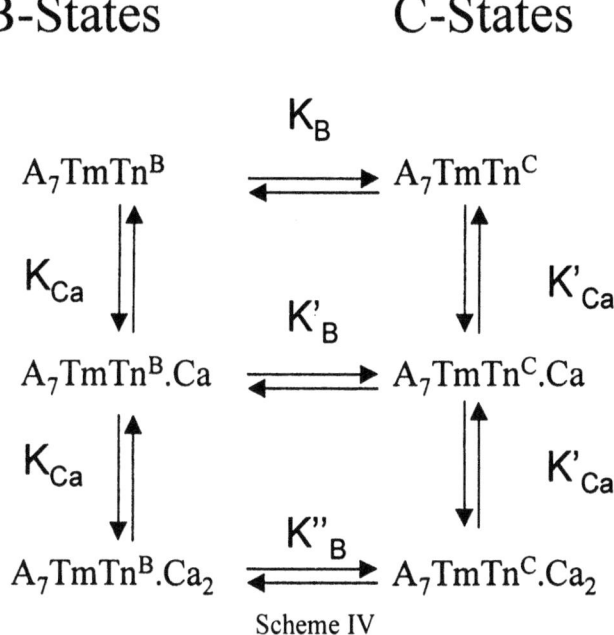

Scheme IV

The above argument does not reveal how many actin sites are switched on in the Ca^{2+} induced B to C transition. It can be assumed to be 7 actin subunits if there is no Tn-Tn interaction. This was tested by adding an excess of S1 to filaments in the absence of Ca^{2+} and the kinetic transient recorded (Fig 4b). In this case S1 occupies all of the actin sites in the filament. Initial binding is limited by the fraction of the filament in the blocked state but as S1 binds it switches the filament on and the rate of binding accelerates. The fraction of actin sites in the two states (B & C) at the start of the transient is defined by K_B at each Ca^{2+} (Fig 4a). The intrinsic rate constant for S1 binding is defined by the exponential at high Ca^{2+}. Thus the only unknown in the model is n, the number of actin sites turned on by each S1 bound and therefore the sigmoid binding transients can be fitted by varying only one parameter, n. The data were shown to be compatible with $n = 7$ for the B to C transition. i.e. the switch from B to C involves the switching of a single A_7TmTn unit. However, it should be pointed out that the model may not be very sensitive to large values of n but it is certainly not smaller than 7 and the data requires no value larger than 7.

4.3. Equilibrium Binding of S1 to Actin Filaments and the C to M Transition

Steady state and equilibrium measurements of S1 binding to thin filaments revealed a different picture to that shown by kinetic studies. Sedimentation and fluorescence methods have been used to estimate the fraction of occupancy of actin filaments with S1 and S1-nucleotide complexes at equilibrium, as the S1 concentration is varied (Greene, & Eisenberg, 1980; Geeves, & Halsall, 1987; Ishii, & Lehrer, 1985,

Maytum et al 1999). The binding of S1 to actin is non-cooperative in the absence of TmTn, i.e. S1 binds to each actin site on the filament as if it were an isolated monomer with no interaction between actin sites. However, for ATmTn plus or minus Ca^{2+} the binding is sigmoidal, indicative of cooperative binding of S1 to actin (Fig 5). Thus at low [S1] the binding is inhibited but after some fraction of the actin sites are occupied the system changes state (turns-on) to a form which binds S1 more readily. As for the MWC classical cooperative system four parameters are required to define the equilibrium binding isotherm as defined above; K_T, the equilibrium between the two states of the filament (off/on) in the absence of S1; n., the apparent size of the cooperative unit which changes state, K_L, the affinity of S1 for the "off-state" of the filament; c, the ratio of affinities of S1 for the two filament states.

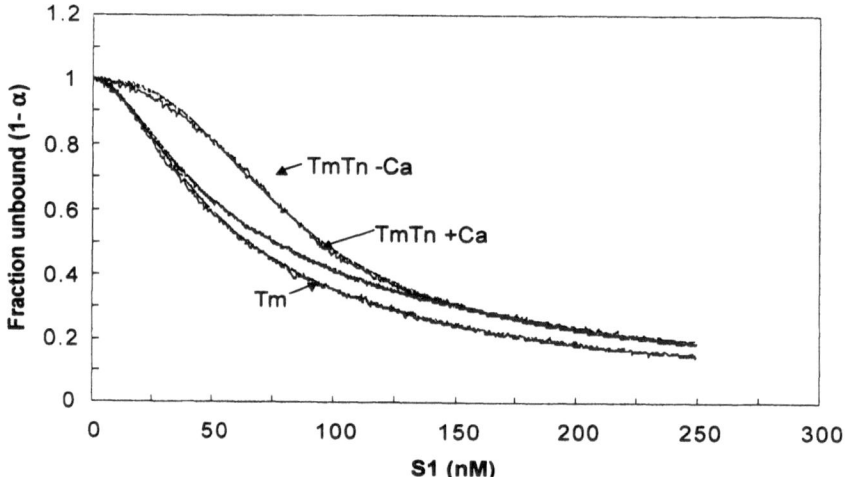

Figure 5. Equilibrium titration of pyrene labelled actin filaments with S1 (Maytum et al 1999). Titration for actin Tm, actin TmTn +/- Ca^{2+} and the data fitted. Data fitted to the model of McKillop & Geeves

$$\alpha = \frac{K_1[M]P^{n-1}(K_T(1+K_2)^n + 1)}{(K_TP^n + Q^n)(1+K_2)^{n-1}}$$

α - fractional saturation of actin
[M] - concentration of free S1
$P = 1 + K_1[M](1 + K_2)$ and $Q = 1 + K_1[M]$.

TABLE 1. *Occupancy of the B/C/M states, Tm position and model parameters for different thin filaments*

Thin filaments	B	C	M	K_B	K_T	n	Tm Position
A.Tm	0	0.8	0.2	-	0.2	7	inner/outer
A.TmTn - Ca^{2+}	0.7	0.25	0.01	0.3	0.03	7	outer domain
A.TmTn + Ca^{2+}	0	0.8	0.2	>10	0.2	12	inner/outer
A.TmTn + myosin	0	0	1.0	-	-	-	inner domain

Based on Lehman et al 2000 and Maytum et al 1999

Two approaches were taken in fitting the cooperative binding isotherms; both using the assumption that n = 7. Hill et al (1980) derived a model which included an extra cooperative parameter to allow the state of the two adjacent units to influence the probability of a change of state. The approach of Geeves & Halsall (1987) and McKillop & Geeves (1993) was a refinement of the Hill et al model. Greene (1982) reported that the binding constants depended on the nucleotide bound to S1. Those constants appeared to correlate with the value of K_2, the value of the equilibrium constant for the isomerization in the two step binding mechanism of S1 to actin from an A (attached) to R-state (rotated or rigor-like) (Geeves, 1991; Geeves, Goody, & Gutfreund, 1984).

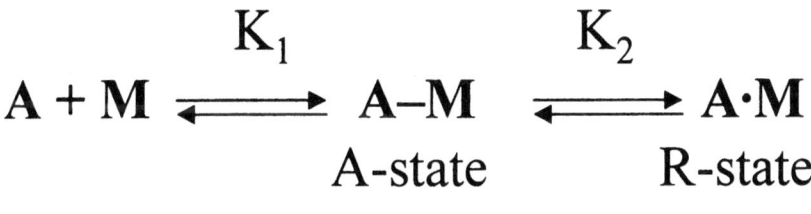

$$A + M \underset{}{\overset{K_1}{\rightleftharpoons}} A\text{–}M \underset{}{\overset{K_2}{\rightleftharpoons}} A\cdot M$$

A-state   R-state

Scheme V

It was proposed that in the "*off-state*" of the thin filament, S1 could bind to form the A- type interaction but that Tm prevented the isomerisation to the R-state. However in the "*on state*", S1 could form the R-state and trap Tm in the "*on-state*" (Fig 6). Thus the A to R transition of actoS1 and the off-on transition form two coupled equilibria. A simple way of imagining this is that Tm blocks only the part of the actin-myosin interface which is involved in forming the strongly bound R-state. Thus in this model $K_L = K_1$ and $c = K_2$.

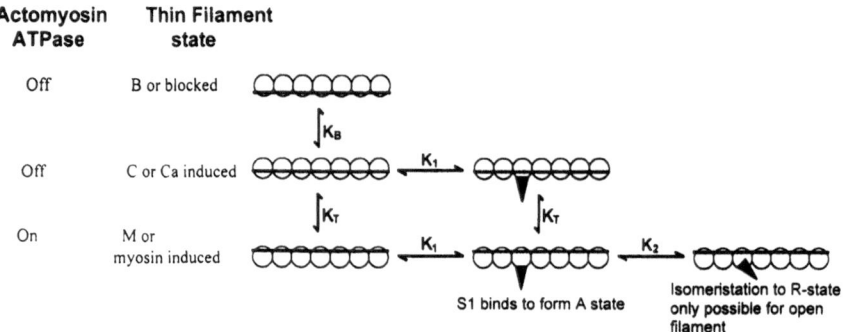

Figure 6. Schematic diagram of the relationship of the three S1-binding states of actinTmTn, (B, C and M) with the ATPase activity states and the A- and R-states of the S1-actin complex.

B-state   Myosin binding blocked. No productive S1-binding and no activation of S1 ATPase

C-state   Closed or Ca²⁺ induced state. S1 binding to form the attached or A-state only but no activation of the ATPase or force generation.

M-state   Open or myosin induced state. Binding of S1 to form the A-state and the rigor like complex or R-state ATPase is accelerated and force can be generated.

The names emphasize the similarity between the states identified in solution binding studies and the 3- structural states identified independently by Vibert et al (1997) see Table1. Adapted from McKillop & Geeves (1993).

A feature of this model (Fig. 6) is that K_T and K_2 are independent of each other but K_T is Ca^{2+} dependent and K_2 is nucleotide dependent. Binding studies in the presence of different nucleotides provided values to fit the Hill et al and McKillop & Geeves models. The most recent and best determined values are those of Maytum et al (1999) which used very low concentration of fluorescently labeled actin (50 nM pyrene-actin) which gives high sensitivity to each of the parameters of the fit. In the absence of Ca^{2+} the analysis gave values of n = 7 and $K_T \approx 0.05$ i.e. less than 5% in the on state. The system appeared more turned off than that seen in kinetic measurements (33 % on). The situation was more marked for the case with TmTn in the presence of Ca^{2+} or Tm alone. In each case the titrations were sigmoid, indicating cooperative binding with $K_T = 0.2$ in each case (80% off). Kinetic data gave 100 % on for both measurements. The differences were too great to be due to experimental error and therefore McKillop & Geeves proposed the data represent two different off states and the two methods have a different sensitivity to the states. The states were named B, C and M states and are defined in Figure 6.

In this model the kinetic measurements are only sensitive to presence of the B state. (Both C & M states can bind myosin at the same rate) whereas the equilibrium titration measurements are sensitive to the M-state, the only one which binds myosin tightly. The parameters that define the model using both kinetic and equilibrium binding data are given in Table 1. Two key points about the model are worth emphasizing. First, the addition of Ca^{2+} does not switch the system on, it eliminates the B-state but the ATPase is still not activated, some tightly bound myosin is required before the ATPase is activated. I.e. Ca^{2+} is an allosteric activator of the ATPase not an on-off switch (Lehrer & Geeves 1998). The second key point is the value of n, the apparent number of actins activated by a single S1 binding is dependent upon Tn. In the absence of Tm it has a value of 6-7, close to the structural repeat A_7Tm. In the presence of Tn and Ca^{2+} the cooperative unit is much larger and almost twice the size of the structural repeat. Thus information about an S1 binding is transmitted to adjacent Tms as in the Hill et al model. In the absence of Ca^{2+} the unit size is again ~7 the size of the structural repeat. Why this cooperative unit should change size can be explained by considering which component of Tn is responsible for the increase in unit size. This is largely due to the N-terminal fragment of TnT known as TnT1. This binds along the Tm and interacts with two adjacent Tms in the N-C overlap region. (The presence of TnI, TnC and the C-terminal part of TnT, TnT2, although conferring Ca^{2+} sensitivity on the S1-actin interaction, had no effect on the apparent cooperative unit size (Schaertl, et al., 1995). Furthermore, use of Tm from the smooth chicken gizzard muscle (smTm) or from fibroblasts (fTm), neither of which normally bind Tn and have stronger Tm-Tm contacts, have a cooperative unit size of 10-12 (Lehrer, et al 1997, Maytum et al 2001). Thus the large cooperative unit size is a property of strong Tm-Tm contacts and the size of the cooperative unit does not relate simply to the length of Tm. This point is emphasized by the fact that smTn and fTm have similar cooperative unit sizes even though the fTm binds to only 6 actin sites compared to the 7 of skeletal and smTm. Surprisingly, in the absence of Ca^{2+} the cooperative unit was equal to 7. Under these conditions the cooperativity is dominated by the B to C transition and is consistent with a strong TnI-A interaction repeating every 7 actins in the B-state limiting any longer range cooperativity along the filament.

The apparent cooperative unit size for the C to M transition has been measured in two other ways. Ishii & Lehrer (1985) demonstrated that the excimer fluorescence of a pyrene label attached to Cys-190 of Tm (Tm*) signaled the C to M transition of ATm and ATnTm. Thus by combining light scattering signals (which measure the fractional occupancy of actin sites by S1) with the Tm* fluorescence (which measures C to M) the average occupancy of actin with S1 required to change the Tm fluorescence was estimated in both equilibrium binding (Ishii & Lehrer, 1985; Ishii & Lehrer, 1990) and kinetic experiments (Geeves, & Lehrer, 1994). The kinetic experiments were performed in both the direction of S1 binding to actin and the ATP induced dissociation of S1 from the actin filament. These results gave cooperative unit sizes in full agreement with those using pyrene actin but are not dependent upon the detailed model for the fitting. Thus both kinetic and equilibrium measurements using pyrene-Tm agree on the apparent cooperative unit size for Tm.

EM studies also addressed the question of how far beyond an S1 bound to a reconstituted thin filament does the Tm retain "memory" of the S1 induced local structure. Partially S1 decorated thin filaments were used. By noting patches of decorated and undecorated areas, the reconstructions showed that Tm moved between two positions (M to C). They obtained a values of about 300 – 500 nm or about 10 Tm molecules (Vibert et al, 1997). These values, which are much greater than the length of 2 Tm's that were obtained in solution studies, can be taken as an upper limit of *n*, in view of the fact that a few S1's in the undecorated area may not be visible which would keep the Tm in the open position.

These results lead to the question of what is the physical meaning of the size of a cooperative unit greater than that of the structural unit. We have discussed two interpretations (Lehrer, et al., 1997, Maytum et al, 1999): one involves the model of Hill et al where the state of a single A_7TmTn unit communicates to adjacent structural units via Tm end-to-end contacts. However this would require major changes in the Tm-Tm contacts when structural units change state. Since the Tm-Tm contacts provide most of the binding energy of Tm for actin a major change in these contacts seems unlikely. The alternative, which we favor, is that when Tm-Tm contacts are strong the Tm forms a continuous dynamic cable along the length of the actin filament (Fig 7) with the apparent cooperative unit size a measure of the stiffness or the persistence length of the Tm filament on actin. Direct evidence for such a dynamic cable has yet to be produced.

'Discontinuous' Tm model

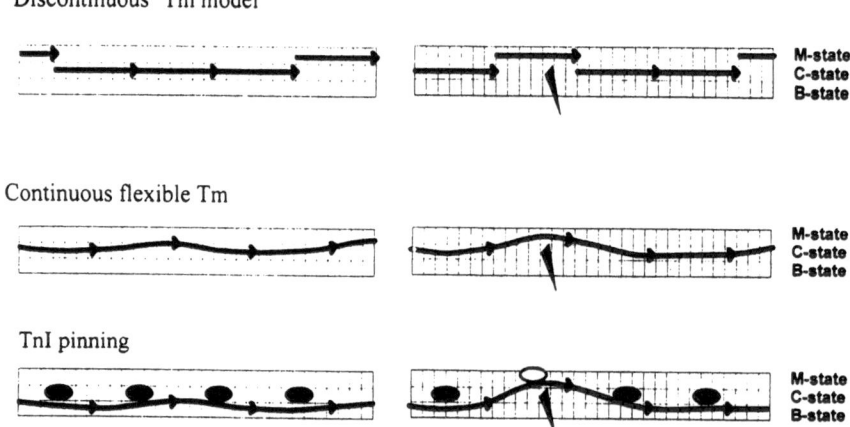

Continuous flexible Tm

TnI pinning

Figure 7. Discontinuous cooperative & continuous flexible models of tropomyosin. Lehrer et al (1997), Maytum et al (1999)

5. STRUCTURAL AND SPECTROSCOPIC EVIDENCE OF 3 STATES

Structural evidence for a three state regulatory system has been obtained from X-ray studies on intact muscle (Holmes, 1995) and electron microscopy studies on reconstituted thin filaments (Vibert et al 1997; for a review see Squire, & Morris, 1998). Both methods detect Ca^{2+} and S1-induced changes in structure. EM data suggests that Tm interacts with a large region of the myosin binding site of actin in the absence of Ca^{2+} but that addition of Ca^{2+} moves Tm but still leaves it in a position to sterically hinder the formation of the myosin rigor-like complex. In the rigor complex the Tm is moved yet further away from the myosin binding site. This interpretation agrees with the modeling of the thin filament by Holmes and his colleagues (Holmes 1995). However, neither method (EM or X-ray) considered the possible influence of Tn reorganization on the interpreted changes (Ishii & Lehrer,

1990; Lehrer, 1994; Squire & Morris, 1998). The EM pictures were helically averaged which can dilute out the contribution of the Tn across 7 actin sites. To address this issue Lehman et al (2001) used a genetically engineered short Tm which spans four actin sites rather than the normal seven. This has the effect of increasing the stoichiometry of the filament to A_4TmTn giving more density to the Tn and a more favorable symmetry for the averaging. An alternative approach was taken by Narita et al (2001) who used single particle averaging methods to produce high resolution images. Both approaches confirmed the three positions of Tm on the actin surface and in addition produce images of Tn on the filament. These confirmed the close association between Tn and actin in the absence of Ca^{2+}, consistent with the Tn-actin interaction contributing to a pinning of the Tm into its blocked or B position.

A problem that remains to be resolved is that the EM pictures show the Tm in one dominant position under a particular experimental condition. In contrast, solution studies show a more dynamic state of the Tm moving between three positions on a millisecond time-scale with all three states occupied except in an S1 decorated thin filament (see table 1). Spectroscopic measurements using fluorescent or spin probes may give a better indication of the dynamic behaviour of the thin filament.

Several studies of the probes attached to thin filament proteins have been undertaken. It is desirable that the probes are sensitive to only one of the transitions; B to C or C to M, so that only data on a particular perturbation is monitored and local effects of direct Ca^{2+} binding or myosin head binding should be taken into account. The pyrene excimer fluorescence probe on Tm has proved useful in monitoring the C to M transition (Ishii & Lehrer, 1985, 1990). Probes of Tn subunit conformational changes associated with thin filament transitions have been performed with fluorescence (Zot & Potter, 1987; Trybus & Taylor 1980, Brenner et al 1999) and spin probes (Li & Fajer, 1998) to detect changes in environment and mobility. Proximity information has been obtained from crosslinking studies (Golitsina and Lehrer 1999) and FRET (Luo et al 2000, Miki et al 1998, Bacchiocchi & Lehrer 2000). Fluorescence images from labeled S1 bound to fibrils in rigor at low and high $[Ca^{2+}]$ have revealed that both in the absence and in the presence of Ca^{2+}, S1 is preferentially bound in the region of the sarcomere where myosin heads can bind to actin to trap the system in the open state, i.e., few heads are bound in the region of the sarcomere that only contained actin (Swartz et al, 1996). This verified that Ca^{2+} is not sufficient nor necessary to turn-on the system. The FRET approach has shown that TnI dissociates from actin when TnC binds Ca^{2+} and has been more recently used in showing that TnI elongates on binding to actin in the ternary complex (Luo et al 2000). A variety of FRET probes between Tm and actin using steady state intensity measurements and early lifetime studies have not revealed changes in distance associated with Ca^{2+} binding (Miki et al 1998, Luo et al 2000). More recent studies with a different acceptor position, using high resolution frequency-detected lifetime techniques and distance-distribution global analyses have revealed a rolling movement of Tm over actin associated with Ca^{2+} binding to skeletal TmTn and S1 binding to gizzard Tm (Bacchiocchi & Lehrer, 2000, 2001).

6. ATPASE STUDIES

The regulation system has so far been confined to studies of Ca^{2+} and myosin heads binding to actin filaments. The situation with cycling cross bridges turning over ATP is more complex since we have established that myosin heads (or S1) binding in the tightly bound R-state are required for full activation of the thin filament. Yet during turnover the interaction of each myosin head with actin is only transitory as each new ATP to bind to a myosin head dissociates it from actin. The lifetime of a myosin attachment is a function of many parameters including the concentrations of ATP, S1, actin and Ca^{2+} in addition to solution conditions such as temperature, pH and in particular the ionic strength. A complete description of regulation for a system turning over ATP has not been achieved. However, ATPase studies of myosin heads interacting with the complete thin filament demonstrate features of the complex cooperativity without the additional complexity that is involved with studies of muscle fibers. The advantage of solution studies of the ATPase is that the concentration of actin, S1 or ATP in the presence or absence of Ca^{2+} can be varied, or the Ca^{2+} concentration can be varied at different protein concentrations. In the absence of actin the S1 Mg^{2+}-ATPase is very low with a k_{cat} of 0.05 s^{1}. The presence of pure actin filaments increases the ATPase up to 200 fold, hence our proposal that actin is the catalyst of the breakdown of the myosin products complex (Lehrer & Geeves, 1998). In the presence of Tm the plot of ATPase vs. [S1] becomes cooperative and in the additional presence of Tn the ATPase is Ca^{2+} regulated (Fig. 8). At low [S1] the ATPase of the regulated actin filament (ATmTn) is inhibited compared to pure actin in both the presence and absence of Ca^{2+} but the inhibition is greater when Ca^{2+} is absent. As the [S1] is increased, the ATPase increases sigmoidally both in the presence and absence of Ca^{2+} above the S1 alone values but the increase is greater when Ca^{2+} is present. This behavior is compatible with a model in which the S1 can cooperatively activate the regulated actin and Ca^{2+} acts as a classical allosteric activator of the thin filament activity (Lehrer & Geeves, 1998). In agreement with the properties of allosteric effectors, the thin filament activity is not switched on by Ca^{2+} per se but only influences the concentration at which the S1 is able to turn-on the system Strongly bound myosin heads, the product of the ATPase, switch on the ATPase (Lehrer & Geeves, 1998). This was observed in early ATPase studies that used NEM-S1, a modified S1 which binds strongly to the thin filament and is not dissociated by ATP (Pemrick, & Weber, 1976). The binding of only a few NEM-heads can turn on the ATPase and fewer heads are required for ATmTn in the presence of Ca^{2+}, compared to the absence of Ca^{2+}. The curve for ATm lies in between the two Tn-containing curves in agreement with the equilibrium titration experiments. These results corroborated the pioneering studies in A. Weber's lab which showed activation of acto-S1 ATPase by rigor heads at low ATP concentrations in the presence of Tm and TmTn (Bremel, Murray, & Weber, 1972). The reciprocal relationship between myosin heads and Ca^{2+} binding was also shown in A. Weber's lab whereby myosin binding increases Ca^{2+} binding (Bremel, & Weber, 1972; Murray, Weber, & Bremel, 1975). Thus Ca^{2+}-induced cooperatively is a product of both Ca^{2+} and S1 binding to the regulated actin filament.

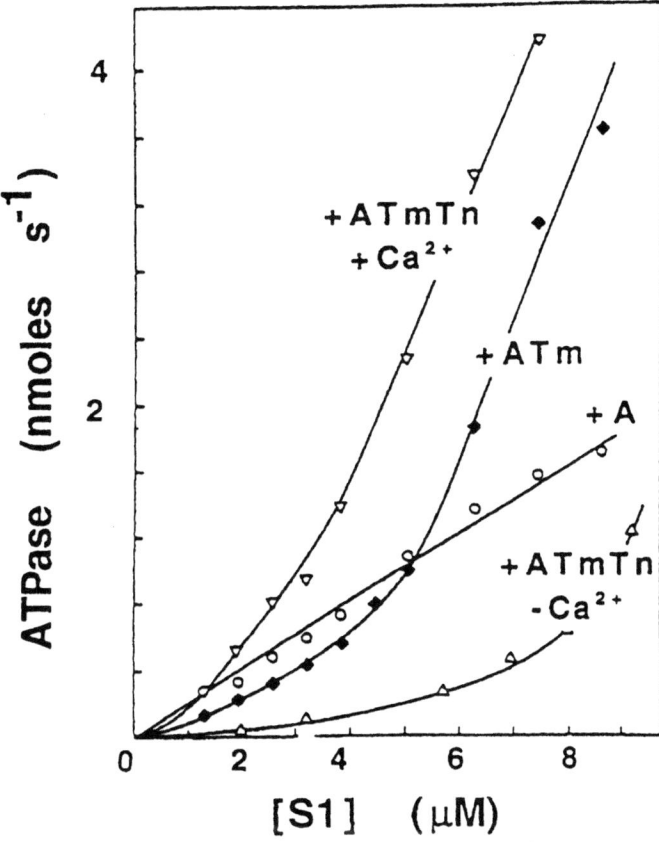

Figure 8. ActoS1 ATPase activity as a function of [S1] for actin, actinTm, actinTmTn+/- Ca²⁺ (Lehrer, & Morris, 1982).

7. RELATIONSHIP OF SOLUTION STATES TO FIBER STUDIES

In this chapter we have shown that the thin filament is a complex assembly of proteins which can exist in a number of structural and functional states. Which state or group of states will be observed depends upon both the methods used to probe the system and the conditions under which it is examined. The overall state of the system is a function of a series of allosteric interactions between the proteins, the principal effectors are calcium and R-type myosin heads bound to the thin filament. We have shown that a self-consistent model can be produced from solution studies of the isolated components at equilibrium or kinetically during the approach to equilibrium. At the end of this chapter it is appropriate to consider the extent to which these studies allow us to understand the regulatory process in a muscle fiber. Unfortunately while the models help our understanding of the underlying mechanism we are still missing several key pieces of information to be able to describe the system in a muscle fiber. The single most important piece of information still missing is the number of cross bridges (individual myosin heads) which are strongly attached to actin during isometric contraction and the lifetime of an individual cross bridge force holding state (often referred to as the duty ratio- the fraction of time a cross bridge spends strongly attached to actin). These numbers are still debated among muscle physiologists with numbers of R-type bound cross bridges ranging from 10 to 50 % but with the most recent data tending to favor the lower values.

At the simplest level the model(s) presented can give an explanation of the force pCa curves in that it is the combination of increases in Ca^{2+} reducing the occupancy of the blocked state which leads to the recruitment of cross bridges. The cross bridges themselves then cooperatively activate the filament leading to additional recruitment and hence the large Hill coefficient. Such a mechanism may result in the non-symmetrical shape of the force-pCa curve. However, the extent to which the plots deviate from symmetry is still the subject of debate (Moss, 1992) and the definition of such plots may be subject to distortion due to mechanical artifacts and non homogeneity of the activation process throughout the muscle fibers.

Because of the nature of the allosteric relationship between calcium and myosin cross bridges we expect and indeed observe changes in the numbers of heads contributing to the force to change the Ca^{2+} sensitivity (Zot & Potter, 1989; Hofmann et al, 1987a,b; McDonald et al, 1995; Millar & Homsher 1990, Gordon et al 1988; See also the original solution work of A. Weber). The addition of any effector which acts on myosin to alter the isometric force by reducing the number of bound cross bridges is expected to change the Ca^{2+} sensitivity of the contraction. For example, phosphate acts as a competitive product inhibitor of the actomyosin ATPase and reduces isometric force thereby producing a left shift of the force pCa curve and an increase in the Hill coefficient (Millar, & Homsher, 1990). ADP acts in the opposite way as an activator to increase isometric force and produces a right shift in the plots. Such effectors should be distinguished from those which can also affect the thin filament (e.g. temperature, pH). Phosphorylation of myosin light chains or of Tn components may affect both force levels and regulation even though their primary site of action may be altering the lifetime of the force holding cross bridges (light chains) or changing the Ca^{2+} sensitivity (Tn).

Allowing fibers to shorten is an alternative way of reducing the average number of strongly bound cross bridges. This reduces the average lifetime of the strongly attached cross bridge (Ridgway & Gordon 1984; Edman, 1996). In the case of shortening fibers a higher Ca^{2+} concentration is required to initiate contraction (rightward shift of the pCa curve) and the contraction becomes almost all or none i.e. a very sharp transition and an increase in the Hill coefficient. However a complication of interpretation arises because in the contracting fiber the tip of the thick filament is moving into an area of the actin filament which by definition is not activated as there are no cross bridges present. The extent to which the cooperative activation of the thin filament is propagated ahead of the tip of the moving thick filament remains to be defined.

While the models do help to understand the events in muscle fibers in general terms without an estimate of the number of cross bridges involved in an isometric contraction the models are not *very* helpful in analyzing the events in detail. Turning the problem around, models of regulation can help define the limits of the number of cross bridges involved in contraction. Since Ca^{2+} is limited to being an allosteric effector of myosin binding it will only begin to turn-on contraction when there is a modest range of actin saturation with cross bridges. If a single rigor-like cross bridge within every 12 actin sites (the apparent cooperative unit size) is sufficient to keep Tm in the on-state and if half of the actin sites are occupied during a contraction, removal of Ca^{2+} will be ineffective at switching off the system. Equally, if the maximum occupancy of actin sites is 1 in 20 (5%) then Ca^{2+} will not switch on the system. Note that we are talking here about the fraction of actin sites occupied which is different to the way the number is usually expressed i.e. as the fraction of myosin heads attached to actin, since there are more actin sites than myosin heads in the overlap zone of a muscle fibre. In a mammalian skeletal muscle the ratio of actin to myosin heads in the overlap zone is 3:2 so that the occupancy of actin sites in rigor is 67%. Assuming that at any moment only one of the two heads of each myosin interacts with actin in the presence of ATP the maximal occupancy of actin sites is reduced to 33% (corresponding to 50% of myosin heads). The threshold for cooperative activation by myosin heads is 1 in 12 actin sites (8 % of actin sites, or 12% of myosin heads). Ca^{2+} must be able to increase the number of cross bridges across this 8% occupancy threshold for the system to turn on or off cooperatively in response to changes in Ca^{2+} concentration. Detailed modeling of this system is beyond the scope of the current chapter but an appreciation of the scale of the problem can be obtained in the following, with the use of the cooperative binding equation of McKillop & Geeves (1993).

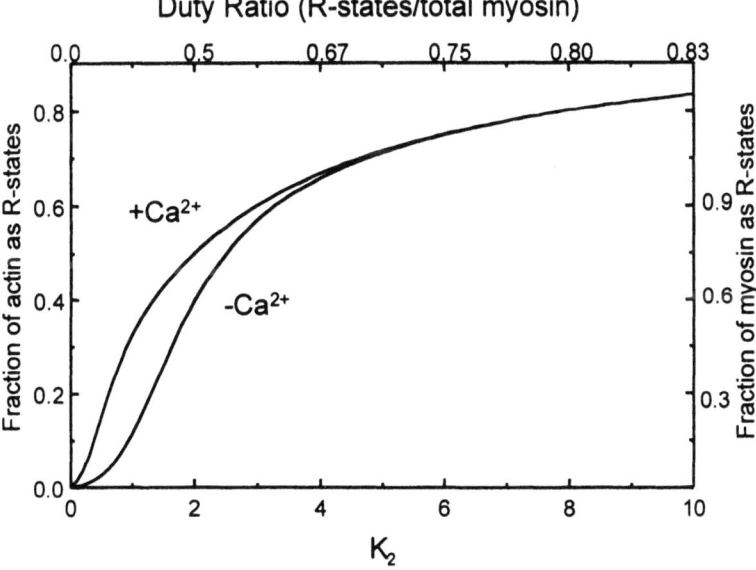

Figure 9. The Ca^{2+} dependence of the number of R-states (force holding crossbridges)

If the values of n, K_T and K_B are used as defined in section 4.3 Table 1 and the value of K_1 [M] is selected such that myosin heads have equal probability of being detached or in the A-state, then we can calculate the fraction of actin sites with a force generating R-state myosin head as a function of K_2 and Ca^{2+}. In the case of a muscle under steady state isometric contraction K_2 provides an approximate value for the fraction of time the cross bridge is in the R-state. A value of $K_2=1$ would be compatible with equal occupancy of the attached A- and R-states; e.g. 33% actin detached, 33% attached in the A-state and 33% in the force holding R-state. Figure 9 shows the relationship between occupancy of R-states and K_2 at high and low Ca^{2+}. As shown, regulation only occurs for a narrow range of K_2 values and therefore, with the fraction occupancy of actin with R-states. With more than 50% of actin occupied the system becomes insensitive to Ca^{2+}. The figure also shows the same data expressed as a Duty Ratio which is an alternate way of expressing the fraction of time a cross bridge remains strongly attached. This is calculated assuming equal occupancy of the detached and A-states and then the Duty Ratio is given by $K_2/(2+K_2)$. The key point is again that the Ca regulation only works for duty ratios of <0.5.

In conclusion the ideas developed in this chapter demonstrate the extent to which the system can be understood from a detailed study of the individual components of the thin filament regulatory system and how the components interact with each other. From the perspective of a solution biochemist this can be seen as a prerequisite for any attempt to understand the complexities of the system. However, the goal is to understand the contracting muscle fiber and this can only be achieved by studying the fiber itself, as this is the only context in which the full subtlety of the interactions can be appreciated.

M. A. GEEVES
University of Kent at Canterbury
Canterbury
Kent
CT2 7NJ
U.K

S. S. LEHRER
Muscle Research Group
Boston Biomedical Research Institute
64 Grove Street
Boston, MA 02472 USA

8. REFERENCES

Bacchiocchi, C. Graceffa, P. & Lehrer, S.S. (2001) Myosin induced movement and actin binding specificity of smooth muscle $\alpha\beta$ tropomyosin. *Biophys. J.* 80: 357a

Bacchiocchi, C. & Lehrer, S.S. (2000) Multi-site fluorescence energy transfer shows Ca^{2+}-induced tropomyosin movement in reconstituted skeletal muscle thin filaments. *Biophys. J.* 78: 364a

Brandt, P.W., Cox, R.N., & Kawai, M. (1980) Can the binding of Ca^{2+} to two regulatory sites on troponin C determine the steep pCa/tension relationship of skeletal muscle? *Proc. Natl. Acad. Sci. USA.* 77: 4717-4720.

Bremel, R.D., Murray, J.M., & Weber, A. (1972) Manifestations of cooperative behavior in the regulated actin filament during actin-activated ATP hydrolysis in the presence of Ca^{2+}. *Cold Spring Harb. Symp Qunat. Biol.* 37: 267-275.

Bremel, R.D. & Weber, A. (1972) Cooperation within actin filament in vertebrate skeletal muscle. *Nature*: 238: 97-101.

Brenner, B., Kraft, T., Yu, L.C., Chalovich, J.M. (1999) Thin filament activation probed by fluorescence of N-((2-(Iodoacetoxy)ethyl)-N-methyl)amino-7-nitrobenz-2-oxa-1, 3-diazole-labeled troponin I incorporated into skinned fibers of rabbit psoas muscle. *Biophys J.* 77: 2677-91.

Edman, K.A. (1996) Fatigue vs shortening-induced deactivation in striated muscle. *Acta Physiol. Scand.* 156: 183-192.

Geeves, M.A. (1991) The dynamics of actin and myosin association and the cross bridge model of muscle contraction. *Biochem J.* 274: 1-14.

Geeves, M.A., Goody, R.S. & Gutfreund, H. (1984) The kinetics of acto-S1 interaction as a guide to a model for the crossbridge cycle. *J.Muscle Res. & Cell Motil.* 5: 351-361.

Geeves, M.A., & Halsall, D.J. (1987) Two step ligand binding and cooperativity. *Biophys. J.* 52: 215-220.

Geeves, M.A., & Lehrer, S.S. (1994) Dynamics of the muscle thin filament regulatory switch: the size of the cooperative unit. *Biophys. J.* 67: 273-282.

Golitsina, N.L., Hitchcock-DeGregori, S.E. & Lehrer, S.S. (1995) Tropomyosin-actin cooperative effects during myosin S1 and heavy meromyosin titrations. Biophys. J. 68: A60.

Golitsina, N.L. and Lehrer, S.S., (1999) Smooth muscle α-tm crosslinks to caldesmon, to actin and to myosin S1 on the muscle thin filament". *FEBS Lett.* 463: 146-150.

Gordon, A.M., Ridgway, E.B., Yates, L.D. & Allen, T. (1988) Muscle cross-bridge attachement: effects of calcium binding and calcium activation. *Adv.Exp. Med. Biol.* 226:89-99.

Grabarek, Z., Grabarek, J., Leavis, P. & Gergely, J. (1983) Cooperative Binding to the Ca^{2+}-Specific Sites of Troponin C in Regulated Actin and Actomyosin. *J. Biol. Chem.* 258, 14098-14102.

Greene, L. (1982) The Effect of Nucleotide on the Binding of Myosin Subfragment 1 to Regulated Actin. *J. Biol. Chem.* 257: 13993-13999.

Greene, L. & Eisenberg, E. (1980) Cooperative Binding of Myosin Subfragment 1 to the Actin-Tropomyosin-Troponin Complex. *Proc. Natl. Acad. Sci. USA.* 77: 2616-1620

Hancock, W.O., Huntsman, L.L., & Gordon, A.M. (1997) Models of calcium activation account for differences between skeletal and cardiac force redevelopment kinetics. *J. Muscle Res & Cell Moltil.* 18: 671-681.

Head, J.G., Ritchie, M.D. & Geeves, M.A. (1995) Characterization of the equilibrium between blocked and closed states of muscle thin filaments. *Eur. J. Biochem.* 227: 694-699.

Hill, T.L. (1960) Introduction to statistical thermodynamics. Addison-Wesley, Reading, MA. 140-143

Hill, T., Eisenberg, E. & Greene, L. (1980) Theoretical Models for the Cooperative Equilibrium Binding of Myosin Subfragment 1 to the Actin-Tropomyosin-Troponin Complex. *Proc. Natl. Acad. Sci. USA.* 77: 3186-3190.

Hoffman, P.A., Greaser, M.L. & Moss, R.L. (1991) C-protein limits shortening velocity of rabbit skeletal muscle fibres at low levels of Ca^{2+} activation. *J. Physiol.* 439: 701-715.

Hoffman, P.A. & Fuchs, F. (1987a) Effect of length and cross-bridge attachement on Ca^{2+} bining to cardiac troponin. *Am. J. Phys.* 253: C90-C96.

Hoffman, P.A. & Fuchs, F. (1987b) Evidence for a force-dependent component of calcium binding to cardiac tropinin. *Am.J.Phys.* 253: C541-C546.

Holmes, K.C. (1995) The actomyosin interaction and its control by tropomyosin. *Biophys. J.* 68: 2s-7s.

Ishii, Y., & Lehrer, S.S. (1985) Fluorescence studies of the conformation of pyrene-labeled tropomyosin. Effects of F-actin and myosin subfragment 1. *Biochem.* 24: 6631-6638.

Ishii, Y. & Lehrer, S.S. (1990) Excimer Fluorescence of Pyrenyliodoacetamide-Labeled Tropomyosin: A Probe of the State of Tropomyosin in Reconstituted Muscle Thin Filaments. *Biochem.* 29: 1160-1166.

Iwamoto, H. (1998) Thin filament cooperativity as a major determinant of shortening velocity in skeletal muscle fibers. *Biophys J.* 74: 1452-1464.

Koshland, D.E., Nemethy, G. & Filmer, D. (1996) Comparison of experimental binding data and theoretical models in proteins containing subunits. *Biochem.* 5, 365-385.

Lehrer, S. S. (1994). the regulatory switch of the muscle thin filament: Ca^{2+} or myosin heads? *J. Mol. Biol.* 15: 232-236.

Lehrer, S. S., & Geeves, M. A. (1998). The muscle thin filament as a classical cooperative/allosteric regulatory system. *J. Mol. Biol.* 277: 1081-1089.

Lehrer, S. S., Golitsina N. L. & Geeves, M. A. (1997) Actin-tropomyosin activation of myosin subfragment 1 ATPase and thin filament cooperativity: The role of tropomyosin flexibility and end to end interactions. *Biochem.* 36: 13449-13454.

Lehrer, S. S., & Morris, E. P. (1982). Dual Effects of Tropomyosin and Tropomyosin-Troponin on Actomyosin Subfragment 1 ATPase. *J. Biol. Chem.* 257: 8073-8080.

Lehman, W., Rosol, M. Tobacman. L. S. & Craig, R. (2001) Troponin organization on relaxed and activated thin filaments revealed by electron microscopy and three-dimensional reconstruction. *J. Mol. Biol.* 307, 739-744.

Li H.C., Fajer P.G., (1998) Structural coupling of troponin C and actomyosin in muscle fibers. *Biochemistry.* 37: 6628-35.

Luo, Y, Leszyk J, Li B, Gergely J, Tao T. (2000) Proximity relationships between residue 6 of troponin I and residues in troponin C: further evidence for extended conformation of troponin C in the troponin complex. *Biochemistry.* 2000 39:15306-15.

Maytum, R., Konrad, M., Lehrer, S.S. & Geeves, M.A. (2001) Regulatory properties of tropomyosin, effects of length, isoform and N-terminal sequence. *Biochem.* 40: 7334-7341.

Maytum, R., Lehrer, S.S., & Geeves, M.A. (1999). Cooperativity and switching within the three state model of muscle regulation. *Biochem.* 38: 1102-1110.

McDonald, K.S., Field, L.J., Parmacek, M.S., Soonpaa, M., Leiden, J.M. & Moss, R.L. (1995) Length dependence of Ca^{2+} sensitivity of tension in mouse cardiac myocytes expressing skeletal troponin C. *J. Phys.* 483: 131-139.

McKillop, D.F.A., & Geeves, M.A. (1993) Regulation of the interaction between actin and myosin subfragment 1: evidence for three states of the thin filament. *Biophys. J.* 65: 693-701.

Miki, M., Miura, T., Sano, K., Kimura, H., Kondo, H., Ishida, H., Maeda, Y. (1998) Fluorescence resonance energy transfer between points on tropomyosin and actin in skeletal muscle thin filaments: does tropomyosin move? *J Biochem (Tokyo).* 123:1104-11.

Millar, N.C. & Homsher, E. (1990) The effect of phosphate and Ca on force generation in glycerinated rabbit skeletal muscle fibers. *J. Biol. Chem.* 265: 20234-20240.

Monod, J., Wyman, J., & Changeux, J.-P. (1965) On the nature of allosteric transitions. *J. Mol. Biol.* 12: 88-118.

Moss, R.L. (1992) Ca^{2+} regulation of mechanical properties of striated muscle. *Circ. Res.* 70: 865-884.

Murray, J.M., Weber, A. & Bremel, R.D. (1975) Could cooperativity in the actin filament play a role in muscle contraction? In E. Carafoli (Ed.), Calcium Transport in Contraction and Secretion North-Holland.

Narita, A., Yasunaga, T., Ishikawa, T., Mayanagi, K. & Wakabayashi, T. (2001) Ca^{2+} induced switching of troponin and tropomyosin on actin filaments as revealed by electron cryomicroscopy. *J. Mol. Biol.* 308: 241-261.

Offer, G. & Knight, P., (1988) Water holding properties of meat Part 1: general principles and water uptake during processing. *Dev. Meat Sci.* 4: 63-171.

Pemrick, S., & Weber, A. (1976) Mechanism of inhibition of relaxation by N-ethyl maleimide treatment of myosin *Biochem.* 15: 5193-5198.

Phillips, G.N., Fillers, J.P. & Cohen, C. (1986) Tropomyosin crystal structure and muscle regulation. *J. Mol. Biol.* 192: 111-131.

Potter, J.D. & Gergely, J. (1975) The calcium and magnesium sites on troponin and theor role in the regulation of myofibrillar adenosine triphosphatase. *J. Biol. Chem.* 250: 4628-4633.

Reiffert, S.U., Jaquet, K., Heilmeyer, L.M.G.J., Ritchie, M.D. & Geeves, M.A. (1996) Bisphosphorylation of cardiac troponin I modulates the Ca-dependent binding of myosin subfragment 1 to reconstituted thin filaments. *FEBS Lett.* 384: 43-47.

Ridgway, E.B. & Gordon, A.M. (1984) Muscle calcium transient. Effect of post stimulus length changes in single muscle fibres. *J. Gen. Physiol.* 83: 75-103.

Schaertl, S., Lehrer, S.S. & Geeves, M.A. (1995) Separation and Characterization of the Two Functional Regions of Troponin Involved in Muscel Thin Filament Regulation. *Biochem.* 34: 15890-15894.

Squire, J.M. & Morris, E.P. (1998) A new look at thin filament regulation in vertebrate skeletal muscle. *Faseb. J.* 12: 761-771.

Swartz, D., Moss, R.L. & Greaser, M.L. (1996) Calcium alone does not fully activate the thin filament for S1 binding to rigor myofibrils. *Biophys. J.* 71: 1891-1904.

Trybus, K.M. & Taylor, E.W. (1980) Kinetics of the Cooperative Binding of Subfragment 1 to Regulated Actin. *Proc. Natl. Acad. Sci. USA.* 77: 7209-7213.

Vibert, P., Craig, R. & Lehman, W. (1997) Steric-model for activation of muscle thin filaments. *J. Mol. Biol.* 266: 8-14.

Zot, H.G. & Potter, J.D. (1987) Calcium binding and fluorescence measurements of dansylaziridine-labelled troponin C in reconstituted thin filaments. *J. Muscle Res Cell Motil,.* 257: 7678-7683.

Zot, H.G. & Potter, J.D. (1989) Reciprocal coupling between TnC and myosin crossbridge attachment. *Biochem.* 28: 6751-6756.

RICHARD L. MOSS, DANIEL P. FITZSIMONS
& MARIA V. RAZUMOVA

REGULATION OF THE RATE OF FORCE DEVELOPMENT IN HEART AND SKELETAL MUSCLES

1. INTRODUCTION

Contractions of heart and skeletal muscle cells *in vivo* are usually well matched to the loads the muscle must move or bear, but despite considerable similarity in motor and regulatory proteins, the detailed mechanisms of regulation differ in significant ways in the two muscle types. Other chapters in this volume address the molecular components of the thick and thin filaments in striated muscles and address the specific processes by which Ca^{2+} and cross-bridges regulate the activation state of the thin filament. This chapter deals with the physiological manifestations of these regulatory processes during force development in heart and skeletal muscles, with particular focus on the regulation of the kinetics of force development.

The fundamental contractile event of heart and skeletal muscles *in vivo* is the twitch, which occurs in response to a single membrane action potential. Twitch contractions of skeletal muscle fibers involve a transient increase in intracellular Ca^{2+} concentration that is sufficient to saturate thin filament regulatory sites on troponin-C. Thus, the force and power developed by the fiber depend upon the duration of the Ca^{2+} transient and the time available for cross-bridge binding to actin, which can be prolonged by tetanic stimulation. In cardiac muscle cells, the amount of Ca^{2+} released from intracellular stores during a twitch is typically not sufficient to saturate thin filament sites, so that force and power are sub-maximal. Because of this, increases in twitch force and power can be achieved by increasing the delivery of Ca^{2+} to the myoplasm, as well as by agonist-induced acceleration of cross-bridge cycling kinetics.

Besides this difference in the ability to modulate force and power, cardiac and skeletal muscles differ in the speed, or kinetics, of the onset of contraction. Skeletal muscles (especially fast-type) undergo impulsive twitch contractions that exhibit little variability in a given muscle fiber, at least under physiological conditions. In contrast, myocardial twitches are finely graded on a continuum that encompasses very slow kinetics under resting conditions and fast kinetics under stressful conditions. The intrinsic kinetic properties of the contractile protein myosin

R.J. Solaro and R.L. Moss (eds), Molecular Control Mechanisms in Striated Muscle
Contraction, 271-290

contribute significantly to twitch kinetics (Schiaffino and Reggiani, 1996), i.e., the rate of rise of force, or dF/dt, but the responsiveness of the thin filaments to Ca^{2+} and cross-bridge binding also has an important role. There is emerging consensus that the activation of force is the product of both Ca^{2+} and cross-bridge binding (Lehrer, 1994; Solaro and Rarick, 1998; Gordon, et al., 2000), but the relative contributions of these two processes differ in cardiac and skeletal muscles. In both muscle types, Ca^{2+} binding initiates cross-bridge binding, and the activating effects of bound cross-bridges in large part determine the rate of rise of force. In living fast-twitch skeletal muscle fibers, the fact that the thin filaments are saturated with Ca^{2+} during a twitch results in fast kinetics due to the saturating, or at least near saturating, effects of Ca^{2+} and cross-bridge binding. As a result, the rate constant of force development in living skeletal muscle is at or near maximal. During the cardiac twitch, the thin filaments are rarely saturated with Ca^{2+}, so that neither Ca^{2+} nor cross-bridge binding is saturated and kinetics are consequently sub-maximal. By varying the amount of activator Ca^{2+} that is released to the myoplasm during a twitch, it is possible to vary myocardial twitch kinetics such that the rate constant of force development becomes faster when myoplasmic $[Ca^{2+}]$ is increased. From this brief introduction, it appears that the regulatory processes in heart and skeletal muscles are well designed for their functions in vivo, i.e., twitches of skeletal muscle fibers are relatively invariant, while myocardial twitches are graded on a beat-to-beat basis to provide the variable force and power that are required to accommodate the variations in load on the heart when venous return changes.

2. ISOMETRIC FORCE

Regulation of isometric force requires the binding of Ca^{2+} to the TnC subunit of the thin filament regulatory protein troponin (reviewed by Tobacman, 1996; Gordon, et al., 2000). Through a series of molecular interactions involving troponin subunits, tropomyosin and actin, Ca^{2+} binding activates the thin filament to allow myosin cross-bridge binding to actin and subsequent force development. Using permeabilized muscle preparations, many investigators have shown that isometric force increases as the concentration of Ca^{2+} in the bathing solution is increased. The relationship between force and pCa (i.e., $-\log[Ca^{2+}]$) is sigmoidal (Figure 1) and can be fit with the Hill equation (Shiner and Solaro, 1982):

$$\frac{P}{P_o} = \frac{1}{1 + 10^{n(pCa - pCa_{50})}} \qquad (1)$$

where P is force, P_o is maximum force, n is the Hill coefficient, and pCa_{50} is pCa of the activating solution at half-maximal activation. The Hill coefficient provides an estimate of the minimum number of binding sites involved in the regulation of force and is generally greater in fast skeletal muscle than in cardiac muscle. Skeletal TnC has two Ca^{2+}-specific binding sites and cardiac TnC has just one, and yet the Hill coefficient is greater than 3 in both muscle types (Figure 1). The discrepancy

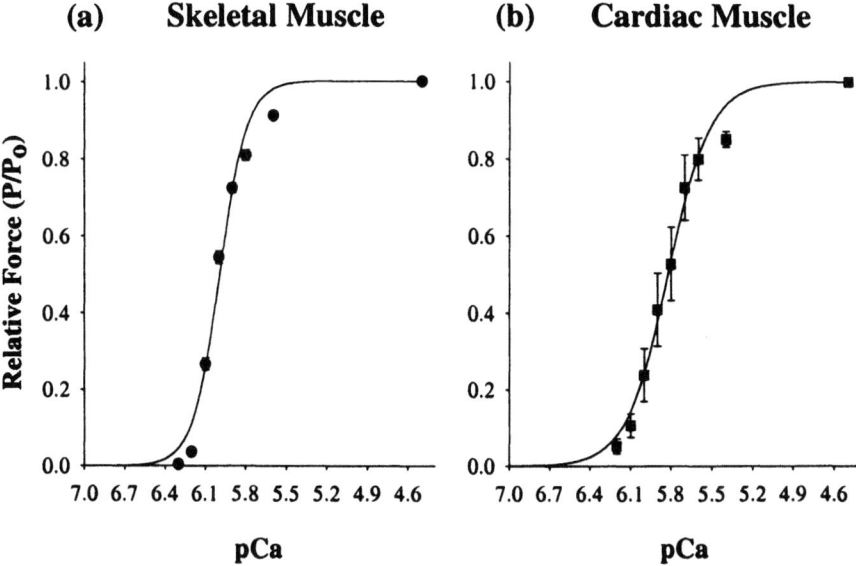

Figure 1. Tension-pCa relationships from (a) rabbit skinned skeletal muscle fibers and (b) myocardium at 15°C. Hill coefficients and pCa$_{50}$ values were 7.9 ± 0.4 and 6.00 ± 0.01 for skeletal muscle and 4.7 ± 0.4 and 5.85 ± 0.05 for myocardium. Force values at submaximal pCa were scaled to maximum force at pCa 4.5.

between the Hill coefficient and the number of regulatory binding sites on TnC has been taken to mean that there is significant intermolecular cooperation in the activation of force. The greater value of the Hill coefficient in skeletal muscle, i.e., the greater steepness of its force-pCa relationship implies that the cooperativity of thin filament activation is greater than in cardiac muscle. Likely mechanisms include positive cooperativity in Ca^{2+} binding to TnC and/or cross-bridge binding to the thin filament or cross-bridge-induced increases in the Ca^{2+} binding affinity of TnC.

Ca^{2+} binding to TnC initiates muscle contraction, but complete activation of tension and the kinetics of tension development appears to involve cooperative effects due to cross-bridge binding to actin (reviewed by Lehrer, 1994; Tobacman, 1996; Moss, 1999; Gordon, et al., 2000). Biochemical data suggest that activation is a positive cooperative process (Williams, et al., 1984, 1988), such that activation of a thin filament cooperative group (estimated to be ± 10-14 actin monomers, e.g., Geeves and Lehrer, 1994; Swartz, et al., 1996) by Ca^{2+} and/or strong binding cross-bridges influences the activation of neighboring functional groups (Lehrer, 1994). Cooperation is apparent in contracting muscle in the greater than expected (on the basis of the numbers of Ca^{2+} binding sites on TnC) steepness of the tension-pCa relationship, especially at low levels of Ca^{2+} (reviewed by Moss, 1992), and observed effects of strong-binding cross-bridges to increase sub-maximal force developed in skinned skeletal muscle fibers (Swartz and Moss, 1992; Fitzsimons, *et al.*, 2001a; Lu, *et al.*, 2001) and myocardium (Fitzsimons, *et al.*, 2001b).

2.1 Cooperativity In Ca^{2+} Binding to TnC

One possibility to explain greater than expected forces at a given $[Ca^{2+}]$ is that the affinity of TnC binding sites for Ca^{2+} increases as either the number of cross-bridges bound to actin or the amount of Ca^{2+} bound to regulatory sites on the thin filament increases. Experimental evidence does indeed suggest that these phenomena occur in striated muscles, although to a greater degree in cardiac than in fast- or slow-twitch skeletal muscles (Fuchs, 1995). Ca^{2+} binding to regulated thin filaments exhibits positive cooperativity in both skeletal (Grabarek, et al., 1983) and cardiac (Tobacman and Sawyer, 1990) muscles. Bremel and Weber (1972) were the first to propose that cross-bridge binding increases Ca^{2+} binding affinity based on their solution experiments in which cross-bridge binding to skeletal muscle regulated thin filaments was found to increase Ca^{2+} binding. Subsequent work showed that strong-binding cross-bridges also increased the apparent Ca^{2+} binding affinity of TnC in skinned cardiac muscle (Pan and Solaro, 1987). Consistent with these results, several studies have reported conformational changes in troponin C in response to strong cross-bridge binding to the thin filaments of both cardiac and skeletal muscles (Guth and Potter, 1987; Hannon, *et al.*, 1992; Martyn, *et al.*, 1999).

Considerable support for the physiological relevance of cross-bridge induced increases in Ca^{2+} binding affinity of TnC has come from mechanical experiments in both skeletal and cardiac muscles in which intracellular $[Ca^{2+}]$ concentration was assessed with highly specific Ca^{2+} indicators. For example, in invertebrate striated muscles, shortening of the muscle during the relaxation phase was found to give rise to an extra Ca^{2+} transient which the authors interpreted to be the release of Ca^{2+} from TnC subsequent to the shortening-induced reduction in numbers of strongly bound cross-bridges (Gordon and Ridgway, 1987 and references therein). Similarly, Allen and Kentish (1988) observed that extra Ca^{2+} appeared in the myoplasm of ferret ventricular muscle when the muscle was allowed to shorten, thereby reducing force and the number of cross-bridges bound to the thin filament.

Of course, the mechanism for shortening-induced release of Ca^{2+} from thin filament sites could involve the length change *per se* or could come about as a result of the decrease in numbers of strongly bound cross-bridges. This issue was addressed by Fuchs in a critical series of studies showing that decreases in force induced by vanadate reduced the Ca^{2+} sensitivity of force in cardiac muscle independent of changes in length (Hofmann and Fuchs, 1987). Fuchs used these and other findings to account for the length-dependence of Ca^{2+} sensitivity observed in cardiac muscle, i.e., Ca^{2+}-sensitivity of force increases at long lengths as does Ca^{2+} binding at any given free $[Ca^{2+}]$ (see chapter by Fuchs in this volume). Importantly, skinned skeletal muscle fibers do not have as great a length-dependence of Ca^{2+} sensitivity as cardiac muscle and exhibit virtually no length-dependence of Ca^{2+} binding (Fuchs and Wang, 1991; Wang and Fuchs, 1994). However, this result should not be taken to mean that Ca^{2+} binding affinity in skeletal muscle is insensitive to the number of cross-bridges bound to the thin filament, since biochemical studies have shown that Ca^{2+} binding to TnC in the troponin complex is a cooperative process that is made more so by strong binding of myosin S1 (Grabarek, *et al.*, 1983). Furthermore, electron probe studies on intact myofilaments have shown that Ca^{2+} preferentially binds in the region of overlap between thick and thin filaments (Cantino, *et al.*, 1993).

While it is clear from these and other studies (reviewed by Gordon, *et al.*, 2000) that cross-bridge binding to thin filaments can increase the apparent Ca^{2+} binding affinity of TnC in both cardiac and skeletal muscles, it remains to be determined whether this increase in affinity significantly increases the force at any given $[Ca^{2+}]$ or is simply the result of increased numbers of cross-bridges bound to the thin filament. The finding that cardiac muscle has greater length-dependence of force and Ca^{2+} binding than skeletal muscle (Fuchs, 1995) strongly suggests that this mechanism is physiologically important at least in heart muscle.

2.2 Positive Cooperativity in Cross-bridge Binding to Actin

Several studies have shown that there is positive cooperativity in cross-bridge binding to thin filaments (reviewed by Gordon, *et al.*, 2000) and that this contributes to the steepness of the relationship between force and $[Ca^{2+}]$. Positive cooperativity in cross-bridge binding was first shown in biochemical experiments in which myosin binding to regulated actin was facilitated by the presence of rigor complexes (Bremel and Weber, 1972). Cooperative binding also occurs in the intact filament lattice, which has been demonstrated in a variety of ways including direct fluorescence imaging of myofibrils in the presence of rhodamine-labeled myosin S1 (Swartz, *et al.*, 1990, 1996). The importance of this process to force development in striated muscles has been shown in experiments in which variations in nucleotide concentration or inorganic phosphate were used to manipulate the number of strongly bound cross-bridges and thereby change the Ca^{2+}-sensitivity of force. For example, an increase in [MgADP] increases the Ca^{2+} sensitivity of force (Lu, *et al.*, 1993, 2001;Thirlwell, *et al.*, 1994), presumably by increasing the number of strongly-bound cross-bridges at each $[Ca^{2+}]$ and thereby facilitating cooperative recruitment of additional cross-bridges to force generating states (Dantzig, *et al.*, 1991). Conversely, an increase in [Pi] reduces the Ca^{2+}-sensitivity of force by reversal of the power-stroke and the consequent decrease in strongly-bound cross-bridges (Dantzig, *et al.*, 1992; Walker, *et al.*, 1992).

While the use of chemical interventions such as changes in [MgADP] or [Pi] to alter the distributions of cross-bridges between weakly and strongly bound states has provided insights into the cooperative activating effects of strongly-bound cross-bridges, they do not provide quantitative information due to the fact that each of these interventions also has direct effects on force, the regulated variable being measured. One experimental tool that has overcome this complication is N-ethylmaleimide-modified myosin subfragment-1, NEM-S1, which can be infused into skinned fibers (Swartz and Moss, 1992) where it binds strongly to the thin filament but does not develop force. In biochemical experiments, NEM-S1 facilitates cross-bridge binding to regulated thin filaments (Nagashima and Asakura, 1982) and potentiates ATPase activity in the presence of Ca^{2+} (Williams, *et al.*, 1988), results which are consistent with the idea that cross-bridge binding exhibits positive cooperativity. Studies using NEM-S1 provide convincing evidence in support of cross-bridge-induced cross-bridge binding to actin, but also show that there are pronounced differences between heart and skeletal muscles in the responsiveness of their thin filaments to strong cross-bridge binding.

2.2.1 *Skeletal Muscle*

In skeletal muscle, increasing the number of strongly-bound cross-bridges using NEM-S1 results in an increase in force at sub-maximal $[Ca^{2+}]$ and no change in maximum force (Swartz and Moss, 1992; Fitzsimons, *et al.*, 2001a), i.e., strong binding cross-bridges increase the Ca^{2+}-sensitivity of force (Figure 2a). Since Fuchs and colleagues (Fuchs and Wang, 1991) have shown that there is no change in the

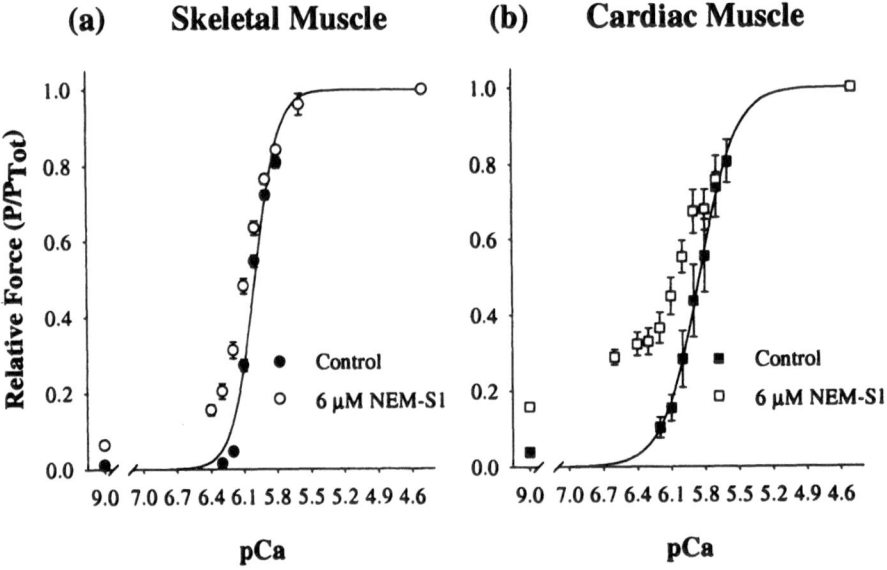

Figure 2. Effects of 6 μM NEM-S1 on force and the Ca^{2+} sensitivity of force in rabbit (a) skinned skeletal muscle fibers and (b) rat myocardium. In both types of muscle, increasing the number of strongly bound cross-bridges increased the Ca^{2+} sensitivity of force (seen as an increased pCa_{50}) and increased the Ca^{2+}-independent force at pCa 9.0, and these effects were greatest in cardiac muscle. Maximum force at pCa 4.5 was unaffected by NEM-S1. Forces at sub-maximal pCa were expressed as a fraction of maximum force developed under the same conditions at pCa 4.5.

Ca^{2+}-binding affinity of troponin-C in skeletal muscle thin filaments as force increases, the increase in Ca^{2+}-sensitivity of force is presumably a manifestation of positive cooperativity in cross-bridge binding, i.e., as predicted in the models of Geeves (McKillop and Geeves, 1993) and Campbell (1997). The mechanism of this cooperation is not known in detail but the results of various studies (reviewed by Tobacman, 1996 and by Gordon, et al., 2000) have suggested that cross-bridge binding has allosteric effects on the regulatory strand, i.e, tropomyosin plus troponin, such that there is a spread of activation within and between functional groups, i.e., 7 actins plus the associated tropomyosin and troponin molecules (Butters, et al., 1997; Lehrer and Geeves, 1998). Importantly, NEM-S1 has virtually no effect on the maximum force developed in the presence of saturating $[Ca^{2+}]$, which implies that the combined activating effects of Ca^{2+} and strong-binding cross-bridges are maximal under these conditions. Thus, in terms of force development the thin filaments of skeletal muscles are fully switched on at saturating $[Ca^{2+}]$.

Another interesting effect of NEM-S1 is that it activates tension at Ca^{2+} concentrations where virtually no Ca^{2+} is likely to be bound to the regulatory sites of TnC. This activating effect in the absence of Ca^{2+} is evident as active tension at pCa 9.0 (Figure 2a)—this tension is certainly active as muscles treated in this way actively shorten and develop force (see section 3.2.1, below). If activation of contraction is indeed a synergistic process involving both Ca^{2+} and cross-bridge binding (Lehrer, 1994), this result indicates that it is possible to activate the thin filament in the absence of Ca^{2+} if the number of strong-binding cross-bridges is sufficiently high. Such a phenomenon is non-physiological since the number of strongly bound cross-bridges is never this high in vivo, although the active shortening of muscles that is sometimes observed soon after death may involve cooperative activation of the thin filament by populations of nucleotide-deficient cross-bridges.

The fact that contraction can be activated by strong-binding cross-bridges in the absence of Ca^{2+} implies that the skeletal muscle thin filament is not completely switched off in the absence of Ca^{2+}. In the activation model proposed by Geeves and colleagues (McKillop and Geeves, 1993), the thin filament without Ca^{2+} bound is in a "blocked" state in which it cannot bind cross-bridges, but Ca^{2+} binding converts the filament to a "closed" state in which it can bind cross-bridges. In the context of this model, the present results suggest that the skeletal muscle thin filament is not fully blocked in the absence of Ca^{2+}.

2.2.2 Cardiac Muscle

Isometric force development in cardiac muscle is also sensitive to increased numbers of strongly-bound cross-bridges (Figure 2b). Addition of NEM-S1 to the bathing solution increases force at each sub-maximal $[Ca^{2+}]$ and thereby increases the Ca^{2+}-sensitivity of force in skinned myocardium (Fitzsimons, et al., 2001b). While this response is qualitatively very similar to the response seen in skeletal muscle, the increase in Ca^{2+} sensitivity induced by a given amount of NEM-S1 is much greater in cardiac muscle. Clearly, both muscle types exhibit positive cooperativity in cross-bridge binding, but skeletal muscle is the more cooperative of the two since it requires a greater number of strongly-bound cross-bridges to achieve a similar potentiation of force. Viewed another way, cardiac thin filaments are much more responsive to the activating effects of even a small number of cross-bridges, which presumably contributes to fine gradations in twitch force observed on a beat-to-beat basis in living myocardium. Just as in skeletal muscle, there is no effect of NEM-S1 on maximum force at pCa 4.5, implying that activation of force is saturated by cross-bridge and Ca^{2+} binding under these conditions.

NEM-S1 also increases force in the absence of Ca^{2+} in skinned myocardium (Figure 2b), but here the effects are much greater than those observed in skeletal muscle. For example, at 6 μM NEM-S1 the force developed in skinned myocardium was ~20% of the maximum active force at pCa 4.5, while in skeletal muscle under the same conditions force was ~5% of maximum. This difference in cross-bridge activation of force is consistent with the greater responsiveness of myocardium to strong-binding cross-bridges observed at sub-maximal Ca^{2+} concentrations and also supports the notion that the cardiac thin filament is not completely switched off ("blocked") in the absence of Ca^{2+}.

3. REGULATION OF THE KINETICS OF FORCE DEVELOPMENT

The primary determinants of the rate of rise of twitch force in living skeletal and cardiac muscles appear to be the rate of delivery of Ca^{2+} to the myoplasm and the rate of cross-bridge cycling during the twitch. Since other chapters in this volume address the mechanisms of Ca^{2+} delivery in heart and skeletal muscles, the discussion here will focus on the contributions of myofibrillar proteins to force development. Most studies of the regulation of the rate of force development have been done using skinned preparations of heart and skeletal muscle in which it is possible to record force development in response to photolysis of caged Ca^{2+} (reviewed by Gordon, *et al.*, 2000) or following a mechanical maneuvers to reduce force to near zero (Brenner and Eisenberg, 1986). The rate constants of force development (k_{tr}) measured in skinned preparations at 15°C are ~18 s^{-1} in fast-twitch skeletal muscle fibers and ~10 s^{-1} in rat skinned myocardium (Figure 3). In both skeletal (Figure 3a) and cardiac (Figure 3b) muscles, the rate of cross-bridge cycling is activation dependent, i.e., k_{tr} increases by an order of magnitude in both muscle types as activation is increased from threshold to maximal levels. Such results imply that the rate of force development in living muscle would increase simply as a consequence of increased delivery of Ca^{2+} to the myoplasm. However, the relationship between k_{tr} and pCa differs from the force-pCa relationship in that the pCa_{50} for k_{tr} is less than the pCa_{50} for force, i.e., a higher $[Ca^{2+}]$ is required to achieve half-maximal activation of k_{tr}.

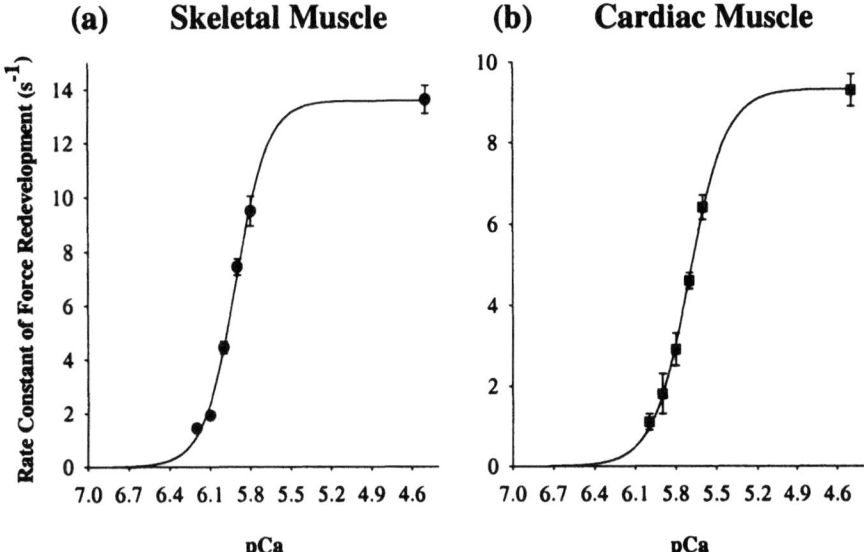

Figure 3. The activation-dependence of the rate constant of force redevelopment (k_{tr}) at 15°C in rabbit skinned (a) skeletal muscle fibers and (b) rat myocardium. k_{tr} was estimated by fitting a single exponential equation to the time course of force redevelopment following a release/re-stretch maneuver to reduce force to near zero (Brenner and Eisenberg, 1986). Data is replotted from Fitzsimons, et al. (2001a, 2001b).

The basis for activation dependence of k_{tr} is a central question in the regulation of contraction given the potential importance of this mechanism in determining the rate of rise of force in living muscles. One possibility is that variations in $[Ca^{2+}]$ *per se* mediate the activation dependence of k_{tr}, which would be expected if cross-bridge cycling kinetics were directly regulated by Ca^{2+} binding to thick (e.g., regulatory light chain) or thin filament (troponin C) proteins. Another possibility is that cross-bridge kinetics are regulated by strong binding of cross-bridges to the thin filament, such that cycling kinetics are cooperatively accelerated by cross-bridge binding. In the latter case, the activation dependence of k_{tr} would be an indirect consequence of alterations in the amount of activator Ca^{2+} delivered to the myoplasm. At the present time, neither of these possibilities has been eliminated and there is evidence in favor of both mechanisms, suggesting that both mechanisms might be operative in striated muscles *in vivo*.

3.1 Ca^{2+} Activation of Cross-bridge Cycling Kinetics

The ~10-fold increase in k_{tr} as $[Ca^{2+}]$ is increased in skinned skeletal or cardiac muscles could be interpreted in terms of a mechanism in which Ca^{2+} binding to myofibrillar regulatory proteins directly modulates cross-bridge cycling kinetics. When Brenner (1988) observed the activation dependence of k_{tr} in skinned preparations, he interpreted his results in terms of a model in which Ca^{2+} directly influenced the rate constant of cross-bridge attachment (f_{app}), assuming that force development was coincident with attachment. The molecular mechanism by which this might occur was not specified in Brenner's model but subsequent modeling by Landesberg and Sideman (1994) was able to account for the activation-dependence of k_{tr} by treating Ca^{2+} binding to TnC as a rate-determining step in the activation of contraction. In such a scheme, the rate of Ca^{2+} binding to TnC would be a second-order determinant of f_{app}, such that f_{app} would increase as $[Ca^{2+}]$ was increased. This mechanism could certainly contribute to the increase in k_{tr} at high levels of activation but it does not account for the effects of strong-binding cross-bridges on k_{tr}, especially the NEM-S1-mediated acceleration of k_{tr} at very low levels of Ca^{2+} (Swartz and Moss, 1992; Fitzsimons, et al., 2001a, b) discussed in section 3.2 of this chapter. Still, there are regulatory phenomena that are particularly well explained by the Landesberg and Sideman model. Foremost among these is the invariance of k_{tr} measured in skinned skeletal muscle fibers when the level of activation at a given Ca^{2+} concentration is varied by partial extraction of TnC (Metzger and Moss, 1991) or replacement of endogenous TnC with a constitutively active TnC mutant (Chase, et al., 1994). If k_{tr} was solely determined by the number of strongly-bound cross-bridges, k_{tr} measured at a fixed $[Ca^{2+}]$ should decrease when some fraction of TnC is removed from the thin filament thereby reducing the number of strongly-bound bridges. Experimental results show that this is not the case.

There is at least one other myofibrillar protein that might be involved in Ca^{2+}-mediated variations in k_{tr}. Foremost among these is the so-called regulatory light chain (RLC), which is a 20 kDa subunit of myosin and is located at the junction between the myosin head (sub-fragment 1) and the myosin rod (sub-fragment 2). By virtue of its location spanning a region of the myosin molecule that is thought to be flexible, the state of the RLC might influence the kinetics of cross-bridge binding by regulating the flexibility of the head/rod junction and thus the probability of myosin binding to actin. Partial extraction of RLC from skinned skeletal muscle fibers also increased k_{tr} at submaximal (but not maximal) levels of Ca^{2+} activation and also increased the Ca^{2+} sensitivity of force (Hofmann, et al., 1990). In a related study, phosphorylation of RLC in skinned skeletal muscle fibers (Metzger, et al., 1989) increased the rate constant of force development (k_{tr}) at submaximal levels of activation, which these authors explained in terms of a model in which phosphorylation of RLC caused charge repulsion between the myosin head and thick filament backbone thereby driving the head toward the thin filament. Subsequent work showed that phosphorylation of RLC in isolated thick filaments induced a radial movement of cross-bridge heads away from the thick filament backbone (Levine, et al., 1996).

An interesting feature of the Ca^{2+} dependence of k_{tr} is that it is affected by changes in $[Mg^{2+}]$, suggesting that there is involvement of a Ca^{2+}/Mg^{2+} binding site as opposed to the low-affinity Ca^{2+}-specific binding site(s) on TnC involved in the regulation of contraction in skeletal and cardiac muscles. An increase in $[Mg^{2+}]$ was found to reduce the Ca^{2+}-sensitivity of k_{tr}, an effect which was unaltered by partial extraction of TnC but was eliminated (or at least substantially reduced) by partial extraction of RLC (Metzger and Moss, 1992). Such a result implies that RLC does indeed play a role in regulating k_{tr}, but its importance in relation to other potential mechanisms (see section 3.2 below) remains to be determined. A puzzling feature of the effects observed due to extraction of RLC, phosphorylation of RLC, or changes in $[Mg^{2+}]$ is that the effects of these interventions are restricted to submaximal Ca^{2+} concentrations, i.e., none of these interventions alters k_{tr} measured at saturating $[Ca^{2+}]$. While additional work is required to understand the basis for this phenomenon, it is possible that the activating effects of the combination of cross-bridge binding and Ca^{2+} binding at saturating Ca^{2+} concentrations is sufficient to yield maximal k_{tr}. In such a case, phosphorylation or extraction of RLC would have no further activating effects on k_{tr}.

3.2 Cross-bridge Activation of Cross-bridge Cycling Kinetics

There is a growing body of evidence that strong binding of myosin cross-bridges to the thin filament cooperatively accelerates the rate of force development in both skeletal and cardiac muscles. For example, increasing the concentration of MgADP in skinned skeletal muscle fibers increases the rate constant of force development (k_{tr}) at low and intermediate levels of Ca^{2+}-activation (Lu, et al., 1993, 2001), presumably by increasing the number of strongly-bound MgADP cross-bridges. Conversely, reducing the number of strongly bound cross-bridges with vanadate (Dantzig and Goldman, 1985) slows k_{tr} at submaximal Ca^{2+} concentrations (Lu, Moss and Walker, unpublished result). As might be expected from results such as these, cross-bridge effects on the rate of force development comprise a key element of current models of muscle contraction (McKillop and Geeves, 1994; Campbell, 1997; Razumova, et al., 2000).

Considerable effort has been directed toward describing the effects on contraction kinetics due to strong-binding of cross-bridges and understanding the importance of cooperativity in cross-bridge binding in determining twitch kinetics in living heart and skeletal muscles. Just as in studies of the regulation of isometric force, NEM-S1 has also been used in both cardiac and skeletal muscles to assess possible cooperative activating effects of strong-binding cross-bridges on the rate of force development.

3.2.1 Skeletal Muscle

In skinned fast-twitch skeletal muscle fibers (Figure 4), NEM-S1 increases the rate constant of force development (k_{tr}) at all submaximal Ca^{2+} concentrations (Swartz and Moss, 1992; Fitzsimons, et al., 2001a). At very low concentrations of Ca^{2+}, NEM-S1 increases k_{tr} to values equivalent to those measured at maximal $[Ca^{2+}]$ in the absence of NEM-S1, while at intermediate $[Ca^{2+}]$ k_{tr} is less than maximal but greater than the value measured at the same Ca^{2+} concentrations in the absence of NEM-S1. These results give rise to at least two additional questions, which we have

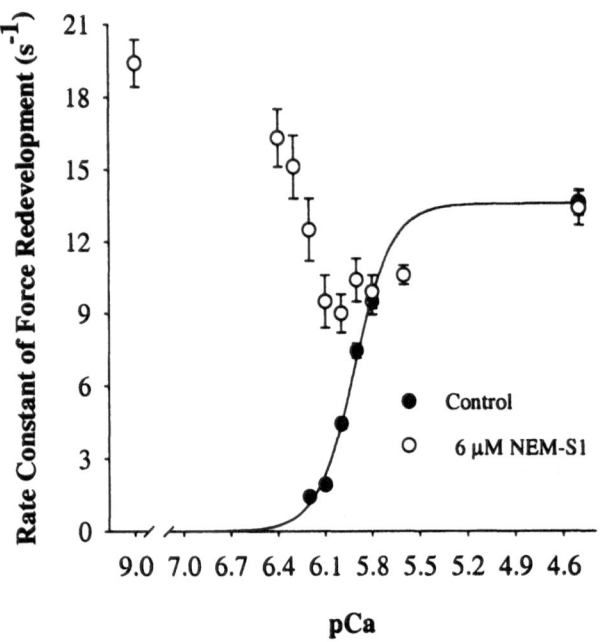

Figure 4. Effects of 6 μM NEM-S1 on the rate constant of force redevelopment (k_{tr}) assessed in rabbit skinned skeletal muscle fibers at 15°C. Data are replotted from Fitzsimons, et al. (2001a).

worked to address. First, is the maximum k_{tr} observed at saturating $[Ca^{2+}]$ in the absence of NEM-S1 a true maximum, i.e., are there conditions under which k_{tr} is supramaximal? Second, what is the basis for the apparent minimum in k_{tr} observed at intermediate levels of activation in skeletal muscle fibers activated in the presence of NEM-S1?

Regarding the first question, experiments in skinned skeletal muscle fibers have shown that it is possible to potentiate k_{tr} at all levels of activation using an ATP derivative, de-oxy ATP, that accelerates the rate of cross-bridge cycling (Regnier, *et al.*, 1998). De-oxy ATP appears to increase the maximum rate of cross-bridge cycling and thereby increases k_{tr} across the entire range of activations, and thus, its effects are directly upon the enzymatic properties of myosin. It is also possible to increase k_{tr}, at least at low $[Ca^{2+}]$, by increasing the concentration of NEM-S1 (Swartz and Moss, 1992; Fitzsimons, *et al.*, 2001a). When this is done, k_{tr} at the lowest levels of activation (where active tension is near zero) increases to values greater than the maximum measured at saturating $[Ca^{2+}]$ (Figure 4). However, the dip in the k_{tr}-pCa relationship is still seen at intermediate levels of Ca^{2+} activation and further increasing the amount of NEM-S1 did not remove the dip or increase k_{tr} at saturating Ca^{2+}. Importantly, the ability to increase k_{tr} to supramaximal values under any conditions leads to the conclusion that in terms of the kinetics of force development the thin filaments of living skeletal muscles are not completely activated during a twitch, or even during a tetanus in which intracellular $[Ca^{2+}]$ is maintained at saturating levels.

The molecular basis for the dip in the k_{tr}-pCa relationship is unclear, although the phenomenon has been observed in both skeletal (Swartz and Moss, 1992; Fitzsimons, et al., 2001a) and cardiac muscles (Fitzsimons, et al., 2001b) and thus appears to be real. Additional experiments in skeletal muscle do provide some insight into the underlying mechanism, in that partial extraction of TnC together with added NEM-S1 completely eliminated the dip, i.e., this combination of interventions eliminated the activation dependence of k_{tr} by increasing k_{tr} to the same supramaximal value at every level of activation (Fitzsimons, et al., 2001a). This is a dramatic and surprising result since TnC extraction by itself was found to have very little effect on the activation dependence of k_{tr}.

It is likely that a variety of models or a combination of models could be used to explain the Ca^{2+}-activation dependence of k_{tr} and the changes in this relationship caused by the addition of NEM-S1 to the bathing solution. When first described by Brenner (1988), the activation dependence of k_{tr} was explained with a model in which Ca^{2+} influences the rate constant of cross-bridge attachment to actin, f_{app}, i.e., f_{app} decreases when $[Ca^{2+}]$ is lowered and this has been extended in more recent work (Landesberg and Sideman, 1994; Brenner and Chalovich, 1999). As discussed above, such a mechanism could certainly be operative in skeletal muscle, but it does not account for the effect of strong-binding cross-bridges (NEM-S1) to accelerate k_{tr}. Because of this we favor a model developed by Campbell (1997), which is a variation of one developed earlier by Geeves and colleagues (McKillop and Geeves, 1993). According to Campbell, the Ca^{2+}-dependence of k_{tr} is due to cooperativity-induced slowing of force development during submaximal activation (Campbell, 1997). In this model, cross-bridges are distributed between cycling and non-cycling populations: cycling cross-bridges undergo repeated transitions between non-force-bearing and force-bearing states, under the influence of the rate constants f_{app} and g_{app}, while non-cycling cross-bridges are recruited to the cycling population as a result of Ca^{2+} binding to troponin or cooperative effects of strong binding cross-bridges to enhance activation of the thin filament. At lower levels of Ca^{2+}, a small fraction of cross-bridges is initially recruited into the cycling population as a direct result of Ca^{2+} binding to the thin filament, so that most cross-bridges are in the non-cycling pool and are thus available for cooperative recruitment to the cycling pool. Progressive recruitment of cross-bridges from the non-cycling pool would then slow the rate of force development. In contrast, at high levels of Ca^{2+}, the rate constant of force development is much greater because most cross-bridges are recruited to the cycling pool when Ca^{2+} binds to troponin, which leaves few cross-bridges available in the non-cycling pool for subsequent cooperative recruitment. At a saturating $[Ca^{2+}]$ of pCa 4.5, the rate constant of force development would thus approach the sum of the forward and reverse rate constants for the force-generating transition, i.e., $f_{app} + g_{app}$.

In the context of this model, the effects of NEM-S1 to accelerate k_{tr} at submaximal Ca^{2+} concentrations can be explained as a cooperative activation of the thin filament by the strong-binding NEM-S1. The observation that k_{tr} is maximal or supramaximal at the very lowest levels of activation implies that there is little cooperativity in cross-bridge binding in this activation range, perhaps due to reduced interactions between neighboring functional groups (Razumova, *et al.*, 2000). The decrease in k_{tr} as $[Ca^{2+}]$ is raised to intermediate levels would then be explained on the basis of an increase in cooperativity in cross-bridge binding, and the increase in k_{tr} at the highest Ca^{2+} would be due to reduced cooperativity due to saturation of Ca^{2+} and cross-bridge binding to the thin filaments.

It is likely that the cooperative activation of the thin filament by NEM-S1 is greater than the effects of cross-bridge binding in skeletal muscles *in vivo*, since k_{tr} measured in control fibers in the absence of NEM-S1 does not increase substantially until Ca^{2+}-activated isometric forces are greater than half-maximal (Swartz and Moss, 1992). This is a potentially important point because the potent activating effects of NEM-S1 could actually mask other activating processes such as direct effects of Ca^{2+} on contraction kinetics, discussed in section 3.1 above.

While NEM-S1 dramatically accelerates k_{tr}, strong-binding cross-bridges do not entirely account for the activation of cross-bridge kinetics, since NEM-S1 alone was insufficient at intermediate levels of activation to increase k_{tr} to maximal or to completely eliminate the activation dependence of k_{tr}. Instead, elimination of activation dependence required both NEM-S1 and partial extraction of TnC from the thin filament (Fitzsimons, *et al.*, 2001a). Since partial extraction of TnC disrupts near-neighbor communication between functional groups in the thin filament (Brandt, *et al.*, 1984; Moss, *et al.*, 1985), the activation dependence of k_{tr} involves effects of strongly-bound cross-bridges to cooperatively recruit additional cross-bridges within the same and neighboring regions of the thin filament. In fact, a modification of Campbell's (1997) model to include near-neighbor interactions that cooperatively recruit cross-bridges to strongly-bound states predicts these results (Razumova, et al., 2000). By extracting TnC, these interactions were presumably disrupted and cooperative recruitment of cross-bridges from neighboring functional groups was reduced or eliminated, thereby speeding the rate of force development.

3.2.2 *Cardiac Muscle*

The rate constant of force development in cardiac muscle is also Ca^{2+}-dependent (Wolff, *et al.*, 1995; Palmer and Kentish, 1998; Fitzsimons, et al., 2001b), increasing approximately 10-fold as Ca^{2+} is increased from threshold to maximal concentrations (Figure 3b). Just as in skeletal muscles, application of NEM-S1 accelerates k_{tr} at submaximal Ca^{2+} concentrations and has no effect on maximal k_{tr} (Figure 5). Unlike skeletal muscles, it is not possible to further increase maximal

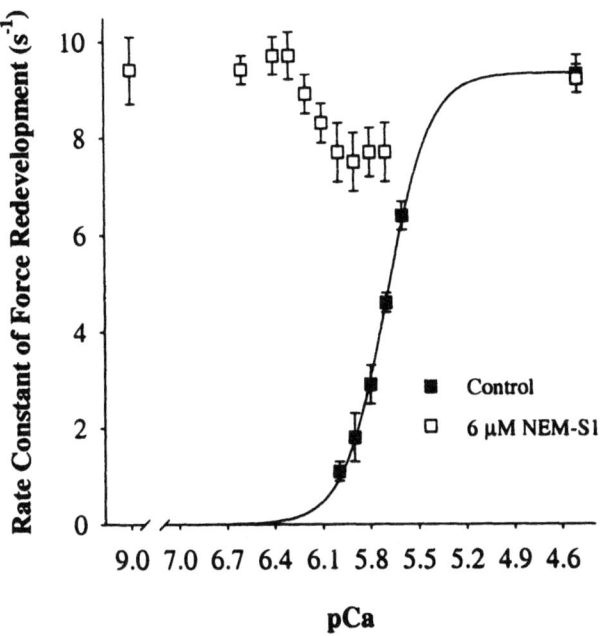

Figure 5. Effects of 6 μM NEM-S1 on the rate constant of force redevelopment (k_{tr}) assessed in rat skinned myocardium at 15°C. Data are replotted from Fitzsimons, et al. (2001b).

k_{tr} by partial extraction of TnC, which implies that the cardiac thin filament is fully activated at saturating Ca^{2+} concentrations with respect to the kinetics of force development. A logical conclusion from this observation is that cardiac muscle is less cooperative than skeletal muscle in the sense that fewer cross-bridges are required to cooperatively activate the cardiac thin filament. These results should not be taken to mean that it is impossible to increase k_{tr} beyond the maximum measured in skinned fibers, since preliminary experiments (Patel, Fitzsimons and Moss, unpublished) have shown that the combination of NEM-S1 and PKA-mediated phosphorylation of myofibrillar proteins in skinned myocardium increases k_{tr} at submaximal activation by as much as 20%, i.e., under these conditions, k_{tr} at the lowest force levels is supramaximal. This finding implies that such phosphorylations have the potential to enhance the kinetics of force development in sub-maximally activated cardiac muscle. Given that the Ca^{2+} activation of living myocardium is typically submaximal, such a mechanism is a potentially important way to increase the rate of rise of force (and pressure) of myocardium *in vivo*.

Compared to skeletal muscles, much less work has been done to determine the mechanisms of activation in skinned myocardial preparations, primarily because of the technical difficulties of working with this tissue. The potential importance of such work is great since these mechanisms might be altered in diseases such as heart failure in which the rate of force (or pressure) development ($+dP/dt_{max}$) can be substantially lower than in healthy tissue. Furthermore, the findings of familial hypertrophic cardiomyopathy mutations in thin filament regulatory proteins (Thierfelder, *et al.*, 1994; Bonne, *et al.*, 1998) indicate that it is plausible to propose that the mechanisms contributing to the cooperative activation of the thin filament are altered or disrupted.

4. SUMMARY

Regulation of contraction in cardiac and skeletal muscles is a cooperative process involving the activating effects of both Ca^{2+} binding to TnC and cross-bridge binding to the thin filament. In terms of isometric force development, skeletal muscle is the more cooperative of the two because (1) more strongly bound cross-bridges are required to activate force and the kinetics of force development and (2) it is not fully activated in terms of kinetics when cross-bridge binding and Ca^{2+} are saturated. Cardiac muscle is also cooperative but less so than skeletal muscle in the sense that fewer cross-bridges are required to completely activate the system. Consistent with this interpretation, both force and the kinetics of force development are maximal in cardiac muscle at saturating concentrations of Ca^{2+}.

These differences in the mechanisms of activation fit well with our notions of how these two muscle types are activated *in vivo*. When stimulated by its motor nerve, a skeletal muscle fiber undergoes a twitch contraction, which is a relatively explosive phenomenon involving switch-like, on-off activation of the thin filament. Sufficient Ca^{2+} is released during excitation-contraction coupling to saturate the thin filament, or nearly so. In order to increase the force developed by a single muscle fiber, stimulation must be repetitive and closely spaced (i.e., a tetanus) so as to prolong the Ca^{2+} transient and assure that a greater proportion of cross-bridges bind to actin. The mechanisms used by motor control systems to vary the force developed by a whole muscle include increases in action potential frequency along the motor nerve and also the recruitment of additional motor units.

In cardiac muscle, tetanus is not a workable mechanism for varying force due to the refractoriness of the membrane during the plateau and repolarization phases of the cardiac action potential. Instead, all myocardial cells contract during a given beat and the force developed by any one cell depends a range of factors. Force is modulated by variations in the amount of trigger Ca^{2+} entering the cell, the amount of activator Ca^{2+} released from the sarcoplasmic reticulum, and the sensitivity of the thin filaments to Ca^{2+}. Changes in these variables are mediated by agonist binding to membrane receptors, e.g., β-adrenergic receptors, and in most cases involve phosphorylations due to activation of cAMP—dependent protein kinase. Phosphorylations of myofibrillar proteins, particularly troponin-I and myosin binding protein-C, also speed the rate of cross-bridge cycling, and thus, treatment with a β-agonist would accelerate myocardial twitch kinetics both as a result of increased Ca^{2+} delivery to the myoplasm and direct effects of second messengers on cross-bridge cycling rates. Sarcomere length is another variable that influences twitch force, in that the Ca^{2+} sensitivity of force is increased at longer lengths. Overall, this combination of regulatory mechanisms makes it possible to achieve exquisite gradations of myocardial twitch force and kinetics on a beat-to-beat basis and thereby match the characteristics of contraction to the preload (end-diastolic volume) and afterload (peak systolic pressure) in the left and right chambers of the heart.

While strong binding cross-bridges have substantial effects on force generation and kinetics in cardiac and skeletal muscles, there are still questions about the relative contributions of strong-binding cross-bridges compared to direct kinetic effects of Ca^{2+} binding to thick or thin filament proteins. In cardiac muscle, the application of exogenous strong-binding cross-bridges is sufficient to increase the rate constant of force development (k_{tr}) to its maximum but not beyond, which implies that the strong-binding cross-bridge component of activation becomes saturated over the physiological range of Ca^{2+} concentrations. In skeletal muscle, application of exogenous strong-binding cross-bridges (i.e., NEM-S1) in sufficient amounts can actually increase k_{tr} at low $[Ca^{2+}]$ to values greater than the maximum seen at saturating Ca^{2+} concentrations in the absence of NEM-S1. Thus, while strong-binding cross-bridges have an accelerating effect on the kinetics of force development in skeletal muscle, this effect is not saturated under physiological conditions, which begs the question about how much of the activation-dependent acceleration of k_{tr} is due to strong binding cross-bridges. However, it is clear that strong-binding cross-bridges contribute significantly to the activation of k_{tr} in fast-twitch skeletal muscle since a reduction in the number of strong binding cross-bridges with vanadate significantly reduces the value of k_{tr} at all levels of activation, including saturating concentrations of Ca^{2+}. In both cardiac and skeletal muscles there is a marked difference in the $[Ca^{2+}]$-dependencies of force and k_{tr}, i.e., force increases at lower concentrations and therefore has greater Ca^{2+} sensitivity than k_{tr}. The difference in Ca^{2+} dependencies of force and k_{tr} can be explained by several different models, none of which has been excluded or uniquely supported by experimental evidence. For example, k_{tr} could be smaller at low $[Ca^{2+}]$ due to the slowing effects of progressive cooperative recruitment of cross-bridges into the cycling pool (Campbell, 1997); k_{tr} could be accelerated at high Ca^{2+} due to the activating effects of increased numbers of strongly bound cross-bridges (McKillop and Geeves, 1993); or k_{tr} could be increased at high Ca^{2+} due to direct effects of increased Ca^{2+} on f_{app}, the cross-bridge attachment rate constant (Brenner, 1988; Landesberg and Sideman, 1994). The nature of the regulatory mechanism underlying the activation-dependence of the rate of force development remains a central question in the biology of striated muscles.

Department of Physiology
University of Wisconsin Medical School
Madison, WI 53706 USA

5. REFERENCES

Allen, D.G., and J.C. Kentish. 1988. Calcium concentration in the myoplasm of skinned ferret ventricular muscle following changes in muscle length. *J. Physiol.* 407:489-503.

Bonne, G., L. Carrier, P. Richard, B. Hainque, and K. Schwartz. 1998. Familial hypertrophic cardiomyopathy. From mutations to functional defects. *Circ. Res.* 83:580-593.

Brandt, P.W., M.S. Diamond and F.H. Schachat. 1984. The thin filament of vertebrate skeletal muscle co-operatively activates as a unit. *J. Mol. Biol.* 180:379-384.

Bremel, R.D., and A. Weber. 1972. Cooperation within actin filament in vertebrate skeletal muscle. *Nature* 238:97-101.

Brenner, B. 1988. Effect of Ca^{2+} on cross-bridge turnover kinetics in skinned single rabbit psoas fibers: implications for regulation of muscle contraction. *Proc. Natl. Acad. Sci. USA.* 85:3265-3269.

Brenner, B., and J.M. Chalovich. 1999. Kinetics of thin filament activation probed by fluorescence of N-((2-iodoacetoxy)ethyl)-N-methyl)amino-7-nitrobenz-2-oxa-1,3-diazole-labeled troponin I incorporated into skinned fibers of rabbit psoas muscle: Implications for regulation of muscle contraction. *Biophys. J.* 77:2692-2708.

Brenner, B., and E. Eisenberg. 1986. Rate of force generation in muscle: correlation with actomyosin ATPase activity in solution. *Proc. Natl. Acad. Sci. USA* 83:3542-3546.

Butters, C.A., J.B. Tobacman, and L.S. Tobacman. 1997. Cooperative effect of calcium binding to adjacent troponin molecules on the thin filament-myosin subfragment 1 MgATPase rate. *J. Biol. Chem.* 272:13196-13202.

Campbell, K. 1997. Rate constant of muscle force redevelopment reflects cooperative activation as well as cross-bridge kinetics. *Biophys. J.* 72:254-262.

Cantino, M.E., T.StC. Allen, and A.M. Gordon. 1993. Subsarcomeric distribution of calcium in demembranated fibers of rabbit psoas muscle. *Biophys. J.* 64:211-222.

Chase, P.B., D.A. Martyn, and J.D. Hannon. 1994. Isometric force redevelopment of skinned skeletal muscle fibers from rabbit activated with and without Ca^{2+}. *Biophys. J.* 67:1994-2001.

Dantzig, J.A., and Y.E. Goldman. 1985. Suppression of muscle contraction by vanadate. *J. Gen. Physiol.* 86:305-327.

Dantzig, J.A., Y.E. Goldman, N.C. Millar, J. Lacktis, and E. Homsher. 1992. Reversal of the cross-bridge force-generating transition by photogeneration of phosphate in rabbit psoas muscle fibres. *J. Physiol.* 451:247-278.

Dantzig, J.A., M.G. Hibberd, D.R. Trentham, and Y.E. Goldman. 1991. Cross-bridge kinetics in the presence of MgADP investigated by photolysis of caged ATP in rabbit psoas muscle fibers. *J. Physiol.* 432:639-680.

Fitzsimons, D.P., J.R. Patel, K.S. Campbell, and R.L. Moss. 2001a. Cooperative mechanisms in the activation dependence of the rate of force development in rabbit skinned skeletal muscle fibers. *J.Gen. Physiol.* 117:133-148.

Fitzsimons, D.P., J.R. Patel, and R.L. Moss. 2001b. Cross-bridge interaction kinetics in rat myocardium are accelerated by strong binding of myosin to the thin filament. *J. Physiol.* 530:263-272.

Fuchs, F. 1995. Mechanical modulation of the Ca^{2+} regulatory protein complex in cardiac muscle. *NIPS* 10:6-12.

Fuchs, F., and Y.-P. Wang. 1991. Force, length, and Ca^{2+}-troponin C affinity in skeletal muscle. *Am. J. Physiol.* 261:C787-C792.

Gordon, A.M., and E.B. Ridgway. 1987. Extra calcium on shortening in barnacle muscle. Is the decrease in calcium binding related to decreased cross-bridge attachment, force, or length. *J. Gen. Physiol.* 90:321-340.

Geeves, M. A., and S. S. Lehrer. 1994. Dynamics of the muscle thin filament regulatory switch: The size of the cooperative unit. *Biophys. J.* 67:273-282.

Gordon, A.M., E. Homsher and M. Regnier. 2000. Regulation of contraction in striated muscle. *Physiol. Rev.* 80:853-924.

Grabarek, Z., J. Grabarek, P.C. Leavis, and J. Gergely. 1983. Cooperative binding to the Ca^{2+} specific sites of troponin C in regulated actin and actomyosin. *J. Biol. Chem.* 258:14098-14102.

Guth, K., and J.D. Potter. 1987. Effect of rigor and cycling cross-bridges on the structure of troponin C and on the Ca^{2+} affinity of the Ca^{2+} specific regulatory sites in skinned rabbit psoas fibers. J. Biol. Chem. 262:13627-13635.

Hannon, J.D., D.A. Martyn, and A.M. Gordon. 1992. Effects of cycling and rigor crossbridges on the conformation of cardiac troponin C. *Circ. Res.* 71:984-991.

Hofmann, P.A., and F. Fuchs. 1987. Effect of length and cross-bridge attachment on Ca^{2+} binding to troponin C. *Am. J. Physiol.* 253:C90-C96.

Hofmann, P.A., J.M. Metzger, M.L. Greaser and R.L. Moss. 1990. Effects of partial extraction of light chain 2 on the Ca^{2+} sensitivities of isometric tension, stiffness, and velocity of shortening in skinned skeletal muscle fibers. *J. Gen. Physiol.* 95:477-498.

Landesberg, A., and S. Sideman. 1994. Coupling calcium binding to troponin C and cross-bridge cycling in skinned cardiac cells. *Am. J. Physiol.* 266:H1260-H1271.

Lehrer, S. S. 1994. The regulatory switch of the muscle thin filament: Ca^{2+} or myosin heads. *J. Muscle Res. Cell Motil.* 15:232-236.

Lehrer, S.S., and M.A. Geeves. 1998. The muscle thin filament as a classical cooperative/allosteric regulatory system. *J. Mol. Biol.* 277:1081-1089.

Levine, R.J., R.W. Kensler, Z. Yang, J.T. Stull, and H.L. Sweeney. 1996. Myosin light chain phosphorylation affects the structure of rabbit skeletal muscle thick filaments. *Biophys. J.* 71:898-907.

Lu, Z., R.L. Moss, and J.W. Walker. 1993. Tension transients initiated by photogeneration of MgADP in skinned skeletal muscle fibers. *J. Gen. Physiol.* 101:867-888.

Lu, Z., D.R. Swartz, J.M. Metzger, R.L. Moss, and J.W. Walker. 2001. Regulation of force development studied by photolysis of caged ADP in rabbit skinned psoas fibers. *Biophys. J.* 81:334-344.

Martyn, D.A., C.J. Freitag, P.B. Chase, and A.M. Gordon. 1999. Ca2+ and cross-bridge-induced changes in troponin-C in skinned skeletal muscle fibers: effects of force inhibition. *Biophys. J.* 76:1480-1493.

McKillop, D. F. A., and M. A. Geeves. 1993. Regulation of the interaction between actin and myosin subfragment 1: evidence for three states of the thin filament. *Biophys. J.* 65:693-701.

Metzger, J. M., M. L. Greaser, and R. L. Moss. 1989. Variations in cross-bridge attachment rate and tension with phosphorylation of myosin in mammalian skinned skeletal muscle fibers. *J. Gen. Physiol.* 93:855-883.

Metzger, J. M., and R. L. Moss. 1991. Kinetics of a Ca^{2+}-sensitive cross-bridge state transition in skeletal muscle fibers. *J. Gen. Physiol.* 98:233-248.

Metzger, J.M., and R.L. Moss. 1992. Myosin light chain 2 modulates calcium-sensitive cross-bridge transitions in vertebrate skeletal muscles. *Biophys. J.* 63:460-468.

Moss, R.L. 1999. Plasticity in the dynamics of myocardial contraction: calcium, cross-bridge kinetics or molecular cooperation? *Circ. Res.* 84:862-865.

Moss, R.L., G.G. Giulian, and M.L. Greaser. 1985. The effects of partial extraction of TnC upon the tension-pCa relationship in rabbit skinned skeletal muscle fibers. *J. Gen. Physiol.* 86:585-600.

Nagashima, H., and S. Asakura. 1982. Studies on cooperative properties of tropomyosin-actin and tropomyosin-troponin-actin comples by the use of N-ethylmaleimide-treated and untreated species of myosin subfragment 1. *J. Mol. Biol.* 155:409-428.

Palmer, S., and J. C. Kentish. 1998. Roles of Ca^{2+} and crossbridge kinetics in determining the maximum rates of Ca^{2+} activation and relaxation in rat and guinea pig skinned trabeculae. *Circ. Res.* 83:179-186.

Pan, B.-S., and R.J. Solaro. 1987. Calcium-binding properties of troponin C in detergent-skinned heart muscle fibers. *J. Biol. Chem.* 262:7839-7849.

Razumova, M.V., A.E. Bukatina and K.B. Campbell. 2000. Different myofilament nearest-neighbor interactions have distinctive effects on contractile behavior. *Biophys. J.* 78:3120-3137.

Regnier, M., D.A. Martyn and P.B. Chase. 1998. Calcium regulation of tension redevelopment kinetics with 2-deoxy-ATP or low [ATP] in skinned rabbit psoas fibers. *Biophys. J.* 74:2005-2015.

Schiaffino, S., and C. Reggiani. 1996. Molecular diversity of myofibrillar proteins: gene regulation and functional significance. *Physiol. Rev.* 76:371-423.

Shiner, J.S., and R.J. Solaro. 1984. The Hill coefficient for the Ca^{2+}-activation of striated muscle contraction. *Biophys. J.* 46:541-543.

Solaro, R.J., and H.M. Rarick. 1998. Troponin and tropomyosin. Proteins that switch on and tune in the activity of cardiac myofilaments.

Swartz, D. R., and R. L. Moss. 1992. Influence of a strong-binding myosin analogue on calcium-sensitive mechanical properties of skinned skeletal muscle fibers. *J. Biol.Chem.* 267:20497-20506.

Swartz, D.R., M.L. Greaser, and B.B. Marsh. 1990. Regulation of binding of subfragment 1 in isolated rigor myofibrils. *J. Cell Biol.* 111:2989-3001.

Swartz, D. R., R. L. Moss, and M. L. Greaser. 1996. Calcium alone does not fully activate the thin filament for S1 binding to rigor myofibrils. *Biophys. J.* 71:1891-1904.

Thierfelder, L., H. Watkins, C. MacRae, R. Lamas, W. McKenna, H.-P. Vosberg, J.G. Seidman, and C.E. Seidman. 1994. α-Tropomyosin and cardiac troponin T mutations cause familial hypertrophic cardiomyopathy: a disease of the sarcomere. *Cell* 77:701-712.

Thirlwell, H., J.E.T. Corrie, G.P. Reid, D.R. Trentham, and M.A. Ferenczi. 1994. Kinetics of relaxation from rigor of permeabilized fast-twitch skeletal fibers from rabbit using a novel caged ATP and apyrase. *Biophys. J.* 67:2346-2447.

Tobacman, L.S. 1996. Thin filament-mediated regulation of cardiac contraction. *Ann. Rev. Physiol.* 58:447-481.

Tobacman, L.S., and D. Sawyer. 1990. Calcium binds cooperatively to the regulatory sites of the cardiac thin filaments. *J. Biol. Chem.* 265:931-939.

Walker, J.W., Z. Lu, and R.L. Moss. 1992. Effects of Ca^{2+} on the kinetics of phosphate release in skeletal muscle. *J. Biol. Chem.* 267:2459-2466.

Wang, Y.P., and F. Fuchs. 1994. Length, force, and Ca^{2+}-troponin C affinity in cardiac and slow skeletal muscle. *Am J. Physiol.* 266:C1077-C1082.

Williams, D. L., L. E. Greene, and E. Eisenberg. 1988. Cooperative turning on of myosin subfragment 1 adenosinetriphosphatase activity by the troponin-tropomyosin-actin complex. *Biochem.* 27:6987-6993.

Wolff, M.R., K.S. McDonald, and R.L. Moss. 1995. Rate of tension development in cardiac muscle varies with level of activator calcium. *Circ. Res.* 76:154-160.

R. JOHN SOLARO, BEATA M. WOLSKA, GRACE ARTEAGA,
ANNE F. MARTIN, PETER BUTTRICK
& PIETER DETOMBE

MODULATION OF THIN FILAMENT ACTIVITY IN LONG AND SHORT TERM REGULATION OF CARDIAC FUNCTION

1. MODULATION OF CARDIAC MYOFILAMENT RESPONSE TO CA^{2+} IS A SIGNIFICANT REGULATOR OF THE INTENSITY AND DYNAMICS OF CARDIAC ACTIVITY

The contraction and relaxation of heart muscle is exquisitely controlled to match the venous return to the cardiac output. If the control mechanisms work, they ensure that output of the right or left ventricle occurs over a broad range with little change in end diastolic volume (EDV). If the control mechanisms fail and increases in cardiac output occur with relatively large changes in EDV, then homeostasis is threatened by edema and by relatively inefficient conversion of wall tension into pressure. Moreover, chronic elevations in EDV and the associated stretch of the myocardium engage pathways for hypertrophic signaling. With hypertrophic adaptation, homeostasis is restored, but this is likely to be temporary; with continued strain on the cells eventually maladapation occurs and a viscous cycle of cardiac cell growth and depressed function leads to the syndrome of heart failure.

The manifestation of the control mechanisms that ensure homeostasis occurs in the form of changes in the extent of contraction and of changes in the rates of relaxation and contraction that are tuned to the prevailing heart rate and metabolic demands. Figures 1 and 2 illustrate these concepts in graphs that show the changes in cardiac function in the transition from rest to moderate exercise. Figure 1 depicts cardiac function as the dynamics of left ventricular volume at rest, during exercise, and in a resting state of heart failure. The data in Figure 1 demonstrate that both the end systolic volume (ESV) and overall cycle time of contraction/relaxation have both decreased with exercise. However, at this level of exercise in which cardiac output has increased several fold, EDV remained essentially the same in the healthy heart. In the failing heart, however, while stroke volume is maintained at rest, EDV is elevated. Figure 2 depicts cardiac function as the relation between volume (Vol) and pressure (P) during a beat of the left ventricle at rest, during exercise, in a resting state of heart failure. The line through the end systolic pressure (ESP)

291

R.J. Solaro and R.L. Moss (eds), Molecular Control Mechanisms in Striated Muscle Contraction, 291-327
©2002 Kluwer Academic Publishers, Printed in the Netherlands

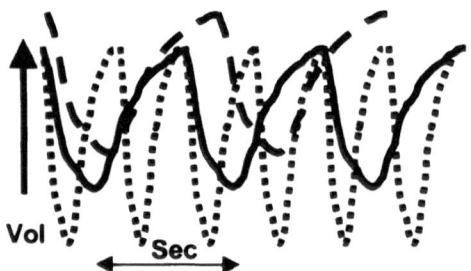

Figure 1. Dynamics of left ventricular volume in a resting state, during exercise, and during heart failure in the resting state. Note that during exercise, increases in stroke volume and heart rate occur with little change in the end diastolic volume. In heart failure, stroke volume is maintained but at an elevated end diastolic volume and with slower kinetics of contraction and relaxation. (see text for details).

provides a measure of contractility. Again, these data illustrate that the increase in stroke volume during exercise has occurred with little change in EDV and EDP, whereas EDV and EDP rise in heart failure. Thus, the data in Figures 1 and 2 provide evidence of the ability of the heart to match its output to the increase in venous return during exercise with no change in EDV. However, with heart failure, shown by the depression of the slope of the Vol-ESP relation note that the increase in stroke volume occurs but with the cost of operating at an elevated EDV.

An appealing aspect of the Vol-ESP relation is that it reflects the length tension properties of heart muscle. Even though the dependence tension on fiber length is related in a complex way to ventricular geometry, it is instructive to relate ventricular volume to fiber and sarcomere length and ventricular pressure to fiber tension. This gets us from the realm of chamber properties of pressures, flows, and volumes to the myocyte properties of tension, shortening, and sarcomere length. The next level of organization is the assembly of molecules that make up the cellular organelles and the proteins responsible for the ability of the myocytes to develop tension and shorten. Thus, the Vol-ESP relation serves as a useful context to discuss molecular mechanisms and their translation into cardiac function.

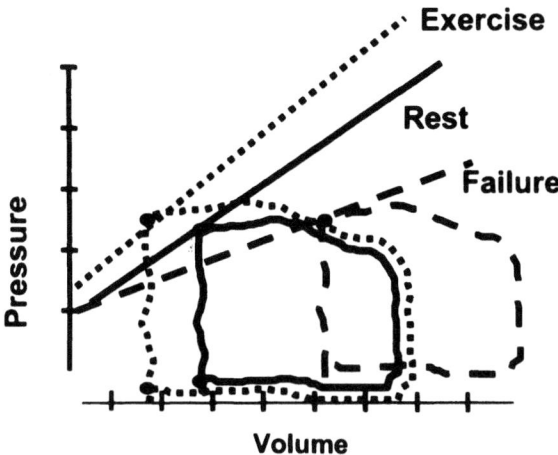

Figure 2.   Relations between left ventricular volume and left ventricular pressure during one beat in a basal state, during exercise, and during heart failure in the basal state. See text for discussion.

Molecular mechanisms regulating cardiac function as illustrated in Figures 1 and 2 ultimately control the reaction of the molecular motor, myosin or crossbridge, with actin. This reaction promotes a translation of the actin-containing thin filaments past the thick myosin-containing filaments and thus induces an ability of the myofilaments to generate force and shorten. The number of crossbridges reacting with actin and the force generated/crossbridge determine the level of tension (pressure) developed by the myofilaments and the extent of fiber shortening (related to volume change during ejection) against a particular load (related to arterial pressure).   In diastole, processes in the thin filament that impede the reaction of crossbridge with actin keep contraction turned off. In systole, release from this inhibition is initiated by Ca^{2+}-binding to the thin filament receptor, troponin C (TnC). Delivery of Ca^{2+} to the myofilament space is determined by membrane-controlled fluxes mainly across the sarcolemma and sarcoplasmic reticulum (SR). Binding of Ca^{2+} to TnC triggers activation, but activation of the myofilaments is sustained by a complex process involving long and short range cooperative interactions elicited along the thin and thick filaments and by the actin-myosin reaction itself (Tobacman, 1996; Geeves and Lehrer, 1994; Solaro and Rarick, 1998).

Inasmuch as the chambers of the heart operate as a functional syncytium, regulation of the intensity of contraction occurs not by recruitment of motor units as is the case in skeletal muscle, but by control mechanisms in the cells themselves. In basal physiological states, there is a reserve of crossbridges that can be recruited to interact with the thin filaments; sarcomeres appear to operate at about 25% of maximum activity (Fabiato, 1981). General mechanisms for variations in the population of active crossbridges include: 1) control of the amounts of Ca^{2+} released into the myofilament space as determined by net fluxes of Ca^{2+} into and out of the myofilament space and 2) control of the myofilament response to Ca^{2+}. In the short term, the regulation of Ca^{2+} movements into and out of the myofilament space may occur by alterations in neuro-humoral signaling, the prevailing chemical and physical environment, and sarcomere length. In the long term the regulation of these Ca^{2+} movements may occur by altered expression of the channels, pumps, and exchangers that control Ca^{2+} fluxes in the cells. There is ample evidence that these mechanisms controlling the delivery of Ca^{2+} to the myofilaments are significant as regulators of the changes in cardiac function illustrated in Figures 1 and 2 (Bers, 2001).

Here, however, we focus on mechanisms that control the myofilament response to Ca^{2+}. We will demonstrate that myofilaments are not passive participants in regulation of cardiac function and not entirely slaves to Ca^{2+} released into the myofilament space. In fact, there is now substantial evidence that neuro-humoral regulation and Starling's Law, the most prominent mechanisms for beat-to-beat control of cardiac function, involve modulation of the myofilament response to Ca^{2+} (Solaro, 2001; deTombe and Solaro, 2000, Fuchs, this monograph). The intracellular environment especially pH, redox state, inorganic phosphate and nucleotide concentration also modifies myofilament response to Ca^{2+}. It is not surprising that the myofilament response to Ca^{2+} is a variable in the control system. Variations in the myofilament response to Ca^{2+} offer the possibility of reducing the energetic cost of transporting Ca^{2+} as well as reducing the threat of the pathology of Ca^{2+} overload. Long term changes in hemodynamic demand associated with normal cardiac development and exercise, as well as pathologies of the cardiac and vascular systems also involve changes in myofilament response to Ca^{2+}. The mechanisms for these changes may involve protein isoform switching, phosphorylation, or proteolysis (Solaro, 2000). These changes may be of particular significance with regard to the transition from the compensated state to failure (deTombe and Solaro, 2000).

2. ACTIVATION OF CARDIAC MYOFILAMENTS IS A COMPLEX PROCESS WITH MULTIPLE SITES FOR MODULATION

In this chapter, we track evidence revealing the functional significance of modulation of the activity of cardiac thin filaments by molecular mechanisms downstream of Ca^{2+}-binding to troponin C (TnC). Consideration of the detailed mechanisms of thin filament control of the activation of the actin-myosin reaction reveals the many potential targets for modulation of cardiac myofilament response to Ca^{2+}. Figures 3 and 4 illustrate a fundamental structural unit of the thin filament during diastole (closed or blocked state) and during systole (open state). The transition between these states involves complex allosteric, steric, and cooperative

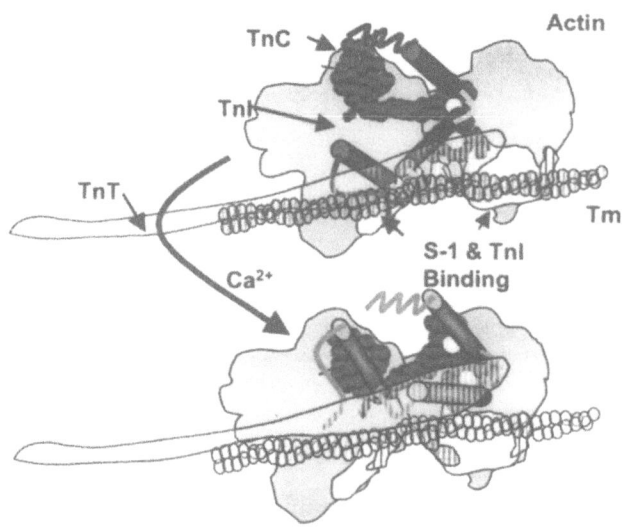

Figure 3. Schematic illustration of effect of Ca^{2+} binding to cTnC on thin filament structure. Tropomyosin (Tm) is shown blocking sites for crossbridge interaction with actin (hatched areas) when Ca^{2+} is not bound to troponin C (TnC). Ca^{2+}-binding to TnC sets into a motion a series of protein-protein interactions leading to movement of Tm away from the sites where crossbridges may interact. See text for discussion.

interactions in which there are many points that are subject to control by intrinsic mechanisms, such as Starling's Law, and extrinsic mechanisms such as neural and hormonal regulation (Solaro 2001, Solaro and Rarick, 1998; Tobacman, 1996). Regions of actin along the thin filament with which the crossbridges react are illustrated by cross hatches in Figures 3 and 4. In diastole crossbridges do not react with actin in a force generating reaction. In diastole when Ca^{2+} is low, crossbridges are either in a blocked state in which they are physically impeded from interacting with actin, or in a closed state in which there are relatively weak, rapid on-off interactions between actin and myosin that do not generate force (Chen et al. 2001; McKillop and Geeves, 1993; Squire and Morris, 1998). In systole, when Ca^{2+} is released into the myofilament space, the crossbridges react with actin in a strong state of binding and high rate of MgATP hydrolysis (open state). Early alternative models indicated that Ca^{2+} either: 1) simply recruits blocked crossbridges to

Figure 4. Schematic illustration of effect of crossbridge binding on thin filament structure. As illustrared in Figure 3, Ca^{2+} alone cannot fully expose sites on actin where crossbridges interact. Crossbridge binding is shown moving Tm further away from the crossbridge binding sites on actin. Thus, crossbridge binding fully activates the thin filament and as shown moves Tm along the thin filament permitting cooperative crossbridge interactions at a near-neighbor functional unit.

participate in force generation in an all or none manner (Podolsky and Teichholz, 1970) or 2) that Ca^{2+} modulates the rate that crossbridges enter their force generating states (Julian, 1969) with no change in the number of actively cycling crossbridges. Data demonstrating the Ca^{2+} dependence of the time course of force redevelopment (quantified by a rate constant designated as ktr) following a quick release and unloaded shortening (to disengage the crossbridges) provided evidence supporting the concept of rate modulation rather than pure recruitment. In addition, theoretical models such as that described by Hill et al. (1980) or by Brenner and Chalovich (1999) and by Brenner et al. (1999) fit the idea that a blocked or closed state is not necessary to explain either steady-state or presteady state crossbridge binding to the thin filament. Thus, even with so much structural, biochemical, and biophysical data on activation of striated muscle contraction, the precise mechanisms remain in dispute. Yet, structural data summarized below and showing steric hindrance of crossbridge binding as well as a movement of Tm in two steps, first with Ca^{2+} binding and then with crossbridge binding provide support for the existence of three states.

As illustrated in Figures 3 and 4, thin filament regulatory proteins control the state of the actin-crossbridge reaction. These proteins act as a molecular switch and as modulators of the state of activation of the thin filament. Associated with the actin helix is tropomyosin (Tm) and troponin (Tn), a hetero-trimer of distinct gene products. Tm is composed of similar but not identical repeating modules that interact with sites on 7 actins along the axis of the thin filament. The Tm molecules interact end to end and form a long strand winding around the actin helix. Troponin

(Tn) components include cardiac TnC (cTnC), the Ca^{2+} receptor; cTnI, an inhibitor of the actin-myosin reaction that toggles between binding to actin and to cTnC; and, TnT, a Tm binding protein that transduces the Ca^{2+}-binding signal by interacting with cTnI and cTnC. cTnC contains a single regulatory binding site for Ca^{2+} in its N-terminal lobe and two slowly exchanging Ca^{2+}/Mg^{2+} sites in its C-terminal lobe. An alpha-helical stalk connects these lobes. At rest Tm, which appears immobile and possibly inflexible on the thin filament, sterically hinders the actin-crossbridge reaction. This is illustrated in Figure 3, which shows Tm covering the crossbridge binding sites on actin. Tm immobility is associated with tight binding of the C-lobe of TnC to TnI and TnT, weak binding of the N-lobe of TnC to TnI and relatively strong binding and tethering of TnI to actin.

Our focus here is on modulation involving cTnI and slow skeletal TnI (ssTnI; the embryonic/neonatal isoform in the heart). Primary structures of cTnI and ssTnI are compared in Figure 5. C-terminal residues of TnI and a stretch of highly basic amino acids in the inhibitory peptide (Ip) make the tight contact with actin in resting muscle. Although results of early experiments indicated that the Ip was the most critical domain for establishing the blocked or closed state, it is now clear that C-terminal regions outside the Ip are critical for inhibition of the actin-myosin reaction and for its activation by Ca^{2+}. In the case of fsTnI, the C-terminus contains "second" regions of interaction with actin-Tm ($fsTnI_{140-148}$) and with TnC ($fsTnI_{116-131}$) that are outside the Ip. These regions are relatively conserved in cTnI and ssTnI, but do contain charge differences that we consider below in discussing modulation of response to Ca^{2+} in myofilaments containing different isoforms of

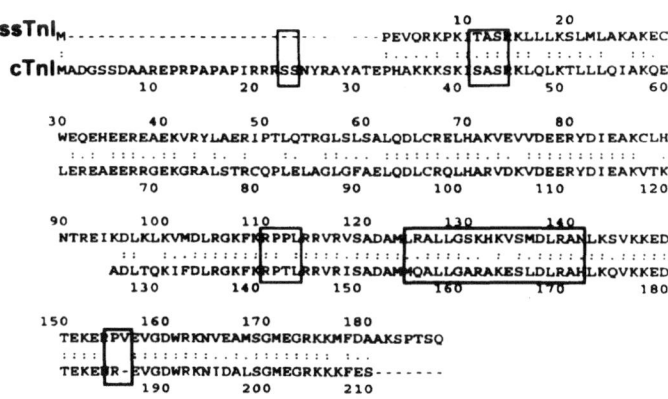

Figure 5. Comparison of primary structures of cardiac and slow skeletal TnI. Regions enclosed in the rectangles are discussed in the text.

TnI. The data of Rarick et al (1997) indicated that $cTnI_{154-199}$ represents a minimal C-terminal sequence for inhibition and Ca^{2+} regulation. Using a monoclonal antibody to an epitope in the far C-terminal region of TnI, Jin et al. (2001) reported evidence that this domain forms an exposed structure.   In addition, Jin et al. demonstrated that the binding of the antibody to the C-terminal TnI epitope was unaffected by binding of TnC and TnT at low Ca^{2+}.   However, when Ca^{2+} was increased, antibody binding increased greatly.   Jin et al. concluded that the C-terminus is essential in the Ca^{2+}-induced switching mechanism.   The C-terminal region of TnT also has interactions with TnI, TnC, and Tm that are important in maintaining the diastolic state.   The position of Tn on the thin filament may also sterically block crossbridges from reacting with actin (Figures 3 and 4).

With Ca^{2+}-binding to the N-lobe of TnC and exposure of its hydrophobic surface Spyracoupoulos et al. 1997), the tether between actin and TnI is released as TnI moves as much as 1.5 nm from its position in relaxed muscle and binds tightly to TnC (Tao et al. 1990 ; Lehman et al.1995) The Ca^{2+} trigger and signal involves exposure of a hydrophobic patch of amino acids in the N-lobe of cTnC.   An important and distinctive feature of the cardiac isoform is that exposure of the hydrophobic patch does not occur unless cTnC exists in a binary complex with cTnI. In the case of fsTnC, the hydrophobic patch is exposed (open state) upon Ca^{2+} binding to the two N-terminal regulatory sites even without fsTnI bound. However, Ca^{2+}-binding to the single regulatory site of cTnC does induce the exposure of hydrophobic residues.   Induction of the open state of cTnC requires the presence of either cTnI (Dong et al.1999) or a peptide comprised of residue 147-163 (Li et al. 1999).   Thus, whereas Ca^{2+} binding to both cTnC and fsTnC both result in similar structural states, there are important differences in how these states are achieved and modulated by neighboring proteins. The positions and movements of the proteins in the Tn complex suggest a lever-like action with the fulcrum consisting of C-terminal regions of cTnT and cTnC and N-terminal regions of cTnI.   With Ca^{2+}-binding to the N-terminal regulatory site, there is a rotation of TnC, as the cTnC N-lobe and cTnI C-terminal and Ip regions move toward each other.   A pivot point for these movements is the C-lobe of cTnC and the N-terminal and C-terminal regions of cTnT.   This pivot region contains sites for PKC-dependent phosphorylation on both cTnT and cTnI (Solaro, 2001).   Phosphorylation of these sites is known to inhibit maximum ATPase activity of reconstituted preparations.   In additions the unique N-terminal extension of cTnI, which is shown in Figures 3 and 4 situated close to the N-terminal regulatory site of cTnC, contains serial serines at positions 23 and 24 that are sites of phosphorylation by PKA.   The Ip of cTnI also contains a unique Thr at position 144, which is a substrate for PKC.   The position of these phosphorylation sites provides the potential for control of the Tn molecular switch at two critical regions: the pivot point at the cTnC- C-lobe and sites of interaction between the cTnI and the N-lobe of cTnC.   These changes in state of cTnC and of cTnI are sensed by cTnT in a signaling cascade ultimately responsible for movement of Tm. Whether sub-domains of actin also move in the activation process remains controversial (Squire and Morris, 1999; Gerson et al. 2001).   Whatever the case, Tm is now free to move on the thin filament and it is apparent that the seven actins under the control of Tm may make a concerted transition to the activated state (Hill et al. 1980, 1981). The movement of Tm is likely to involve a two step process in which Ca^{2+} binding to TnC induces an initial movement of Tm away from the region on actin where crossbridges bind. Figure 4 illustrates a second process in which strong crossbridge binding to the thin filament moves Tm further toward the groove formed by the actin helix. Figure 4 also shows that this crossbridge-induced movement of Tm eases the reaction of near-neighbor crossbridges with the thin filament. This long-range crossbridge dependent movement of Tm is essential for full myofilament activation and appears largely responsible for the highly cooperative nature of the activation process (Moss, 2001). Thus, strong binding of crossbridges moves Tm, inducing activation by a crossbridge-dependent process rather than a Ca^{2+}-dependent process. Direct binding of TnI to Tm may also restrict the movement of Tm; crossbridge binding to the thin filament not only pushes Tm into the groove of the actin helix, but also overcomes allosteric inhibition by TnI. In addition, the hyper-variable N-terminal half of cTnT, which anchors the cTn complex to Tm independently of Ca^{2+}, may also modify the strength of the interactions between adjacent Tm molecules on the thin filament, and thus affects the size of the cooperative unit. The number of actins under the control of one Tn may be more than presumed from the stoichiometry of 1 Tn: 7 actins. Current data indicate that one Tn may control as many 14 actins and that this cooperative unit may be a variable regulating tension (Moss, 2001).

3. MYOFILAMENT RELATED PROCESSES MODULATE DYNAMICS OF RELAXATION

Relaxation of the myofilaments and the return to the diastolic state in the intact ventricle is a poorly understood and an apparently complex process. Processes by which myofilament kinetics might modulate cardiac function include the following: Ca^{2+}-binding and dissociation from cTnC, kinetics of crossbridge cycling, kinetics of protein-protein interactions associated with activation and relaxation of the thin filament, and kinetics of the cooperative activating effect of strong crossbridges and the active state of the thin filament. Ca^{2+}-binding to the thin filament appears relatively fast compared to the other processes and is not rate limiting. The off rate for Ca^{2+}-release from cTnC is much slower than the on rate and may be a regulated variable (Robertson et al. 1982). Evidence indicates that steps that limit the rate of relaxation at the level of the myofilament crossbridge detachment rather than release of Ca^{2+} from TnC. Even so agents that enhance the affinity of TnC for Ca^{2+} and thereby slowing Ca^{2+} release are able to slow down relaxation (Regnier et al. 1996).

The dominance of myofilament related processes, as a determinant of relaxation kinetics is evident in measurements of intracellular Ca^{2+} and tension in isometric twitches of cardiac muscle. Intracellular Ca^{2+} falls well ahead of tension. This is what one would expect in a system in which strong crossbridges generated by Ca^{2+} binding to TnC sustain activation of the thin filament. In unloaded myocytes the mechanical event of shortening occurs with few cross-bridges reacting strongly with the thin filament, and, in this case, the falling phase of the Ca^{2+} transient coincides more closely with relaxation (re-lengthening). Hunter (2000) characterized two determinants of the relaxation process and its relation to myofilament properties. Hunter viewed one phase of relaxation as "load dependent" as the number of cross-bridges reacting with the thin filament (and thus cross-bridge dependent activation) depend on the load.. He viewed a second phase as "displacement dependent" as strong cross-bridges are strained with shortening and cannot reconnect easily. For example, in the pressure changes occurring during the cardiac cycle, the first phase is isovolumic (isometric) and the number of strong cross-bridges reacting with the thin filament is determined by the load (arterial pressure). With opening of the aortic valve, filament sliding occurs with rapid sarcomere shortening (Covell and Ross, 2001) and the point of relaxation becomes more displacement dependent. With enhanced rates of cross-bridge cycling, one might therefore expect that not only would there be an enhancement of cross-bridge detachment rate, but also with an increased shortening rate, enhanced displacement-related rate of relaxation. Of course the control of relaxation by myofilament dependent processes will diminish in conditions where Ca^{2+} removal from the cytoplasm is relatively slow compared to myofilament processes. However, if cross-bridge cycling is slowed or if cross-bridge dependent activation is diminished, these myofilament-related processes could remain rate limiting for relaxation.

Pharmacological and transgenic approaches have provided a powerful means of testing the hypothesis that modulation of the myofilament response to Ca^{2+} is a significant determinant of cardiac function. We next discuss how these approaches have provided evidence and insights into the physiological and patho-physiological implications of altered activity of the myofilament and their control by Ca^{2+}.

4. SPECIFIC MODIFICATIONS IN THE RESPONSE OF CARDIAC MYOFILAMENTS TO CA^{2+} BY PHARMACOLOGICAL AGENTS ALTER CARDIAC FUNCTION

Data in Figure 6 provides a clear illustration of how pharmacological manipulation of the myofilament response to Ca^{2+} may vary over a broad range with little or no change in the movements of Ca^{2+} to and from the myofilaments. The data in Figure 6 are records of length changes and intracellular Ca^{2+}-transients of single rat cardiac myocytes in a control condition and after exposure to two agents that affect myofilament response to Ca^{2+}. After exposure to BDM (butane dione monoxime = DAM or diacetyl monoxime), which is known to inhibit the actin crossbridge reaction (Li et al. 1985 ; Wolska et al. 1996), the Ca^{2+} transient remained constant, but the cell could no longer shorten. The inhibition of shortening was overcome by adding CGP 48506, a benzodiazocine derivative that directly activates the myofilaments (Herold et al. 1995; Wolska et al. 1996). Finally with washout of the drugs, the full effect of CGP 48506 to increase the extent of shortening is evident before the control response is restored. Throughout these wide swings in extent of cell shortening, the Ca^{2+} transient remained essentially unchanged. These data show that major changes in cellular shortening need not involve major changes in

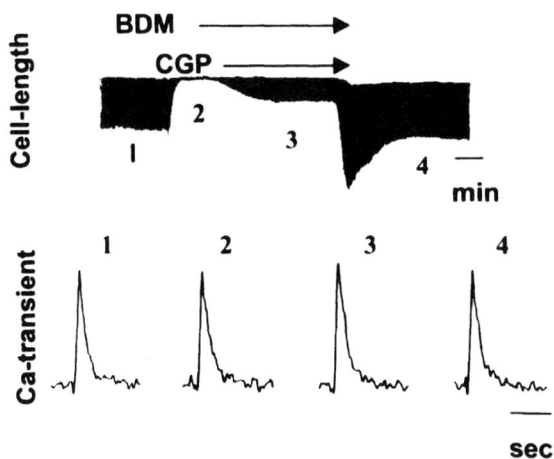

Figure 6. Effects of inotropic interventions on shortening of rat heart myocytes with no change in Ca²⁺ transients. 1. Control condition. 2. Effect of BDM which inhibits the actin-crossbridge interaction. 3. Effect of CGP 48506, a myofilament Ca²⁺-sensitizer to overcome BDM inhibition. 4. Return to control state following washout of BDM and CGP revealing the full inotropic effect of CGP. See text for further discussion.

intracellular fluxes of Ca^{2+}. *In vitro* experiments have identified that this response of the cells is most likely due to direct interactions of both BDM (Li et al. 1985) and CGP 48506 (Wolska et al. 1996) with the myofilaments. In contrast to these effects of CGP 48506, a (+) enantiomer, the (-) enantiomer (CGP 48508) is without effects on myofilament response to Ca^{2+}. The stereo-specificity reveals that subtle and specific alterations of sites (receptors) in the array of myofilament proteins can induce substantial changes in myofilament response to Ca^{2+}. Data shown in Figure 6 thus indicate that modifications in myofilament response to Ca^{2+} are able to enhance contractility. The site of action of CGP 48506 may be directly at the actin-myosin interface as it promotes tension development of crossbridges activated in the absence of TnI-TnC. CGP 48506 also activates rigor tension development at low Ca^{2+} in the absence of MgATP. It is also evident in the data shown in Figure 6 that agents such as CGP 48506 that enhance the force and shortening properties of cardiac myofilaments may also affect dynamics of contraction/relaxation especially by slowing down relaxation. Here it is apparent that with no change in the Ca^{2+}-transient, the effects of CGP on the myofilaments slow relaxation. Despite this prediction, there is no strong evidence that CGP 48506 significantly slows down relaxation of the intact ventricle (Slinker et al. 1997). Slowing of unloaded myocyte relaxation may reflect an effect of CGP on cross-bridge detachment rate. The lack of effect of relaxation of the intact ventricle may be related to the ability of CGP to promote cross-bridge dependent activation of the thin filament, therefore making relaxation more dependent on Ca^{2+}-removal from the cytoplasm. Although CGP shifts the Ca^{2+}-tension relation to the left, the relation in the presence of CGP is less steep than controls. We interpreted these data as indicating that in the presence of CGP, activation is more dependent on Ca^{2+} than on strong cross-bridges. Ca^{2+} binding to TnC in cardiac myofilaments is only modestly cooperative (Pan and Solaro, 1987). A similar increase in Ca^{2+}-sensitivity occurs with a change in slope of the Ca^{2+}-tension relation occurs when levels of inorganic phosphate are reduced from relatively high levels to relatively low levels (Westfall et al. 1993).

Another class of agents exemplified by EMD 57033 (a thiadiazinone) has properties similar, but not identical to those of CGP 48506. EMD 57033 is a potent Ca^{2+}-sensitizer *in vitro* and, as with CGP compounds, the actions of EMD 57033 are stereospecific. Its counterpart, the (-) enantiomer, is EMD 57439, which has no activity as a Ca^{2+}-sensitizer but does inhibit phosphodiesterase (PDE). On the other hand EMD 57033, as is the case with both enantiomers of CGP, has essentially no PDE inhibitory activity. When we (Wolska et al. 1996) compared concentrations of EMD 57033 and CGP 48506 that had the same effect on half-maximally activating Ca^{2+} in skinned fiber bundles, there were differences in the threshold for the onset of tension activation as Ca^{2+} was increased. This suggested to us that CGP might have less of an effect in prolonging relaxation than EMD. This was borne out in our studies of relaxation of isolated cells contracting under zero load. Slinker et al (1997) provided more convincing evidence for this difference in lusitropic effects of EMD and CGP. They directly compared the effects of CGP 48506 and EMD 57033 on the relationships between both the time of onset of relaxation and the rate of relaxation and wall stress in isolated perfused rabbit hearts. Their data showed that there was a negative lusitropic effect of EMD-57033 (relaxation was prolonged), and a negligible lusitropic of CGP 48506. A positive lusitropic effect (relaxation was abbreviated) of dobutamine served as a standard. Although this difference between CGP and EMD suggests a different target site on the myofilaments, the site(s) of action of both CGP and EMD are not clear. Both CGP and EMD activated crossbridge activity in the absence of the Tn-Tm complex and both appear not to affect Ca^{2+}-binding to TnC (Soalro et al. 1993; Wolska et al. 1996). We (Solaro et al. 1993) also reported that EMD 57033 stimulated actin filament sliding velocity on myosin heads attached to cover slips in the motility assay. In view of our results showing that EMD 57033 had no effect on myosin ATPase activity of pure myosin, we concluded that the receptor for EMD 57033 might reside on actin. Using sinusoidal analysis, Zhao and Kawai (1996) studied the effect of EMD-53998 (the racemic mixture) on elementary steps of the cross-bridge cycle in skinned pig heart preparations. Their data showed that EMD increased the number of attached crossbridges in association with a suppression of nucleotide binding to crossbridges and a resistance to the effects of phosphate accumulation to shift the population of crossbridges to weak binding states. The association constants of MgATP and MgADP, and Pi decreased substantially in the presence of EMD 53998.

Direct binding measurements (Pan and Johnson, 1996), and   NMR structural data (Wang et al. 2001; Kleerekoper and Putkey, 1999) indicated that the receptor for EMD may be located at the C-terminal metal binding domain (so-called structural sites) of cTnC.   Hydrophobic regions of the cleft of the C-domain of cTnC interact with the hydrophobic EMD 57033. EMD 57033 was demonstrated to be oriented so that the chiral group fit deeply into the cleft, making several contacts with the C-domain of cTnC. This may explain why EMD 57439 is inactive as a myofilament Ca^{2+} sensitizer. Why this enantiomer has activity as a PDE inhibitor remains unclear. Binding of the N-terminus of cTnI (34-71) completely displaced EMD 57033 from its binding site. Whether or not this is the functionally significant target for EMD 57033, these results indicate that alterations in the interactions of cTnI (34-71) with the C-domain of cTnC may have important effects on the sensitivity of the myofilaments to Ca^{2+}. As discussed below, PKC-dependent phosphorylation of the cTnI N-terminus at Ser 43 and Ser 45 provides a mechanism by which interaction of this region of TnI with the C-lobe of TnC may be varied.

There are other agents with both phosphodiesterase inhibitory and Ca^{2+}-sensitizing activities that are used therapeutically, and target to effects on Ca^{2+} binding to cTnC.   One of these agents is pimobendan (Acardi, UDCG 115BS), a benzodiazepine derivative, and another is levosimendan (Simdax), a pyridazinone-dinitrile derviative. Pimobendan is a racemic mixture and, although, there is some enantiomeric separation of the Ca^{2+} sensitizing activities, both enantiomers of sensitize the myofilaments to Ca^{2+} (Fujino et al.1988); both are also phosphdiesterase inhibitors.   Both pimobendan and levosimendan increase Ca^{2+} binding to cTnC. However, levosimendan binds directly and enhances Ca^{2+}-binding to cTnC (Sorsa et al. 2001). This activity accounts for the ability of levosimendan to increase Ca^{2+}-sensitivity of skinned fiber bundles of guinea pig heart (Edes et al. 1995).   Although pimobendan increases Ca^{2+}-binding to cTnC in skinned fiber bundles (Fujino et al. 1988), it is not clear whether this effect involves direct binding to cTnC. Experiments of Ohte et al (1997) addressed the question as to which of the two effects --myofilament Ca^{2+} sensitization or PDE inhibitory activity – is important in enhancing contractility of failing hearts. They compared the inotropic effect of pimobendan and amrinone (a PDE inhibitor with no Ca^{2+} sensitizing activity) on the left ventricular end systolic volume- end systolic pressure (Vol-ESP) relation in conscious dogs before and after pacing induced heart failure. Before the induction of heart failure both agents had similar effects to reduce end systolic pressure and increase the slope of the of the Vol-ESP relation. After the induction of heart failure, pimobendan was able to increase the slope of the Vol-ESP relation to a greater extent than amrinone. Both agents had similar effects on vasodilation. In studies on failing human myocardium, Bohm et al. (1991) also concluded that the positive inotropic effect of pimobendan involved both Ca^{2+}-sensitization and PDE inhibition.   In the case of levosimendan, Janssen et al. (2000) reported an improvement of diastolic and systolic function of failing human myocardium. Their analysis involved determination of force-frequency relations, which go from a positive slope in normal myocardium to a negative slope in the failed myocardium. Levosimendan improved this relation and induced a positive slope at therapeutic concentrations and relevant frequencies. At the same time diastolic force was decreased by levosimendan. This observation fits with a special feature of the effects of levosimendan on the Ca^{2+}-binding properties of cTnC. Binding of drug requires that Ca^{2+} be bound to the regulatory site of cTnC. The hypothesis is that when levosimendan is bound, the lifetime of the open state of the hydrophobic cleft in cTnC is prolonged. A corollary is that when Ca^{2+} is removed from cTnC levosimendan would be expected to dissociate. Thus the combination of this property together with PDE inhibition appears to enhance both systolic and diastolic function of the failing heart.

These data, which show that molecular changes in the myofilament proteins induced by binding of small molecules translate to functional effects, provide strong evidence regarding the potential that modulation of myofilament response to Ca^{2+} may affect cardiac function. We next consider evidence for this regulatory device based on studies with myofilaments in which there are natural or imposed (by transgenesis) variations in the isoform population of the myofilament proteins. We focus on troponin I, which is not only a key protein in myofilament Ca^{2+} -signaling, but also undergoes a complete shift from the slow skeletal isoform (ssTnI) to the adult cTnI isoform during cardiac development.

5. ISOFORM SWITCHING OF TNI IN HEART MUSCLE ALTERS MYOFILAMENT RESPONSE TO CA^{2+} AND ITS MODULATION BY pH

Early evidence that the state of the thin filament proteins other than troponin C may determine the response to Ca^{2+} came from studies comparing activity of cardiac myofilaments at different developmental stages, and from comparisons of myofilaments from fast, slow, and cardiac muscle. Donaldson and Hermanson (1978) reported that Ca^{2+}-sensitivity and pH induced shifts in the Ca^{2+} -force relation of skinned fiber bundles was different among these muscle types. Comparisons between slow and cardiac muscle myofilaments were particularly interesting inasmuch as both of these muscle types express the same isoform of TnC. Moreover, Solaro et al (1986) reported that myofilaments from neonatal and adult hearts also demonstrated a difference in Ca^{2+}-sensitivity and response to acidic pH. This study also presented evidence that these differences are not due to shifts in isoforms of TnC or myosin heavy or light chains. Although Solaro et al. (1986) noted that adult cardiac TnI was missing in neonatal myofilaments analyzed by polyacrylamide gel electrophoresis, the identification of the neonatal TnI as slow skeletal TnI (ssTnI) was made by Saggin et al. (1989). Subsequently, Martin et al. (1991) reported a close correlation between pH sensitivity and expression of ssTnI in developing rat heart myofilaments. The apparent difference in pH modulation of myofilament activity by isoform shifts of thin filament proteins was particularly significant in that there was evidence that the deactivation of contraction in intact muscle by acidic pH was largely due to a desensitization of the myofilaments to Ca^{2+} rather than an impairment in Ca^{2+} delivery to the myofilaments (Allen and

Orchard, 1983). This finding offered the opportunity to test whether deactivation by acidic pH is different in neonatal versus adult heart cells. We (Solaro et al. 1988) tested this hypothesis by comparing Ca^{2+}-transients, measured using the photo protein aequorin as a Ca^{2+} indicator in muscle preparations from adult and one-day old neonatal rat hearts. Results of these studies showed that the peak amplitude of Ca^{2+}-transients in both adult and neonatal heart cells were either unaffected or actually increased upon a reduction in pH by hypercapnic acidosis. On the other hand the early phase of the fall in peak isometric twitch tension, in response to high CO_2 in the perfusion medium was significantly blunted in the neonatal preparations compared to the adult. These data were the first to clearly point to the possibility that a specific change in a thin filament protein could translate to a change in tension in intact heart cells. Beyond this, the data indicated that modifications of the adult isoform of TnI might be a significant factor determining the inotropic state.

To test the functional significance of specific changes in the TnI isoform population on cardiac function, we generated a transgenic mouse (ssTnI-TG) in which cTnI was completely replaced with ssTnI (Fentzke et al. 1999). The top panel of Figure 7 shows results of PAGE analysis of myofilaments from control and TG hearts. Note the absence of cTnI in the TG myofilaments, and the similar stoichiometry of the other myofilament proteins in the control and TG preparations. The bottom panel shows that the myofilaments from hearts of the TG-ssTnI are more sensitive to Ca^{2+} than the controls. Moreover, whereas the wild-type myofilaments are desensitized to Ca^{2+} following PKA-dependent phosphorylation, the TG myofilaments were unaffected by PKA. We have also compared effects of a drop from pH 7.0 to pH 6.5 on Ca^{2+}-activated tension generated by detergent-extracted (skinned) fiber bundles dissected from papillary muscles of control and TG mouse hearts (Wolska et al. 2001). Skinned fiber bundles from hearts of non-transgenic controls and TG-cTnI (which expressed the native cardiac isoform) mice had identical properties (Fig. 8). In both these control heart myofilaments, the relation between Ca^{2+} and tension was significantly and identically right shifted with a drop from pH 7.0 to pH 6.5 (Fig. 9). Compared to controls, the TG-ssTnI fibers, demonstrated increased Ca^{2+} sensitivity at pH 7.0 and a reduced rightward shift of the Ca^{2+}-tension relation with a change from pH 7.0 to pH6.5 (Figure 9, top panel). Moreover, as shown in the bottom panel of Figure 9, twitch tension of the TG-ssTnI papillary muscle preparations fully recovered from a transient fall in tension as CO_2 was introduced into the bath, whereas the control heart preparations had a sustained and significant depression in tension. These results clearly showed the significance of specific modifications in TnI in altered cardiac function induced by acidic pH. We speculated that relative resistance to acidic pH may be of benefit to embryonic/neonatal heart cells, which operate in a relatively hypoxic environment and with a relatively low abundance of Ca^{2+} transport membranes (Solaro et al. 1988; Wolska et al. 2001).

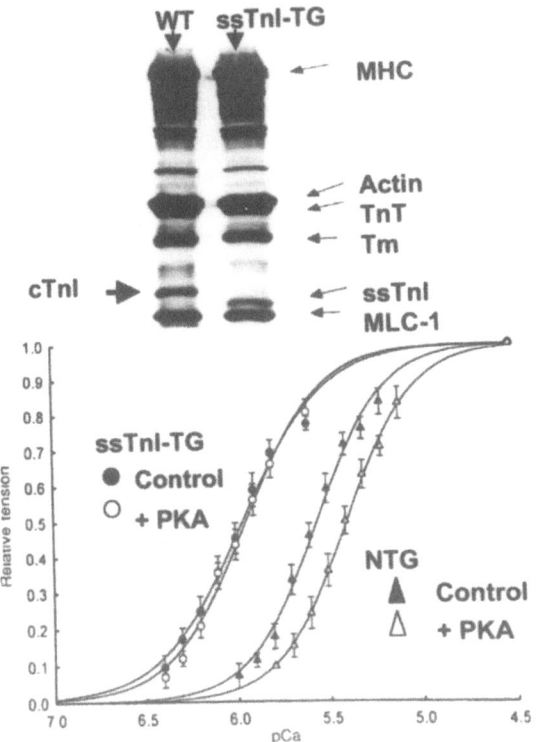

Figure 7. Relation between Ca²⁺ and tension of detergent extracted myocytes from hearts of wild type controls and hearts in which cTnI has been replaced with ssTnI by transgenesis. The top panels shows SDS-PAGE analysis of isolated myofilaments and demonstrates complete replacement of cTnI with ssTnI. Note that the ssTnI containing myofilaments are more sensitive to Ca²⁺ and unaffected by PKA dependent phosphorylation. The inset shows that cTnI is completely replaced with ssTnI in this transgenic model. Data redrawn from Fentzke et al. 2000.

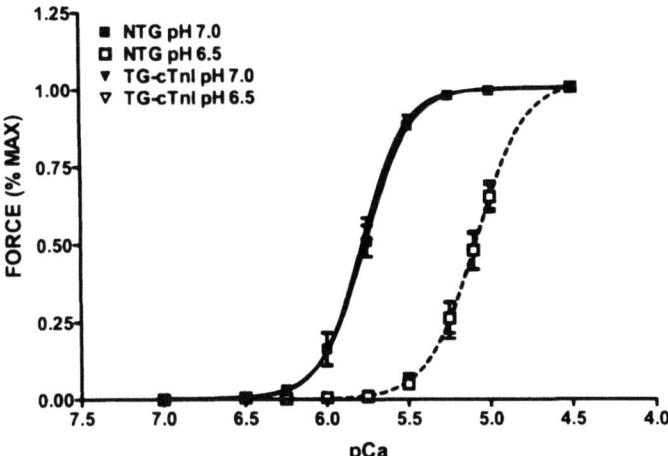

Figure 8. Relation between Ca²⁺ and tension of detergent extracted fiber bundles from hearts of wild type controls and hearts harboring a transgene expressing cTnI and in which native cTnI has been replaced with cTnI by transgenesis. The data illustrate that expression of cTnI in the transgenic model does not affect relations between Ca²⁺ and tension in the detergent treated fiber bundles at neutral and acidic pH.

We (Li et al. 2001) and others (Westfall and Metzger, 2001) have addressed the question of the localization of the domain variations between cTnI and ssTnI that might account for the differences in effects of acidic pH on myofilament function. These studies generated chimeras of cTnI, ssTnI, and fast skeletal TnI in which the regions comprised of the N-terminal domain, the Ip, and the C-terminal domain were varied. Results of these studies support the hypothesis that differences in charged amino acids in C-terminal regions of cTnI outside the Ip are the main determinants of the differential effects of acidic pH on cardiac and skeletal myofilaments. Figure 5 compares the primary structures of ssTnI and cTnI and highlights a number of differences in charged amino acids that could account for the importance of the C-terminal region in response to pH changes. In previous studies we (El-Saleh and Solaro, 1988) reported that when pH is reduced from 7.0 to 6.5 a region surrounding Cys133 of fsTnI, which links the actin-Tm and TnC binding regions, undergoes a conformational change. Cys 133 of fsTnI corresponds to Ser

135 in ssTnI and Ser 166 in cTnI. Flanking this region is a number of potentially significant amino acid differences. One residue, in a region critical for interactions with TnC, is Gln 157 of cTnI, which is Lys 124 in fsTnI and Arg 124 in ssTnI.

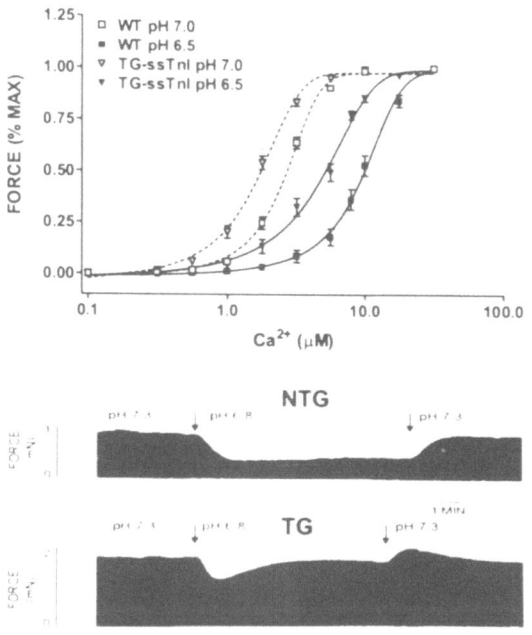

Figure 9.   Data demonstrating that specific replacement of cTnI with ssTnI in heart myofilaments blunts the effect of acidosis on heart myofilaments.   Top panel. Ca²⁺-tension relation in detergent extracted fiber bundles at pH 7.0 and pH 6.5.   Bottom panel. Twitch tension of papillary muscle preparations before and after hypercapnic acidosis.   Data redrawn from Wolska et al. 2001.

Other significant charge differences exist at cTnI-Ala$_{164}$ (His in ssTnI and fsTnI), cTnI-Glu$_{166}$ (Val in ssTnI and fsTnI), and cTnI-His$_{172}$ (Asn in ssTnI and fsTnI). Despite identity of these amino acids in fsTnI and ssTnI, there remains a significant difference in deactivation of fast skeletal and slow skeletal myofilaments by acidic pH (Morimoto et al. 1999; Li et al. 2001).

One possibly significant difference in a charged amino acids between fsTnI and ssTnI is at a location homologous with Arg$_{186}$ in cTnI, which is Pro in ssTnI and Asp in fsTnI. Arg$_{186}$ also flanks a region of cTnI (residues 193-198) that appears critical in regulation of myofilament activation. We localized this as a critical region on the basis of our experiments (Rarick et al. 1997) showing that deletion of cTnI amino acids 198-210 had no effect on myofilament Ca^{2+} regulation, whereas truncation of amino acids 193-210 has a significant effect of myofilament activation (Murphy et al. 2000). The region of fsTnI and ssTnI homologous to cTnI$_{193-198}$ also have Ala/Glu and Met/Lys substitutions that appear likely to be important in explaining the differences between effects of pH on Ca^{2+} activation of fast skeletal and slow skeletal myofilaments. Definitive experiments involving point mutations to test these ideas have not been carried out. It will also be important to test whether charge differences between ssTnI and cTnI in the three amino acids upstream of the Ip may account for functional differences. These charge differences are at Arg92, Lys 95, and Lys 98 of ssTnI, which in cTnI are Thr 123, Ala 126, and Thr 129. As will be discussed below, there is evidence that this region of cTnI may be important in activation of the myofilaments by strong crossbridges.

6. ISOFORM SWITCHING OF TNI IN HEART MUSCLE ALTERS LENGTH DEPENDENT ACTIVATION.

We (Arteaga et al. 2000) have also reported evidence that expression of the embryonic/neonatal isoform, ssTnI, in adult hearts alters length-dependent activation of the myofilaments. Previous studies had indicated that length dependent activation might be modified by alterations in the isoform population of TnT (Akella et al., 1995) or by phosphorylation of myofilament proteins (Komukai and Kurihara, 1997). Our data extended these findings by the demonstration that a specific exchange of cTnI with ssTnI in the thin filament induced a reduction in length dependent activation. This blunting of length dependent activation that occurred with TnI isoform switching is not an inevitable consequence of the associated increase in Ca^{2+}-sensitivity. In a previous investigation (Wolska et al., 1999), we found skinned fibers from a TG-∃-Tm heart, in which more than half the native ∀-Tm was replaced with ∃-Tm, were more sensitive to Ca^{2+} than controls containing ∀-Tm. Yet the sensitization of the cardiac myofilaments to Ca^{2+} and a reduction in maximum tension did not alter length dependent activation.

How does a specific change in the TnI isoform population alter length dependence of myofilament activation? Current theories regarding the mechanism of length dependent activation have at least two components. These are that variations in strong crossbridge binding that occur with a change in length either increase the affinity of cTnC for Ca^{2+} (Allen and Kentish, 1985; Hofmann & Fuchs, 1988) or alter spread of activation to near neighbor functional units (Fitzsimmons & Moss, 1998; McDonald & Moss, 1995; McDonald *et al.*, 1997). Convincing evidence (reviewed in Solaro and Van Eyk, 1996) indicates that length dependence of activation in fast skeletal muscle may occur with no change in Ca^{2+} binding to TnC (Fuchs and Wang, 1991; Wang and Fuchs, 1994). Thus, we favor a scheme in which, apart from the mechanism determining the relation between sarcomere length and strong crossbridge binding, variation in the ability of strong crossbridge to bring near neighbors to the open state is a critical mechanism. In support of this idea, length dependence of myofilament activation has been significantly attenuated experimentally by artificially promoting crossbridge binding either with Ca^{2+} sensitizing agents (Arteaga et al. 2000), MgADP (Fukuda et al. 2000), or titration with strongly binding (NEM-S1) myosin heads (Fitzsimmons and Moss 1998). These data indicate that mechanisms by which a change in thin filament proteins may modulate length dependent activation are: 1) by shifts in the steady-state relation between the blocked, closed, and open state of the myofilaments, and 2) by modulating the interaction energy among functional units. As suggested by Lehrer (1994), the number of crossbridges in the closed state may be modulated by the state of TnI. Moreover, TnI and TnT, both of which are asymmetric molecules extending across 2-3 actins, could in fact contribute significantly to steric blocking of the actin-crossbridge reaction monomers (Squire and Morris, 1998; Solaro and Rarick, 1998). Although it is not known whether their extended molecular length is different, there are substantial structural differences between cTnI and ssTnI. One of the most prominent structural differences between cTnI and ssTnI is that the cardiac variant possesses an additional mass of 32 amino acids forming an N-terminal extension not present in either slow skeletal or fast skeletal variants. This N-terminal extension of cTnI contains serial Ser residues that are sites of PKA dependent phosphorylation (Solaro and Rarick, 1998). Importantly, phosphorylation of these sites by activation of PKA-dependent signaling cascades *in situ* and *in vitro* have been reported to alter length dependent activation of cardiac muscle preparations (Komukai and Kurihara, 1997; Konhilas et al. 2000). When these sites are phosphorylated *in vitro* in the binary cTnI-cTnC complex, there are global changes in TnI structure that may alter the ability of TnI to affect the closed state (Dong et al. 1997). Phosphorylation of the sites also reduces affinity of TnC for Ca^{2+} and could also affect length dependence of activation by attenuating the ability of strong crossbridges to influence Ca^{2+} binding to TnC.

Alterations in cTnI either by covalent modifications or by isoform switching may influence length-dependent activation by altering interaction energies between cTnC and the crossbridge binding sites on actin. There is evidence that the state of TnC can influence crossbridge kinetics and that the crossbridge-actin interaction feeds back to affect the state of cTnC. Regnier et al. (1996) reported that calmidazolium (CDZ), an agent that binds to cTnC (El-Saleh and Solaro, 1987), not only increases myofilament Ca^{2+}-sensitivity, but also increases transition of crossbridges to the force generating state. CDZ did not directly affect crossbridge kinetics. Moreover, by measuring linear dichroism of a fluorescent probe attached to cTnC, Martyn and Gordon (2001) demonstrated effects of changes in sarcomere length on cTnC structure in cardiac myofilaments. However, sarcomere length dependent shifts in fsTnC structure did not occur in fast skeletal skinned fibers. The transmission of these signals between crossbridges and cTnC is likely to involve cTnI

One important aspect of these experimental approaches toward modification is that they provide a means of testing the significance of length dependent activation as an important determinant of the Vol-ESP relation. For example, we would predict that with blunting of length dependent activation, the Vol-ESP would be dominated by changes in filament overlap and/or altered Ca^{2+}-release. A preliminary test of this hypothesis was carried out in experiments comparing the Vol-ESP relation. We compared the Vol-ESP relation in wild type hearts beating *in situ* with hearts of TG-ssTnI mice following beta-adrenergic stimulation. This approach provided two extremes that would aid in detecting a difference in the slope of the Vol-ESP relation. Our previous studies (Arteaga et al. 2000) predicted that the length dependence of activation would be suppressed in the TG-ssTnI and unaffected by beta-adrenergic stimulation, whereas in the wild-type heart length-dependence of activation would be amplified. Indeed our data indicated a reduction in the slope of the Vol-ESP relation, a change predicted from the change in length dependence of activation as indicated in Figure 2 (Novak et al. Unpublished Observations).

Interestingly, Fukuda et al. (2001) reported that acidosis enhanced length dependence of tension in rat skinned cardiac muscle preparations. This results indicates that at acidic pH the ability of strong crossbridges to activate the thin filament is impaired. In related findings, Morimoto et al. (2001) reported that the effect of acidic pH to depress crossbridge dependent thin filament activation at pCa 9.0 was attenuated when cTnI-cTnT-cTnC was exchanged with ssTnI-cTnT-cTnC in skinned fiber bundles from rat heart. In these experiments the population of strong crossbridges reacting with the thin filament was varied by reducing the concentration of MgATP in the bathing medium of skinned fiber bundles at pCa 9.0. Morimoto et al. (2001) found that whereas strong crossbridges could activate myofilaments containing either cTnI-cTnT or ssTnI-cTnT, but lacking cTnC, the activation was the same at pH 7.0 and pH 6.2. These results indicate: 1) that strong crossbridges can reverse the inhibition of the thin filament by TnI, presumably by releasing the cTnI Ip and C-terminal binding sites from actin in a mechanism that is unaffected by a change from pH 7.0 to pH 6.2. and 2) that activation of the thin filament by strong crossbridges depends on a differential pH dependent interaction of the N-terminal domain of ssTnI and cTnI with the C-terminal lobe of cTnC. It remains unknown whether the N-terminal differences in charged residues at Arg 92, Lys 95 and Lys 98 of ssTnI (Thr 123, Ala 126, and Thr 129 of cTnI) are important in this difference.

7. PROTEIN KINASE A DEPENDENT PHOSPHORYLATION OF TROPONIN I
MODULATES RELAXATION KINETICS

Figure 10 summarizes the PKA and PKC phosphorylation sites on cTnI and indicates their location with regard to the other thin filament proteins. Evidence that phosphorylation of cTnI may be an important factor determining the enhanced relaxation during β-adrenergic stimulation of the heart began with the discovery that the N-terminus of cTnI contains a unique stretch of amino acids at its N-terminus (see Solaro, 2001 for review). As shown in Figure 10, this region, which contains consensus sequences for cAMP-dependent protein kinase (PKA) sites at Ser 22 and Ser 23 (Solaro 2001), is situated close to the N-lobe of cTnC. Mutagenesis and phospho-peptide maps such as those shown in Figure 10, demonstrated that the Ser

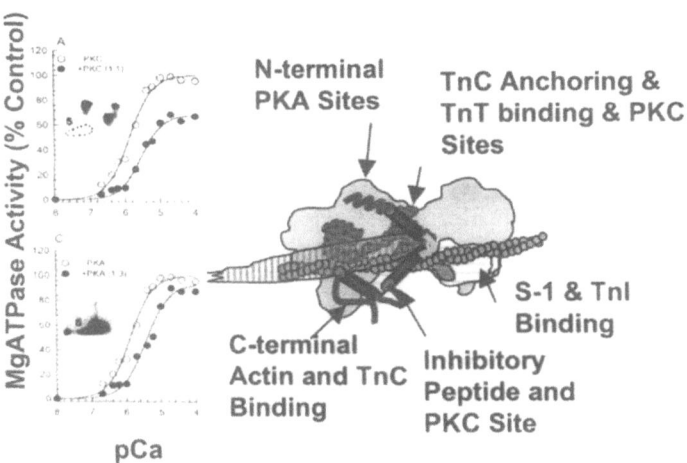

Figure 10. Location of PKA and PKC sites on cTnI and dissection of effects on Ca²⁺-dependence of actin-heavy-meromyosin MgATPase activity of preparations of thin filaments reconstituted with cTnI phosphorylated by PKA or PKC. Data on the left side of the figure are reproduced from Noland et al. 1995, and indicate that PKC dependent phosphorylation depresses maximum ATPase acticity, whereas PKA dependent phosphorylation depresses Ca²⁺ sensitivity. Insets in the figures depict phospho-peptide mapping of the cTnI used in reconstitution. The spot labeled "5" arises from the N-terminal PKA sites. See text for details.

residues in the unique N-terminal extension of cTnI are the only sites phosphorylated by PKA (Noland et al. 1995). *In vitro* studies, such as that shown in the left panel of Figure 10, demonstrated that the phosphorylation of these sites is responsible for the desensitization of the myofilaments to Ca^{2+} with no effect on maximum tension or ATPase rate. These studies also showed that PKA-dependent phosphorylation of cTnI reduced the affinity of cTnI for cTnC, and reduced the affinity of cTnC for Ca^{2+}. Pre-steady state measurements showed that phosphorylation of cTnI increased the off rate for Ca^{2+} exchange. Early studies also showed that cTnI is phosphorylated in isolated and *in situ* beating hearts in association with Ǝ-adrenergic stimulation. *In vitro* effects clearly indicated that phosphorylation of TnI was more likely to be important in the relaxant effect rather than the positive inotropic effects of β-adrenergic stimulation. To test this hypothesis was however difficult in that cTnI is not the only substrate for PKA in the cardiac myocyte. With β-adrenergic stimulation it is well known that membrane proteins regulating Ca^{2+} fluxes are phosphorylated. Phospholamban appears to be the most prominent among these proteins as a determinant of relaxation rate. Moreover, in addition to cTnI, cardiac myofilaments contain sites of PKA-dependent phosphorylation on myosin binding protein C (MyBPC) and titin. To determine the specific role of cTnI phosphorylation in cardiac function, we investigated transgenic mice (ssTnI-TG) expressing slow skeletal TnI (ssTnI) in the heart. (Fentzke et al. 1999). ssTnI is not a substrate for PKA and lacks the N-terminal domain unique to cTnI. Studies comparing *in situ* function of ssTnI-TG hearts demonstrated a significant blunting of the relaxant effect of Ǝ-adrenergic stimulation. Measurements of levels of protein phosphorylation in myocytes showed that during β-adrenergic stimulation, there were essentially normal levels of phospholamban and MyBP-C phosphorylation. Studies on Ca^{2+} uptake rates of vesicles of the SR showed no differences between preparations from ssTnI-TG hearts and controls. Taken together these data derived from studies on the ssTnI-TG mouse hearts provide direct evidence that the state of TnI is an important determinant of relaxation kinetics. Studies with mutant mice (PKB-KO) hearts in which phospholamban was ablated also support the conclusion that myofilament phosphorylation is an important determinant of the relaxant effect of Ǝ-adrenergic stimulation. Relaxation of isometric tension of myocardial preparations was significantly hastened even in the absence of phospholamban (Li et al. 2000). Interestingly the relaxant effect does not appear to be present in the free floating myocyte contracting with zero load. Under these conditions crossbridge cycling rate may be near maximum and unaffected by cTnI phosphorylation.

General mechanisms by which phosphorylation of cTnI could translate into an effect on cardiac function include an increase in Ca^{2+} release from cTnC and an increase in crossbridge cycling rate. An effect of cTnI phosphorylation on crossbridge cycling rate is appealing as evidence indicates that the rate of crossbridge detachment, for example, may be a more important factor in relaxation that removal of Ca^{2+} release from cTnC (Palmer and Kentish, 1998; Gordon et al. 2000). To test the effect of β-adrenergic stimulation, we (Kentish et al. 2001) assessed the kinetics of cross-bridge cycling by measuring f_{min} (the frequency for minimum dynamic stiffness) in muscles during tetanic contractions. Associated with Ǝ-adrenergic stimulation with isoproterenol, f_{min} increased from 1.9 to 3.1 s^{-1} in muscles from wild-type mice, but was without effect in muscles from transgenic mice. We also compared the intrinsic rate of relaxation of control and ssTnI-TG myofilaments with and without pre-incubation with PKA. To rapidly decrease the Ca^{2+} within skinned cardiac muscle preparation, we used flash photolysis of diazo-2, a caged compound that releases a Ca^{2+}-chelator. The intrinsic rate of myofibrillar relaxation nearly doubled with PKA incubation of the control preparation but did not change significantly in preparations from ssTnI-TG mice. These data strongly suggested that the acceleration of myofibrillar relaxation rate by PKA is due to phosphorylation of TnI, rather than MyBP-C, and that this is due, at least in part, to faster cross-bridge cycling kinetics. In general agreement with these data, Strang et al. (1994) reported an increase in unloaded shortening velocity of rat heart myocytes treated stimulated with a beta-adrenergic agonist before skinning. Saeki et al (1990) showed that stimulation of rat heart preparations with adrenaline also increased rate of cross bridge cycling. However, there is no clear agreement as to whether phosphorylation of the myofilament by PKA alters crossbridge cycling rate. Janssen et al (1997) reported no change in unloaded shortening velocity in skinned rat trabeculae treated with PKA, and Hoffman et al (1995) could find no change in unloaded shortening velocity of isolated myocytes from rat hearts.

8. PROTEIN KINASE C DEPENDENT PHOSPHORYLATION OF TROPONIN I MODULATES MYOFILAMENT CA²⁺-SENSITIVITY AND MAXIMUM NUMBER OF ACTIN-MYOSIN INTERACTIONS

Although both cTnI and cTnT are substrates for phosphorylation by protein kinase C (PKC), we will focus on cTnI phosphorylation, in which some clearer connections have been made to cardiac function (see Solaro 2001 for review).   cTnI is phosphorylated at Ser 43 and Ser 45 in the N-terminal region that interacts with the C-lobe of cTnC and the C-terminal region of cTnT. Thr 144, located in the Ip of cTnI is also a substrate for PKC. Figure 10 shows the location of these sites in relation to the disposition of the thin filament proteins. Ser 43 and Ser 45 are located in what we have described as a pivot for movements of cTnI, cTnC, and cTnT. Investigations employing reconstituted thin filament-myosin S-1 preparations containing either unphosphorylated or fully phosphorylated cTnI (Katoh et al. 1983; Noland et al. 1989; Noland and Kuo, 1993; Noland et al. 1995) demonstrated that PKC dependent phosphorylation of TnI inhibits the actin-crossbridge reaction. The left panel of Figure 10 shows this depression in maximum ATPase activity for such preparations. The mechanism of the inhibitory effects of the phosphorylation of cTnI by PKC appears to involve promotion of the blocked state as described in Figures 2 and 3. It is apparent that PKC-dependent phosphorylation induces increased competition of cTnI with myosin S-1 for binding to actin. Noland and Kuo (1993) reported that although phosphorylation of TnI always induced a decrease in the Ca^{2+}-dependent ATPase rate of reconstituted actomyosin, the inhibition was partially overcome as the concentration of myosin or S-1 increased. This was not the observation in the case of phosphorylated TnT in which inhibition was the same at all concentrations of myosin or S-1.   As discussed above, phosphorylation of cTnI at Ser 43 and 45 may also affect interactions of the N-terminal region of cTnI with the C-lobe of cTnC and act to inhibit strong crossbridge connections. This idea is based on evidence that the binding site of EMD 57003 is located at the C-lobe of cTnC and EMD is displaced by an N-terminal cTnI peptide containing Ser 43 and 45.

Mutational analysis of the sites of PKC dependent phosphorylation on cTnI led us to the conclusion that phosphorylation of Ser 43 and Ser 45 ismore prominent than phosphorylation of Thr 144 in eliciting the inhibitory effect on actomyosin ATPase activity and apparent affinity of myosin S-1 for reconstituted (regulated) actin (Noland et al 1995). Moreover, when both Ser-23/Ser-24 and Ser-42/Ser-44 were phosphorylated, there was an additive effect at sub-maximal Ca^{2+} concentration resulting in a severe depression in ATPase activity as a result of a combination of a decreased in Ca^{2+}-sensitivity and Vmax. We (Noland et al. 1996) also reported site specificity of PKC isoenzymes resulting in different alterations in activation of the reconstituted actomyosin. Both free cTnI and cTnI in the Tn complex exhibited discrete specificity for phosphorylation by the PKC isozymes-alpha and delta. PKC-delta, but not PKC-alpha that had a propensity to phosphorylate the consensus PKA sites (Ser-23/Ser-24), resulting in a reduced Ca^{2+}-sensitivity of the reconstituted actomyosin S-1 MgATPase. Thus, PKC-delta appears to function as a hybrid of PKC-alpha and PKA. On the other hand, PKC-alpha preferred to phosphorylate Ser-42/Ser-44, and thus reduced the maximal Ca^{2+}-stimulated activity of the MgATPase.

We (Noland et al. 1996) phosphorylated wild type and mutant forms of cTnI by the PKC isozymes alpha and delta and by PKA in experiments investigating the regulation of Ca^{2+}-stimulated MgATPase activity of reconstituted actomyosin S-1 PKC-delta was able to phosphorylate both the PKC and PKA sites on cTnI, where as PKC-alpha (which gave results similar to PKC-epsilon) did not phosphorylated the PKA sites. As predicted by this finding, N32 (a mutant missing the N-termnal PKA sites on cTnI) was a much poorer substrate for PKC-delta than for PKC-alpha. These results provided the first evidence for site-selective preferences of PKC-alpha and PKC-delta for phosphorylating a single substrate (cTnI) in the myocardium that may lead to distinct functional consequences. Jideama et al. (1996) have shown further evidence that the PKC isozymes alpha, delta, epsilon, and zeta, which are expressed in adult rat cardiomyocytes, show specificity for the phosphorylation of cTnI and cTnT. Bovine cTnI was the preferred substrate for PKC-alpha, -delta, and -epsilon when compared to cTnT, whereas the opposite was true for PKC-zeta.

Whatever the path of phosphorylation, our results indicate that Ser-42, Ser-44 are the sites largely responsible for the inhibitory effect of TnI PKC-phosphorylation on actomyosin ATPase activity, and furthermore, that phosphorylation of Thr-144 by PKC is relatively unimportant in modulating the activity of the Ip. Although earlier work (Noland and Kuo, 1991) had indicated that Thr 144 in cTnI is important for the observed decrease of Ca^{2+}-stimulated actomyosin MgATPase activity, results of studies (Noland et al. 1995,1996) with site directed mutants provide strong evidence that this may not be the case. A different conclusion was reached by Malhotra et al. (1997) in a study of regulated actomyosin preparations containing S22A/23A and

TI44A. Although they agree with the work of Noland et al. (1995,1996) that PKA sites are exclusively phosphorylated by PKA with a resulting decrease in Ca^{2+} sensitivity, they indicate a more prominent role of Thr-144 phosphorylation than for Ser-42/Ser-44. In contrast to the data reported by Noland et al. (1995), the curve relating Ca^{2+} to the ATPase activity was substantially shifted to the right, when wild type TnI, which was phosphorylated by PKC, was employed for regulation. Moreover, this shift was markedly attenuated when T144A was incorporated into the preparations. Although not analyzed a residual reduction in maximum ATPase activity was still seen in these preparations containing the T144A mutant. It is apparent that this could be due to phosphorylation of Ser 42/Ser44.

Physiological effects of PKC dependent phosphorylation of myofilaments have been difficult to demonstrate clearly. Moreover, there are conflicting data. This may not be surprising inasmuch as the role of multiple isoforms of PKC and targeting and compartmentation of PKC on RACKS (Pucéat and Vassort, 1996), introduce complexities to the system. . In addition, there are at least four major sarcomeric substrates for PKC including cTnI, cTnT, MyBP-C and MLC-2 (Solaro, 2001). In the case of cTnI and cTnT, there are multiple sites of phosphorylation and interactions among the Tn components appear to affect phosphorylation by PKC. For example, we (Montgomery et al. 2001) found that phorbol ester induced phosphorylation of cTnI in myofilaments from hearts of transgenic mice expressing fast skeletal TnT (which lack PKC phosphorylation sites) was reduced compared to non-transgenic controls. Interactions among the phosphorylation sites may be expected as the PKC sites are clustered on cTnI and cTnT at regions of interactions between the two proteins. In some cases phosphorylation of these sarcomeric proteins have offsetting effects. Phosphorylation of MLC2 and MyBP-C increase myofilament Ca^{2+} sensitivity with no effect on maximum activity, whereas phosphorylation of cTnI and cTnT decrease maximum activity and decrease Ca^{2+}-sensitivity. These differential effects may underlie data demonstrating both negative (Capogrossi et al. 1990; Gwathmey and Hajjar, 1990; Johnson and Mochly-Rosen, 1995) and positive inotropic effects (Terzic et al. 1992) with activation of the PKC pathway. Unfortunately these studies did not correlate activity with specific changes in phosphorylation of cellular proteins. There are no studies directly comparing inotropic state of cardiac myocytes to levels of phosphorylation of specific PKC sites. However, Talosi and Kranias (1992) reported no change in cTnI phosphorylation in rabbit hearts perfused with phenylephrine concentrations that were sufficient to increase +dp/dt and redistribute PKC. Whether this negative result is peculiar to the rabbit is not clear, but it may also be related to the difficulty of ^{32}P labeling of the ATP pool to high specific activity in isolated perfused hearts. Venema and Kuo (1993) were able to demonstrate in rat heart myocytes a relatively specific increase in cTnI phosphorylation at Ser 43 and Ser 45 following treatment with phenylephrine or phorbol esters. Myofibrillar preparations isolated from these myocytes also showed a depressed maximum ATPase rate compared to the controls. In view of evidence indicating that Ser 43 and Ser 45 are important substrates for PKC, we generated a transgenic mouse in with approximately 50% of the native cTnI was replaced with cTnI in which Ser 43 and Ser 45 were mutated to Ala (MacGowan et al. 2001). Ca^{2+}-transients and ventricular pressures were measured simultaneously in isolated perfused hearts. Activation of the PKC pathway in the transgenic hearts by increasing perfusate Ca^{2+} concentration resulted in a significant alteration in the Ca^{2+}-pressure relation and susceptibility to ischemic contracture. These data indicated that Ser 43 and Ser 45 may be phosphorylated in the non-transgenic hearts. Papillary muscles from the transgenic mouse hearts also showed a blunted response to phenylephrine as well as blunted depression of maximum tension following treatment with a phorbol ester.

9. MYOFILAMENT RELATED PROCESSES ARE INTEGRAL ELEMENTS IN SIGNALING OF CARDIAC HYPERTROPHY.

A hypothesis guiding our investigations is that modulation of myofilament activity is an important element in signaling pathways leading to compensatory hypertrophy and potentially to decompensation. Figure 11 illustrates a minimal model of such a signaling process. Hypertension exemplifies a stress extrinsic to the heart that promotes myocyte hypertrophy. With the increase in afterload, there is increased likelihood that cardiac output regulation occurs with relative elevation in end diastolic volume. There is ample evidence that the associated stretch and increased strain on the myocyte leads to release of peptides, angiotension II (Ang II) and endothelin-1 (ET-1). The peptides bind to G protein coupled receptors (GPCR), activate G proteins and phospholipases. The result is the release of signaling molecules IP3 and DAG. Increased IP3 may release a special pool of cellular Ca^{2+} targeted to the activation of calmodulin dependent processes. Either or both release of DAG and Ca^{2+} are important in activation of PKC. Factors that regulate transcription are in turn controlled by Ca^{2+}-calmodulin and PKC dependent processes. There is also recent evidence that Ang II increases protein synthesis by increasing translation in a process involving Ca^{2+}-calmodulin dependent phosphorylation of elongation factor-2 (Everett et al. 2001). In the case of Ca^{2+}-calmodulin dependent regulation of the phosphatase, calcineurin, dephosphorylation of the transcription factor, N-FAT, has been demonstrated to be an important determinant of its translocation into the nucleus (De Windt et al. 2001). There is also evidence for Ca^{2+}-calmodulin dependent translocation of the transcription factor, MEF-2 (Nicol et al. 2000). Although a variety of PKC isoforms are potentially important for the hypertrophic process, PKC beta appears to be especially important in hypertrophy. Bowling et al. (1999) reported that samples of hearts with dilated or ischemic myopathies had increased activity and membrane expression of PKC-alpha, and PKC Beta I and II. Bowman et al. (1997) investigated effects of activation of the PKC beta pathway using a mouse in which a tetracycline dependent promoter in the transgene controlled expression of PKC beta.

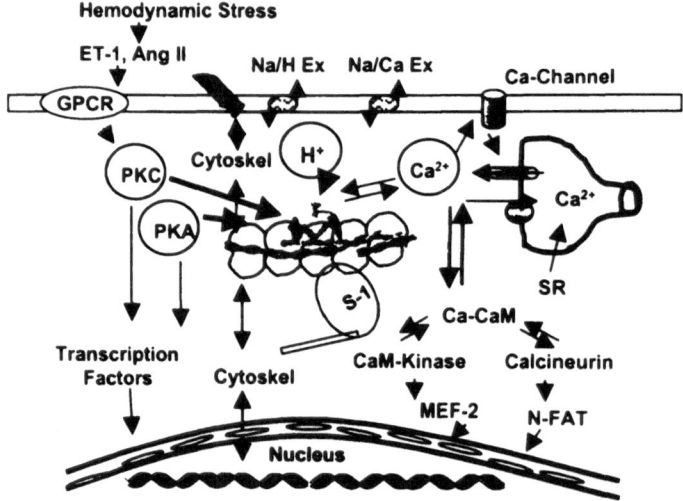

Figure 11. Integration of cardiac myofilaments in the flow of signals for hypertrophy and failure. A prevailing hemodynamic stress such as hypertension leads to myocardial stretch and release of peptides that bind to G protein coupled receptors (GPCR). The transduction of this signal promotes activation of kinases that not only increase protein synthesis but affect myofilament activity and Ca^{2+} buffering. See text for further discussion.

This conditional expression in the adult resulted in progressive hypertrophy and impaired relaxation of the hearts of the transgenic mice. Takeishi et al. (1998) reported that a transgenic mouse model over expressing high levels of PKC beta II had heart cells that demonstrated a decrease in contraction with no change in the Ca^{2+}-transient, and an increase in cTnI phosphorylation. Superfusion of these cardiac myocytes with a specific inhibitor of PKC beta II reversed these effects. D'Angelo et al. (1997) and Sakata et al. (1998) investigated a transgenic mouse model over expressing G alpha-q and also reported increased phosphorylation of cTnI associated with the hypertrophy. Using the same model studied by Bowman et al. (1997), we (Huang et al. 2001) determined effects of conditional expression of PKC beta on myocyte and myofilament function. These experiments confirmed that expression of PKC beta at relative low levels was able to induce hypertrophy. However, myocytes isolated early on the hypertrophic process had a relative increase in contractility associated with an increase in the amplitude of intracellular Ca^{2+} transients. The source of this increase is most likely due to an effect of PKC on L-type Ca^{2+} channels. However, we cannot rule out the possible activation of the calcineurin-dependent pathway, which has also been demonstrated to enhance Ca^{2+} channel currents in a transgenic model of cardiac specific expression of activated calcineurin (Yatani et al. 2001). Cingolani et al. (1998) also reported that stretch of myocardium is associated with a change in intracellular pH. The mechanism for this change in pH apparently involves PKC dependent alteration in the activity of the Na/H exchanger. Changes in the activity of the Na/H exchanger might indirectly affect intracellular Ca^{2+} by altering intracellular Na concentrations and therefore the activity of the Na/Ca exchanger.

It is now clear that the hypertrophic signals are sensed and possibly modified by the myofilaments (De Tombe and Solaro, 2001). For, example a change in intracellular pH associated with myocyte stretch strongly affects the Ca^{2+}-sensitivity of the myofilaments, and would be expected to alter myofilament Ca^{2+}-buffering. The change in sarcomere length associated with stretch would also be expected to alter myofilament Ca^{2+}-sensitivity and Ca^{2+}-buffering. Phosphorylation of the myofilament proteins that occurs in association with effects on cascades of kinases and phosphatases during hypertrophy adds to the complexity of signaling. As discussed above, phosphorylation of the myofilaments, modifies Ca^{2+}-sensitivity and Ca^{2+}-buffering, maximum tension, and length-dependent activation. In fact, there is excellent evidence that changes in myofilament phosphorylation occur in cardiac hypertrophy. Compared to controls, an increase in myofilament Ca^{2+}-sensitivity has been reported in skinned myocytes isolated from explanted failed human hearts. The increase in Ca^{2+}-sensitivity is thought to occur in association with down regulation of beta-adrenergic receptors and signal transduction, which results in a depressed PKA activation and relative dephosphorylation of cTnI (Wolff et al. 1996; Bartel et al. 1996; Bodor et al. 1997). There is also evidence for increased protein phosphatase activity associated with down regulation of beta-adrenergic receptors (Boknik et al. 2000; 2001). On the other hand, we (Huang et al. 2001) have found increased phosphorylation of cTnI in mouse hearts responding to conditional expression of PKC beta. Myofilaments from these hearts also had reduced maximum tension compared to controls. There was no change in Ca^{2+}-sensitivity indicating no change in phosphorylation of the PKA sites. Yet, the exact sites of cTnI phosphorylation in this study and in other have not been defined. Moreover, there have been no measurements of phosphorylation of PKC sites in myofilament proteins from hearts in end-stage heart failure.

In summary the perception of the hypertrophic process illustrated in Fig. 11 is a complex signaling process in which cascades of hypertrophic signals increase cell size, remodel cell structures, and also affect the activity of pumps, channels and exchangers that alter fluxes of intracellular Ca^{2+}. We think a key element in this process is the alteration of the activity and Ca^{2+} buffering by the myofilaments. For example, PKC dependent phosphorylation of the myofilaments would be expected to lead to a vicious cycle in which signals, which are geared toward compensation of a stress such as hypertension, also lead to depression of tension, which is geared toward decompensation. Moreover, alterations in transmission of force through the cytoskeleton may be important for altered regulation of diverse membrane, cytoplasmic and nuclear processes that determine cell size and function. As the main buffer for Ca^{2+} in the myocyte it is likely that alteration in the myofilaments may affect Ca^{2+} signaling of key calmodulin-dependent mechanisms affecting transcription and translation.

Strong support for the idea that modulation of myofilament activity and Ca^{2+} buffering may be especially important in hypertrophy/failure comes from analyses linking mutations in sarcomeric proteins to hypertrophic and dilated myopathies (Palmiter and Solaro, 1997). In this case the hypertrophic process *begins* with an intrinsic defect in the sarcomere. It seems reasonable to extrapolate this to a stiuation described above in which the myofilament alteration is a consequence of an extrinsic stess. Many of the functional changes associated with the sarcomere in heart failure also occur in sarcomeres containing mutant proteins. For example, a highly penetrant mutation in cardiac myosin heavy chain has been shown to be associated with abnormal diastolic levels of Ca^{2+} that have been proposed to induce the hypertrophic response (Fatkin et al. 2000). Mutations in thin filament proteins, which are also highly penetrant, include a missense mutation in cTnT. In this case, we have reported alterations not only in myofilament sensitivity to Ca^{2+}, but also an altered response to acidic pH and sarcomere length (Chandra et al. 2001; Solaro et al. 2002). It will be important also in these models to determine whether effects of protein phosphorylation are modified by the mutations in the sarcomeric proteins. In any case, the genetic linkage of sarcomeric mutations to hypertrophic and dilated myopathies has provided exciting new insights and directions into the question of the role of modulation of myofilament activity in both long and short-term regulation of the heart.

10. ACKNOWLEDGEMENTS

This work described in this review was supported in part by grants from the NIH NHLBI and AHA.

Departments of Physiology and Biophysics, Pediatrics, Medicine (Cardiology Section) and Program in Cardiovascular Sciences
College of Medicine
University of Illinois at Chicago
835 South Wolcott Avenue
Chicago, IL 60612-7342

Address for Correspondence:
R. John Solaro, Ph.D.
Department of Physiology and Biophysics (M/C 901)
College of Medicine
University of Illinois at Chicago
835 South Wolcott Avenue
Chicago, IL 60612-7342
TEL (312) 996-7620
FAX (312) 996-1414
email: solarorj@uic.edu

11. REFERENCES

Akella, A.B., Ding, X.L., Chen, R., and Gulati, J. (1995) Diminished Ca2+ sensitivity of skinned cardiac muscle contractility coincident with troponin T-band shifts in the diabetic rat. *Circ. Res.* **76**:600-606.

Allen, D.G. and Kentish, J.C. (1985) The cellular basis of the length-tension relation in cardiac muscle. *J. Mol. Cell Cardiol.* **17**:821-840.

Allen, D.G. and Orchard, C.H. (1983) The effects of changes of pH on intracellular calcium transients in mammalian cardiac muscle. *J. Physiol.* **335**:555-67.

Arteaga, G.M., Palmiter, K.A., Leiden, J.M., and Solaro, R.J. (2000) Attenuation of length-dependent activation in myofilaments of transgenic mouse hearts expressing slow skeletal troponin I. *J. Physiol.* (London) **526**:541-549.

Bartel, S., Stein, B., Eschenhagen, T., Mende, U., Neumann, J., Schmitz, W., Krause, E.G., Karczewski, P., and Sholz, H. (1996) Protein phosphorylation in isolated trabeculae from nonfailing and failing human hearts. *Mol. Cell Biochem.* **157**:171-179.

Bers, D.M. (2001) Excitation Contraction Coupling and Cardiac Contractile Force, 2nd Edition, Klluwer Academic Publishers).

Bodor, G.S., Oakely, A.E., Allen, P.D., Crimmins, D.L., Ladenson, J.H., and Anderson, P.A. (1997) Troponin I phosphorylation in the normal and failing adult human heart. *Circulation* **96**:1495-1500.

Bohm, M., Morano, I., Pieske, B., Ruegg, J.C., Wankerl, M., Zimmermann, R., and Erdmann, E. (1991) Contribution of cAMP-phosphodiesterase inhibition and sensitization of the contractile proteins for calcium to the inotropic effect of pimobendan in the failing human myocardium. *Circ. Res.* **68**:689-701.

Boknik, P., Fockenbrock, M., Herzig, S., Knapp, J., Linck, B., Luss, H., Muller, F.U., Muller, T., Schmitz, W., Schroder, F., and Neumann, J. (2000) Protein phosphatase activity is increased in a rat model of long-term beta-adrenergic stimulation. *Naunyn Schmiedebergs Arch Pharmacol* **362**:222-31.

Boknik, P., Heinroth-Hoffmann, I., Kirchhefer, U., Knapp, J., Linck, B., Luss, H., Muller, T., Schmitz, W., Brodde, O., and Neumann, J. (2001) Enhanced protein phosphorylation in hypertensive hypertrophy. *Cardiovasc. Res.* **51**:717-28.

Boknik, P., Khorchidi, S., Bodor, G.S., Huke, S., Knapp, J., Linck, B., Luss, H., Muller, F.U., Schmitz, W., and Neumann J. (2001) Role of protein phosphatases in regulation of cardiac inotropy and relaxation. *Am. J. Physiol. Heart Circ. Physiol.* **280**:H786-94.

Bowling, N., Walsh, R.A., Song, G., Estridge, T., Sandusky, G.E., Fouts, R.L., Mintze, K., Pickard, T., Roden, R., Bristow, M.R., Sabbah, H.N., Mizrahi, J.L., Gromo, G., King, G.L., and Vlahos, C.J. (1999) Increased protein kinase C activity and expression of Ca^{2+}-sensitive isoforms in the failing human heart. *Circulation* **99**:384-91.

Bowman, J.C., Steinberg, S.J., Jiang, T., Geenen, D.L., Fishman, G.I., and Buttrick, P.M. (1997) Expresson of protein kinase C beta in the heart causes hypertrophy in adult mice and sudden death in neonates. *J. Clin. Invest.* **100**:2189-2195.

Brenner, B. and Morano, I. (1990) Effect of myosin lihgt chanin pshosphorylation on isometric force and crossbridge turnover kinetics. *Pflugers Arch* **415**: R73.

Brenner, B., Kraft, T., Yu, L.C., and Chalovich, J.M. (1999) Thin filament activation probed by fluorescence of N-((2-(Iodoacetoxy)ethyl)-N-methyl)amino-7-nitrobenz-2-oxa-1, 3-diazole-labeled troponin I incorporated into skinned fibers of rabbit psoas muscle. *Biophys. J.* **77**:2677-91.

Brenner, B. and Chalovich, J.M. (1999) Kinetics of thin filament activation probed by fluorescence of N-((2-(Iodoacetoxy)ethyl)-N-methyl)amino-7-nitrobenz-2-oxa-1, 3-diazole-labeled troponin I incorporated into skinned fibers of rabbit psoas muscle: implications for regulation of muscle contraction. *Biophys J* **77**:2692-708.

Capogrossi, M.C., Kaku, T., Filburn, C.R., Pelto, D.J., Hansford, R.G., Spurgeon, H.A., and Lakatta, E.G. (1990) Phorbol ester and dioctanoylglycerol stimulate membrane association of protein kinase C and have a negative inotropic effect mediated by changes in cytosolic Ca^{2+} in adult rat cardiac myocytes. *Circ. Res.* **66**:1143-1155.

Chandra, M, Rundell, V.L., Tardiff, J.C., Leinwand, L.A., de Tombe, P.P., and Solaro, R.J. (2001) Ca(2+) activation of myofilaments from transgenic mouse hearts expressing R92Q mutant cardiac troponin T. *Am. J. Physiol. Heart Circ. Physiol.* **280**:H705-13.

Chen, Y., Yan, B., Chalovich, J.M., and Brenner, B. (2001) Theoretical kinetic studies of models for binding myosin subfragment-1 to regulated actin: Hill model versus Geeves model. *Biophys. J.* **80**:2338-49.

Cingolani, H.E., Alvarez, B.V., Ennis, I.L., and Camilion de Hurtado, M.C. (1998) Stretch-induced alkalinization of the feline papillary muscle. An autocrine-paracrine system. *Circ. Res.* **83**:775-780.

Covell, J.W. and Ross, J. Jr. Systolic and diastolic function (mechanics) of the intact heart. In. *Handbook of Physiology: Section 2. The Cardiovascular System. Vol 1. The Heart* (E. Page, H. Fozzard, R. J. Solaro, Eds.) Oxford University Press, New York, pp 741-772.

D'Angelo, D.D., Sakata, Y., Lorenz, J.N., Boivin, G.P., Walsh, R.A., Liggett, S.B., and Dorn, G.W., II. (1997) Trangenic G□q overexpression induces cardiac contractile failure in mice. *Proc. Natl. Acad. Sci. USA* **94**:8121-8126.

de Tombe, P.P. and Solaro, R.J. (2000) Integration of Cardiac Myofilament Activity and Regulation with Pathways Signaling Hypertrophy and Failure. *Annals of Biomed. Eng.* **28**:991-1001.

De Windt, L.J., Lim, H.W., Bueno, O.F., Liang, Q., Delling, U., Braz, J.C, Glascock, B.J., Kimball, T.F., del Monte, F., Hajjar, R.J., and Molkentin, J.D. (2001) Targeted inhibition of calcineurin attenuates cardiac hypertrophy in vivo. *Proc. Natl. Acad. Sci. USA* **98**:3322-7.

Donaldson, S.K., Hermansen, L., and Bolles, L. (1978) Differential, direct effects of H+ on Ca2+ - activated force of skinned fibers from the soleus, cardiac and adductor magnus muscles of rabbits. *Pflugers Arch.* **376**:55-65

Dong, W.J., Xing, J., Villain, M., Hellinger, M., Robinson, J.M., Chandra, M., Solaro, R.J., Umeda, P.K., and Cheung, H.C. (1999) Conformation of the regulatory domain of cardiac muscle troponin C in its complex with cardiac troponin I. *J. Biol. Chem.* **274**:31382-90.

Edes, I., Kiss, E., Kitada, Y., Powers, F., Papp, J.G., Kranias, E.G., and Solaro, R.J. (1995) Effects of levosimendan, a cardiotonic agent targeted to troponin C, on cardiac function and on phophorylation and Ca^{2+}-sensitivity of cardiac myofibrils and sarcoplasmic reticulum in guinea pig heart. *Circ. Res.* **77**:107-113.

El-Saleh, S. and Solaro, R.J. (1987) Calmidazolium, A calmodulin antagonist, stimulates Ca-calmodulin and Ca-troponin C dependent activation of striated myofilaments. *J. Biol. Chem.* **262**:17240-17246.

El-Saleh, S. and Solaro, R.J. (1988) Troponin I enhances acidic pH induced depression of Ca-binding to the regulatory sites in skeletal troponin C. *J. Biol. Chem.* **263**:3274-3278.

Everett, A.D., Stoops, T.D., Nairn, A.C., and Brautigan, D. (2001) Angiotensin II regulates phosphorylation of translation elongation factor-2 in cardiac myocytes. *Am. J. Physiol. Heart Circ. Physiol.* **281**:H161-7.

Fabiato, A. (1981) Myoplasmic free calcium concentration reached during the twitch of an intact isolated cardiac cell and during calcium-induced release of calcium from the sarcoplasmic reticulum of a skinned cardiac cell from the adult rat or rabbit ventricle. *J. Gen. Physiol.* **78**:457-97.

Fatkin, D., McConnell, B.K., Mudd, J.O., Semsarian, C., Moskowitz, I.G., Schoen, F.J., Giewat, M., Seidman, C.E., and Seidman, J.G. (2000) An abnormal Ca(2+) response in mutant sarcomere protein-mediated familial hypertrophic cardiomyopathy. *J. Clin. Invest.* **106**:1437-9.

Fentzke, R.C., Buck, S.H., Patel, J.R., Lin, H., Wolska, B.M., Stojanovic, M.O., Martin, A.F., Solaro, R.J., Moss, R.L., and Leiden, J.M. (1999) Impaired cardiomyocyte relaxation and diastolic function in transgenic mice expressing slow skeletal troponin I in the heart. *J. Physiol. (Lond.)* **517**:143-157.

Fitzsimons, D.P. and Moss, R.L. (1998) Strong binding of myosin modulates length-dependent Ca2+ activation of rat ventricular myocytes. *Circ. Res.* **83**:602-607.

Fuchs, F. and Wang, Y.P. (1991) Force, length, and Ca(2+)-troponin C affinity in skeletal muscle. *Am. J. Physiol.* **261**:C787-C792.

Fuchs, F. and Wang, Y.P. (1996) Sarcomere length versus interfilament spacing as determinants of cardiac myofilament Ca2+ sensitivity and Ca2+ binding. *J. Mol. Cell Cardiol.* **28**:1375-1383.

Fuchs, F. and Wang, Y.P. (1997) Length-dependence of actin-myosin interaction in skinned cardiac muscle fibers in rigor. *J. Mol. Cell Cardiol.* **29**:3267-3274.

Fujino, K., Sperelakis, N., and Solaro, R.J. (1988) Sensitization of dog and guinea pig cardiac myofilaments to Ca^{2+} activation and inotropic effect of pimobendan: Comparison with milrinone. *Circ. Research* **63**:911-922.

Fukuda, N., Kajiwara, H., Ishiwata, S., and Kurihara, S. (2000) Effects of MgADP on length dependence of tension generation in skinned rat cardiac muscle. *Circ. Res.* **86**:E1-6.

Gao, L., Kennedy, J.M., and Solaro, R.J. (1995) Differential expression of TnI and TnT isoforms in rabbit heart during the perinatal period and during cardiovascular stress. *J. Mol. Cell Cardiol.* **27**:541-550.

Geeves, M.A. and Lehrer, S.S. (1994) Dynamics of the muscle thin filament regulatory switch: the size of the cooperative unit. *Biophys J.* **67**:273-282.

Gerson, J.H., Kim, E., Muhlrad, A., and Reisler, E. (2001) Tropomyosin-troponin regulation of actin does not involve subdomain 2 motions. *J. Biol. Chem.* **276**:18442-9

Gordon, A.M., Homsher, E., and Regnier, M. (2000) Regulation of contraction in striated muscle. *Physiological Reviews* **80**:853-924.

Gwathmey, J.K. and Hajar, R.J. (1990) Effect of protein kinase C activation on sarcoplasmic reticulum function and apparent myofibrillar Ca^{2+} sensitivity in intact and skinned muscles from normal and diseased human myocardium. *Cir. Res.* **67**:744-652.

Herold, P., Herzig, J.W, Leutert, T., Zbinden, P., Fuhrer, W., Stutz, S., Schenker, K., Meier, M., and Rihs, G. (1995) 5-Methyl-6-phenyl-1,3,5,6-tetrahydro-3,6-methano-1,5-benzodiazocine-2,4-dione (BA 41899): representative of a novel class of purely calcium-sensitizing agents. *J. Med. Chem.* **38**:2946-54.

Hill, T.L., Eisenberg, E., and Chalovich, J.M. (1981) Theoretical models for cooperative steady-state ATPase activity of myosin subfragment-1 on regulated actin. *Biophys. J.* **35**:99-112.

Hill, T.L., Eisenberg, E., and Greene, L. (1980) Theoretical model for the cooperative equilibrium binding of myosin subfragment 1 to the actin-troponin-tropomyosin complex. *Proc. Natl Acad Sci USA* **77**:3186-90.

Hoffman, P.A .and Lange, J.H., III. (1995) Effect of phosphorylation of troponin I and C protein on isometric tension and velocity of unloaded shortening in skinned single cardiac myocytes from rats. *Circ. Res.* **74**:718-726.

Hofmann, P.A. and Fuchs, F. (1988) Bound calcium and force development in skinned cardiac muscle bundles: effect of sarcomere length. *J. Mol. Cell Cardiol.* **20**:667-677.

Huang, L., Wolska, B.M., Montgomery, D.E., Burkart, E., Buttrick, P.M., and Solaro, R.J. (2001) Increased contractility and altered Ca^{2+}-transients of mouse heart myocytes conditionally expressing PKC-beta. *Am. J. Physiol. (Cell)* **280**:C1114-C1120.

Hunter, W.C. (2000) Role of myofilaments and calcium handling in left ventricular relaxation. *Cardiol. Clin.* **18**:443-57.

Janssen, P.M., Datz, N., Zeitz, O., and Hasenfuss, G. (2000) Levosimendan improves diastolic and systolic function in failing human myocardium. *Eur. J. Pharmacol.* **404**:191-9.

Janssen, P.M.L. and deTombe, P.P. (1997) Protein kinase A does not alter unloaded velocity of sarcomere shortening in skinned rat cardiac trabeculae. *Am. J. Physiol.* H2415-H2422.

Jideama, N.M., Noland, T.A., Jr., Raynor, RL., Blobe, G.C., Fabbro, D., Kazanietz, M.G., Blumberg, P.M,. Hannun, Y.A., and Kuo, J.F. (1996) Phosphorylation specificities of protein kinase C isozymes for bovine cardiac troponin I and troponin T and sites within these proteins and regulation of myofilament properties. *J. Biol. Chem.* 271:23277-23283.

Jin, J.P., Yang, F.W., Yu, Z.B., Ruse, C.I., Bond, M., and Chen, A. (2001) The highly conserved COPH terminus of troponin I forms a Ca^{2+}-modulated allosteric domain in the troponin complex. *Biochemistry* 40:2623-31.

Johnson, J.A. and Mochly-Rosen, D. (1995) Inhibition of the spontaneous rate of contraction of neonatal cardiac myocytes by protein kinase C isozymes. A putative role for the □ isozyme. *Circ. Res.* 76:654-663.

Julian, F.J. (1969). Activation in a skeletal muscle model with a modification for insect fibrillar muscle. *Biophys J.* 9:547-570.

Katoh, N., Wise, B.C., and Kuo, J.F. (1983) Phosphorylation of cardiac troponin inhibitory subunit (troponin I) and tropomyosin-binding subunit (troponin T) by cardiac phospholipid-sensitive Ca^{2+}-dependent protein *Biochem. J.* 209:189-195.

Kentish, J.C., McCloskey, D.T., Layland, J., Palmer, S., Leiden, J.M., Martin, A.F., and Solaro, R.J. (2001) Phosphorylation of troponin I by protein kinase A accelerates relaxation and cross-bridge cycle kinetics in mouse ventricular muscle. *Circ. Res.* 88:1059-1065.

Kleerekoper, Q. and Putkey, J.A. (1999) Drug binding to cardiac troponin C. *J Biol. Chem.* 274:23932-9.

Komukai, K. and Kurihara, S. (1997) Length dependence of Ca(2+)-tension relationship in aequorin-injected ferret papillary muscles. *Am. J. Physiol.* 273:H1068-H1074.

Konhilas, J.P., Wolska, B.M., Martin, A.F., Solaro, R.J., and de Tombe, P.P. (2000) PKA modulates length-dependent activation in murine myocardium. *Biophys. J.* 78:108A.

Kraft, T. and Brenner, B. (1997) Force enhancement without changes in cross-bridge turnover kinetics: the effect of EMD 57033. *Biophys. J.* 72:272-281.

Laemmli, U.K. (1970) Cleavage of structural proteins during the assembly of the head of bacteriophage T4. *Nature* 227:680-685.

Lehman, W., Vibert, P., Uman, P., and Craig, R. (1995) Steric-blocking by tropomyosin visualized in relaxed vertebrate muscle thin filaments. *J. Mol. Biol.* 251:191-196.

Li, G., Martin, A.F., and Solaro, R.J. (2001) Localization of regions of troponin I important in deactivation of cardiac myofilaments by acidic pH. *J. Mol. Cell Cardiol.* 33:1309-1320.

Li, L., Desantiago, J., Chu, G., Kranias, E.G., and Bers, D.M. (2000) Phosphorylation of phospholamban and troponin I in beta-adrenergic-induced acceleration of cardiac relaxation. *Am. J. Physiol. Heart Circ. Physiol.* 278:H769-79.

Li, M.X., Spyracopoulos, L., and Sykes, B.D. (1999) Binding of cardiac troponin-I147-163 induces a structural opening in human cardiac troponin-C. *Biochemistry* 38:8289-98.

L,i T., Sperelakis, N., TenEick, R.E., and Solaro, R.J. (1985) Effects of diacetyl monoxime (DAM) on cardiac excitation-contraction coupling. *J. Pharm. Exp. Ther.* 232:688-695.

MacGowan, G.A., Du, C., Cowan, D.B., Stamm, C., McGowan, F.X., Solaro, R.J., Koretsky, A.P., and Del Nido, P.J. (2001) Ischemic dysfunction in transgenic mice expressing troponin I lacking protein kinase C phosphorylation sites. *Am. J. Physiol. Heart Circ. Physiol.*

Martin, A.M., Ball, K., Gao, L., Kumar, P.K., and Solaro, R.J. (1991) Identification and functional significance of troponin I isoforms in neonatal rat heart myofibrils. *Circ. Res.* 69:1244-1252.

Martyn, D.A. and Gordon, A.M. (2001) Influence of length on force and activation-dependent changes in troponin c structure in skinned cardiac and fast skeletal muscle. *Biophys. J.* 80:2798-808.

Martyn, D.A., Regnier, M., Xu, D., and Gordon, A.M. (2001) Ca^{2+} - and cross-bridge-dependent changes in N- and C-terminal structure of troponin C in rat cardiac muscle. *Biophys. J.* 80:360-70.

Mcdonald, K.S. and Moss, R.L. (1995) Osmotic compression of single cardiac myocytes eliminates the reduction in Ca2+ sensitivity of tension at short sarcomere length. *Circ. Res.* 77:199-205.

Mcdonald, K.S, Field, L.J., Parmacek, M.S., Soonpaa, M., Leiden, J.M., and Moss, R.L. (1995) Length dependence of Ca2+ sensitivity of tension in mouse cardiac myocytes expressing skeletal troponin C. *J. Physiol. (Lond.)* 483:131-139.

Mcdonald, K.S., Wolff, M.R., and Moss, R.L. (1997) Sarcomere length dependence of the rate of tension redevelopment and submaximal tension in rat and rabbit skinned skeletal muscle fibres. *J. Physiol. (Lond.)* **501**:607-621.

McKillop, D.F. and Geeves, M.A. (1993) Regulation of the interaction between actin and myosin subfragment 1: Evidence for three states of the thin filament. *Biophys. J.* **65**:693-701.

Montgomery, D.E., Chandra, M., Huang, Q.-Q., Jin, J.-P., and Solaro, R.J. (2001) Transgenic incorporation of fast skeletal troponin T into cardiac myofilaments blunts PKC-mediated depression of force. *Am. J. Physiol. (Heart)* **280**:H1011-H1018.

Morimoto, S., Harada, K., and Ohtsuki, I. (1999) Roles of troponin isoforms in pH dependence of contraction in rabbit fast and slow skeletal and cardiac muscles. *J. Biochem. (Tokyo)* **126**:121-9.

Morimoto, S., Ohta, M., Goto, T., and Ohtsuki, I. (2001) A pH-sensitive interaction of troponin I with troponin C coupled with strongly binding cross-bridges in cardiac myofilament activation. *Biochem. Biophys. Res. Commun.* **282**:811-5.

Moss, R.L. and Buck, S.H. Regulation of Cardiac Contraction by Calcium In: *Handbook of Physiology: Section 2. The Cardiovascular System. Vol 1. The Heart* (E. Page, H. Fozzard, R. J. Solaro, Eds.) Oxford University Press, New York, pp 420-454.

Murphy, A.M., Kogler, H., Georgakopoulos, D., McDonough, J.L., Kass, D.A,, Van Eyk, J.E., and Marban, E. (2000) Transgenic mouse model of stunned myocardium. *Science* **287**:488-91.

Nicol, R.L., Frey, N., and Olson, E.N. (2000) From the sarcomere to the nucleus. Role of Genetics and Signaling in Structural Heart Disease. *Annu. Rev. Genomics Hum. Genet.* **1**:179-223.

Noland, T.A., Jr., Raynor, R.L., and Kuo, J.F. (1989) Identification of sites phosphorylated in bovine cardiac troponin I and troponin T by protein kinase C and comparative substrate activity of synthetic peptides containing the phosphorylation sites. *J. Biol. Chem.* **264**:20778-20785.

Noland, T.A., Jr. and Kuo, J.F. (1993a) Protein kinase C phosphorylation of cardiac troponin I and troponin T inhibits Ca^{2+}-stimulated MgATPase activity in reconstituted actomyosin and isolated myofibrils, and decreases actin-myosin interactions. *J. Mol. Cell. Cardiol.* **25**:53-65.

Noland, T.A. Jr., Raynor, R.L., Jideama, N.M., Guo, X., Kazanietz, M.G., Blumberg, P.M., Solaro. R.J., and Kuo, J.F. (1996) Differential regulation of cardiac actomyosin S-1 MgATPase by protein kinase C isozyme-specific phosphorylation of specific sites in cardiac troponin I and its phosphorylation site mutants. *Biochemistry* **35**:14923-14931.

Noland, T.A., Guo, X., Raynor, R.L., Averyhart-Fullard, V., Jideama, N.M., Solaro, R.J., and Kuo, J.F. (1995) Cardiac troponin I mutants: phosphorylation by protein kinases C and A and regulation of Ca^{2+}-stimulated MgATPase of reconstituted actomyosin S-1. *J. Biol. Chem.* **43**:25445-25454.

Ohte, N., Cheng, C.P., Suzuki, M., and Little, W.C. (1997) The cardiac effects of pimobendan (but not amrinone) are preserved at rest and during exercise in conscious dogs with pacing-induced heart failure. *J. Pharmacol. Exp. Ther.* **282**(1):23-31.

Palmer, S. and Kentish, J.C. (1997) Differential effects of the Ca2+ sensitizers caffeine and CGP 48506 on the relaxation rate of rat skinned cardiac trabeculae. *Circ. Res.* **80**:682-7.

Palmer, S. and Kentish, J.C. (1998) Roles of Ca^{2+} and crossbridge kinetics in determining the maximum rates of Ca^{2+} activation and relaxation in rat and guinea-pig skinned trabeculae. *Circ. Res.* **83**:179-186

Palmiter, K.A. and Solaro, R.J. (1997) Molecular mechanisms regulating the myofilament response to Ca^{2+}: Implications of mutations causal for familial hypertrophic cardiomyopathy. *Basic Res. Cardiol.* **92**:63-4.

Pan, B.S. and Johnson, R.G.J. (1996) Interaction of cardiotonic thiadiazinone derivatives with cardiac troponin C *J. Biol. Chem.* **271**, 817-823.

Pan B.-S. and Solaro, R.J. (1987) Calcium binding properties of troponin C in detergent skinned heart muscle fibers. *J. Biol. Chem.* **262**:7339-7349.

Podolsky, R.J. and Teichholz, L.E. (1970) The relation between calcium and contraction kinetics in skinned muscle fibres. *J. Physiol. (Lond)* **211**:19-35.

Pucéat, M. and Vassort, G. (1996) Signaling by protein kinase C isoforms in the heart. *Mol. Cell Biochem.* **157**:65-72.

Rarick, H.M., Tu, X., Solaro, R.J., and Martin, A.M. (1997) The C-terminus of cardiac troponin I is essential for full inhibitory activity and Ca^{2+}-sensitivity of rat myofibrils. *J. Biol. Chem.* **272**:26887-26892.

Regnier, M., Martyn, D.A., and Chase, P.B. (1996) Calmidazolium alters Ca^{2+} regulation of tension redevelopment rate in skinned skeletal muscle. *Biophys. J.* **71**:2786-94.

Robertson, S.P., Johnson, J.D., Holroyde, M,J., Kranias, E., Potter, J.D., and Solaro, R.J. (1982) The effect of TnI phosphorylation on static and kinetic Ca binding by cardiac TnC. *J. Biol. Chem.* **257**:260-263.

Saeki, Y., Shiozawa, K., Yanagisawa, K., and Shibata, T. (1990) Adrenaline increases the rate of cross-bridge cycling in rat cardiac muscle. *J. Mol. Cell Cardiol.* **22**:453-460.

Saggin, L., Gorza, L., Ausoni, S., and Schiaffino, S. (1989) Troponin I switching in the developing heart. *J. Biol. Chem.* **264**:16299-302.

Sakata, Y., Hoit, B.D., Liggett, S.B., Walsh, R.A., and Dorn, G.W., II. (1998) Decompensation of pressure-overload hypertrophy in G□q overexpressing mice. *Circulation* **97**:1488-1495.

Slinker, B.K., Green, H.W. 3rd, Wu Y, Kirkpatrick RD and Campbell KB. (1997) Relaxation effect of CGP-48506, EMD-57033, and dobutamine in ejecting and isovolumically beating rabbit hearts. *Am. J. Physiol.* **273**:H2708-20.

Solaro, R.J., Kumar, P., Blanchard, E.M., and Martin, A.M. (1986) Differential effects of pH on Ca²⁺ activation of myofilaments of adult and perinatal dog hearts: Evidence for developmental differences in thin filament regulation. *Circ. Res.* **58**:721-729.

Solaro, R.J., Lee, J., Kentish, J., and Allen, D.A. (1988) Differences in the response of adult and neonatal heart muscle to acidosis. *Circ. Res.* **63**:779-787.

Solaro, R.J., Gambassi, G., Warshaw, D.M., Keller, M.R., Spurgeon, H.A., Beier. N., and Lakatta, E.G. (1993) Stereoselective actions of thiadiazinones on canine cardiac myocytes and myofilaments. *Circ. Res.* **73**:981-990.

Solaro, R.J. and Van Eyk, J. (1996) Altered interactions among thin filament proteins modulate cardiac function. *J. Mol. Cell Cardiol.* **28**:217-230.

Solaro, R.J. and Rarick, H.M. (1998) Troponin and tropomyosin: proteins that switch on and tune in the activity of cardiac myofilaments. *Circ. Res.* **83**: 471-480.

Solaro, R.J. (2000) Mechanisms Regulating Cardiac Myofilament Response to Ca²⁺. In: *Heart Physiology and Pathophysiology*, 4ᵗʰ Edition (Sperelakis, N, Kurachi, Y, Terzik, A, and MV. Cohen, eds) pp. 519-526.

Solaro, R.J. (2001) Modulation of cardiac myofilament activity by protein phosphorylation. *Handbook of Physiology: Section 2. The Cardiovascular System. Vol 1. The Heart* (E. Page, H. Fozzard, R. J. Solaro, Eds.) Oxford University Press, New York, pp 264-300.

Solaro, R.J., Varghese, J., Marian, A.J., and Chandra, M. (2002) Molecular Mechanisms of Cardiac Myofilament Activation: Modulation by pH and a Troponin T Mutant R92Q. *Cardiovascular Res.* (In Press).

Sorsa, T., Heikkinen, S., Abbott, M.B., Abusamhadneh, E., Laakso, T., Tilgmann, C., Serimaa, R., Annila, A., Rosevear, P.R., Drakenberg T, Pollesello P and Kilpelainen I. (2001) Binding of levosimendan, a calcium sensitizer, to cardiac troponin C. *J. Biol. Chem.* **276**:9337-43.

Squire, J.M. and Morris, E.P. (1998) A new look at thin filament regulation in vertebrate skeletal muscle. *FASEB J.* **12**:761-771.

Strang, K.T., Sweitzer, N.K., Greaser, M.L., and Moss, R.L. (1994) □-adrenergic receptor stimulation increases unloaded shortening velocity of skinned single ventricular myocytes from rats. *Circ. Res.* **74**:542-549.

Takeishi, Y., Chu, G., Kirkpatrick, D.M., Li, Z., Wakasaki, H., Kranias, E.G., King, G.L., Walsh, R.A. (1998) In vivo phosphorylation of cardiac troponin I by protein kinase C □2 decreases cardiomyocyte calcium responsiveness and contractility in transgenic mouse hearts. *J. Clin. Invest.* **102**:72-78.

Tao, T., Gong, B.-J., and Leavis, P.C. (1990) Calcium-induced movement of troponin-I relative to actin in skeletal muscle thin filaments. *Science* **247**:1339-1341.

Tardiff, J.C., Hewett, T.E., Factor, S.M., Vikstrom, K.L., Robbins, J., and Leinwand, L.A. (2000) Expression of the beta (slow)-isoform of MHC in the adult mouse heart causes dominant-negative functional effects. *Am. J. Physiol. Heart Circ. Physiol.* **278**:H412-9.

Terzic, A., Puceat, M., Clement, O., Scamps, F., and Vassort, G. (1992) □₁-Adrenergic effects on intracellular pH and calcium and on myofilaments in single rat cardiac cells. *J. Physiol. (London)* **447**:275-292.

Tobacman, L.S. (1996) Thin filament-mediated regulation of cardiac contraction. *Ann. Rev. Physiol.* **58**:447-481.

Tobias, A.H., Slinker, B.K., Kirkpatrick, R.D., and Campbell, K.B. (1996) Functional effects of EMD-57033 in isovolumically beating isolated rabbit hearts. *Am. J. Physiol.* **271**, H51-H58.

Venema, R.C. and Kuo, J.F. (1993) Protein kinase C-mediated phosphorylation of troponin I and C-protein in isolated myocardial cells is associated with inhibition of myofibrillar actomyosin ATPase. *J. Biol. Chem.* **268**:2705-2711.

Wang, X., Li, M., Spyracopoulos, L., Beier, N., Chandra, M., Solaro, R.J., and Sykes, B.D. (2001) Structure of the C-domain of human cardiac troponin C in complex with the Ca^{2+}-sensitizing drug EMD 57003. *J. Biol. Chem.* **276**:25456-25466.

Wang. Y.P, and Fuchs, F. (1994) Length, force, and Ca(2+)-troponin C affinity in cardiac and slow skeletal muscle. *Am. J. Physiol.* **266**:C1077-C1082.

Westfall, M.V. and Metzger, J.M. (2001) Troponin I isoforms and chimeras: tuning the molecular switch of cardiac contraction. *News Physiol. Sci.* **16**:278-81.

Westfall, M.V., Wahler, G.M., and Solaro, R.J. (1993) A highly specific benzimidazole pyridazinone reverses phosphate-induced changes in cardiac myofilament activation. *Biochemistry* **32**:10464-70.

Wolff, M.R., Buck, S.H., Stoker, S.W., Greaser M, and Mentzer, R.M. (1996) Myofibrillar calcium sensitivity of isometric tension is increased in human dilated cardiomyopathies: role of altered beta-adrenergically mediated protein phosphorylation. *J. Clin. Invest.* **98**:167-176.

Wolska, B.M., Kitada, Y., Palmiter, K.A., Westfall, M.V., Johnson, M.D., and Solaro, R.J. (1996) CGP-48506 increases contractility of ventricular myocytes and myofilaments by effects on actin-myosin reaction. *Am. J. Physiol.* **270**:H24-H32.

Wolska, B.M., Vijayan, K., Arteaga, G.M., Konhilas, J.P., Phillips, R.M., Kim, R., Naya, T., Leiden, J.M., Martin, A.F., de Tombe, P.P., and Solaro, R.J. (2001) Expression of slow skeletal troponin I in adult transgenic mouse heart muscle reduces the force decline observed during acidic conditions. *J. Physiol.* **536**:863-70.

Wolska, B.M., Keller, R.S., Evans, C.C., Palmiter, K.A., Phillips, R.M., Muthuchamy, M., Oehlenschlager, J., Wieczorek, D.F., de Tombe, P. and Solaro, R.J. (1999) Correlation between myofilament response to Ca^{2+} and altered dynamics of contraction and relaxation in transgenic cardiac cells that express □-tropomyosin. *Circ. Res.* **84**:745-751.

Yatani, A., Honda, R., Tymitz, K.M., Lalli, M.J., and Molkentin, J.D. (2001) Enhanced Ca2+ channel currents in cardiac hypertrophy induced by activation of calcineurin-dependent pathway. *J. Mol. Cell Cardiol.* **33**:249-59.

Zhao, Y. and Kawai, M. (1996) Inotropic agent EMD-53998 weakens nucleotide and phosphate binding to cross bridges in porcine myocardium. *Am. J. Physiol.* **271**:H1394-406.

PAGE ANDERSON

THIN FILAMENT REGULATION IN DEVELOPMENT

1. THE HEART AND DEVELOPMENT

The basic function of the heart is to eject blood. This function is supported throughout development by the sarcomere and the interaction of its thick and thin filaments. In the face of this constancy, the ability of the heart to work, modulate its output, and respond to pathophysiologic states undergoes marked changes with development. The potential contributions of troponin I (TnI) cardiac (c) TnT, actin, and tropomyosin isoforms to these changes are the focus of this chapter.

We will also consider, in the context of myofilament response to calcium, the effects of development on other systems basic to heart function. These include the systems that regulate cytosolic calcium concentration ($[Ca]_i$) and those that transduce adrenoreceptor stimulation. Their effects on thin filament modulation of heart function will be related to birth and the pathophysiologic states associated with birth and heart disease.

1.1 Cardiac Contractility and Development

The force of myocardial contraction and the ability of the heart to work increase with development. For example, the early stage embryonic ventricle develops a peak systolic pressure of about 0.3 mm Hg, the adult left ventricle generates pressures in excess of 100 mm Hg, and the stroke volume of the adult heart is over 1000 fold greater than that of the embryonic heart (Hu and Clark, 1989). Peak tension, peak rate of rise of tension, and velocity of shortening against a load increase with development. These positive effects of development on the cardiac contraction have been described in avian and mammalian myocardium (e.g. Friedman, 1973; Godt et al., 1991; Reiser et al., 1994; Anderson et al., 1984).

When sarcomere shortening and re-extension are examined in the cardiac myocyte, isolated from the heart and so free of the extracellular matrix and its

R.J. Solaro and R.L. Moss (eds), Molecular Control Mechanisms in Striated Muscle Contraction, 329-377
©2002 Kluwer Academic Publishers, Printed in the Netherlands

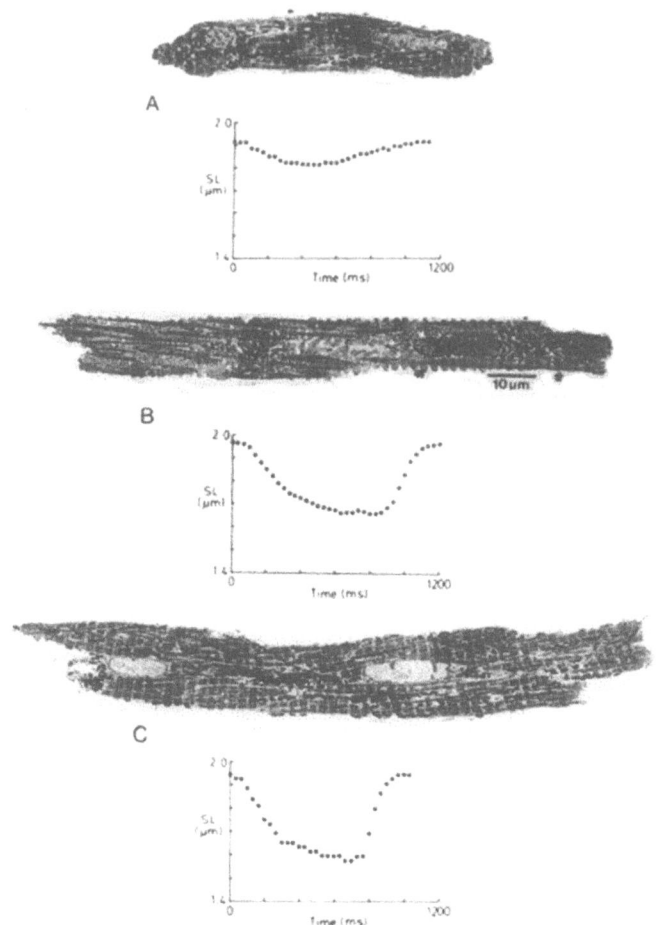

Figure 1. Comparison of the ultrastructure and sarcomere contraction waveforms from adult and neonatal rabbit cardiac myocytes. The contraction waveforms and longitudinal central sections of three isolated cardiac myocytes are illustrated. Sarcomere contractions were measured, and the cells then prepared for electron microscopy. Sarcomere length (SL) is plotted as a function of time (1 mM extracellular calcium). All cells in the figure are shown at identical magnification. A. An average-sized myocyte from a neonatal rabbit. B. A relatively small adult cell. C. An average-sized cell from an adult heart. (from Nassar, R., Reedy, M.C. and Anderson, P.A.W: (1987) Developmental changes in the ultrastructure and sarcomere shortening of the isolated rabbit ventricular myocyte. Circulation Research 61, 465-483.)

constraints, the positive effect of development on the cardiac contraction are still seen. The amount and velocity of sarcomere shortening of the adult cardiac myocyte are greater than those of the myocyte from the neonatal heart (see Figure 1, Nassar et al., 1987). The velocity of sarcomere re-extension is greater and the duration of the contraction shorter in the adult cardiac myocyte than in the myocyte from the neonatal heart (Nassar et al., 1987).

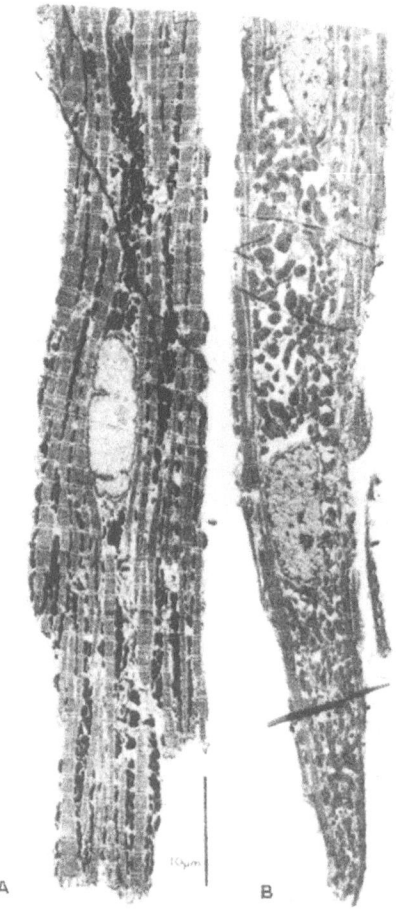

A   B

Fig. 2. Electron micrographs of a longitudinal section through (A) a myocyte isolated from an adult rabbit and (B) a myocyte isolated from a neonatal rabbit show the maturational differences in cell shape, myofibril distribution, and the cytoskeleton, relative to mitochondria and nuclei. The second nucleus of the adult cell lies just off the top of the page (from Nassar, R., Reedy, M.C., and Anderson, P.A.W.: (1987) Developmental changes in the ultrastructure and sarcomere shortening of the isolated rabbit ventricular myocyte. Circulation Research 61, 465-483.)

These functional changes in the myocyte, the myocardium, and the intact heart are a consequence of developmental changes in many components of the systems that affect cardiac contraction. These include the amount and organization of the myofibrils, the cytoskeleton, and the membrane systems that control [Ca]$_i$ (Figure 2, Nassar et al., 1987). The percentage of the cell volume made up of myofibrils and the relative amount, organization, and ATPase activity of the sarcoplasmic reticulum (SR) increase, and the peak [Ca]$_i$ transient increases with development. For example, peak [Ca]$_i$ is approximately 150 nM and 400 nM in the neonatal and adult myocyte respectively, while diastolic [Ca]$_i$ does not differ (Anderson et al., 1993; Figure 3). As will be discussed below, the contractile response to these developmental changes in calcium regulation depends on thin filament isoform expression and their post-translational modifications.

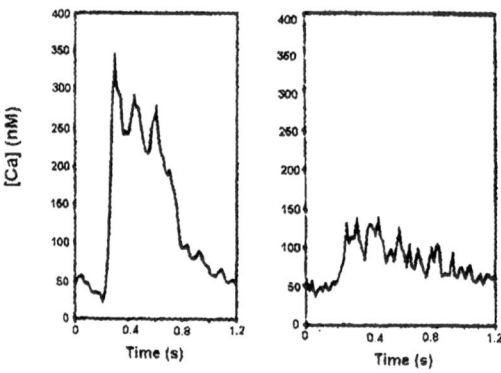

Figure 3. Response of an adult (left) and immature (right) myocyte to a depolarization from a holding potential of –40 mV to 0. Note that the [Ca]$_I$ transient is threefold higher in the adult myocyte, while diastolic [Ca]$_I$ is the same.

1.2. Frank-Starling Relation and Development

A fundamental property of the heart, essential for altering cardiac output, is the Frank-Starling relation. This relation is seen in the working heart, where an increase in ventricular end-diastolic volume increases stroke volume (Starling, 1914), and in the heart contracting isovolumetrically, where an increase in volume increases peak systolic pressure (Frank, 1895). The positive effect of an increase in ventricular end-diastolic volume on ejection of blood and development of pressure resides, in part, within the myofilaments. An increase in sarcomere length and decrease in filament spacing increases the sensitivity of the myofilaments to calcium shifting the force-pCa relation toward lower calcium concentrations (Wang and Fuchs, 1995). This sarcomere-length induced shift will result in the same peak [Ca]$_i$ transient having a markedly different effect on tension and so ventricular output.

The immature heart was once thought to lack a Frank-Starling relation or to have a blunted one. However, fetal left and right ventricular output and pressure development have been found to be quite sensitive to changes in ventricular end-diastolic volume (Anderson et al., 1986, 1987). The importance of the Frank-Starling relation in the immature heart is emphasized by the increase in left ventricular end-diastolic volume and cardiac output that occur with birth (Anderson et al., 1984; Lister et al., 1979; Breall et al., 1984). Birth is marked by an abrupt change in circulation from that of the fetus, where the cardiac output is the sum of the right and left ventricular outputs to that of the neonate, where the right and left ventricles function in series, and systemic output is supported only by the left ventricle. With birth, left ventricular end-diastolic volume increases, and left ventricular output increases two-to-threefold (Anderson et al., 1984; Breall et al., 1984). Corrected for body weight, left ventricular output of the neonate is three times that of the adult (Klopfenstein and Rudolph, 1978). This marked neonatal increase in cardiac output found in the mammal is also seen in the hatchling chick (Tazawa et al., 1983).

Left ventricular output in the neonate must be further increased, often several fold, if the ductus arteriosus does not close with birth (Clyman et al., 1987, 1988; Shimada et al., 1994). The ductus arteriosus, the arterial connection between the aorta and pulmonary artery, is an essential component of the circulation *in utero* where it directs blood away from the pulmonary arteries to the descending aorta and the placenta. Following birth, blood flow through the ductus arteriosus is reversed, flowing from the aorta to the pulmonary artery. When the ductus stays open, as it usually does for hours following birth, left ventricular output must support both blood flow to the body and flow that passes through the ductus into the pulmonary circulation. A failure to generate this needed output can lead to death (Nguyen et al., 1997).

The presence of the Frank-Starling relation in the neonatal heart will allow it to respond to the neonatal increase in left ventricular end-diastolic volume with an increase in stroke volume and cardiac output. While increasing sarcomere length enhances the sensitivity of the myofilaments to calcium in the immature heart, developmental changes in contractile protein isoform expression and their post-translational modifications may alter quantitatively this relation (see below). These various effects remain to be fully assessed in the immature myocardium.

1.3. The Sympathetic System and Development

Catecholamine stimulation of the adrenergic receptors (AR's) enhances peak force of the cardiac contraction and can significantly affect thin filament regulation of contraction (see below) and the membrane systems that regulate $[Ca]_i$ (Kranias and Solaro, 1982). For example, β-adrenergic stimulation results in cAMP-dependent protein kinase A (PKA) phosphorylation of proteins essential in $[Ca]_i$ regulation. Phosphorylation of the L-type calcium channel increases calcium current density, and phospholamban phosphorylation removes its inhibitory effect on SR ATPase activity (Osaka and Joyner, 1992; Kranias and Solaro, 1982). The phosphorylation of cTnI decreases the sensitivity of myofilaments to calcium The combined effects result in an enhancement of the $[Ca]_i$ transient and an acceleration in relaxation.

1.3.1 Acquisition of the AR's

The cardiac myocyte acquires AR's during fetal development. β-AR receptors have been identified in the human heart following the 12th week of gestation (Coltart and Spilker, 1972). The myocardial density of the β AR's increases during fetal and neonatal life while that of the α-AR's increases and falls (Felder et al., 1982; Yamada et al., 1980; Chen et al., 1979; Schumacher et al., 1984). At birth, the α-AR's have been described in some species as being at their highest density and relatively greater in number than the β-AR's. This differential expression of the α and β-AR's with development will affect the response of the heart to endogenous catecholamines, norepinephrine and epinephrine.

1.3.2 Myocardial Innervation

Sympathetic innervation of the heart follows the acquisition of the β-AR receptors. The intramyocardial release of norepinephrine in response to sympathetic nervous stimulation depends on the presence of the synaptic vesicles that synthesize and store the transmitter (Bareis and Slotkin, 1980). This delivery system is incomplete until the sympathetic nerves grow into the myocardium.

The developmental timing of cardiac innervation varies among species. For example, myocardial nerves appear during postnatal life in the rat, late gestation in the rabbit, and at mid-gestation in the fetal lamb (Lebowitz et al., 1972; Lipp and Rudolph, 1972; DeChamplain et al., 1970; Hoar and Hall, 1970; Papka, 1981).

1.3.3 Myocardial Response to AR-Agonists

The myocardial response to exogenous catecholamines changes with development and varies among species and from study to study. Based on studies of sheep myocardium, the immature heart was found to be more sensitive to β-agonists stimulation than is the adult (Friedman, 1972). In contrast to this observation, other studies, including those of the lamb and rabbit, have demonstrated a developmental increase in the *in vivo* and *in vitro* sensitivity of the myocardium to isoproterenol, a β1 agonist (Park et al., 1980; Assali et al., 1978). In the neonatal and adult dog, active tension and the maximum and minimum first derivatives of tension are affected in a same manner by an increase in β1 stimulation (Park et al., 1982). However, the effect of β1 stimulation on time-to-peak tension and relaxation time differ between the neonate and the adult. In the two day old, these measures were unaffected by an increase in β1 stimulation, while both were shortened in the adult (Park et al., 1982).

The complexity of the effects of β-adrenergic stimulation are likely to follow from changes in thin filament isoform expression, as will be discussed below, and in the regulation of $[Ca]_i$. For example, expression of phospholamban, SR Ca^{2+} ATPase, and the L-type calcium channel increase with maturation (Szymanska et al., 1995; Osaka and Joyner, 1991). Of note, the expression of phospholamban relative to that of the SR calcium ATPase is lowest in the fetal heart and highest in the adult, suggesting an increase in phospholamban regulation of SR function with development (Szymanska et al., 1995). The expression of another regulator of $[Ca]_i$, the Na^{2+}/Ca^{2+} exchanger, is also affected by development, rising and falling during perinatal life in the human and other species (Artman et al., 1995; Boerth et al., 1994). The functional effects of these developmental changes in $[Ca]_i$ regulators is further complicated by the changes in thin filament protein isoform expression and their post-translational modifications (see below).

1.3.4 Catecholamine Stimulation During Perinatal Life

Perinatal life is marked by a transient increase in sympathetic stimulation (Padbury and Martinez, 1988). At birth, the fetus and newborn have markedly elevated blood levels of epinephrine and norepinephrine. These levels are likely to result from catecholamine release from extracardiac sympathetic ganglia and the adrenals. Their effects will be further enhanced by intramyocardial release of norepinephrine. In the newborn infant who appears unstressed at birth, epinephrine and norepinephrine are two-fold those of the adult who has exercised to exhaustion (Padbury et al., 1982; Padbury and Martinez, 1988; Galbo et al., 1975). In the stressed newborn, the levels of norepinephrine and epinephrine are 10-fold or greater than those achieved in the adult exercised to exhaustion. As a point of reference, these neonatal levels equal or exceed those achieved in the infant receiving an intravenous catecholamine infusion to treat severe low cardiac output.

The importance of this neonatal increase in epinephrine and norepinephrine in opposing the negative inotropic effects of perinatal acidosis, ischemia, and hypoxia is evidenced by the lethal effect of β-adrenergic receptor blockade in the stressed infant (Buechler and Palmer, 1982; Joelsson and Barton, 1969). As will be considered below, how these normal and pathophysiologic events may affect thin filament regulation of heart function depend upon thin filament composition and the systems transducing catecholamine stimulation of the AR's.

1.4. Development and Myocardial Response to Calcium

1.4.1 Intact Myocardium and Extracellular Calcium ($[Ca]_o$).

Myocardial sensitivity to $[Ca]_o$, as measured by peak tension, decreases with development. In the chick heart, $[Ca]_o$ sensitivity of the myocardium has been examined in membrane intact preparations from the embryonic and post-hatchling heart. The sensitivity to $[Ca]_o$ decreased by more than 1 log unit with development (Marsh and Allen, 1989; Godt et al., 1991). A similar decrease in the response to $[Ca]_o$ was found during postnatal development in the heart and isolated myocardium of the rabbit (Park et al., 1980; Jarmakani et al., 1982). These differences in the response to $[Ca]_o$ have multiple interpretations. They include: the mechanisms that regulate the $[Ca]_i$ transient are less affected by $[Ca]_o$ with maturation, and the sensitivity of the myofilaments to $[Ca^{2+}]$ decreases with maturation.

The postnatal fall in Na^+/Ca^{2+} exchanger expression, observed in some species, could explain a developmental decrease in the response to $[Ca]_o$ (Artman et al., 1995; Boerth et al., 1994). On the other hand, the maturational increase in calcium current (I_{Ca}) density would enhance the ability of the myocyte to respond to an increase in $[Ca]_o$ (Fabiato, 1982; Osaka and Joyner, 1991; Marsh and Allen, 1989). These findings suggest that a maturational fall in the sensitivity of myofilaments to calcium contributes to the developmental fall of the sensitivity in the myocardium to $[Ca]_o$.

1.4.2 Membrane Skinned Myocardium and Calcium

The effect of development on the sensitivity of cardiac myofilaments to calcium has been examined in avian and mammalian species. In the rat, the force-pCa relation has been examined in isolated membrane-extracted cardiac myocytes from the two-day prepartum fetus and the adult (Figure 4, Fabiato, 1982). The $[Ca^{2+}]$ that generated half-maximal force in the myocytes from the fetal heart was approximately 0.3 µM lower than in the adult myocyte. In a subsequent study of membrane-skinned myocardium from the neonatal and adult rat, ventricular trabeculae demonstrated a similar maturational shift in the sensitivity of the myofilaments to calcium (Reiser et al., 1994).

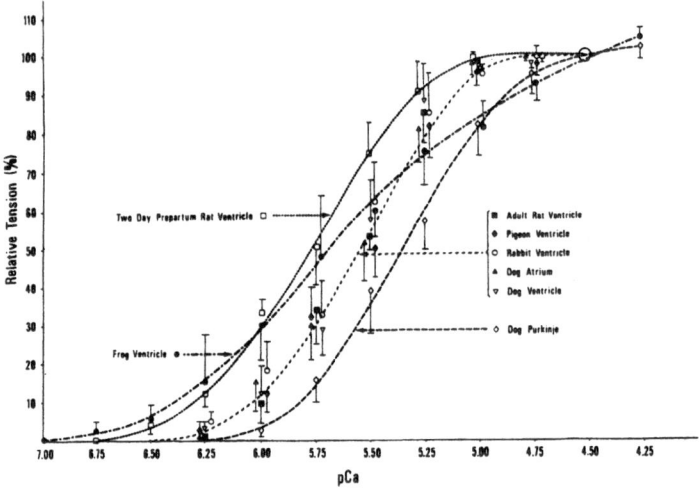

Figure 4. Relationship between tension and pCa ($-log_{10}$ [free Ca^{2+}]) in skinned cardiac cells from two developmental stages of the rat ventricle (open squares, 2 day preparation; closed squares, adult), the ventricle of different adult animal species, and different cardiac tissues of the adult dog heart. Each point corresponds to the mean of 6-15 determinations, and each vertical bar corresponds to the SD represented in one direction only. For each skinned cardiac cell the tension obtained at a given pCa value was expressed as a percentage of the tension induced in the same cell by pCa 4.50. (from Fabiato, A. (1982) Calcium release in skinned cardiac cells: variations with species, tissues, and development. Federation proceedings 41, 2238-2244.)

In the mouse and the rabbit, the relations between $[Ca^{2+}]$ and force or myofibrillar ATPase activity demonstrate a maturational shift similar to that observed in the rat. Skinned myocardium from adult female mice and mouse fetuses (approximately day 17 post-coitum) demonstrated a developmental decrease in myofilament sensitivity to calcium as measured by the relation between force and calcium (see Figure 5, Metzger et al., 1994). *In vitro* differentiation of mouse embryonic stem cells demonstrated a similar fall in myofilament sensitivity (Metzger et al., 1994). However, species differences appear to occur in the effect of development on the Hill coefficient of the force-pCa relation. In the rabbit, the Hill coefficient fell with development (McAuliffe et al., 1990), while in the mouse the Hill coefficient increased (Metzger et al., 1994).

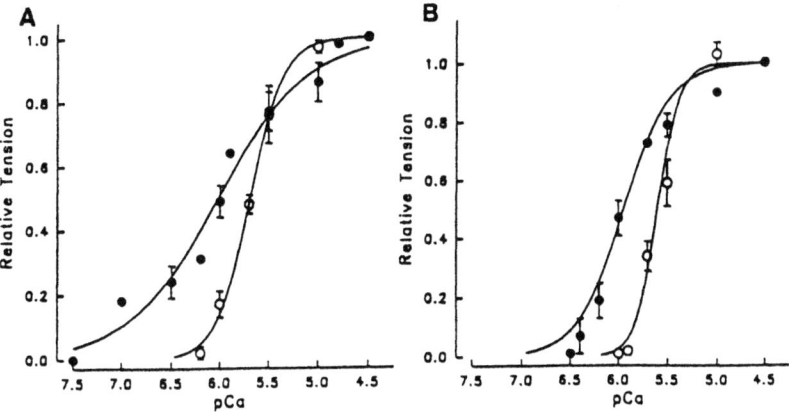

Figure 5. Summary of tension-pCa relationships in cardiac myocytes obtained during development in vitro from embryonic stem (ES) cells (A), and in vivo from mouse cardiac myocytes (B). (A) Filled circles are ES cell-derived cardiac myocyte preparations that had been contracting for an average of 6 ± 1 d in vitro, and open circles are preparations that had been contracting for an average of 37 ± 1 d in vitro. (B) Filled circles are fetal mouse cardiac myocytes (obtained on average at day 17 post coitum), and open circles are cardiac myocytes from adult mice. For each plot the best fit curve is shown. Values are mean ± SEM, average n of 6. In some instances error bars were covered by the symbol. (from Metzger, J.M., Lin, W.I., and Samuelson, L.C. (1994). Transition in cardiac contractile sensitivity to calcium during the in vitro differentiation of mouse embryonic stem cells. Journal of Cell Biology 126, 701-11.)

In membrane-extracted preparations from the chicken heart, maximum calcium activated force increased from embryonic day 7 to embryonic day 19 and then remained unchanged through seven weeks post-hatching (Godt et al., 1991). The pCa_{50} of the relation between force and calcium shifted toward higher $[Ca^{2+}]$ with maturation, similar to the change observed in the rat (Godt et al., 1991; Reiser et all, 1994).

The maturational fall in the sensitivity of myofilaments to calcium is opposite in direction to the maturational increase in the peak of the $[Ca]_i$ transient (Figure 3).

Teleologically, the greater sensitivity of the myofilaments in the immature heart would appear to compensate for the lower $[Ca]_i$ transient of the immature myocyte. The $[Ca]_i$ transient of the immature myocyte would generate relatively little force in the adult myocardium but in the neonatal myocardium this transient will generate a twitch that can exceed 50% of maximum active force.

1.5. Development and Myocardial Response To Acidosis

1.5.1 Intact Myocardium and Acidosis
Acidosis depresses heart function, and hypercarbic acidosis, sometimes accompanied by metabolic acidosis, often occurs during birth. As noted above, the left ventricle must be able to markedly increase its output with birth to support the increased metabolic needs of the newborn in the presence of this acidosis.

Teleologically, consistent with the need of the heart to be resistant to acidosis during perinatal life, the immature myocardium is initially more resistant to the depressive effects of acidosis on function than is adult myocardium (Solaro et al., 1988). In the rat, acidosis was found to cause an initial greater fall in peak tension in adult myocardium as compared to that of the neonate. Continued exposure to acidosis caused a further fall in developed tension in both age groups with the effect being greatest in the neonatal myocardium. The complexity of these time dependent effects of acidosis on myocardial function suggests that developmental changes in multiple systems, including those that regulate $[Ca]_i$ and myofilament response to $[Ca^{2+}]$, play significant roles in the response to acidosis.

1.5.2 Myofilament Function and Acidosis
The negative effects of acidosis on myocardial function with maturation follow, in part, from alterations in the myofilament's response to $[H^+]$. A maturational increase in the negative effects of acidosis on the sensitivity of myofilaments to calcium occurs in the dog and rat (Solaro et al., 1986; Martin et al., 1991). In the fetal and neonatal dog heart, calcium activation of preparations was unaffected by a decrease in pH from 7.0 to 6.5. In contrast, in the adult dog heart, this decrease in pH shifted the $[Ca^{2+}]$ required for half-maximal myofibrillar ATPase activity by 0.4 log units toward a higher $[Ca^{2+}]$. The biochemical properties of troponin C were the same in the adult and perinatal dog myocardium. The developmental shift in myosin heavy chain isoform expression cannot explain the relative resistance to acidosis of the perinatal heart (Solaro et al., 1986). In the rat, myofilaments from the adult heart responded to a reduction in pH from 7.0 to 6.5 with a shift in the pCa_{50} of approximately 0.6 log units toward a higher $[Ca^{2+}]$, while in the neonatal heart, this pH change shifted the force-pCa relation by only 0.3 log units. In chick myocardium from developmental stages that range from embryonic day 7 through young adulthood, acidosis depressed the sensitivity of the myofilaments to calcium. However, the extent of the shift in the force-pCa relation was three-fold higher in the adult. Acidosis also had a greater depressive effect on maximum force development in the adult preparation. In summary, across species, the function of

myofilaments from the immature heart is more resistant to acidosis than those from the adult.

In the following sections, the maturational changes in isoform expression of the thin filament proteins and their post-translational modifications of these proteins will be related to the maturational changes in the relations between pCa and force or myofibrillar ATPase activity and how these relations are modulated by acidosis, adrenoreceptor stimulation, and sarcomere length.

1.6. Development and Thick Filament Proteins

Although the focus of this chapter is the thin filament, maturational changes in thick filament isoform expression and post-translational modifications may contribute to developmental differences in myofilament function.

Although myofilaments sensitivity to calcium, as defined by the force-pCa relation, is not thought to be affected by the myosin heavy chain (MHC) isoforms, an effect may be present. Inducing a hypothyroid state to generate myofilaments containing only the β-MHC isoform has revealed an apparent MHC isoform effect on the force-pCa relation (Metzger et al., 1999). The pCa_{50} of preparations containing β-MHC was 5.51 ± 0.03, while that of α-MHC containing preparations was 5.68 ± 0.05. The relative stiffness-pCa relation was also shifted toward a higher $[Ca^{2+}]$ in β-MHC expressing myocytes ($pCa_{50}=5.47\pm0.05$ versus 5.76 ± 0.05 in α-MHC expressing myocytes), and no difference in the steepness of the tension-pCa relation or in maximum active force was observed.

The effect of development on the expression of the MHC isoforms has been extensively examined in multiple species (Swynghedauw, 1986). In the rodent, α-MHC is the predominantly expressed isoform, while β-MHC is predominantly expressed in the fetus. In the human, β-MHC is the predominant isoform expressed throughout development (Miyata et al., 2000). As regards the rodent heart, the positive effects of α-MHC on myofilament sensitivity to $[Ca^{2+}]$, the negative effect of maturation on the sensitivity of myofilaments to calcium, and the developmental increase in α-MHC isoform expression make it unlikely that a switch in MHC isoform expression contributes to the maturational fall in the sensitivity of the myofilaments to calcium.

Studies assessing myofilament sensitivity and the effects of maturation usually make use of ventricular myocardium. The myosin light chain (MLC) 1 isoforms (MLC1A and MLC1V) are both expressed in the ventricle prior to adult life. With maturation, MLC1V becomes the only MLC1 isoform expressed in the ventricle. MLC1A appears to confer on myocardium faster cross-bridge kinetics and greater force generation at a given $[Ca^{2+}]$ (Schaub et al., 1998). Replacement of MLC1V with MLC1A, using transgenesis, resulted in an increase in V_{max} in skinned myocardium and the velocity of translocating actin filaments in vitro (Fewell et al., 1998). Although the developmental fall in ventricular expression of MLC1A is consistent with the observed developmental decrease in the sensitivity of myofilaments to calcium, the time courses of the switch in MLC1 isoform expression and the maturational fall in myofilament sensitivity suggest that this switch in expression of MLC1 isoforms is insufficient to explain the maturational changes in myofilament function.

The two MLC2 isoforms (MLC2V and MLC2A) are restricted to the ventricular and atrial compartments, respectively, in contrast to the expression pattern of the MLC1 isoforms. MLC2 phosphorylation shifts the pCa_{50} toward a lower $[Ca^{2+}]$ (Kelly and Buckingham, 1997), so a developmental decrease in MLC2V phosphorylation could bring about a decrease in the sensitivity of the myofilaments to calcium. Such a developmental fall in MLC2V phosphorylation remains to be demonstrated.

2. THIN FILAMENT PROTEINS AND DEVELOPMENT

2.1 Intact Troponin and Development

Cardiac troponin isolated from the human fetus and adult are functionally different based on an *in vitro* motility assay (Purcell et al., 1999). Troponin isolated from myocardium at these two stages of development regulated actin-tropomyosin filament movement over a bed of immobilized heavy meromyosin, and, at pCa 9, troponin from both stages of development reduced the fraction of moving filaments from 90% to <5% with a modest decrease in velocity. However, at pCa 5, troponin from normal adult heart increased filament velocity by approximately 47%, while fetal troponin increased velocity by only 4%. Interestingly, troponin isolated from the failing adult human heart had an intermediate effect on filament velocity (Purcell et al., 1999). As will be reviewed below, the isoform content and post-translational modifications of troponin isolated from the fetal, adult, and failing human heart are likely to differ and so contribute to these functional differences.

3. TROPONIN I

3.1 Troponin I Isoforms

Troponin I (TnI) inhibits thick and thin filament interaction. This inhibition is suppressed by the binding of calcium to TnC (Solaro and Van Eyk, 1996). Here, we will focus on the TnI isoforms, their modulation of myofilament function, and how these effects may produce significant changes in function with cardiac development.

Two TnI isoforms, slow skeletal (ss) TnI and cardiac (c) TnI, are expressed in the heart during development. The heart is the only organ in which cTnI is expressed (Hastings et al., 1991). The structural and functional differences between slow ssTnI and cTnI are reviewed in detail in the Chapter by Solaro. The differences include an amino-terminal extension of 24 to 32 residues present in cTnI and absent in ssTnI, the peptide length varying among species. Phosphorylation of two serines in this extension by PKA or PKC and that of other PKC substrates in cTnI, absent in ssTnI, have been shown to modulate function (Wattanapermpool, 1995; Noland and Kuo, 1991, 1993). Other structural differences underlie the differential responses to calcium and acidosis of myofilaments containing ssTnI or cTnI (see below, Response to Acidosis).

3.1.1. TnI Isoforms and Development

All the chambers of the heart express the two isoforms at sometime during development. Their relative levels of expression change throughout maturation in a species-dependent manner (Gorza et al., 1993; Saggin et al., 1989; Sabry and Dhoot, 1989; Murphy et al., 1991). In the rat, ssTnI is dominantly expressed during embryonic life. The level of ssTnI mRNA falls during fetal and early neonatal development reaching an undetectable level in the adult rat heart. In the chicken, ssTnI and cTnI isoforms were found in approximately equal amounts in the heart throughout embryonic development. The expression of ssTnI disappears in the rat, mouse, and chicken heart by two weeks following hatching or birth. Only ssTnI is expressed in the human heart in the 12th week of gestation (Figure 6, Bodor et al., 1997). The subsequent developmental shift in expression to only cTnI in the human has a time course with some similarity to that seen in the rat (Bhavsar et al., 1991; Hunkeler et al., 1991). For example, in the human and rat heart, ssTnI and cTnI mRNA and protein are both expressed in the neonate, and only cTnI is expressed in the adult heart. However, ssTnI expression persists in the human ventricle well into the first year of life while its expression in the rat heart disappears in the first two weeks of life (Bhavsar et al., 1991; Hunkeler et al., 1991).

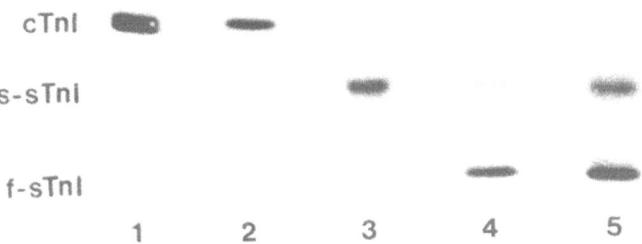

Figure 6. Western blot. Expression of TnI isoforms in fetal myocardium and adult failing and control myocardium. Myofibrillar proteins from fetal and adult hearts were resolved in a 7.5% SDS-Page gel and transferred to a PVDF membrane. TnI was detected using TnI Mab 3C5.10, which is equally reactive with cTnI and both skeletal muscle isoforms. The skeletal muscle preparations did not contain cTnI; the adult cardiac preparations contained only cTnI. The two bands of cTnI in the adult failing heart result from different levels of phosphorylation. Numbered lanes indicate the following myofibrillar preparations: 1, failing myocardium; 2, control myocardium; 3, human fetal heart (14 weeks of gestation); 4, fetal human skeletal muscle; and 5, adult human skeletal muscle. (from Bodor, G.S., Oakeley, A.E., Allen, P.D., Crimmins, D.L., Ladenson, J.H., and Anderson, P.A.W. (1997) Troponin I phosphorylation in the normal and failing adult human heart. Circulation 96, 1495-1500)

The maturational acquisition of cTnI expression follows a temporal and regional pattern. The regional differences in the timing of the onset of cTnI expression and the shift in expression from ssTnI to cTnI have been described in detail in the rat heart (Gorza et al., 1993). At embryonic day 10, only ssTnI mRNA was detected in the cardiac tube. At embryonic day 11, cTnI mRNA was rarely detected in atrial myocytes and was absent in the outflow tract and the primitive right ventricle. The appearance of cTnI transcripts and the disappearance of ssTnI transcripts started in the atria and the caudal regions of the heart and extended rostrally to the ventricles and into the outflow tract. Except for continued expression of ssTnI in adult myocytes within the atrioventricular node and conduction system, this temporal spatial process results in the adult heart expressing only cTnI. The heart is not rescued by an increase in ssTnI expression if cTnI expression does not occur (Huang et al., 1999). Mice null for cTnI demonstrate the irreversible nature of the developmentally driven suppression of ssTnI expression. In mice null for cTnI, the neonatal fall in expression of ssTnI results in death.

3.1.2 Thyroid Hormone

Thyroid hormone modulates the developmental time course of the expression of the TnI isoforms. In the rat, hypothyroidism delays the switch in expression from ssTnI to cTnI (Averyhart-Fullard et al., 1994; Dieckman and Solaro, 1990). The hypothyroid effect was associated with a lower percentage of cTnI expression throughout embryonic life. In hypothyroid cTnI null mice, ssTnI expression was prolonged following birth with the survival of the mice to day 21 (Huang et al., 2000; Averyhart-Fullard et al., 1994). Consistent with the differential effects of the TnI isoforms on myofilament response to acidosis (see below, Response to Acidosis) myocardium from hypothyroid rats expressed a higher level of ssTnI and were more resistant to acidosis (Dieckman and Solaro, 1990). In hyperthyroid cTnI null mice, the duration of ssTnI expression was abbreviated.

The developmental fall in ssTnI expression and the reciprocal increase in cTnI expression in the perinatal heart may be modulated by the abrupt increase in thyroid stimulating hormone at birth and the subsequent marked increase in thyroxine levels (Behrman et al., *Nelson Textbook of Pediatrics*, 2000). The levels of the thyroid hormones then gradually fall over the first weeks of life. Regardless of how this transient hyperthyroid state would enhance the expression of cTnI, ssTnI is expressed in the human heart for many months after the thyroid hormone level is no longer elevated. These findings suggest the importance of other factors in the developmental increase in the expression of cTnI in the human heart.

3.2 Functional Comparison of ssTnI and cTnI

Both ssTnI and cTnI serve the same basic role in regulating contraction through their inhibition of actomyosin interaction (Solaro and Van Eyk, 1996). However, the quantitative responses to $[Ca^{2+}]$ and to $[H^+]$, as measured by force and ATPase activity, are modulated by the TnI isoform composition of the thin filaments. These effects of the TnI isoforms have been examined *in vitro* and *in vivo* (for example, Westfall et al., 1997; Guo et al., 1994; Wattanapermpool et al., 1995; Fentzke et al., 1999).

3.2.1. Response to Calcium

Myofilaments containing ssTnI are more sensitive to calcium than are those containing cTnI (Figure 7). Adenovirus-induced expression of ssTnI in cardiac myocytes *in vitro* (Westfall et al., 1997) has been used to almost completely suppress cTnI expression. Maximum activated tension was found not to differ between cTnI and ssTnI expressing myocytes. However, the threshold for calcium activated force development was shifted toward a lower calcium concentration in the ssTnI expressing myocytes. The Hill coefficient of this relation was also lower in ssTnI expressing myocytes (1.6 ± 0.2 versus 3.0 ± 0.2). Although the Hill coefficient of *in vitro* ssTnI expressing myocytes and that of native muscle expressing ssTnI, for example soleus fibers and fetal and neonatal myocardium, were the same, the pCa_{50} of the ssTnI expressing cardiac myocytes and native slow skeletal muscle fibers differed (Westfall et al., 1997). The difference in myofilament sensitivity to calcium among these fibers expressing only ssTnI indicates that other contractile proteins contribute to the functional differences between slow skeletal muscle and cardiac muscle.

A transgenesis study has examined the effect of ssTnI expression in the adult mouse heart (Fentzke et al., 1999). As might be anticipated, the rate of rise of $[Ca]_i$, peak $[Ca]_i$, and the velocity of shortening did not differ in myocytes from the transgenic and wild type hearts. The enhanced *in vivo* expression of ssTnI was associated with a reciprocal fall in cTnI expression, similar to that observed *in vitro* in the presence of adenovirus-induced ssTnI expression (Westfall et al., 1997). The membrane intact myocytes from the transgenic animal shortened more than those from the wild type. In addition, the half-time of the $[Ca]_i$ transient and the relengthening time of the transgenic myocytes was prolonged. Permeabilized single cardiac myocytes, isolated from the transgenic heart, had an increased sensitivity to calcium as measured by the relation between $[Ca^{2+}]$ and tension. *In vivo* diastolic function, measured by ventricular diastolic relaxation rate, was slower in the transgenic animals consistent with the functional effects of ssTnI expression in the skinned preparation and the intact myocyte. The functional effects of enhanced ssTnI expression on myocardial function are reminiscent of differences between immature myocardium, where ssTnI is expressed, and adult myocardium, where only cTnI is expressed.

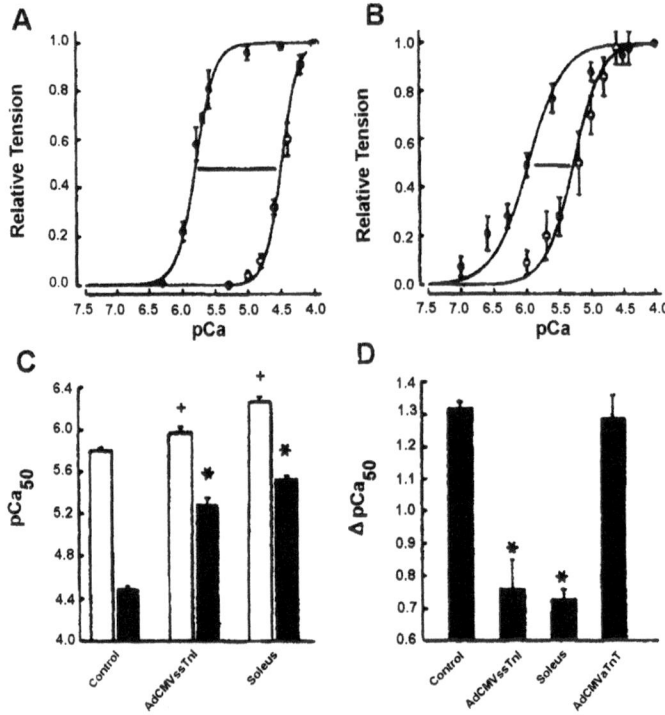

Figure 7. In vitro study of adrenovirus-induced expression of ssTnI in adult cardiac myocytes. A summary of the tension-pCa relationships are shown for control (A) and adrenovirus ssTnI (AdCMVssTnI)-treated (B) myocytes at pH 7.0 (closed symbols) and pH 6.20 (open symbols). Active tension (P) at each submaximal pCa is expressed as a fraction of the maximum active isometric tension at pCa 4.0 (P_o). The shape and position of the tension-pCa relationship in controls cultured for 6 days are the same as those obtained from acutely isolated myocytes. (C) Summary of pCa_50 in control and AdCMVssTnI-treated cardiac myocytes and in single soleus skeletal muscle fibers at pH 7.00 (empty bars) and pH 6.20 (filled bars). (D) Summary of ΔpCa_50 (pCa_50 pH 7.00 – pCa_50 at pH 6.20) in control, AdCMVssTnI-treated, and AdCMVaTnT-treated cardiac myocytes and in single soleus skeletal muscle fibers. Values expressed as mean ± SEM. Cross indicates significantly different from control at pH 7.00 (C; P < 0.01). Asterisk indicates significantly different (P <0.001) from control (D) or different from control at pH 6.20 (C). (from Westfall, M.V., Rust, E.M. and Metzger, J.M. (1997) Slow skeletal troponin I gene transfer, expression, and myofilament incorporation enhances adult cardiac myocyte contractile function. Proceedings from the National Academy of Science USA 94, 5333-5449).

3.2.2 Response to Acidosis

Acidosis has a more negative effect on the function of intact myocardium from the adult heart as compared to that of the immature heart, as reviewed above. Replacement of native neonatal troponin-tropomyosin with adult troponin-tropomyosin yielded data that support the hypothesis that the resistance of myofilaments from the immature heart to an acidic pH resides in troponin (Martin et al., 1991). The native troponin isolated from these two stages of development are likely to have different TnI and cTnT isoforms.

The correlation between the developmental fall in ssTnI expression and the developmental increase in the negative effects of acidosis has been tested by altering ssTnI expression *in vitro* and *in vivo* and by *in vitro* replacement experiments. Acidosis had a more negative effect on the relation between force and $[Ca^{2+}]$ in *in vitro* myocytes that express cTnI than those expressing ssTnI (Westfall et al., 1997). The acidosis induced shift in the pCa_{50} of ssTnI expressing cardiac myocytes was similar to the acidosis induced shift in ssTnI expressing soleus fibers (Westfall et al., 1997). Transgenic expression of ssTnI has also been shown to confer onto the myofilaments a relative resistance to acidosis (Fentze et al., 1999). Of note, in the adult cardiac myocyte, *in vitro* expression of a chimera, consisting of the cTnI amino-terminus and the ssTnI carboxyl-terminus, and replacement of endogenous cTnI with a mutant cTnI lacking the amino-terminal extension have the same protective effect against acidosis as those of ssTnI expression (Westfall et al., 2000). These results demonstrate the increased pH sensitivity conferred onto the myofilaments by cTnI is a consequence of residues outside the cTnI amino-terminal extension (Guo et al., 1994). Together, the *in vitro* and *in vivo* results demonstrate that, while ssTnI does not prevent the negative effect of acidosis on myofilament function, this effect is much decreased in myofilaments containing ssTnI.

3.2.3. Response to Adrenergic Stimulation

β-adrenergic receptor stimulation results in cAMP-dependent phosphorylation of cTnI but does not affect ssTnI. The functional consequence of PKA phosphorylation of cTnI has been directly addressed in extraction-reconstitution experiments (Wattanapermpool et al., 1995). The pCa-actomyosin Mg ATPase activity relations obtained from nonphosphorylated native myofibrils, myofibrils reconstituted with nonphosphorylated cTnI, and PKA treated myofibrils reconstituted with a mutant cTnI, lacking the PKA substrates, were indistinguishable. In contrast, the pCa_{50} of this relation was shifted rightward by an identical amount in PKA phosphorylated native myofibrils and myofibrils in which endogenous cTnI was replaced with PKA phosphorylated cTnI. In summary, cAMP-dependent phosphorylation of cTnI is sufficient to cause the decrease in sensitivity of myofilaments to calcium induced by PKA treatment of the myofilaments (A potential contribution from C protein phosphorylation was not found. C protein was removed from the myofibrils with the extraction process and not replaced, Wattanapermpool, 1995). Consequently, cAMP-dependent PKA phosphorylation of cTnI will decrease the sensitivity of myofilaments to calcium, while this effect will be absent in the presence of ssTnI.

A positive effect of catecholamines on cardiac contraction is mediated by α-AR stimulation in some species. In that PKC phosphorylation has both positive and negative effects, the PKC dependent pathways that transduce α-AR stimulation into this positive effect remain to be resolved. Among the proteins phosphorylated by PKC are the contractile proteins, cTnI, cTnT, C protein, and MLC2 ((Noland and

Kuo, 1991, 1992, 1993; Venema et al., 1993; Venema and Kuo, 1993). The overall negative effect of PKC phosphorylation on Ca^{2+} actomyosin Mg ATPase activity suggest that the positive effect of MLC2 phosphorylation on myocardial sensitivity is blunted by the negative effects of cTnI and cTnT phosphorylation. These effects appear to have biological significance based on the effects of *in situ* phosphorylation of isolated intact cardiac myocytes incubated with a α1AR agonist (Venema and Kuo, 1993). The extent of this phosphorylation may be less in the intact heart (Talosi and Kranias, 1992). This negative effect of PKC phosphorylation of cTnI will be absent in developing myocardium expressing only ssTnI.

3.2.4. Frank-Starling Relation and the TnI Isoforms

The contributions of the TnI isoforms to the sarcomere length-induced increase in the sensitivity of the myofilaments to calcium has been examined *in vivo* and *in vitro* (Arteaga et al., 2000). At a given sarcomere length, the replacement of cTnI with ssTnI through transgenesis enhances the sensitivity of the myofilaments to calcium. An increase in sarcomere length shifts the force-pCa relation of preparations containing ssTnI and cTnI or cTnI alone toward a lower calcium concentration (Figure 8). However, the extent of this enhancement in myofilament sensitivity was greater in preparations containing only cTnI.

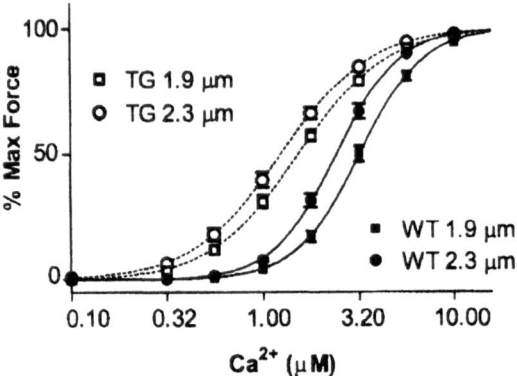

Figure 8. Ca^{2+}-force relations of wild-type (WT) and transgenic (TG) skinned cardiac trabaculae over-expressing ssTnI at sarcomere lengths (SL) of 1.9 and 2.3 μm. the half-maximal activating Ca^{2+} concentration (EC_{50}) for WT skinned myofilaments was 3.20 ± 0.24 μm at SL 1.9 μm (n=28, ■) and 2.40 ± 0.16 μm at SL 2.3 μm (n=25, ●); for TG skinned myofilaments, the EC_{50} was 1.53 ± 0.88 μm at SL 1.9 μm (n=28, □) and 1.25 ± 0.09 μm at SL 2.3 μm (n=25, O) (P < 0.05). (from Arteaga, G.M>, Palmiter, K.A., Leiden, J.M., and Solaro R..J. (2000). Attention of length dependence of calcium activation in myofilaments of transgenic mouse hearts expressing slow skeletal troponin I. Journal of Physiology 526, 541-549.)

The effect of PKA phosphorylation on the sarcomere length-dependence of myofilament sensitivity to calcium has been examined in cardiac myofilaments *in vitro* by phosphorylating native myofilaments and by replacing endogenous cTnI with phosphorylated and nonphosphorylated cTnI (Kajiwara et al., 2000). The shift toward a lower [Ca^{2+}] with increasing muscle length was present only when cTnI was not phosphorylated (Figure 9, Kajiwara et al., 2000).

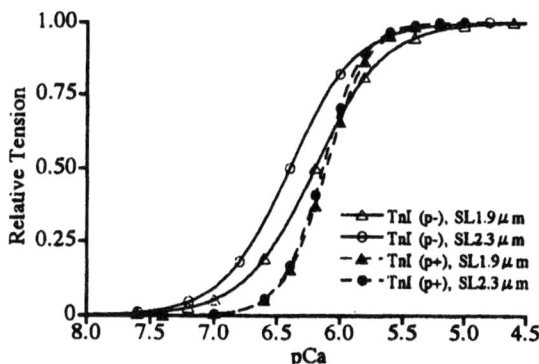

Figure 9. Effect of TnI reconstitution with PKA and non-PKA treated cTnI on sarcomere length-dependent shift of pCa-tension relation. A shift does not occur in the presence of phosphorylated TnI. Skinned preparations were reconstituted with phosphorylated or dephosphorylated cTnI. pCa-tension relation at SL 1.9 μm (Δ, ▲), and SL 2.3 μm (●, O) after replacement with dephosphorylated cTnI (open symbols) or phosphorylated cTnI (closed symbols). (from Kajiwara H, Morimoto S, Fukuda N, Ohtsuki I, and Kurihara S. (2000) Effect of Troponin I Phosphorylation by protein kinase A on length-dependence of tension activation in skinned cardiac muscle fibers. Biochemical and Biophysical Research Communications 272, 104-110)

This effect of cAMP-dependent phosphorylation of cTnI was not apparent in a study of donor and failing human cardiac myocytes where increasing sarcomere length shifted the pCa the same amount in myocytes from the donor and failing heart (van der Velden et al., 2000). The phosphorylation state of the two groups was different in that PKA phosphorylation of the donor myocytes had little effect on pCa_{50}, while PKA treatment of the failing myocytes significantly shifted the pCa_{50} rightward. This difference in cTnI phosphorylation between donor and failing human myocardium is consistent with previous studies (Bodor et al., 1997). The negative effect of cAMP-dependent phosphorylation on Frank-Starling relation in rat myocardium and the absence of any effect in human myocytes may reflect differences in the expression of contractile proteins between these two species.

3.3 The Functional Consequences of the TnI Isoforms in the Developing Heart

3.3.1. Acidosis

Acidosis is a frequent event during *in utero* life and birth. The ability to maintain cardiac function and output in the presence of acidosis is essential to the survival of the newborn. The ssTnI isoform-mediated resistance of the immature myocardium to acidosis is likely to be important in the ability of the immature heart to respond to pathophysiologic stress. Although the expression of only cTnI in the mature heart might be considered a disadvantage as regards the negative effects of acidosis on myofilament function, cTnI expression is essential for enhancing the modulation of ventricular function through adrenergic stimulation (see below).

3.3.2. β-Adrenoreceptor Stimulation

The sympathetic system of the heart is acquired with maturation (see above). The effects of sympathetic stimulation or a β-AR agonist will be absent in the embryonic and early fetal heart where β-AR's are not expressed. With the developmental acquisition of the components of this system, β-AR agonists will increase $[Ca]_i$ and heart rate. These responses must be accompanied by an increase in the rate of ventricular diastolic relaxation to avoid compromise of ventricular end-diastolic volume and so stroke volume. Although phosphorylation of phospholamban is likely to play a major role in enhancing the rate of relaxation (Li et al., 2000), PKA phosphorylation of cTnI will enhance relaxation by decreasing the sensitivity of myofilaments to calcium as measured by force, myofibrillar ATPase activity, and cross-bridge cycle kinetics (see Figure 10, Wattanapermpool, 1995; Robertson et al., 1982; Kentish et al., 2001). The cTnI phosphorylation-mediated shift in these processes toward higher calcium concentrations will decrease contraction duration and preserve ventricular diastolic filling. Of note, neither ssTnI nor cTnT are PKA substrates. This catecholamine-induced effect on myofilament function will be present in the adult heart, where only TnI is expressed, and absent in those developmental stages where only ssTnI is expressed in the heart.

Ventricular diastolic filling in the ssTnI expressing heart, relative to that of the heart expressing only cTnI, might thus be slowed, potentially decreasing end-diastolic volume, stroke volume, and so cardiac output. Recall that circulating catecholamine levels are high at birth and with perinatal stress, increasing the $[Ca]_i$ transient and heart rate. However, a potential negative effect of ssTnI expression does not appear to occur in that perinatal survival of ssTnI transgenic mouse pups appears to be the same as that of wild type pups expressing predominantly cTnI (Fentzke et al., 1999).

Maturational changes in myocardial response to catecholamines, the acquisition of myocardial sympathetic innervation, and the fall in cardiac ssTnI expression differ among species, suggesting that the relation between TnI isoform expression and these processes is complex. For example, sympathetic innervation of the rat heart occurs during postnatal life, while in the human, nerve structures are present in the fetal myocardium in the second trimester (Chen et al., 1979; Baker and Potter, 1980; Schumacher et al., 1984; Gennser and von Studnitz, 1972). In contrast, the developmental fall in ssTnI expression occurs earlier in the rat with ssTnI essentially disappearing by two weeks of life (Sabry and Dhoot, 1989), while in the human, ssTnI expression persists for months following birth (Hunkeler et al., 1991).

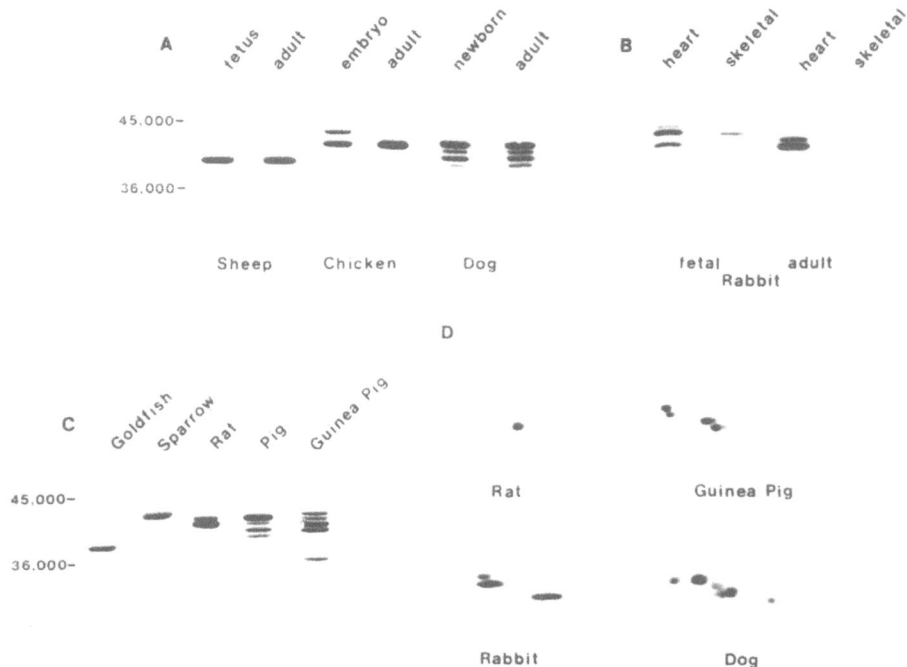

Figure 10. Phylogenic and ontogenic expression of cTnT isoforms. Panel A shows a cTnT specific antibody (mAb 13-11) recognizes its epitope in Western blots of cardiac preparations from newborn dog, adult dog, fetal sheep, adult sheep, embryonic chicken, and adult chicken. Sheep myocardium demonstrates little effect of development on cTnT isoform expression, in contrast to chicken, dog and rabbit (illustrated in panel C) myocardium. The two size markers are from Sigma: egg albumin, 45,000 Da and glyceraldehydes-3-phosphate dehydrogenase, 36,000 Da. Panel B shows that cTnT is expressed in rabbit fetal skeletal muscle but not in adult rabbit skeletal muscle. The two other lanes were loaded with fetal and adult rabbit cardiac preparations to illustrate the developmental changes in rabbit cTnT expression. Panel C shows cTnT isoform expression in juvenile pig, .adult rat, adult guinea pig, goldfish, and adult swamp sparrow (Melospiza Georgiana). Panel D shows Western blots, probed with mAb 13-11, of two-dimensional gels, 6.5% polyacrylamide concentration, loaded with cardiac preparations from adult rat, adult guinea pig, 2-day postnatal rabbit, and adult dog. The more basic part of the gel is to the right. In a study of human cTnT (Anderson et al, 1991), we found that the more acidic protein in pairs of troponin spots with the same M_r was phosphorylated. (from Malouf, N.N., McMahon, D., Oakeley, A.E., and Anderson, P.A.W. (1992) A cardiac troponin T epitope conserved across phyla. Journal of Biological Chemistry 267, 9269-9274.)

3.3.3. α-Adrenoreceptor Stimulation

The effects of development on expression of thin filament protein isoforms and the α-AR's may result in PKC isoforms having different effects on myocardial contraction at different stages of development. Many PKC isoforms are expressed in the heart, and their developmental pattern of expression varies among studies (Pucéat et al., 1994; Rybin and Steinberg, 1994). PKC epsilon, which localizes to the myofibrils, has been shown to be most abundant in the neonatal heart and to decrease calcium actomyosin ATPase activity (Rybin and Steinberg, 1994; Disatnik et al., 1994). Transgenic overexpression of PKC epsilon results in concentric cardiac hypertrophy, suggesting that α_1-AR stimulation in the neonatal heart contributes to the normal physiological neonatal increase in cardiac myocyte size and myocardial mass (Takeishi et al., 2000). *In vivo*, PKC β2-induced phosphorylation of cTnI decreases cardiac myocyte responsiveness to calcium and may cause depressed cardiac myocyte function (Takeishi et al., 1998). Although the maturational effects on α1AR and PKC isoform expression suggest that α-AR stimulation will have its greatest effects on cardiac function during perinatal life, ssTnI expression in the fetal and neonatal heart would appear to protect against negative effects of α-AR stimulation on maximum myofibrillar ATPase activity, mediated through PKC phosphorylation of cTnI.

3.3.4. Frank-Starling Relation

The potential value of the Frank-Starling relation (see above) in enhancing cardiac function in the neonate is supported by the marked increase in left ventricular end-diastolic volume and output with birth. How the multiple developmental changes in thin filament isoform composition and their post-translational modifications affect the Frank-Starling relation has only been partially examined.

The sarcomere length-dependence of the sensitivity to myofilaments to calcium has been found to be greater in myofilaments containing cTnI (Arteaga et al., 2000). This finding suggests that ssTnI expression in the fetal and neonatal heart could decrease the positive effect of an increase in left ventricular end-diastolic volume on cardiac output. On the other hand, sympathetic tone is increased at birth. The resultant cAMP-dependent phosphorylation of cTnI may blunt the effect of an increase in sarcomere length on myofilament function (Kajiwara et al., 2000; Komukai and Kurihara, 1997). In that ssTnI is not phosphorylated by this system, the expression of ssTnI in the neonatal heart will protect against a blunting of the Frank-Starling relation through enhanced sympathetic tone and PKA phosphorylation of cTnI. These potential effects on the Frank-Starling relation in the *in vivo* immature heart remain to be established.

4. CARDIAC TROPONIN T

Troponin T binds the troponin complex to tropomyosin and is essential for calcium regulated force development and actomyosin ATPase activity (Solaro and Van Eyk, 1996; Zot and Potter, 1987; Potter et al., 1995). Cardiac troponin T (cTnT) is expressed in the heart throughout development. Similar to the expression of cTnC, cTnT is the only TnT isoform expressed in the heart (except for the identification of ssTnT mRNA in the ventral interventricular groove of the embryonic mouse heart; Saggin et al., 1990). Unlike the cTnC and cTnI genes, which express only one isoform of TnC and TnI, the cTnT gene expresses multiple isoforms through combinatorial alternative splicing of the primary transcript (Anderson et al., 1988; Anderson and Oakeley, 1989; Greig et al., 1994). In the following section, we consider the cTnT isoforms, their molecular basis, and the effect of development on isoform expression. We will consider the findings that support a modulating role of the cTnT isoforms and their post-translational modifications on myofilament function.

4.1. cTnT Isoform Expression

The expression of more than one cTnT isoform was first identified in the chicken where a single cTnT gene generates two isoforms through alternative splicing (Cooper and Ordahl, 1984). They searched for other isoforms and found none, proposing their failure was a consequence of other transcripts being present at very low abundanc or at other stages of development. Only two cTnT isoforms have been identified at the protein level in the chick and quail heart (Malouf et al., 1992). Two cTnT isoforms were subsequently identified in the rat, bovine, and sheep heart (Jin and Lin, 1988; Lesyzk et al., 1987; McAuliffe and Robbins, 1991).

The expression of only two cTnT isoforms is not consistent with the rabbit and chicken fast skeletal (fs) TnT genes generating potentially dozens of isoforms through combinatorial alternative splicing of 5′ and 3′ exons (Breitbart et al., 1985, Wilkinson et al., 1984). Subsequently, multiple cTnT isoforms were found to be expressed at the protein level in the rabbit, rat, mouse, guinea pig, cat, and dog (Figure 11; Anderson et al., 1988; Anderson and Oakeley, 1989; Malouf et al., 1992).

The expression of human cTnT isoforms has been examined at the protein level in heart and skeletal muscle (Figure 12; Anderson et al., 1991). Four isoforms of cTnT ($cTnT_1$-$cTnT_4$, based on increasing electrophoretic mobility) were identified. In the human, similar to the rabbit, rat, and chicken heart, the larger isoforms were more acidic. All the isoforms undergo post-translational modifications in that each was resolved as at least two spots with different isoelectric points (Anderson et al., 1991; Malouf et al., 1992).

A

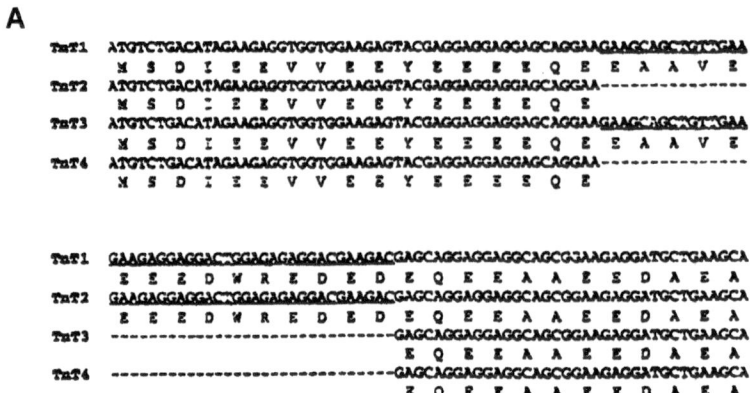

B

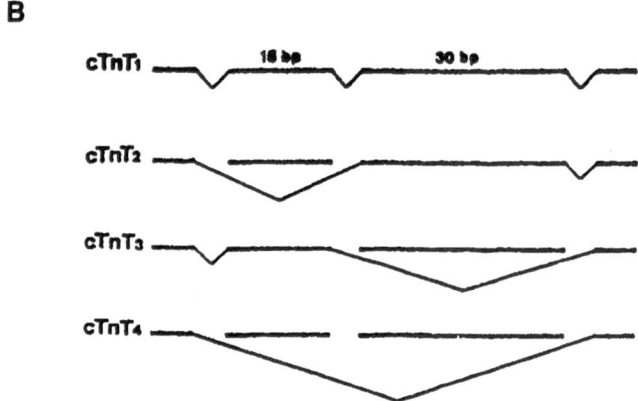

Figure 11. Alternatively spliced exons in the 5' end of rabbit cTnT (TnT) mRNA. A, Nucleotide and deduced amino acid sequences of the 18- and 30-nt alternatively spliced exons are underlined. They are separated by a codon that encodes a glutamate present in all the rabbit cDNA isoforms and the cardiac TnT isoform sequences of sheep, rat, cow, and chicken. B, Schematic depiction of the splicing pattern of the alternatively spliced 5' exons that gives rise to four cardiac TnT isoforms. Cardiac TnT₁, mRNA retains both exons; cardiac TnT₂ and cardiac TnT₃ retain either the 30- or the 18-nt exons, respectively; and cardiac TnT₄ lacks both exons. (from Greig, A., Hirschberg, Y., Anderson, P.A.W., Hainsworth, C., Malouf, N.N., Oakeley, A.E., and Kay, B.K. (1994) Molecular basis of cardiac troponin T isoform heterogeneity in rabbit heart. Circulation Research 74, 41-47.)

Evidence at the protein level for expression of more than five cTnT isoforms is found in the dog heart and in other species (Malouf et al., 1992). In the dog, at least eight isoforms are expressed, with the larger ones comigrating with cTnT isoforms identified in other species. The developmental change in the relative amounts of the dog cTnT's suggest they are not products of proteolysis.

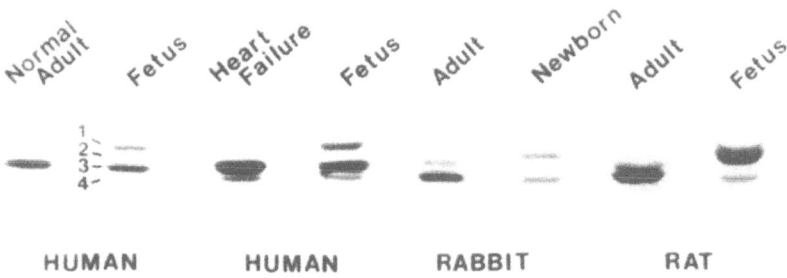

Figure 12. Western Blots. Variation in cTnT isoform expression across species and development is demonstrated by probing myofibrillar preparations from normal adult and fetal human hearts, failing adult (secondary to ischemic heart disease) and fetal human hearts, newborn and adult rabbit hearts, and adult and fetal rat hearts with a cTnT specific antibody. (from Anderson, P.A.W., Greig, A., Mark, T.M., Malouf, N.N., Oakeley, A.E., Ungerleider, R.M., Allen, P.D., Kay, B.K. (1995) Molecular basis of human cardiac troponin T isoforms expressed in the developing, adult, and failing heart. Circulation Research 76, 681-686.

4.2. The Molecular Basis of the cTnT Isoforms

Cooper and Ordahl (1984) found that a 30 nt exon, exon 5 of the chicken cTnT gene, is alternatively spliced yielding two isoforms that differ by a peptide, containing 8 negatively charged residues. Alternative splicing of a 5' exon was also identified in the rat and sheep. In the rat, similar to the chicken, alternative splicing of a 30 nt exon was found, while in the sheep, alternative splicing of a 5' 15 nt exon was found (Jin and Lin, 1988; McAuliffe and Robbins, 1991). The two bovine cTnT isoforms also differ by a 5 residue acidic amino-terminal peptide (Leszyk et al., 1987).

Combinatorial alternative splicing of 5' 18 and 30 nt exons yields four cTnT isoforms in the rabbit (Figure 11; Greig et al., 1994). The clones also demonstrated variable expression of a 3' 9 nt exon. They contained the entire exon, encoding QAQ, or its 3' 6 nt or lacked the entire exon. The product of the rabbit 30 nt exon has eight acidic residues as does that of the rat and chicken variably expressed 30 nt exon. The variably expressed amino-terminal 6 and 5 residue peptides of the rabbit, sheep, and bovine cTnT are also acidic. Using four rabbit cDNAs that contained the 3' 9 nt sequence and that differed only through the combinatorial alternative splicing of the two 5' exons, *in vitro* transcription and translation demonstrated that the four rabbit isoforms comigrated with the endogenously expressed cTnT isoforms.

However, the variably expressed 3′ 9 nt exon suggests the potential for 12 cTnT isoforms (Greig et al., 1994).

4.2.1. Human cTnT Isoforms

The first molecular evidence for multiple human cTnT isoforms was provided by Townsend et al. (1994). The sequence of the full-length cTnT cDNA, which they described, was very similar in sequence to one published earlier by Mesnard et al. (1993). Analysis of PCR products of the 5′ region supported the presence of alternative splicing. The products differed by a 27 nt sequence encoding an amino-terminal acidic peptide similar to those identified previously in chicken, rat, and rabbit cTnT.

The molecular basis of the human cTnT isoforms, identified at the protein level in the human heart (Anderson et al., 1991), was more fully characterized by subsequent studies of Anderson et al. (1995), Mesnard et al. (1995), and Townsend et al. (1995). In addition to the alternatively spliced exon described by Townsend et al. (1994), combinatorial alternative splicing of 5′ 30 and 15 nt sequences was identified along with the variably expressed 3′ 9 nt exon, previously identified in the rabbit (Greig et al., 1994). The previously described 27 nt sequence and the 30 nt sequence differ by a 5′ codon for glutamate. Mesnard et al. (1995) also found another variably expressed codon for glutamate.

In vitro expression of the four cDNAs that variably contained the 30 nt and 15 nt sequences yielded products that comigrated with the native cTnT isoforms (Anderson et al., 1995). In addition to these four cTnT isoforms, variable expression of the 3′ 9 nt exon and the two codons, identified by Townsend et al. (1995) and Mesnard et al. (1995), could yield at least 36 cTnT isoforms. Most of the products of this combinatorial alternative splicing would differ by only one or two residues, potentially precluding their separation by SDS-PAGE. Polymorphisms of cTnT have also been identified (Farza et al., 1998).

4.2.2. Molecular Basis of Other cTnT Isoforms

Additional cTnT isoforms may be present, based on the number of fsTnT isoforms (Wilkinson et al., 1984; Breitbart et al, 1985). This potential is supported by the identification of eight or more cTnT isoforms at the protein level in the dog heart (Malouf et al., 1992). Alternative splicing in the highly conserved central portion of the cTnT molecule is suggested by the RT-PCR results of Townsend et al. (1995). However, a comparison across species demonstrates that, although the amino-terminal 67 residues of cTnT show significant sequence variability, the rest of the molecule is highly conserved.

4.3. Development and cTnT Isoform Expression

The same cTnT isoforms are expressed in the fetal heart and skeletal muscle in the human with cTnT gene expression disappearing in skeletal muscle with maturation.

In some species, the developmental shift in cTnT isoform expression leads to essentially a single isoform being expressed in the adult heart, e.g. the human and sheep (Anderson et al., 1991; Godt et al., 1991; McAuliffe and Robbins, 1991). In other species, multiple isoforms expressed early in development continue to be expressed in the adult with the smallest least acidic isoforms being expressed predominantly in the adult.

4.4. Heart Disease and cTnT Isoform Expression

Heart failure and hypertrophy have been found to alter cTnT isoform expression in the human, guinea pig, rabbit, and rat (Anderson et al., 1991, 1995; Townsend et al., 1995; Mesnard et al., 1995; Gulati et al., 1994; Akella et al., 1995; Peterson et al., 2002). In the adult human, the failing heart expresses three isoforms, while in the donor heart, two of these isoforms that differ by the presence or absence of the 5′ 15 and 30 nt combinatorially spliced exons are not expressed or are expressed at very low levels (Anderson et al., 1991, 1995; Mesnard et al., 1995; Townsend et al., 1995). The expression of the smallest of the human cTnT isoforms, which lacks both of these 5′ sequences, is significantly higher in the adult failing heart than in the donor heart (Anderson et al., 1991). In addition to changes in isoform expression, proteolysis of cTnT has been identified in human cardiomyopathic hearts (Margossian et al., 1999).

Alterations in cTnT isoform expression have also been characterized in myocardium obtained from infants and children undergoing surgery for congenital cardiac defects that cause heart failure or hypoxemia (Saba et al., 1996). The expression of the cTnT isoform containing both the 10 residue and 5 residue amino-terminal peptides ($cTnT_1$) was highest in the neonate and disappeared in the first months of life regardless of the congenital defect. In contrast to the developmental fall in the expression of this isoform, the expression of the smallest cTnT isoform ($cTnT_4$), which lacks both of the amino-terminal acidic peptides, persisted during development in the infants with heart failure (Figure 13). The level of its expression was directly correlated with the severity of heart failure, the post-operative duration of inotropic support, and Intensive Care Unit stay (Saba et al., 1996). These results are consistent with heart failure at all stages of development shifting cTnT isoform expression in the human heart (see below, cTnT Isoforms and Myofilament Function). The functional consequences of the disease-induced switch in cTnT isoform expression in the infant, child, and adult and the developmentally regulated pattern of cTnT isoform expression remain to be established.

As described above, perinatal stress is associated with high circulating levels of norepinephrine and epinephrine in the human infant. A maternal injection of phenylephrine, an α1AR agonist (Gao et al., 1995), was used as a means for inducing fetal stress. The developmental shift in cTnT isoform expression was accelerated in the hearts of fetuses whose mothers received an injection of phenylephrine, while the perinatal switch from ssTnI to cTnI was unaffected. The potential explanations for these fetal effects in response to a maternal injection of an α1AR agonist remain unresolved until additional data are obtained.

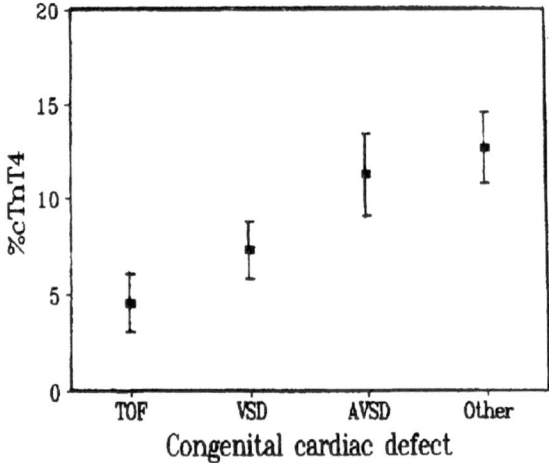

Figure 13. Mean percent (±SEM) cTnT₄ of total (see Figure 11) expressed in myocardium obtained from patients with congenital cardiac defects undergoing surgical repair (age 3 weeks to 18 months). The different diagnostic groups: tetrology of Fallot (TOF), ventricular septal defect (VSD), atrioventricular septal defect (AVSD), other (complex congenital cardiac defects). The difference in percent cTnT₄ between TOF preparations and the average of the VSD, AVSD, and other groups was statistically significant (P=.018). The patients with AVSD and in the Other group had marked heart failure. (from Saba, Z., Nassar, R., Ungerleider, R.M., Oakeley, A.E., and Anderson, P.A.W. (1996). Cardiac troponin T isoform expression correlates with pathophysiological descriptors in patients who underwent corrective surgery for congenital heart disease. Circulation 94, 472-476.)

4.5. cTnT Isoforms, Development, and Electrophysiology

Regional differences in cTnT isoform expression have been described in the adult and newborn canine right atria , tissue essential for maintaining heart rhythmicity (Spach et al., 1989). These regional differences in cTnT isoform expression appear unique in that regional differences in cTnT isoform expression have been sought and not been found in the ventricle. This variation in thin filament isoform expression in electrophysiologically important tissue is reminiscent of ssTnI expression persisting in the Purkinje fibers of the mature heart (Gorza et al., 1993).

The regional changes in cTnT isoform expression and electrophysiological characteristics differed between the neonatal and adult heart (Spach et al., 1989). In both the adult and neonatal dog atrium, the expression of the largest isoform was greatest in the area of the sinus node. Although its expression fell with increasing distance from the sinus node in the adult, the pattern of cTnT isoform expression remained unchanged in the neonate.   As regards the electrophysiological characteristics, in both the adult and the neonate, the action potentials were longest in the area of the sinus node.   Although action potential duration decreased with increasing distance from the sinus node in the adult, these regional changes in action potential characteristics did not occur in the newborn.   The maturational and regional changes in the electrophysiological properties and in cTnT isoform expression suggest that the expression of molecules that regulate ionic mechanisms and myofilament function are coordinated.

4.6. cTnT Isoforms And Myofilament Function

4.6.1. Development
Development affects the sensitivity of myofilaments to calcium and the expression of the contractile proteins, for example, the cTnT and cTnI isoforms (see above). The maturation decrease in the sensitivity of the myofilaments to calcium and the increase in expression of cTnT isoforms with a more basic amino-terminal region suggest a correlation.   In rat and bird, development is associated with a fall in the sensitivity of the myofilaments to calcium and a fall in the expression of the larger acidic cTnT isoforms (Fabiato and Fabiato, 1979; Reiser et al., 1994; Godt et al., 1991; Jin and Lin, 1988; Cooper and Ordahl, 1984).   Similarly, the pCa actomyosin ATPase activity relations of the five day and 22 day old rabbit demonstrate a maturational fall in calcium sensitivity (McAuliffe et al., 1990) associated with a decrease in the expression of the more acidic cTnT isoforms (Greig et al., 1994). Correlations between function and TnT isoform expression in vertebrate fast skeletal muscle (Schachat et al., 1987) and in the indirect flight muscle of the dragonfly (Fitzhugh and Marden, 1997) are consistent with these results.

A potential modulating effect of the cTnT isoforms on calcium sensitivity of the myofilaments is supported by other studies (Ventura-Clapier et al., 1990; Nassar et al., 1991).   A positive correlation was found between the sensitivity of rabbit cardiac myofilaments to calcium and the expression of the cTnT isoform containing the variably expressed amino-terminal 10 residue peptide (Figure 14; Nassar et al., 1991; called $cTnT_2$ in Greig et al., 1994; see Figure 11).   This correlation is consistent with the hypothesis that the charge of the cTnT amino-terminus that extends over the amino and carboxyl terminis of adjoining tropomyosin molecules modulates myofilament function (Perry, 2000).

4.6.2. Heart Failure
A potential effect of cTnT isoform expression on myofilament function comes from the study of myofibrillar ATPase activity and cTnT isoform expression in the normal and failing adult human heart (Anderson et al., 1991).   Myofibrillar ATPase activity of the failing adult human heart is depressed, while myosin ATPase activity is similar in the normal and failing heart (Pagani et al., 1988; Alpert and Gordon,

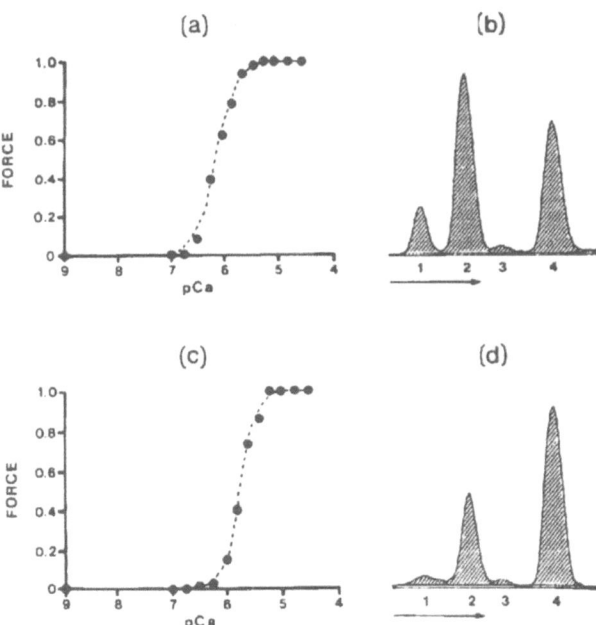

Figure 14. The force-pCa relations of two neonatal rabbit ventricular strands and the densitometric scans of the Western blots of these two strands. The strand proteins were resolved with SDS-PAGE and probed with a cTnT specific monoclonal antibody. Force was normalized with respect to force at maximum activation. The cTnT isoforms are numbered 1-4 (see Figure 11). The force-pCa curve in Panel a is shifted to the left relative to the curve shown in Panel c. The strand that was more sensitive to calcium (Panel a) had a greater proportion of cTnT₂ (Panel b) and less of cTnT₄ than the less sensitive strand (Panel d). The arrows indicate the direction of electrophoresis. (from Nassar, R., Malouf, N.N., Kelly, M.B., Oakeley, A.E., and Anderson, P.A.W. (1991) Force-pCa relation and troponin T isoforms of rabbit myocardium. Circulation Research 69, 1470-1475.)

1962). A study of myofibrillar ATPase activity and cTnT isoform expression in these hearts (Pagani et al., 1988; Anderson et al., 1991) examined whether contractile proteins, other than the MHC isoforms, are the basis of this disease-induced fall in myofibrillar ATPase activity. The expression of the smallest human cTnT isoform, lacking both the 5 and 10 residue amino-terminal peptides, was found to be correlated in an inverse manner with maximum myofibrillar ATPase activity (Anderson et al., 1991). This correlation is reminiscent of a mutational analysis of fsTnI that found deletion of the TnT amino-terminus resulted in a decrease in maximal Ca^{2+} actomyosin activity (Figure 15; Pan et al., 1991).

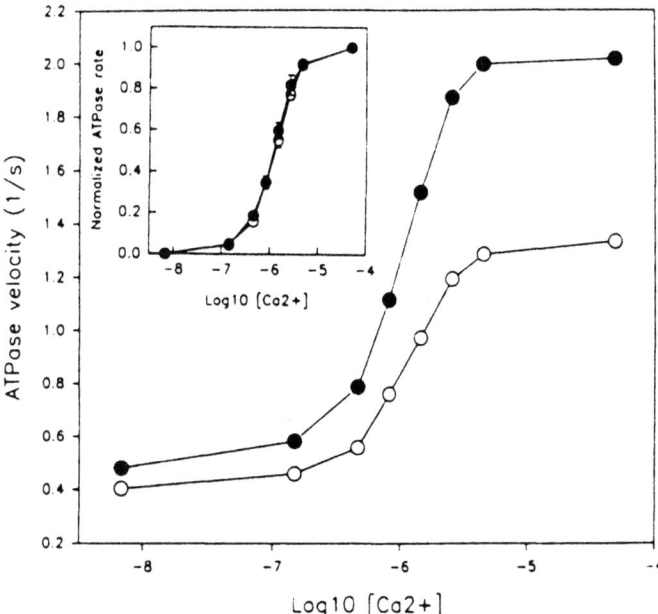

Figure 15. Effect of substitution of a 26,000 dalton TnT fragment (26-TnT), which corresponds to residues 46-259 of TnT₂f, for TnT₂f on Ca²⁺ dependence of actin-activated S1 MgATPase. Shown are the data from a representative experiment with reconstituted thin filaments containing TnT₂f (closed circles) or 26K TnT (open circles). Note the lower maximum MgATPase in the presence of 26K-TnT. Insert: averaged normalized data from four experiments. The same symbols are used. The lines are drawn through the data points. (from Pan, B.S., Gordon, A.M., and Potter, J.D. (1991) Deletion of the first 45 NH2-terminal residues of rabbit skeletal troponin T strengthens binding of troponin to immobilized tropomyosin. Journal of Biological Chemistry 266, 12432-12438.)

4.6.3. Hypertrophy

A comparison of cross-bridge kinetics was made between pressure overloaded rabbit myocardium expressing 100% β-myosin heavy chain and a control group that also expressed 100% β-myosin heavy chain (Peterson et al., 2001). Cross-bridge force-time interval and expression of $cTnT_2$, which contains the 10 residue acidic peptide (Figure 11), were significantly greater in the hemodynamically overloaded myocardium.

4.6.4. Mutational Analysis of the NH₂-Region of TnT

The effects of the TnT amino-terminal region on function has been directly assessed using mutational analysis. A cTnT mutant, lacking a 38 residue amino-terminal peptide, was found to shift the pCa_{50} of Mg-ATPase activity of myosin S-1-actin-tropomyosin toward a higher calcium concentration (Tobacman, 1988). The calcium specific binding site of TnC was 50% weaker in the presence of the amino-terminal truncated cTnT. In a study of intact and mutant fsTnT, deletion of the first 45 residues of TnT_{2F} decreased maximal ATPase activity by approximately 40% while having no effect on the pCa_{50} or Hill coefficient of this relation (Figure 15; Pan et al., 1991). In a subsequent study, a cTnT mutant lacking the first 76 residues of cTnT was studied in detergent skinned tension-bearing cardiac fibers. Binding of the truncated cTnT to tropomyosin was 3-4 fold greater than that of intact cTnT, while the pCa_{50} and the Hill coefficient of the pCa-force relation were unaffected (Chandra et al., 1999). However, the maximal force in the presence of the truncated mutant was approximately two-thirds of that developed by fibers containing intact cTnT. The various functional consequences of the TnT amino-terminal deletion mutants may reflect differences in the extent of the deletion, the isoforms that were mutated, and the assays applied to the analysis. The results, however, are consistent with the importance of the cTnT amino-terminus in modulating function. Transgenesis experiments will provide further insights into the functional differences that result from the amino-terminal differences among the cTnT isoforms.

4.7. Phosphorylation of cTnT, Functional Consequences

All the cTnT isoforms contain the substrates for PKC. PKC has been shown to phosphorylate isolated cTnT (Noland et al., 1989), while cTnT is not phosphorylated to a significant extent by PKA. Phosphorylation of cTnT can be induced in cardiac myocytes by activation of PKC with a phorbol ester (Liu et al., 1987). In reconstituted bovine cardiac actomyosin systems, PKC phosphorylation of cTnT decreased the maximum Ca^{2+} Mg-ATPase activity but did not affect calcium sensitivity (Noland and Kuo, 1991, 1992, 1993). In contrast, PKC α phosphorylation of cTnT *in vitro* decreased both calcium sensitivity and maximum actomyosin Mg-actomyosin ATPase activity (Jideama et al., 1996).

A functional consequence of PKC phosphorylation of cTnT is supported by transgenesis experiments in which cTnT expression was partially replaced by fsTnT (Montgomery et al., 2001). The fsTnT isoform does not contain the cTnT residues phosphorylated by PKC. Phorbol ester treatment of the wild type preparation reduced maximum tension by approximately 30% but had no affect on maximum tension in transgenic fsTnT containing preparations.

Phorbol ester treatment of the fsTnT transgenic preparations did induce a significant increase in calcium sensitivity compared to the wild type controls, suggesting the positive effect of PKC phosphorylation of other contractile proteins (Montgomery et al., 2001). This potential modulatory effect of other contractile proteins was demonstrated in a reconstituted system where phosphorylation of cTnT resulted in a 60% decrease in maximum actomyosin Mg-ATPase activity, phosphorylation of cTnI produced a much smaller decrease in maximum activation, and phosphorylation of both cTnI and cTnT produced an intermediate decrease in activity (Noland and Kuo, 1993).

Although cTnT phosphorylation has been described in the human at different stages of development and with disease (Anderson et al., 1991), variations in the level of phosphorylation with development or disease processes that could result in functional differences remain to be identified.

4.8. cTnT Isoforms, Development, and Function

The developmental changes in cTnT isoform expression with a relative increase in expression of isoforms with more basic amino-terminal sequences, the correlations among myofilament function and isoform expression with development, and the mutational analysis of the functional effects of the cTnT amino-terminus all suggest that filament function is modulated by developmental changes in cTnT isoform expression. The exact nature of these functional effects, however, remains to be established.

The functional effect of PKC phosphorylation of cTnT suggest that the perinatal increase in cardiac α-adrenoreceptors and circulating catecholamine levels will have a depressive effect on neonatal ventricular function (Montgomery et al., 2001). However, a positive effect on survival of the transgenic fsTnT mouse in which this negative effect would be decreased was not described (Montgomery et al., 2001). As regards the modulatory effects that follow from post-translational modification of cTnI (Montgomery, 2001; Noland and Kuo, 1993), this positive effect will be absent in early cardiac development where only ssTnI is expressed in the heart. When cTnI is expressed later in development, PKC phosphorylation of cTnI (see above 4.7) could decrease the negative effects of PKC phosphorylation of cTnT on myofilament function.

5. ACTIN

Actin is a multifunctional protein that participates in a large number of cell processes. These include contractility, motility, division, and muscle contraction. In striated muscle, actin provides the backbone of the thin filament and is essential for contraction.

5.1. α-actin Isoforms

Six actin isoforms are expressed in higher vertebrates (Vandekerckhove and Weber, 1978). The expression of four actin isoforms is limited to muscle. In the adult, skeletal and cardiac α-actins are expressed in striated muscle, smooth muscle α-actin in vascular smooth muscle, and smooth muscle gamma-actin in gastrointestinal and genital tracts (Vandekerckhove and Weber, 1979, 1981). These isoforms are completely conserved from bird to the human. Although skeletal and cardiac α-actin differ by only four residues, three of the residues are found in the myosin binding region (Sutoh, 1982), suggesting a potential functional difference.

5.2. α-actin Isoforms and Cardiac Development

Cardiac, skeletal, and smooth muscle α-actin are expressed in the myocardium at different stages of development and are incorporated into the cardiac sarcomere (Sugi and Lough, 1992; Sawtell and Lessard, 1989; Sassoon et al., 1988; Ruzicka and Schwartz, 1988; McHugh et al., 1991; Boheler et al., 1991). In the adult chicken, mouse, rat, and bovine heart, cardiac α-actin is the predominant isoform. In contrast, in the adult human heart, the level of skeletal α-actin mRNA is equal to or greater in amount than that of cardiac α-actin (Boheler et al., 1991; Gunning et al., 1983). At the protein level, skeletal α-actin is found at a much higher level in the human heart than in the rodent (Bennetts et al., 1986; Vandekerckhove and Weber, 1978).

Sequential expression of vascular smooth muscle α-actin, skeletal α-actin, and cardiac α-actin has been described in the developing heart (Ruzicka and Schwartz, 1988). Vascular smooth muscle α-actin is the first α-actin isoform expressed in the embryonic heart with either skeletal or cardiac α-actin being the second isoform to be expressed (Sugi and Lough, 1992; Ruzicka and Schwartz, 1988; Sawtell and Lessard, 1989; McHugh, 1995). Vascular smooth muscle α-actin expression disappears during early neonatal life (Sawtell and Lessard, 1989).

The developmental expression of the actin isoforms also follows a regional distribution in the heart. In the chick heart, vascular smooth muscle α-actin expression disappears first in the atrium and ventricle, becoming localized to the sinus venosus and the outflow tract of the heart (Sugi and Lough, 1992). A similar pattern of vascular smooth muscle α-actin expression was observed in the rat (Sawtell and Lessard, 1989).

The developmental time course of skeletal muscle and cardiac α-actin expression is, in general, the same in the chicken, mouse, rat, pig, and bovine heart. Skeletal α-actin expression falls, and cardiac α-actin becomes the predominant isoform in the adult heart. In contrast, in the human heart, cardiac α-actin is the predominant isoform in the fetus rather than skeletal α-actin, and cardiac α-actin expression falls with maturation, reaching its lowest level of expression in the adult heart (Boheler et al., 1991).

5.3. Pathophysiologic Processes and α-Actin Isoform Expression

Pathophysiologic processes affect the cardiac expression of the α-actin isoforms. In the rat, hemodynamic overload rapidly increases the level of skeletal α-actin mRNA (Schwartz et al., 1986). However, the accumulation is transient disappearing in the presence of persistent hemodynamic overload (Schwartz et al., 1986). In the human heart, no differences were found among the levels of skeletal and cardiac α-actin expression in donor, cardiomyopathic, or ischemic hearts (Boheler et al., 1991). This apparent absence of a difference in expression in the α-actin isoforms may result from the chronic nature of the disease process, as compared to the animal model, patient-to-patient variability in response, and different disease processes.

Actin isoform expression is affected by hyperthyroid and hypothyroid states. Thyroid hormone administration rapidly increases skeletal α-actin expression and, to a much smaller degree, cardiac α-actin (Winegrad et al., 1990). *In vitro*, α1AR agonists increase skeletal α-actin expression through the PKC pathways (Bisphoric et al., 1987; Simpson et al., 1989). The positive effects of these agents on skeletal α-actin expression are reminiscent of the neonatal increase in thyroid hormone and circulating catecholamines and the transient increase in skeletal α-actin expression found in the neonatal rat heart (Carrier et al., 1992).

5.4. α-Actin Isoforms and Function

The developmentally regulated pattern of expression of the three α-actin isoforms in the heart suggests that small differences in sequence among these isoforms have a functional consequence. A naturally occurring model for testing these functional effects is the BALB/c mouse that expresses high levels of skeletal α-actin that vary among adult animals. In the isolated working heart, a correlation was found between skeletal α-actin mRNA (ranging from 20% to 50% of total sarcomeric actin mRNA) and ventricular function: the higher the level of skeletal α-actin mRNA, the higher was the peak of rise of left ventricular pressure, the shorter was the time to peak pressure, and the more rapid was left ventricular diastolic relaxation (Hewett et al., 1994). In contrast, an *in vitro* study of isolated cardiac myocytes from BALB/c mice and control mice found no difference in baseline function as measured by the force-pCa relation (Metzger, 1997). A pathophysiological intervention, acidosis, did depress maximum tension to a greater degree in the BALB/c myocytes.

The apparent functional advantage of skeletal α-actin expression, observed in the *in vitro* BALB/c heart, was not found *in vivo* in a study that made use of homologous recombination to disrupt the cardiac α-actin gene and transgenesis to express smooth muscle gamma-actin in the heart (Kumar et al., 1997). In the heterozygous cardiac α-actin null mouse, cardiac α-actin was significantly decreased, while vascular smooth muscle and skeletal α-actin expression was increased. These heterozygous null mice had normal survival, and their *in vivo* ventricular function was the same as that of the wild type animal. The hearts of the heterozygous null mice demonstrated no hypertrophy or ultrastructural abnormalities. Thus, similar to an *in vitro* motility study that demonstrated no functional difference between smooth muscle and skeletal α-actin (Harris and Warshaw, 1993), this *in vivo* study of cardiac α-actin deficient mice suggest that smooth muscle, skeletal, and cardiac α-actin are equivalent in supporting cardiac function.

The homozygous cardiac α-actin null mice also demonstrated an increase in smooth muscle and skeletal α-actin expression. However, total actin was only about 50% of that found in the wild type, and the majority of these homozygous null mice did not survive to term. This morbid effect appears to be related to the depressed level of total α-actin expression rather than that of the individual isoforms.

Smooth muscle gamma-actin does not appear to be equivalent to the other three muscle α-actin isoforms, based on attempts to rescue homozygous cardiac α-actin null mice with transgenic expression of gamma-actin. Transgenic expression of gamma-actin rescued only about one-third of the homozygous null mice. Systolic arterial pressure, dP/dt_{max}, and dP/dt_{min} were significantly lower in the survivors than in heterozygous cardiac α-actin null mice and the wild type, and transgenic expression of gamma-actin was also associated with dilation and hypertrophy of the heart. This apparent negative effect of gamma-actin on ventricular function is supported by the gamma-actin transgene positive animals having depressed ventricular function, as compared to wild type controls. The absence of smooth muscle gamma-actin expression in the heart throughout development appears teleologically reasonable given this negative effect (Sawtell and Lessard, 1989).

5.5 Actin Isoforms, Heart Function, and Development

In that the functional differences among the α-actin isoforms remain to be resolved, any consideration of the functional consequences of the developmental changes in expression is speculative. For example, if skeletal α-actin expression does confer onto the heart an enhancement of ventricular function, as seen in the BALB/c mouse, the neonatal increase in hemodynamic load may be partially supported by the transient increase in skeletal α-actin expression, observed in the rat (Carrier et al., 1992) and the sustained increase in the human neonate (Schwartz et al., 1986). To extend the speculation about the functional consequences of the neonatal increase in skeletal α-actin expression in the human infant, this normal developmental increase in skeletal α-actin expression may play an important role in supporting ventricular function in the infant with a congenital cardiac defect where the hemodynamic load increases progressively over the first weeks and months of life.

6. TROPOMYOSIN

Tropomyosin is a fibrous molecule composed of two α-helical chains arranged as a coiled coil associated with actin (Perry, 2001). Tropomyosin provides structural stability and modulates filament function. In striated muscle, calcium induces tropomyosin movement on actin from a position that blocks interaction to one that exposes myosin-binding sites of actin. The role of tropomyosin in tension development is supported by findings in reconstituted actin filaments (Fujita and Ishiwata, 1999) and motility assays (Bing et al., 2000a; Van Buren et al., 1999).

6.1. Tropomyosin Isoforms

Four tropomyosin genes have been identified in mammals. In the human, they have been named TPM1, TPM2, TPM3, and TPM4 (Cuticchia and Pearson, 1993). TMP1 and TPM2 encode the α and β tropomyosin isoforms found in striated muscle. The TPM3 gene (gamma TM) codes for fibroblast TM5 and a slow skeletal muscle α-tropomyosin. The TPM4 gene (delta or TM4 gene) codes for fibroblast TM4.

6.2. Tropomyosin Isoforms and Development

In the heart, α and β-tropomyosin are predominantly expressed. Their relative expression changes with time and differs among species. In the mouse heart, the α-tropomyosin/β-tropomyosin ratio increases from approximately 5:1 during early embryonic life to 60:1 in the adult (Muthuchamy et al., 1993). In the mouse, βTM transcripts have been identified in the embryonic heart but not the neonatal and adult heart, while a βTM3 isoform has been identified at all stages of heart development (Muthuchamy et al., 1993; Wang and Rubenstein, 1992). Similar to the mouse, the rat heart expresses multiple tropomyosin isoforms at different levels during development with α-tropomyosin being dominantly expressed in the adult heart (L'ecuyer et al., 1991). In the embryonic rat heart, thin filament preparations contained α and β-tropomyosin and nonmuscle isoforms of TM3, 4, and 5. TM4 and 5 expression persisted throughout rat heart development, while TM3 disappeared during neonatal life (L'ecuyer et al., 1991).

In the adult human heart, α-tropomyosin is predominantly expressed (Humphreys and Cummins, 1984). However, in the human and bovine heart, in contrast to the rodent heart, β-tropomyosin expression increases during development, reaching approximately 10% of the tropomyosin in the adult heart. The differences in α and β-tropomyosin expression among species with markedly different heart rates led to the hypothesis that a faster heart rate requires a higher level of α-tropomyosin expression to enhance the rate of ventricular diastolic relaxation.

6.3. Functional Consequences of Altering α and β-Tropomyosin Expression

The functional consequences of increased β-tropomyosin expression have been examined using transgenesis and studying the isolated heart, isolated cardiac myocyte, and membrane extracted preparation (Muthuchamy et al., 1995; Palmiter et al., 1996; Wolska et al., 1999; Patel et al., 2001). The transgenic increase in β-tropomyosin expression was accompanied by a reciprocal decrease in α-tropomyosin expression.

In the isolated working heart, the only difference was slower ventricular diastolic relaxation in the β-tropomyosin transgenic animal. Importantly, this difference in diastolic relaxation disappeared when ventricular volume was increased, or the hearts were exposed to isoproterenol. After these interventions, measures of ventricular relaxation were indistinguishable between the transgenic and wild type heart (Muthuchamy et al., 1995).

The functional consequences of enhanced β-tropomyosin expression in the myocardium of transgenic animals included an increase in thin filament activation by strongly bound cross-bridges and an increase in calcium sensitivity of steady state force, pCa 5.7 and 5.57 in the transgenic and wild type, respectively (Palmiter et al., 1996). In isolated intact cardiac myocytes, the maximum rate of myocyte shortening and of relaxation were slower in the transgenic myocytes (Wolska et al., 1999). Confirming the previous studies of skinned multicellular preparations (Palmiter et al., 1996), steady state force and ATPase activity of the transgenic preparations were more sensitive to calcium than the wild type (Figure 16). Importantly, the detergent extracted fibers from the transgenic hearts demonstrated significantly less maximum tension and ATPase activity than did the wild type (Figure 16). Surprisingly, an increase in sarcomere length, comparable to increasing end-diastolic volume or preload, affected pCa_{50} of the wild type and transgenic preparations to the same extent. This is in contrast to the working heart model where an increase in preload eliminated any physiological difference in ventricular function between the wild type and transgenic hearts.

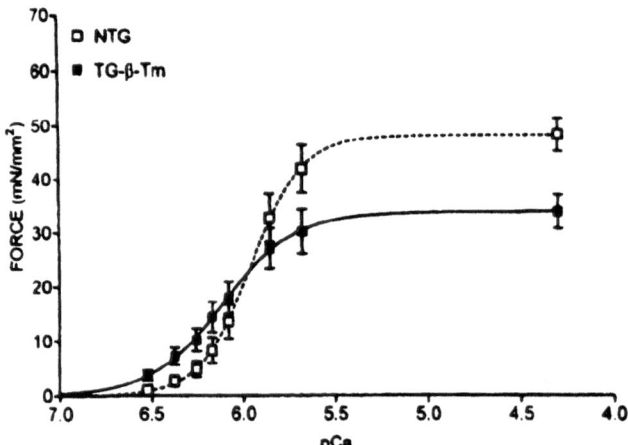

Figure 16. pCa force relation of skinned fiber bundle preparations from non-transgenic (NTG) and transgenic (TG) –beta-tropomyosin mouse hearts at sarcomere length 2.3 μm. The TG bundles were more sensitive to calcium and developed lower maximal force than did NTG bundles. (from Wolska, B.M., Keller, R.S., Evans, C.C., Palmiter, K.A., Phillips, R.M., Muthuchamy, M., Oehlenschlager, J., Wieczorek, D.F., de Tombe, P.P., and Solaro, R.J. (1999) Correlation between myofilament response to Ca^{2+} and altered dynamics of contraction and relaxation in transgenic cardiac cells that express beta-tropomyosin. Circulation Research 84, 745-751.

Transgenesis has been used to overexpress β-tropomyosin to a level where the α:β-tropomyosin ratio was 1:3 (Muthuchamy et al., 1998). These levels of β-tropomyosin expression are far higher than those ever normally observed in the heart *in vivo*. The transgenic animals expressing this high level of β-tropomyosin had severely depressed ventricular function and died by two weeks following birth. This marked impairment of diastolic function is reminiscent of the much milder impairment of diastolic relaxation in the working heart of transgenic animals expressing lower levels of β-tropomyosin (Muthuchamy et al., 1995).

6.4 Tropomyosin Phosphorylation

The striated α and β-tropomyosins contain a serine at 283 that appears to be phosphorylated to different levels in the two isoforms (Stull and Buss, 1977). Tropomyosin kinase phosphorylates both α and β-tropomyosin with phosphorylation of the α-isoform being five times faster than that of the β-tropomyosin (Perry, 2001). A functional role of tropomyosin phosphorylation is supported by the Mg-ATPase activity in reconstituted actomyosin S1 systems being higher in the presence of phosphorylated tropomyosin (Heeley et al., 1989).

The effects of PKA phosphorylation of the myofilaments was examined in β-tropomyosin transgenic and wild type preparations. Previously, cAMP-dependent phosphorylation of cTnI has been demonstrated to be sufficient to induce all of the rightward shift of the calcium-force and ATPase relation observed in the intact preparation (Wattanpermpool et al., 1995). PKA phosphorylation of β-tropomyosin transgenic and wild type preparations lead to different effects on the sensitivity of the myofilaments to calcium (Palmiter et al., 1996). Although the wild type myofilaments responded with the expected rightward shift of the relation, the β-tropomyosin transgenic preparations demonstrated no significant effect of PKA phosphorylation on the pCa_{50}. This differential effect of phosphorylation further magnified the difference in pCa_{50} between nonphosphorylated wild type and transgenic preparations. In contrast to the finding of this study (Palmiter et al., 1996), a subsequent study in wild type and β-tropomyosin transgenic myocardium found that PKA treatment reduced calcium sensitivity of both preparations by the same amount (Patel et al., 2001). Also, no differential effect of PKA treatment was seen in maximum k_{tr} or the rates of submaximal force development (Patel et al., 2001). The basis of these differences remains to be resolved.

6.5. Tropomyosin Phosphorylation and Development

In the rat heart, the amount of tropomyosin phosphorylation appears to decrease with development (L'ecuyer et al., 1991). In late gestation, tropomyosin is resolved as two spots with the more acidic one being the predominant species, while in the adult, the two spots are approximately equal in intensity. In fetal human and bovine heart, the acidic and basic spots of α and β-tropomyosin are approximately equal in intensity, while in the adult hearts, the majority of tropomyosin is nonphosphorylated (Humphreys and Cummins, 1984). The functional differences that arise from such developmental changes in the level of tropomyosin phosphorylation remain to be determined.

6.6. Development, Function, Tropomyosin Isoforms

The developmental fall in β-tropomyosin expression in the rodent may result in a decrease in the sensitivity of myofilaments to calcium, enhancing ventricular diastolic relaxation (Muthuchamy, 1995; Palmiter et al., 1996; Wolska et al., 1999). Thus, the observed developmental fall in myofilament to sensitivity to calcium could be contributed to by a decrease in β-tropomyosin expression as well as the effects of a switch in TnI expression from ssTnI to cTnI. However, such an effect of altered tropomyosin isoform expression would not be apparent in the human heart where β-tropomyosin expression increases with development (Humphreys and Cummins, 1984). Moreover, the subtle changes seen in ventricular function in the transgenic mouse where approximately half of the tropomyosin expressed is β-tropomyosin make it unlikely that the much lower level of β-tropomyosin expression seen in the human heart would have a detectable functional consequence.

The increase in circulating catecholamine levels during birth and with perinatal stress suggests that cAMP-dependent phosphorylation of contractile proteins may be enhanced during those periods of development. Consequently, an increased level of cTnI phosphorylation would be expected. Whether β-tropomyosin expression in the neonate would obscure the usual functional consequences of cTnI phosphorylation remains to be determined (Wolska, 1999; Patel et al., 2001). As regards the Frank-Starling relation and the marked increase in left ventricular end-diastolic volume that accompanies birth, the effect of transgenic β-tropomyosin expression on ventricular diastolic relaxation was eliminated by increasing preload (Muthuchamy, 1995; Patel et al., 2001).

7. SUMMARY

The functional consequences of the developmental changes in thin filament isoform expression remain to be established. The complexity of these effects are likely to be increased through, yet to be described, interactions of the isoforms and the modifications of these interactions by changes in post-translational state.

Department of Pediatric Cardiology
Duke University Medical Center
Durham, NC 27710 USA

8. REFERENCES

Akella, A.B., Ding, X-L., Cheng, R., and Gulati, J. (1995). Dimished Ca^{2+} sensitivity of skinned cardiac muscle contractility coincident with troponin T-band shifts in the diabetic rat. *Circulation Research* **76**, 600-606.

Alpert, N.R., and Gordon, M S. (1962) Myofibrillar adenosine triphosphatase activity in congestive heart failure. *American Journal of Physiology* **202**, 940-946.

Anderson, P.A.W., Moore, G.E., and Nassar, R.N. (1988) Developmental changes in the expression of rabbit left ventricular troponin T. *Circulation Research* **63**, 742-747.

Anderson, P.A.W., Malouf, N.N., Oakeley, A.E., Pagani, E.D., and Allen, P.D. (1991) Troponin T isoform expression in man: A comparison among normal and failing adult heart, fetal heart, and adult and fetal skeletal muscle. *Circulation Research* **69**, 1226-1233.

Anderson, P.A.W. and Oakeley, A.E. (1989) Immunological identification of five troponin T isoforms reveals an elaborate maturational troponin T profile in rabbit myocardium. *Circulation Research* **65**:1087-1093.

Anderson, P.A.W., Glick, K.L., Killam, A.P., and Mainwaring, R.D. (1986) The effect of heart rate on *in utero* left ventricular output in the fetal sheep. *Journal of Physiology* **372**, 557-573.

Anderson, P.A.W., Glick, K.L., Manring, A., and Crenshaw, C. (1984) Developmental changes in cardiac contractility in the fetal and neonatal sheep: *in vitro* and *in vivo*. *American Journal of Physiology* **247**, H371-H379.

Anderson, P.A.W., Henderson, P.M., and Nassar, R. (1993) The calcium transient in cardiac myocytes increases with development. *Circulation* **88**, 1436.

Anderson, P.A.W., Killam, A.P., Mainwaring, R.D., and Oakeley, A.E. (1987) *In utero* right ventricular output in the fetal lamb: the effect of heart rate. *Journal of Physiology* **387**, 297-316.

Anderson, P.A.W., Greig, A., Mark, T.M., Malouf, N.N., Oakeley, A.E., Ungerleider, R.M., Allen, P.D., and Kay, B.K. (1995) Molecular basis of human cardiac troponin T isoforms expressed in the developing, adult, and failing heart. *Circulation Research* **76**, 681-686.

Arteaga, G.M., Palmiter, K.A., Leiden, J.M., and Solaro, R.J. (2000) Attenuation of length dependence of calcium activation in myofilaments of transgenic mouse hearts expressing slow skeletal troponin I. *Journal of Physiology* **526**, 541-549.

Artman, M., Ichikawa, H., Avkiran, M., and Coetzee, W.A. (1995) Na^+/Ca^{2+} exchange current density in cardiac myocytes from rabbit and guinea pigs during postnatal development. *American Journal of Physiology* **268**, H1714-H1722.

Assali, N.S., Brinkman, C.R.III, Woods, R. Jr., Dandavino, A., and Nuwayhid, B. (1978) Ontogenesis of the autonomic control of cardiovascular functions in the sheep, in L.D. Longo and D.D. Reneau (eds.), *Fetal and Newborn Cardiovascular Physiology, Volume 1, Developmental Aspects*. Garland STPM Press, New York, pp 47-91.

Averyhart-Fullard, V., Fraker, L.D., Murphy, A.M., and Solaro, R.J. (1994) Differential regulation of slow skeletal and cardiac troponin I mRNA during development and by thyroid hormone in rat heart. *Journal of Molecular and Cellular Cardiology* **26**, 609-616.

Baker, S.P., and Potter, L.T. (1980) Cardiac β-adrenoceptors during normal growth of male and female rats. *British Journal of Pharmacology* **68**, 65-70.

Bareis, D.L., and Slotkin, T.A. (1980) Maturation of sympathetic neurotransmission in the rat heart. I. Ontogeny of the synaptic vesicle uptake mechanism and correlation with development of synaptic function. Effects of neonatal methadone administration on development of synaptic vesicles. *Journal of Pharmacology Experimental Therapy* **212**, 120-125.

Behrman, R.E., Kliegman, R.M., and Jenson, H.B. (2000) *Nelson Textbook of Pediatrics*. W.B. Saunders Company, Philadelphia, PA: 16th edition.

Bennetts, B.H., Burnett, L., and dos Remedios, C.G. (1986) Differential co-expression of alpha-actin genes within the human heart. *Journal of Molecular and Cellular Cardiology* **18**, 993-996.

Bhavsar, P.K., Dhoot, G.K., Cumming, D.V., Butler-Browne, G.S., Yacoub, M.H., and Barton, P.J. (1991) Developmental expression of troponin I isoforms in fetal human heart. *FEBS Letters* **292**, 5-8.

Bing, W., Knott, A., and Marston, S.B. (2000 a) A simple method for measuring the relative force exerted by myosin on actin filaments in the *in vitro* motility assay: evidence that tropomyosin and troponin increase force in single filaments. *Biochemistry Journal* **350**, 693-699.

Bisphoric, N.H., Simpson, P.C., and Ordahl, C.P. (1987) Induction of the skeletal α actin in α1 adrenoreceptor-mediated hypertrophy of rat cardiac myocytes. *Journal of Clinical Investigation* **80**, 1194-1199.

Bodor, G.S., Oakeley, A.E., Allen, P.D., Crimmins, D.L., Ladenson, J.H., and Anderson, P.A.W. (1997) Troponin-I phosphorylation in the normal and failing adult human heart. *Circulation* **96**, 1495-1500.

Boerth, S.R., Zimmer, D.B., and Artman, M. (1994) Steady-state mRNA levels of the sarcolemmal Na$^+$/Ca^{2+} exchanger peak near birth in developing rabbit and rat hearts. *Circulation Research* **74**, 354-359.

Boheler, K.R., Carrier, L., Chassagne, C., de la Bastie, D., Mercadier, J.J., and Schwartz, K. (1991) Regulation of myosin heavy chain and actin isogenes expression during cardiac growth. *Molecular and Cellular Biochemistry* **104**, 101-107.

Breall, J.A., Rudolph, A.M., and Heymann, M.A. (1984) Role of thyroid hormone in postnatal circulatory and metabolic adjustments. *Journal of Clinical Investigation* **73**, 1418-1424.

Breitbart, R.E., Nguyen, H.T., Medford, R.M., Destree, A.T., Mahdavi, V., and Nadal-Grinard, B. (1985) Intricate combinatorial patterns of exon splicing generate multiple regulated troponin T isoforms from a single gene. *Cell* **41**, 67-82.

Buechler, A.A., and Palmer, S.K. (1982) Intrapartum fetal death associated with propranolol: case report and review of physiology. *Wisconsin Medical Journal* **81**, 23-25.

Carrier, L., Boheler, K.R.. Chassagne, C., de la Bastie, D., Wisnewsky, C., Lakatta, E.G., and Schwartz, K. (1992) Expression of the sarcomeric actin isogenes in the rat heart with development and senescence. *Circulation Research* **70**, 999-1005.

Chandra, M., Montgomery, D.E., Kim, J.J., and Solaro, R.J. (1999) The N-terminal region of troponin T is essential for the maximal activation of rat cardiac myofilaments. *Journal of Molecular and Cellular Cardiology* **31**, 867-880.

Chen, F.C., Yamamura, H.I., and Roeske, W.R. (1979) Ontogeny of mammalian myocardial β-adrenergic receptors. *European Journal of Pharmacology* **58**, 255-264.

Clyman, R.I., Mauray, F., Heymann, M.A., and Roman, C. (1987) Cardiovascular effects of patent ductus arteriosus in preterm lambs with respiratory distress. *Journal of Pediatrics* **111**, 579-587.

Clyman, R.I., Teitel, D., Padbury, J., Roman, C., and Mauray, F. (1988) The role of β-adrenoreceptor stimulation and contractile state in the preterm lamb's response to altered ductus arteriosus patency. *Pediatric Research* **23**, 316-322.

Coltart, D.J., and Spilker, B.A. (1972) Development of human foetal inotropic responses to catecholamines. *Experientia* **28**, 525-526.

Cooper, T.A. and Ordahal, C.P. (1984) A single troponin T gene regulated by different programs in cardiac and skeletal muscle development. *Science* **226**, 979-982.

Cuticchia, A.J., and Pearson, P.L. (1993) *Human Gene Mapping: a Compendium*. Johns Hopkins University Press, Baltimore.

De Champlain, J., Malmfors, T., Olson, L., and Sachs, C. (1970) Ontogenesis of peripheral adrenergic neurons in the rat: Pre- and postnatal observations. *Acta Physiology of Scandinavia* **80**, 276-288.

Dieckman, L.J., and Solaro, R.J. (1990) Effect of the thyroid status on thin filament Ca^{2+} regulation and expression of troponin I in perinatal and adult rat hearts. *Circulation Research* **67**, 344-351.

Disatnik, M-H., Buraggi, G., and Mochly-Rosen, D. (1994) Localization of protein kinase C isozymes in cardiac myocytes. *Experimental Cellular Research* **210**, 287-297.

Fabiato, A. (1982) Calcium release in skinned cardiac cells: variations with species, tissues, and development. *Federation Proceedings* **41**, 2238-2244.

Fabiato, A. and Fabiato, F. (1978) Calcium-induced release of calcium from the sarcoplasmic reticulum of skinned cells from adult human, dog, cat, rabbit, rat, and frog hearts and from fetal and new-born rat ventricles. *Annals of the New York Academy of Science* **307**, 491-522.

Farza, H., Townsend, P.J., Carrier, L., Barton, P.J., Mesnard, L., Bährend, E., Forisser, J-F., Fiszman, M., Yacoub, M.H., and Schwartz, K. (1998) Genomic organization, alternative splicing and polymorphisms of the human cardiac troponin T gene. *Journal of Molecular and Cellular Cardiology* **30**, 1247-1253.

Felder, R.A., Calcagno, P.L., Eisner, G.M., and Jose, P.A. (1982) Ontogeny of myocardial receptors. II. Alpha adrenoceptors. *Pediatric Research* **16**, 340-342.

Fentzke, R.C., Buck, S.H., Patel, J.R., Lin, H., Wolska, B.M., Stojanovic, M.O., Martin, A.F., Solaro, R.J., Moss, R.L., and Leiden, J.M. (1999) Impaired cardiomyocyte relaxation and diastolic function in transgenic mice expressing slow skeletal troponin I in the heart. *Journal of Physiology* 517, 143-157.

Fewell, J.G., Hewett, T.E., Sanbe, A., Klevitski, R., Hayes, E., Warshaw, D., Maughan, D., and Robbins, J. (1998) Functional significance of cardiac myosin essential light chain isoform switching in transgenic mice *Journal of Clinical Investigations* 101, 2630-2639.

Fitzhugh, G. and Marden, J.J. (1997) Maturation changes in troponin T expression, Ca^{2+} sensitivity and twitch contraction kinetics in dragonfly flight muscle. *Journal of Experimental Biology* 200, 1473-1482.

Frank, O. (1895) Zur Dynamik des Herz Muskels. *Zeitschrift für Biologie* 32, 370-447.

Friedman, W.F. (1972) Neuropharmacologic studies of perinatal myocardium. *Cardiovascular Clinic* 4, 43-57.

Friedman, W.F. (1973) The intrinsic physiologic properties of the developing heart, in W.F. Friedman, M. Lesch, and E.H. Sonnenblick (eds.), *Neonatal Heart Disease*. New York: Grune & Stratton, pp 21-49.

Fujita, H. and Ishiwata, S. (1999) Tropomyosin modulates pH dependence of isometric tension. *Biophysical Journal* 77, 1540-1446.

Galbo, H., Holst, J.J., and Christensen, N.J. (1975) Glucagon and plasma catecholamine responses to graded and prolonged exercise in man. *Journal of Applied Physiology* 38, 70-76.

Gao, L., Kennedy, J.M., and Solaro, R.J. (1995) Differential expression of TnI and TnT isoforms in rabbit heart during the perinatal period and during cardiovascular stress. *Journal of Molecular and CellularCardiology* 27, 541-550.

Gennser, G., and von Studnitz, W. (1972) Noradrenaline synthesis in human fetal heart. *Experientia.* 28, 525-526.

Godt, R.E., Fogaca, R.T.H., and Nosek, T.M. (1991) Changes in force and calcium sensitivity in the developing avian heart. *Canadian Journal of Physiology Pharmacology* 69, 1692-1697.

Gorza, L., Ausoni, S., Merciai, N., Hastings, K.E.M., and Schiaffino, S. (1993) Regional differences in troponin I isoform switching during rat heart development. *Developmental Biology* 156, 253-264.

Greig, A., Hirschberg, Y., Anderson, P.A.W., Hainsworth, C., Malouf, N.N., Oakeley, A.E., and Kay, B.K. (1994) Molecular basis of cardiac troponin T isoform heterogeneity in rabbit heart. *Circulation Research* 74, 41-47.

Gulati, J., Akella, A.B., Nikolic, S.D., Starc, V., and Siri, F. (1994) Shifts in contractile regulatory protein subunits troponin T and troponin I in cardiac hypertrophy. *Biochemical and Biophysical Research Communications* 202, 384-390.

Gunning, P., Ponte, P., Blau, H., and Kedes, L. (1983) α-Skeletal and α-cardiac actin genes are coexpressed in adult human skeletal muscle and heart. *Molecular and Cellular Biology* 3, 1985-1995.

Guo, X., Wattanapermpool, J., Palmiter, K.A., Murphy, A.M., and Solaro, R.J. (1994) Mutagenesis of cardiac troponin I. *Journal of Biological Chemistry* 269, 15210-15216.

Harris, D.E. and Warshaw, D.M. (1993) Smooth and skeletal muscle actin are mechanically indistinguishable in the *in vitro* motility assay. *Circulation Research* 72, 219-224.

Hastings, K.E.M., Koppe, R.I., Marmor, E., Bader, D., Shimada, Y., and Toyota, N. (1991) Structure and developmental expression of troponin I isoforms. *Journal of Biological Chemistry* 266, 19659-19665.

Heeley, D., Watson, M.H., Mak, A.S., Dubord, P., and Smillie, L.B. (1989) Effect of phosphorylation on the interaction and functional properties of rabbit striated muscle alpha alpha-tropomyosin. *Journal of Biological Chemistry* 264, 2424-2430.

Hewett, T.E., Grupp, I.L., Grupp, G., and Robbins, J. (1994) α-Skeletal actin is associated with increased contractility in the mouse heart. *Circulation Research* 74, 740-746.

Hoar, R.M., and Hall, J.L. (1970) The early pattern of cardiac innervation in the fetal guinea-pig. *American Journal of Anatomy* 128, 499-508.

Hu, N. and Clark, E.B. (1989) Hemodynamics of the stage 12 to stage 29 chick embryo. *Circulation Research* 56, 1665-1670.

Huang, X., Lee, K.J., Riedel, B., Zhang, C., Lamanski, L.F., and Walker, J.W. (2000) Thyroid hormone regulates slow skeletal troponin I gene in activation in cardiac troponin I null mouse hearts. *Journal of Molecular and Cellular Cardiology* 32, 221-228.

Huang, X.P., Pi, Y.Q., Lee, K.J., Henkel, A.S., Gregg, R.G., Powers, P.A., and Walker, J.W. (1999) Cardiac troponin I gene knockout. A mouse model of myocardial troponin I deficiency. *Circulation Research* **84**, 1-8.

Humphreys, J.E. and Cummins, P. (1984) Regulatory proteins of the myocardium atrial and ventricular tropomyosin and troponin-I in the developing and adult bovine and human heart. *Journal of Molecular and Cellular Cardiology* **16**, 643-657.

Hunkeler, N.M., Kullman, J., and Murphy, A.M. (1991) Troponin I isoform expression in human heart. *Circulation Research* **69**, 1409-1414.

Jarmakani, J.M., Nakanishi, T., George, B.L., and Bers, D. (1982) Effect of extracellular calcium on myocardial mechanical function in the neonatal rabbit. *Developmental Pharmacology & Therapeutics* **5**, 1-13.

Jideama, N.M., Noland, T.A., Raynor, R.I., Blobe, G.C., Fabbro, D., Kazanietz, M.G., Blumberg, P.M., Hannun, Y.A., and Kuo, J.F. (1996) Phosphorylation specificities of protein kinase C isozymes for bovine cardiac troponin I and troponin T and sites within these proteins and regulation of myofilament properties. *Journal of Biological Chemistry* **271**, 23277-23283.

Jin, J.-P. and Lin, J.J.-C. (1988) Rapid purification of mammalian cardiac troponin T and its isoform switching in rat hearts during development. *Journal of Biological Chemistry* **263**, 7309-7315.

Joelsson, I., and Barton, M.D. (1969) The effect of blockade of the beta-receptors of the sympathetic nervous system of the fetus. *Acta Obstetric Gynecology of Scandinavia* **48**, 75-79.

Kajiwara, H., Morimoto, S., Fukuda, N., Ohtsuki, I., and Kurihara S. (2000) Effect of troponin I phosphorylation by protein kinase A on length-dependence of tension activation in skinned cardiac muscle fibers. *Biochemical and Biophysical Research Communications* **272**, 104-110.

Kelly, R. and Buckingham, M. (1997) Manipulating myosin light chain 2 isoforms *in vivo*: a transgenic approach to understanding contractile protein diversity. *Circulation Research* **80**, 751-753.

Kentish, J.C., McCloskey, D.T., Layland, J., Palmer, S., Leiden, J.M., Martin, A.F., and Solaro, R.J. (2001) Phosphorylation of troponin I by protein kinase A accelerates relaxation and crossbridge cycle kinetics in mouse ventricular muscle. *Circulation Research* **88**, 1059-1065.

Klopfenstein, H.S. and Rudolph, A.M. (1978) Postnatal changes in the circulation and responses to volume loading in sheep. *Circulation Research* **42**, 839-845.

Komukai, K., and Kurihara, S. (1997) Length dependence of Ca^{2+}-tension relationship in aequorin-injected ferret papillary muscles. *American Journal of Physiology* **273**, H1068-H1074.

Kranias, E.G., and Solaro, R.J. (1982) Phosphorylation of troponin I and phospholamban during catecholamine stimulation of rabbit heart. *Nature* **298**, 182-184.

Kumar, A., Crawford, K., Close, L., Madison, M., Lorenz, J., Doetschman, T., Pawlowski, S., Duffy, J., Neumann, J., Robbins, J., Boivini, G.P., O'Toole, B.A., and Lessard, J.L. (1997) Rescue of cardiac α-actin-deficient mice by enteric smooth muscle γ-actin. *Proceedings of the National Academy of Sciences USA* **94**, 4406-4411.

L'ecuyer, T.J., Schulte, D., and Lin, J.J.-C. (1991) Thin filament changes during *in vivo* rat heart development. *Pediatric Research* **30**, 232-238.

Lebowitz, E.A., Novick, J.S., and Rudolph, A.M. (1972) Development of myocardial sympathetic innervation in the fetal lamb. *Pediatric Research* **6**, 887-893.

Leszyk, J., Dumaswala, R., Potter, J.D., Gusev, N.B., Verin, A.D., Tobacman, I.S., and Collins, J.S. (1987) Bovine cardiac troponin T: amino acid sequences of the two isoforms. *Biochemistry* **26**, 7035-7042.

Li, L., Desantiago, J., Chu, G., Kranias, E.G., and Bers, D.M. (2000) Phosphorylation of phospholamban and troponin I in β-adrenergic-induced acceleration of cardiac relaxation. *American Journal of Physiology* **278**, H769-H779.

Lipp, J.M., and Rudolph, A.M. (1972) Sympathetic nerve development in the rat and guinea-pig heart. *Biological Neonate* **21**, 76082.

Lister, G., Walter, T.K., Versmold, H.T., Dallman, P.R., and Rudolph, A.M. (1979) Oxygen delivery in lambs: cardiovascular and hematologic development. *American Journal of Physiology* **237**, H668-H675.

Liu, J.D., Wood, J.G., Raynor, R.L., Wang, Y.-C., Noland, T.A. Jr., Ansari, A.A., and Kuo, J.F. (1987) Subcellular distribution and immunochemical localization of protein kinase C in myocardium and phosphorylation of troponin in isolated myocytes stimulated by isoproterenol or phorbol ester. *Biochemical and Biophysical Research Communications* **162**, 1105-1110.

Malouf, N.N., McMahon, D., Oakeley, A.E., and Anderson, P.A.W. (1992) A cardiac tropoin T epitope conserved across phyla. *Journal of Biological Chemistry* **267**, 9269-9274.

Margossian, S.S., Anderson, P.A.W., Chantler, P.D., Deziel, M., Umeda, P.K., Patel, H., Stafford, W.F., Norton, P., Malhotra, A., Yang, F., Aulfield, J.B., and Slayter, H.S. (1999) Calcium regulation in the human myocardium affected by dilated cardiomyopathy: A structural basis for impaired Ca^{2+}-sensitivity. *Molecular and Cellular Biochemistry* **194**, 301-313.

Marsh, J.D. and Allen, P.D. (1989) Developmental regulation of cardiac calcium channels and contractile sensitivity to $[Ca]_o$. *American Journal of Physiology* **256**, H179-H185.

Martin, A.F., Ball, K., Gao, L.Z., Kumar, P., and Solaro, R.J. (1991) Identification and functional significance of troponin I isoforms in neonatal rat heart myofibrils. *Circulation Research* **69**, 1244-52.

McAuliffe, J.J. and Robbins, J. (1991) Troponin-T expression in normal and pressure-loaded fetal sheep heart. *Pediatric Research* **29**, 580-585.

McAuliffe, J.J., Gao, L., and Solaro, R.J. (1990) Changes in myofibrillar activation and troponin C Ca^{2+} binding associated with troponin T isoform switching in developing rabbit heart. *Circulation. Research* **66**:1204-1216.

McHugh, K.M.,Crawford, K., and Lessard, J.L. (1991) A comprehensive analysis of the developmental and tissue-specific expression of the iosactin multigene family in the rat. *Developmental Biology* **148**, 442-458.

McHugh, K.M. (1995) Molecular analysis of smooth muscle development in the mouse. *Developmental Dynamics* **204**, 278-290.

Mesnard, L., Logeart, D., Taviaux, S., Diriong, S., Mercadier, J.-J., and Samson, F. (1995) Human cardiac troponin T: cloning and expression of new isoforms in the normal and failing heart. *Circulation Research* **76**, 687-692.

Mesnard, L., Samson, F., Espinasse, I., Durand, J., Neveux, J.Y., and Mercadier, J.J. (1993) Molecular cloning and developmental expression of human cardiac troponin T. *FEBS Letters* **328**, 139-44.

Metzger, J.M. (1997) Effects of skeletal α-actin isoform expression on contractile function of cardiac myocytes isolated from balb/c mice. *Biophysical Journal* **72**(2), A175.

Metzger, J.M., Wahr, P.A., Michele, D.E., Albayya, F., and Westfall, M.V. (1999) Effects of myosin heavy chain isoform switching on Ca^{2+}-activated tension development in single adult cardiac myocytes *Circulation Research* **84**, 1310-1317.

Metzger, J.M., Lin, W.I., and Samuelson, L.C. (1994) Transition in cardiac contractile sensitivity to calcium during the in vitro differentiation of mouse embryonic stem cells. *Journal of Cell Bioliology* **126**, 701-11.

Miyata, S., Minobe, W., Bristow, M.R., and Lwinwand, L.A. (2000) Myosin heavy chain isoform expression in the failing and nonfailing human heart. *Circulation Research* **86**, 386-390.

Montgomery, D.E., Chandra, M., Huang, Q., Jin, J., and Solaro, R.J. (2001) Transgenic incorporation of skeletal TnT into cardiac myofilaments blunts PKS-mediated depression of force. *American Journal of Physiology* **280**, H1011-1018.

Murphy, A.M., Jones, L. 2nd, Sims, H.F., and Strauss, A.W. (1991) Molecular cloning of rat cardiac troponin I and analysis of troponin I isoform expression in developing rat heart. *Biochemistry* **30**, 707-712.

Muthuchamy, M., Grupp, I., Grupp, G., O'Toole, B., Kier, A., Boivin, G., Neumann, J., and Wieczorek, D.F. (1995) Molecular and physiological effects of overexpressing striated muscle β-tropomyosin in the adult murine heart. *Journal of Biological Chemistry* **270**, 30593-30603

Muthuchamy, M., Pajak, L., Howles, P., Doetschman, T., and Wieczorek, D.F. (1993) Developmental analysis of tropomyosin gene expression in embryonic stem cells and mouse embryos. *Molecular and Cellular Biology* **13**, 3311-3323.

Muthuchamy, M., Boivin, G.P., Grupp, I.L., Wieczorek, D.F. (1998) β-tropomyosin overexpression induces severe cardiac abnormalities. *Journal of Molecular and Cellular Cardiology.* **30**, 1545-1557.

Nassar, R., Reedy, M.C., and Anderson, P.A.W. (1987) Developmental changes in the ultrastructure and sarcomere shortening of the isolated rabbit ventricular myocyte. *Circulation Research* **61**, 465-483.

Nassar, R., Malouf, N.N., Kelly, M.B., Oakeley, A.E., and Anderson, P.A.W. (1991) Force-pCa relation and troponin T isoforms of rabbit myocardium. *Circulation Research* **69**, 1470-1475.

Nguyen, M.T., Camenisch, T., Snouwaert, J.N., Hocks, E., Coffman, T.M., Anderson, P.A.W., Malouf, N.N., and Koller, B.H. (1997) The prostaglandin EP₄ receptor triggers remodeling of the cardiovascular system at birth. *Nature.* **390**, 78-81.

Noland, T.A. and Kuo, J.F. (1991) Protein kinase C phosphorylation of cardiac troponin I or troponin T inhibits Ca^{2+}-stimulated actomyosin MgATPase activity. *Journal of Biological Chemistry* **266**, 4974-4978.

Noland, T.A. and Kuo, J.F. (1992) Protein kinase C phosphorylation of cardiac troponin T decreases Ca^{2+}- dependent actomyosin MgATPase activity and troponin T binding to tropomyosin-F-actin complex. *Biochemistry Journal* **288**, 123-129.

Noland, T.A. and Kuo, J.F. (1993) Protein kinase C phosphorylation of cardiac troponin I and troponin T inhibits Ca^{2+} stimulated MgATPase activity in reconstituted actomyosin and isolated myofibrils and decreases actin-myosin intereactins. *Journal of Molecular and Cellular Cardiology* **25**, 53-65.

Noland, T.A., Raynor, R.L., and Kuo, J.F. (1989) Identification of sites phosphorylated in bovine cardiac troponin I and troponin T by protein kinase C and comparative substrate activity of synthetic peptides containing the phosphorylation sites. *Journal of Biological Chemistry* **264**, 20778-20785.

Osaka, T., and Joyner, R.W. (1991) Developmental changes in calcium currents of rabbit ventricular cells. *Circulation Research* **68**, 788-796.

Osaka, T., and Joyner, R.W. (1992) Developmental changes in the beta-adrenergic modulation of calcium currents in rabbit ventricular cells. *Circulation Research* **70**, 104-115.

Padbury, J.F. and Martinez, A.M. (1988) Sympathoadrenal system activity at birth: integration of postnatal adaption. *Seminars in Perinatology* **12**, 163-172.

Padbury, J.F., Roberman, B., Oddie, T., Hobel, C.J., and Fisher, D.A. (1982) Fetal catecholamine release in response to labor and delivery. *Obstetrics and Gynecology* **60**, 607-611.

Pagani, E.D., Alousi, A.A., Grant, A.M., Older, T.M., Dziuban, S.W. Jr., and Allen, P.D. (1988) Changes in myofibrillar content and Mg-ATPase activity in ventricular tissues from patients with heart failure caused by coronary artery disease, cardiomyopathy, or mitral valve insufficiency. *Circulation Research* **63**, 380-385.

Palmiter, K.A., Kitada, Y., Muthuchamy, M., Wieczorek, D., and Solaro, R.J. (1996) Exchange of β for α-TM in hearts of TG mice induces changes in thin filament response to Ca^{2+}, strong cross-bridge binding, and protein phosphorylation. *Journal of Biological Chemistry* **271**, 11611-11614

Pan, B.S., Gordon, A.M., and Potter, J.D. (1991) Deletion of the first 45 NH2-terminal residues of rabbit skeletal troponin T strengthens binding of troponin to immobilized tropomyosin. *Journal of Biological Chemistry* **266**, 12432-12438.

Papka, R.E. (1981) Development of innervation to the ventricular myocardium of the rabbit. *Journal of Molecular and Cellular Cardiology* **13**, 217-228.

Park, I.-S., Michael, L.H., and Driscoll, D.J. (1982) Comparative response of the developing canine myocardium to inotropic agents. *American Journal of Physiology* **242**, H13-H18.

Park, K.P., Sheridan, P.H., Morgan, W.W., and Beck, N. (1980) Comparative inotropic response of newborn and adult rabbit papillary muscles to isoproterenol and calcium. *Developmental Pharmacological Therapy* **8**, 70-82.

Patel, J.R., Fitzsimons, D.P, Buck, S.H., Muthuchamy M, Wieczorek D.F. and Moss, R.L. (2001) PKA accelerates rate of force development in murine skinned myocardium expressing α- or β-tropomyosin. *American Journal of Physiology* **280**, H2732-H2739.

Perry SV. (2001) Vertebrate tropomyosin: distribution, properties, and function. *Journal of Muscle Research and Cell Motility* **22**, 5-49.

Peterson, J.N., Nassar, R., Anderson, P.A.W., and Alpert, N.R. (2001) Altered cross-bridge characteristics following haemodynamic overload in rabbit hearts expressing V₃ myosin. *Journal of Physiology* **536.2**, 569-532.

Potter, J.D., Sheng, Z., Pan, B-S., and Zhao, J. (1995) A direct regulatory role for troponin T and a dual role in the Ca^{2+} regulation of muscle contraction. *Journal of Biological Chemistry* **270**, 2557-2562.

Pucéat, M., Hilal-Dandan, R., Strulovici, B., Brunton, L.L., and Brown, J.H. (1994) Differential regulation of protein kinase C isoforms in isolated neonatal and adult rat cardiomyocytes. *Journal of Biological Chemistry* **269**, 16938-16944

Purcell, I.F., Bing, W., and Marston, S.B. (1999) Functional analysis of human cardiac troponin by the *in vitro* motility assay: comparison of adult, fetal, and failing hearts. *Cardiovascular Research* **43**, 884-891.

Reiser, P.J., Westfall, M.V., Schiaffino, S., and Solaro, R.J. (1994) Tension production and thin-filament protein isoforms in developing rat myocardium. *American Journal of Physiology*. **267**, H1589-H1596.

Robertson, S.P., Johnson, J.D., Holroyde, M.J., Kranias, E.G., Potter, J.D., and Solaro, R.J. (1982) The Ca^{2+} and Mg^{2+} dependence of Ca^{2+} uptake and respiratory function of porcine heart mitochrondria. Probable physiological significance during the cardiac contraction-relaxation cycle. *Journal of Biological Chemistry* **257**, 260-263.

Ruzicka, D.L. and Schwartz, R.J. (1988) Sequential activation of alpha-actin genes during avian cardiogenesis: vascular smooth muscle alpha-actin gene transcripts mark the onset of cardiomyocyte differentiation. *Journal of Cell Biology* **107**, 2575-2586.

Rybin, V.O. and Steinberg, S.F. (1994) Protein kinase C isoform expression and regulation in the developing rat heart. *Circulation Research* **74**, 299-309.

Saba, Z., Nassar, R., Ungerleider, R.M., Oakeley, A.E., and Anderson, P.A.W. (1996) Cardiac troponin T isoform expression correlates with pathophysiological descriptors in patients who underwent corrective surgery for congenital heart disease. *Circulation* **94**, 472-476.

Sabry, M.A. and Dhoot, G.K. (1989) Identification and pattern of expression of a developmental isoform of troponin I in chicken and rat cardiac muscle. *Journal of Muscle Research Cell Motility* **10**, 85-91.

Saggin, L., Gorza, L., Ausoni, S., and Schiaffino, S. (1989) Troponin I switching in the developing heart. *Journal of Biological Chemistry* **264**, 16299-16302.

Saggin, L., Gorza, L., Ausoni, S., and Schiaffino, S. (1990) Cardiac troponin T in developing, regenerating and denervated rat skeletal muscle. *Development* **110**, 547-554.

Sassoon, D.A., Garner, J., and Buckingham, M. (1988) Transcripts of alpha-cardiac and alpha-skeletal actins are early markers for myogenesis in the mouse embryo. *Development* **104**, 155-164.

Sawtell, N.M., and Lessard, J.L. (1989) Cellular distribution of smooth muscle actins during mammalian embryogenesis: expression of the alpha-vascular but not the gamma-enteric isoform in differentiating striated myocytes. *Journal of Cell Biology* **109**, 2929-2937.

Schachat, F.H., Diamond, M.S., and Brandt, P.W. (1987) Effect of different troponin T-tropomyosin combinations on thin filament activation. *Journal of Molecular Biology* **198,** 551-554.

Schaub, M.C., Hefti, M.A., Zuellig, R.A., and Morano, I. (1998) Modulation of contractility in human cardiac hypertrophy by myosin essential light chain isoforms *Cardiovascular Research* **37**, 381-404.

Schumacher, W., Mirkin, B.L., and Sheppard, J.R. (1984) Biological maturation and β-adrenergic effectors: Development of β-adrenergic receptors in rabbit heart. *Molecular and Cellular Biochemistry* **58**, 173-181.

Schwartz, K., de la Bastie, D., Bouveret, P., Oliviero, P., Alonso, S., and Buckingham, M. (1986) Alpha-skeletal muscle actin mRNA's accumulate in hypertrophied adult rat hearts. *Circulation Research* **59**, 551-555.

Shimada, E., Kasai, T., Konishi, M, and Fujiwara, T. (1994) Effects of patent ductus arteriosus on left ventricular output and organ blood flows in preterm infants with respiratory distress syndrome treated with surfactant. *Journal of Pediatrics* **125**, 270.

Simpson, P.C., Long, C.S., Waspe, L.E., Henrich, C.J., and Ordahl, C.P. (1989) Transcription of early developmental isogenes in cardiac myocyte hypertrophy. *Journal of Molecular and Cellular Cardiology* **21**, 79-89.

Solaro, R.J. and Van Eyk, J. (1996) Altered interactions among thin filament proteins modulate cardiac function. *Journal of Molelcular and Cellular Cardiology* **28**, 217-230.

Solaro, R.J., Kumar, P., Blanchard, E.M., and Martin, A.F. (1986) Differential effects of pH on calcium activation of myofilaments of adult and perinatal dog hearts. Evidence for developmental differences in thin filament regulation. *Circulation Research* **58**, 721-729.

Solaro, R.J., Lee, J.A., Kentish, J.C., and Allen, D.G. (1988) Effects of acidosis on ventricular muscle from adult and neonatal rats. *Circulation Research* **63**, 779-87.

Spach, M.S., Dolber, P.C., and Anderson, P.A.W. (1989) Multiple regional differences in cellular properties that regulate repolarization and contraction in the right atrium of adult and newborn dogs. *Circulation Research* **65**, 1594-1611.

Starling, E.H. (1915) The Linacre Lecture on the Law of the Heart (Cambridge, 1915, monograph). London: Longmans, Green, and Co.

Stull, J.T., and Buss, J.E. (1977) Phosphorylation of cardiac troponin by cyclic adenosine 3'-5' monophosphate dependent protein kinase. *Journal of Biological Chemistry* **252**, 851-857

Sugi, Y. and Lough, J. (1992) Onset of expression and regional deposition of alpha-smooth and sarcomeric actin during avian heart development. *Developmental Dynamics* 193, 116-124.

Sutoh, K. (1982) Identification of myosin-binding sites on the actin sequence. *Biochemistry* 21, 3657-3661.

Swynghedauw, B. (1986) Developmental and functional adaptation of contractile proteins in cardiac and skeletal muscles. *Physiological Reviews* 66, 710-771.

Szymanska, G., Grupp, I.L., Slack, J.P., Harrer, J.M., and Kranias, E.G. (1995) Alterations in sarcoplasmic reticulum calcium uptake, relaxation parameters and their responses to beta-adrenergic agonists in the developing rabbit heart. *Journal of Molecular and Cellular Cardiology* 27, 1819-1829.

Takeishi, Y., Chu, G., Kirkpatrick, D.L., Li, Z., Wakasaki, H., Kranias, E.G., King, G.L., and Walsh, R.A. (1998) *In vivo* phosphorylation of cardiac troponin I by PKC β_2 decreases cardiomyocyte calcium responsiveness and contractility in transgenic mouse hearts. *Journal of Clinical Investigation* 102, 72-78.

Takeishi, Y., Ping, P., Bolli, R., Kirkpatrick, D.L., Hoit, B.D., and Walsh R.A. (2000) Transgenic overexpression of constitutively active protein kinase C (epsilon) causes concentric cardiac hypertrophy. *Circulation Research* 86, 1218-1223.

Talosi, L. and Kranias, E.G. (1992) Effect of α-adrenergic stimulation on activation of protein kinase C and phosphorylation of proteins in intact rabbit hearts. *Circulation Research* 70, 670-678.

Tazawa, H., Visschedijk, A.H.J., Wittmann, J., and Piiper, J. (1993) Gas exchange, blood gases and acid–base status in the chick before, during, and after hatching. *Respiratory Physiology* 53, 173-185.

Tobacman, L.S. (1988) Structure-function studies of the amino-terminal region of bovine cardiac troponin T. *Journal of Biological Chemistry* 263, 2668-2672.

Townsend, P.J., Barton, P.J., Yacoub, M.H., and Farza, H. (1995) Molecular cloning of human cardiac troponin T isoforms: expression in developing and failing heart. *Journal of Molecular and Cellular Cardiology* 27, 2223-2236.

Townsend, P.J., Farza, H., MacGeoch, C., Spurr, N.K., Wade, R., Gahlmann, R., Yacoub, and M.H., and Barton, P.J.R. (1994) Human cardiac trononin T: Identification of fetal isoforms and assignment of the TNNT2 locus to chromosome 1q. *Genomics* 21, 311-316.

Van Buren, P., Palmiter, K.A., and Warshaw, D.M. (1999) Tropmyosin directly modulates actomyosin mechanical performance at the level of a single actin filament. *Proceedings of the National Academy of Sciences USA* 96, 12488-12493.

van der Velden, J., de Jong, J.W., Owen, V.J., Burton, P.B.J., and Stienen, G.J.M. (2000) Effect of protein kinase A on calcium sensitivity of force and its sarcomere length dependence in human cardiomyocytes. *Cardiovascular Research* 46, 487-495.

Vandekerckhove, J. and Weber, K. (1979) The complete amino acid sequence of actins from bovine aorta, bovine heart, bovine fast skeletal muscle and rabbit slow skeletal muscle. *Differentiation* 14, 123-133.

Vandekerckhove, J. and Weber K. (1978) At least six different actins are expressed in a higher mammal: an analysis based on the amino acid sequence of the amino terminal tryptic peptide. *Journal of Molecular Biology* 126, 783-802.

Vandekerckhove, J., and Weber, K. (1981) Actin typing on total cellular extracts: a highly sensitive protein-chemical procedure able to distinguish different actins. *European Journal of Biochemistry* 113, 595-603.

Venema, R.C., Raynor, R.L., Noland Jr., T.A., and Kuo, J.F. (1993) Role of protein kinase C in the phosphorylation of cardiac myosin light chain 2. *Biochemistry Journal* 294, 401-406

Venema, R.C. and Kuo, J.F. (1993) Protein kinase C-mediated phosphorylation of troponin I and C-protein in isolated myocardial cells is associated with inhibition of myofibrillar actomyosin MgATPase. *Journal of Biological Chemistry* 268, 2705-2711.

Ventura-Clapier, R., Hoerter, J., and Murat, I. (1990) Developmental changes in Ca^{2+} sensitivity of skinned rabbit cardiac fibers. *Biophysical Journal* 57, 546A.

Villar-Palasi, C., and Kumon, A. (1981) Purification and properties of dog cardiac troponin T kinase. *Journal of Biological Chemistry* 256, 7409-7415.

Wang, Y.-C. and Rubenstein, P.A. (1992) Splicing of two alternative exon pairs in β-tropomyosin free-mRNA is independently controlled during myogenesis. *Journal of Biological Chemistry* 267, 12004-12010.

Wang, Y.-P. and Fuchs, F. (1995) Osmotic compression of skinned cardiac and skeletal muscle bundles: effects on force generation, Ca^{2+} sensitivity and Ca^{2+} binding. *Journal of Molecular and Cellular Cardiology* **27**, 1235-1244.

Wattanapermpool, J., Guo, X., and Solaro, R.J. (1995) The unique amino-terminal peptide of cardiac troponin I regulates myofibrillar activity only when it is phosphorylated. *Journal of Molecular and Cellular Cardiology* **27**, 1383-91.

Westfall, M.V., Albayya, F.P., Turner, I.I., and Metzger, J.M. (2000) Chimera analysis of troponin I domains that influence Ca^{2+} activated myofilament tension in adult cardiac myocytes. *Circulation Research* **86**, 470-477.

Westfall, M.V., Rust, E.M., and Metzger, J.M. (1997) Slow skeletal troponin I gene transfer, expression, and myofilament incorporation enhances adult cardiac myocyte contractile function. *Proceedings of the National Academy of Science USA* **94**, 5444-5449.

Wilkinson, J.M., Moir, A.J.G., and Waterfield, M.D. (1984) The expression of multiple forms of troponin T in chicken fast muscle may result from differential splicing of a single gene. *European Journal of Biochemistry* **143**, 47-56.

Winegrad, S., Wisnewsky, C., and Schwartz, K. (1990) Effect of thyroid hormone on the accumulation of mRNAs for skeletal and cardiac α-actin in hearts from normal and hypophysectomized rats. *Proceedings of the National Academy of Sciences U.S.A.* **87**, 2456-2460.

Wolska, B.M., Keller, R.S., Evans, C.C., Palmiter, K.A., Phillips, R.M., Muthuchamy, M., Oehlenschlager, J., Wieczorek, D.F., de Tombe, P.P., and Solaro, R.J. (1999) Correlation between myofilament response to Ca^{2+} and altered dynamics of contraction and relaxation in transgenic cardiac cells that express beta-tropomyosin. *Circulation Research* **84**, 745-751.

Yamada, S., Yamamura, H.I., and Roeske, W.R.. (1980) Ontogeny of mammalian cardiac $α_1$-adrenergic receptors. *European Journal of Pharmacology* **68**, 217-221.

Zot, A.S., and Potter, J.D. (1987) Structural aspects of troponin-tropomyosin regulation of skeletal muscle contraction. *Annual Review of Biophysics and Biophysical Chemistry* **16**, 535-559.

FRANKLIN FUCHS

THE FRANK -STARLING RELATIONSHIP: CELLULAR AND MOLECULAR MECHANISMS

"The law of the heart is thus the same as the law of muscular tissue generally, that the energy of contraction, however measured, is a function of the length of the muscle fibre". Starling, E.H., The Linacre Lecture on the Law of the Heart, 1918, Longmans, Green and Co., London.

1. INTRODUCTION

It is probably not an exaggeration to suggest that the modern era of cardiovascular research began with the classical studies of Otto Frank and Ernest Starling on the relationship between end diastolic volume and systolic function in the isolated heart (Frank, 1895; Patterson and Starling, 1914). From the standpoint of integrative physiology, their work provided a mechanism for linking cardiac output to peripheral vascular perturbations (blood volume, venous compliance, skeletal muscle activity, etc.) which can alter central venous pressure and, hence, the degree of stretch of the myocardial muscle fibers. Their findings, still valid, are central to our understanding of the events that take place in such situations as exercise, hemorrhage, and congestive heart failure. For the muscle physiologist, their work provided the challenge of explaining cardiac function in terms of the basic cellular and molecular mechanisms which control force generation and shortening in striated muscle fibers. This effort has continued to the present day. Building on recent advances in the areas of muscle ultrastructure, contractile protein function, and Ca^{2+} regulation, it is now possible to formulate a generally plausible mechanism which accounts, at least in large part, for the linkage between ventricular filling pressure and systolic performance. The intracellular signaling pathways which mediate this process remain to be clarified.

The formulation of the sliding filament model of muscle contraction (Huxley and Hanson, 1954) and the great success of this model in accounting for the isometric force-length curve of tetanically-stimulated skeletal muscle (Gordon, et al, 1966) provided an initial framework for analysis of length-dependent force generation in cardiac muscle. However, it was soon realized that a simple extrapolation of the sliding filament model from skeletal to cardiac muscle had serious limitations. Cardiac muscle normally functions as a "twitch" muscle in which the contractile system is only partially activated. The level of activation under physiological conditions is modulated not only by such well-known factors as neurotransmitters, hormones, and heart rate, but by muscle fiber length as well. Thus the distinction has

379

R.J. Solaro and R.L. Moss (eds), Molecular Control Mechanisms in Striated Muscle Contraction, 379-415
©2002 Kluwer Academic Publishers, Printed in the Netherlands

been made between the contributions of "physical" factors and "activation" factors in the determination of length-dependent force (Jewell, 1977; Allen and Kentish, 1985). Physical factors refer to those force determinants which are solely related to changes in the geometric arrangement of the actin and myosin filaments (filament overlap, interfilament spacing). Activation factors refer to effects of length on any of the steps in the activation process, including Ca^{2+} release into the cytoplasm, binding of Ca^{2+} to troponin C (TnC), or the molecular interactions which are initiated by Ca^{2+} binding. In tetanically-stimulated skeletal muscle the shape of the isometric force-length relation can be largely accounted for by physical factors (a statement not necessarily true for single twitches and contractions at subtetanic stimulus frequencies). In cardiac muscle there is a significant contribution from physical factors but it has become clear that activation factors make the quantitatively more important contribution to length-dependent force generation (Allen and Kentish, 1985). This chapter will focus on the mechanisms whereby a change in muscle length can alter the level of activation, particularly with respect to the effects of length on the interrelationship between the Ca^{2+} regulatory protein complex and the cross-bridges.

Length-dependent activation is a property of both skeletal and cardiac muscle. It is generally considered to play a more important role in force regulation in cardiac muscle and this belief is reflected in the relative amount of space given to the two muscle types in the literature on length modulation. Hence, the emphasis in this chapter will be on cardiac muscle. However, data from skeletal muscle will be discussed where the comparisons may shed light on basic mechanisms relevant to both muscle types. Summaries of the earlier literature can be found in excellent reviews by Jewell (1977), Stephenson and Wendt (1984), Allen and Kentish (1985), and Lakatta (1992), as well as a symposium volume edited by ter Keurs and Noble (1988). There are several more recent reviews which deal with various aspects of length-dependent force production in cardiac muscle (Fuchs, 1995; Crozatier, 1996; ter Keurs, 1996; Solaro and Rarick, 1998; Calaghan and White, 1999;) and skeletal muscle (Rassier, et al, 1999). The reader should consult Gordon, et al (2000) for a comprehensive review of virtually all aspects of Ca^{2+} regulation in both skeletal and cardiac muscle.

2. STRUCTURAL BASIS OF THE FORCE-LENGTH RELATION

A starting point for any discussion of the force-length relation in striated muscle is the classical study by Gordon, et al (1966) of the relation between isometric force and sarcomere length in single frog skeletal muscle fibers. By using a sophisticated servo system to maintain a constant sarcomere length in the central part of the fiber the authors were able to obtain a precise relationship between tetanic isometric force and sarcomere length. With the availability of accurate measurements of the lengths of actin and myosin filaments (Page and Huxley, 1963), it was possible to relate isometric tetanic force to length-dependent changes in filament overlap (Fig 1A). In accordance with the predictions of the sliding filament model, maximum force was recorded at that sarcomere length (2.0-2.2μm) where all of the cross-bridges were overlapped by actin filaments. With further increase in length there was a linear decline in force, with zero force occuring at that sarcomere length where filament overlap disappeared. This is as expected if the cross-bridges function as independent force generators.

For the cardiac physiologist the ascending limb of the force-length curve is clearly of greater interest. The high diastolic stiffness of cardiac muscle ensures that the cardiac myocytes will almost always be functioning in a sarcomere length range (1.7-2.3μm) where an increase in length is associated with an increase in developed force. Returning to the skeletal muscle data of Gordon, et al (1966), in the sarcomere length range 2.0-2.2μm there was a force plateau which was attributed to the fact that the middle of the myosin filament contains a short bare zone which is devoid of cross-bridges (Huxley, 1963). At sarcomere lengths less than 2.0μm there was a decline in force with a relatively shallow slope (Fig. 1A). The actin filaments extend approximately 1.0μm from each side of the Z disk. Hence at a sarcomere length less than 2.0μm the there will be a zone of double overlap in which oppositely polarized actin filaments project into the opposite sides of the sarcomere, presumably interfering with normal cross-bridge interactions. Other factors may also contribute to force reduction in this range but as far as skeletal muscle is concerned it will be assumed that physical factors are of primary importance. At a sarcomere length equal to the length of the myosin filament (1.6μm) the ends of the myosin filaments would be colliding with the Z disks and force is seen to decline sharply. This length is probably outside the range of physiological function for both skeletal and cardiac muscle.

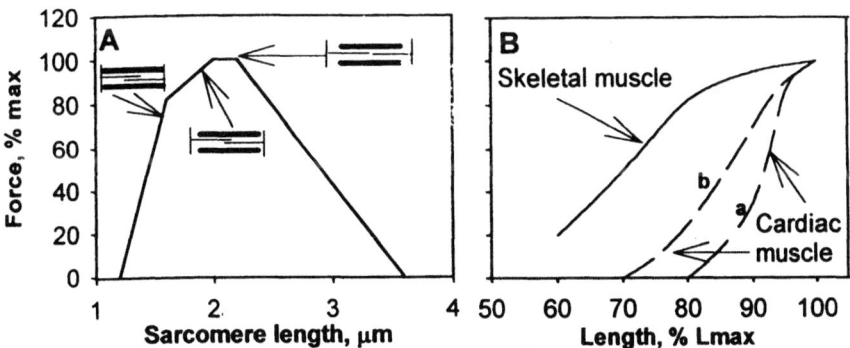

*Figure 1. A. The relationship between maximum tetanic force and sarcomere length in frog skeletal muscle fibers, redrawn from the results of Gordon,et al (1966). Also shown is the approximate disposition of actin and myosin filaments at various points on the ascending limb of the force-length curve. B. Schematic diagram comparing the approximate length dependence of normalized tetanic force developed by skeletal muscle and normalized twitch force developed by cardiac muscle. Although actin and myosin filament lengths are similar in skeletal and cardiac muscle the slope of the force-length relation in the physiological length range (80-100%L$_{max}$) is much steeper for cardiac muscle. The shift from curve **a** to curve **b** illustrates the effect of an increase in contractile state on the position of the cardiac force-length curve.*

The basic sarcomere structure and filament lengths of vertebrate skeletal and cardiac muscle are sufficiently similar that if physical factors were the primary determinants of cardiac force generation the normalized force-length curves for the two muscle types should also be similar (Allen, et al, 1974; Allen and Kentish, 1985). Such is not the case. In cardiac muscle almost the entire ascending limb falls within the range of 80-100% L_{max}, where L_{max} is the length at which maximum force is developed (Allen, et al, 1974; Julian, et al, 1976). Skeletal muscle at 80% L_{max} can still develop more than 80% of maximal tetanic force whereas for cardiac muscle at the same relative length the developed force is less than 10% of the maximum (Fig. 1B). This sharp drop in force is not due to any intrinsic limitations of the cardiac contractile apparatus. Under conditions of maximum activation by Ca^{2+} the force-length relationship of skinned cardiac muscle in the sarcomere length range 1.7-2.3μm has the same shallow slope as seen with fully activated skeletal muscle (Fabiato and Fabiato, 1975; Hibberd and Jewell, 1982; Kentish and Stienen, 1994). Of special significance was the discovery that the position of the cardiac force-length curve is not fixed. Positive inotropic interventions that promote increased Ca^{2+} entry into the muscle cell (increased extracellular Ca^{2+}, increased stimulus frequency) cause a leftward shift (Fig. 1B) and a reduction in the slope of the curve (Allen, et al, 1974; Lakatta and Jewell, 1977; Huntsman and Stewart, 1977; ter Keurs, et al, 1980). That is, the effect of inotropic intervention is greater at short length than it is at L_{max}. These results provided strong evidence that the level of Ca^{2+} activation is less at the shorter length.

It should be noted that a similar phenomenon can be demonstrated in skeletal muscle. With single twitches or subtetanic contractions at low stimulus frequency the ascending limb of the length-force relation is shifted to the right and the peak force occurs at a longer length than peak tetanic force (Close, 1972; Roszek, et al, 1994; Balnave and Allen, 1996). As with cardiac muscle, this shift can be related to a length-dependent change in Ca^{2+} activation (Balnave and Allen, 1996). The physiological role of length-dependent activation in skeletal muscle is discusssed by Rassier, et al (1999).

3. LENGTH, FORCE, AND INTRACELLULAR Ca^{2+}

When a papillary muscle or ventricular trabecula is mounted in a muscle chamber under isometric conditions and stimulated at a fixed frequency an increase in length produces a characteristic biphasic increase in twitch force. Immediately following the length change there is an increase in force, the magnitude of which is related to the length change. This is followed over a period of several minutes by a slow, secondary increase in force (Parmley and Chuck, 1973; Lakatta and Jewell, 1977). The slow force increase is not an artifact of the isolated muscle preparation inasmuch as it can be demonstrated in the intact, blood-perfused ventricle under conditions in which isovolumic developed pressure is measured as a function of end diastolic volume (Tucci, et al, 1984). Most authors equate the Frank-Starling phenomenon with the initial response to stretch and this chapter will be primarily focused on the mechanisms behind the initial force increase, although reference will be made to probable mechanisms underlying the slow response (see below). It seems to be accepted (Nichols, et al, 1988) that the slow force response is a manifestation of the slow increase in contractility (homeometric autoregulation) which is observed in the intact heart following the imposition of an increased afterload (Sarnoff, et al, 1960).

The first systematic study examining Ca^{2+} transients in intact cardiac muscle in relation to length-dependent activation was that of Allen and Kurihara (1982). These investigators microinjected the Ca^{2+}-sensitive photoprotein aequorin into myocytes of rat and cat papillary muscles and trabeculae and recorded both isometric force and Ca^{2+} transients as a function of muscle length. With stimulation at a constant frequency, a sudden increase in length caused a large increase in force without a measureable change in the height of the Ca^{2+} transient. With continued stimulation this initial response was then followed over a period of several minutes by a slow additional increase in force, accompanied by a parallel increase in the height of the Ca^{2+} transient. This general pattern has now been confirmed in a variety of muscle and single myocyte preparations and with other intracellular Ca^{2+} indicators (Bluhm and Lew, 1995; Hongo, et al, 1996; Kentish and Wrzosek, 1998; Alvarez, et al, 1999). A detailed consideration of the mechanism behind the slow increase in force is outside the scope of this review; suffice it to say that prolonged stretch of the cardiac myocyte causes a change in intracellular Ca^{2+} homeostasis which ultimately leads to an increase in the height of the Ca^{2+} transient. How this comes about is still a subject of investigation. Participation of the sarcoplasmic reticulum (Bluhm and Lew, 1995; Kentish and Wrzosek, 1998) and L-type Ca^{2+} channels (Hongo, et al, 1996) appear to have been ruled out. Alvarez et al (1999) have produced strong evidence in favor of an autocrine/paracrine mechanism involving release of angiotensin II and endothelin from the stretched cardiac myocytes. The combined effect of these agents is activation of the Na^+-H^+ and Cl^--HCO_3^- exchangers. The net result is an increase in intracellular Na^+ concentration with no significant change in intracellular pH. The increased intracellular Na^+ is postulated to cause an increased Ca^{2+} influx via the Na^+-Ca^{2+} exchanger operating in the reverse mode.

With respect to the Ca^{2+} transients seen immediately following the stretch, Allen and Kurihara (1982) made two additional observations which anticipated a large body of subsequent research. First, they showed that although the height of the Ca^{2+} transient was not altered immediately following a stretch, there was a more rapid decay of the Ca^{2+} transient at the longer length. Second, if the muscle was subjected to a quick release while developing tension there was a brief increase in the intracellular Ca^{2+} concentration during the decay of the Ca^{2+} transient. The authors hypothesized that at longer length, with greater force being developed, Ca^{2+} was bound to troponin C more tightly. Thus the rate of Ca^{2+} dissociation from the myofilaments during relaxation was reduced relative to the rate of Ca^{2+} removal by the sarcoplasmic reticulum, thereby accounting for the more rapid decay of the Ca^{2+} transient. On the other hand, quick release during contraction would reduce the number of force-generating cross-bridges and promote a more rapid release of Ca^{2+} from the myofilaments, leading to the appearance of "extra Ca^{2+}" in the cytoplasm. The possibility of a feedback between mechanical state and Ca^{2+} binding had been raised previously (Edman and Kiessling, 1971; Kaufmann, et al, 1972). A biochemical basis for such a feedback was provided by the work of Bremel and Weber (1972) which established a role for the cross-bridges in the activation of the skeletal muscle thin filament. The interpretation of Allen and Kurihara was consistent with data showing that rigor bonds between actin and myosin promoted a tighter binding of Ca^{2+} to troponin C (Bremel and Weber, 1972). Studies with Ca^{2+} indicators in intact muscle cells have an intrinsic limitation in that "sources" and "sinks" for Ca^{2+} cannot be directly identified. Thus one could not rule out the possibility, for example, that the extra Ca^{2+} seen upon quick release came from a membrane compartment or from a protein other than troponin C. However, studies with skinned muscle preparations, usually obtained by detergent extraction of the membranes, have provided strong support for the interpretations based on data from intact muscle.

4. SARCOMERE LENGTH, Ca^{2+} SENSITIVITY, AND Ca^{2+} BINDING

An effect of sarcomere length on myofilament Ca^{2+} sensitivity was first described in skinned frog skeletal muscle fibers by Endo (1972) and subsequently confirmed in skinned cardiac myocytes by Fabiato and Fabiato (1978). However, both of these studies examined Ca^{2+} sensitivity only over the descending limb of the force-length curve. Hibberd and Jewell (1982) were the first to study the effect of sarcomere length on the Ca^{2+} sensitivity of skinned cardiac muscle in the physiological sarcomere length range. Comparing skinned rat ventricular muscle of sarcomere lengths 1.9-2.0µm and 2.3-2.5µm, it was shown that at the longer length the force-pCa relation was shifted significantly to the left (~0.2 pCa units) relative to muscles at the shorter length. It was also found that the slope of the normalized force-length curve became flatter as the Ca^{2+} concentration was elevated. These observations were confirmed and extended by Kentish, et al (1986), who demonstrated a general parallelism between intact and skinned muscle with respect to the effects of Ca^{2+} on the force-length relationship. Thus at least part of the Frank-Starling relation could be attributed to an effect of sarcomere length on myofilament Ca^{2+} sensitivity.

An increase in Ca^{2+} sensitivity with stretch could be due to an increased binding of Ca^{2+} at the longer length or a greater recruitment of cross-bridges for a given amount of bound Ca^{2+} (or both). The experiments with intact muscle preparations containing intracellular Ca^{2+} indicators provided indirect evidence for a greater Ca^{2+}-troponin C affinity at the longer sarcomere length. Direct evidence on this point was first obtained in studies of the binding of $^{45}Ca^{2+}$ to detergent-treated bovine ventricular muscle (Hofmann and Fuchs, 1987a,b; 1988). These measurements were made with a double isotope technique in which [^{3}H]glucose was used as a marker for the solvent space within the muscle bundles. With this experimental protocol it was possible to make parallel measurements of force and bound Ca^{2+} in the same preparation (Fuchs and Fox, 1982).

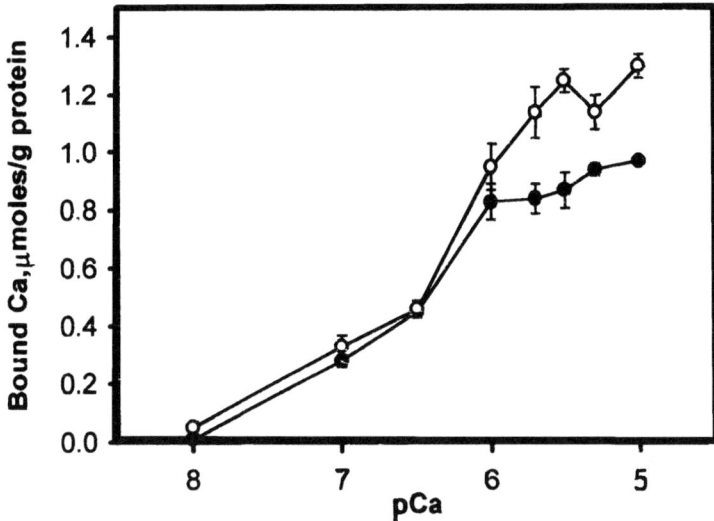

Figure 2. Results of measurements of Ca²⁺ binding to skinned bovine cardiac muscle in the rigor state. Measurements were made at sarcomere lengths 2.3μm (open circles) and 1.8μm (closed circles). Data redrawn from Hofmann and Fuchs (unpublished experiments).

Initial studies on fiber bundles in rigor showed that the binding of Ca^{2+} in the sarcomere length range 1.7-1.8μm was significantly less than in the sarcomere length range 2.3-2.5μm (Fig. 2). Moreover, this difference was confined to the pCa range (6.0-5.0) in which the single regulatory site of cardiac troponin C (cTnC) is titrated (Holroyde, et al, 1980). Essentially the same result was obtained with fiber bundles engaged in active force generation (Hofmann and Fuchs, 1988). Evidence was presented that under the conditions used in these studies cTnC was the only significant Ca^{2+}-binding species (Hofmann and Fuchs, 1988).

Experiments were also carried out to determine the effect of force generation (at constant sarcomere length) in the presence of MgATP on Ca^{2+}-cTnC affinity (Hofmann and Fuchs, 1987b; Wang and Fuchs, 1994). In these experiments the phosphate analog sodium vanadate (Vi) was used to regulate force independent of free Ca^{2+} concentration and Ca^{2+} saturation curves were obtained under conditions of contraction and relaxation. These experiments showed that the amount of Ca^{2+} bound to the regulatory site of cTnC was a function not only of the pCa but of the amount of force developed (Fig. 3). It should be pointed out that the same experimental protocol was used in a previous study (Fuchs, 1985) to determine whether force-generating cross-bridges influenced the amount of Ca^{2+} bound to troponin C in skinned rabbit psoas muscle fibers. No measureable effect could be found. On the other hand, it has been well-documented that rigor cross-bridges in skeletal muscle do promote increased Ca^{2+} binding to troponin C (Bremel and Weber, 1972; Fuchs, 1977; Cantino, et al, 1993). This point will be taken up again in a later section.

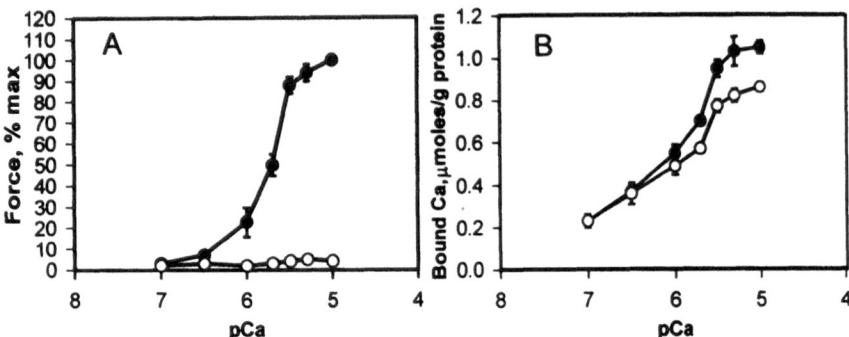

Figure 3. The efffect of sodium vanadate (1mM) on force generation (A) and Ca^{2+} binding (B) by skinned bovine ventricular muscle bundles (sarcomere length 2.2-2.3μm). Control, closed circles; vanadate, open circles. Data redrawn from Wang and Fuchs (1994)

Further evidence has been gathered to support the conclusion that Ca^{2+}-cTnC affinity is modulated by the number of strong-binding cross-bridges attached to actin rather than by variation in sarcomere length. Hofmann and Fuchs (1987a) showed that inhibition of strong-binding interactions between actin and myosin in skinned bovine ventricular muscle eliminated the length-dependence of Ca^{2+} binding to cTnC. Allen and Kentish (1988) immersed skinned ferret ventricular muscle in small volumes of aequorin-containing buffer solution and measured changes in light emission in response to changes in length. They found that the amount of Ca^{2+} released into the solution in response to a step reduction in length was correlated with the change in force rather than with the change in length. Experiments in intact muscles have led to the same conclusion (Saeki, et al, 1993; Kurihara and Komukai, 1995). Of particular interest were studies on tetanic tension in ryanodine-treated fibers loaded with aequorin (Saeki, et al, 1993). The extra Ca^{2+} seen on quick release was eliminated following treatment with the actin-myosin interaction inhibitor 2,3-butanedione monoxime (BDM). Thus in both skinned and intact cardiac muscle length-dependent changes in Ca^{2+}-cTnC affinity are coupled to length-dependent changes in force rather than length itself. A tight coupling between strong-binding cross-bridge interactions, Ca^{2+} sensitivity, and Ca^{2+} binding seems to be a key feature of the cardiac myofilament complex.

The effects of force-generating cross-bridges on Ca^{2+} sensitivity and Ca^{2+} binding are now considered to be manifestations of the general phenomenon of myosin-mediated thin filament activation. Myosin activation of the thin filament can be most easily illustrated by placing skinned fibers or myofibrils in a Ca^{2+}-free relaxing solution and reducing the MgATP concentration to very low levels (~10^{-5}M). The attachment of a critical number of strong-binding cross-bridges (rigor bridges) to actin causes an activation of the thin filament which is manifested as an increase in myosin ATPase activity and the generation of force (Bremel and Weber, 1972; Brandt, et al, 1990; Metzger, 1995). The force-generating cross-bridges formed under physiological conditions (3-5mM MgATP, 10^{-7}-10^{-5} M Ca^{2+}) also activate the thin filament and there is now general agreement that physiological thin filament activation in both skeletal and cardiac muscle is a highly cooperative process involving both Ca^{2+} binding to TnC and myosin binding to actin (see Tobacman, 1996; Gordon, et al, 2000, for reviews).

The findings described above provided the basis for a simple qualitative scheme (Fig. 4) for length-dependent Ca^{2+} activation which took into account the contributions of both physical factors and activation factors (Fuchs, 1995). The basic assumption made is that in the sarcomere length range 1.7-2.3μm physical factors alone would account for a family of force-length curves of relatively shallow slope. The free Ca^{2+} concentration would determine which curve the cardiac myocyte would be operating on at any given moment. If the number of force-generating complexes which can form is a function of sarcomere length as well as pCa then, with only partial Ca^{2+} saturation of cTnC, an increase of sarcomere length would, through myosin-mediated activation of the thin filament, lead to an increase in bound Ca^{2+} and a disproportionate increase in developed force. This would be equivalent to an upward shift of the force-length curve. Thus at a given free Ca^{2+} concentration the actual force-length curve would have a steeper slope than would be expected on the basis of physical factors alone. In subsequent sections evidence will be presented that at a given sarcomere length the developed force is a function of the combined effects of Ca^{2+} and cross-bridges on the level of thin filament activation.

5. INTERFILAMENT SPACING AND LENGTH-DEPENDENT Ca^{2+} ACTIVATION

In considering a role for cross-bridges in the modulation of Ca^{2+} sensitivity one must take into account the fact that in both skeletal muscle (Endo, 1972) and cardiac muscle (Fabiato and Fabiato, 1978) the Ca^{2+} sensitivity is increased with stretch along the descending limb of the force-length curve, where the number of cross-bridge interactions would be expected to diminish with increase in sarcomere length. However, there is now a convincing body of data showing that Ca^{2+} sensitivity is not a function of sarcomere length *per se* but, rather, a function of the lateral spacing between actin and myosin filaments. Actually, this concept is not new. The striated muscle myofilament lattice is an isovolumic system in which the actin and myosin filaments come closer together as the sarcomere length is increased (see Millman, 1998, for review). Earlier studies in which osmotically active polymers (dextran, polyvinylpyrolidone) were used to reduce the interfilament spacing in skinned skeletal muscle fibers had established that Ca^{2+} sensitivity could be reversibly increased by bringing the actin and myosin filaments closer together (Godt and Maughan, 1981; Stienen, et al, 1985). The possibility was raised that the length-dependent increase in Ca^{2+} sensitivity of skinned skeletal muscle fibers (Endo,

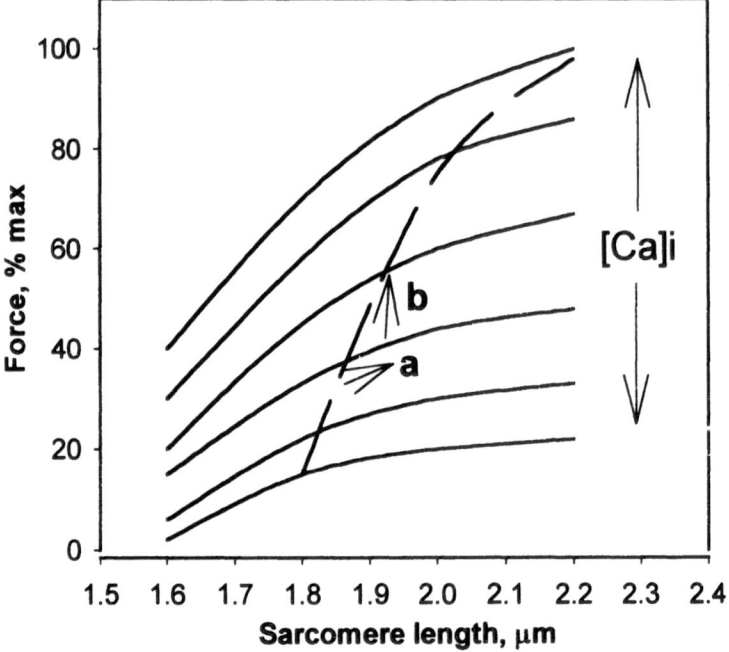

*Figure 4. Schematic drawing indicating how the steep force-length relation of cardiac muscle might result from an interaction between filament overlap and cross-bridge mediated activation of the thin filament. Solid lines represent a family of force-length curves, each characteristic of a given free Ca^{2+} concentration. It is assumed that the shape of these curves is determined solely by physical factors. At any given $[Ca^{2+}]$ an increase in length (**a**) would alter filament geometry and would also promote cooperative activation of the thin filament, leading to greater binding of Ca^{2+} and greater force (**b**). Thus the actual force-length curve (dashed line) would be steeper than that expected on the basis of filament overlap alone (redrawn from Fuchs, 1995).*

1972), was due simply to a length-dependent reduction in interfilament spacing (Stienen, et al 1985). Such a reduction in interfilament spacing would increase the probability that strong-binding interactions between actin and myosin would take place. Harrison, et al (1988) published the first detailed study of the effects of osmotic compression on skinned cardiac muscle. They reported that at short sarcomere length the addition of 3% Dextran T70 enhanced Ca^{2+} sensitivity and they suggested that lattice shrinkage was the basis of the length-dependent increase in Ca^{2+} sensitivity.

More recent studies have provided strong support for this concept. Working with single skinned rat ventricular myocytes, McDonald and Moss (1995) showed that osmotic compression with 2.5% Dextran T-500 at short sarcomere length (~1.85μm) produced an increase in Ca^{2+} sensitivity equivalent to that caused by stretch to sarcomere length ~2.25μm. When comparing the effects of stretch and osmotic compression on myocyte diameter it appeared that most of the effect of stretch could be accounted for by the change in lattice dimensions. Wang and Fuchs (1995) arrived at the same conclusion based on studies with bovine ventricular muscle bundles and they further showed that osmotic compression at short sarcomere length produced an increase in Ca^{2+} binding to the regulatory site of cTnC equivalent to that produced by stretch to long length. In a subsequent study (Fuchs and Wang, 1996), varying concentrations of Dextran T-500 were applied to skinned ventricular muscle bundles at sarcomere lengths 1.7, 2.0, and 2.3µm. Based on measurements of diameter changes at the different sarcomere lengths and with different Dextran T-500 concentrations it was possible to make comparisons of Ca^{2+} sensitivity and Ca^{2+}-cTnC affinity under conditions where the interfilament spacing was either changing with sarcomere length or was relatively constant at all sarcomere lengths. The data clearly showed that Ca^{2+} sensitivity (and Ca^{2+} binding) was strongly correlated with changes in interfilament spacing and not at all with changes in sarcomere length. Thus the increased Ca^{2+} sensitivity with stretch presumably reflects the greater "effective" concentration of myosin heads in the vicinity of the actin filaments.

Although there is now strong evidence that interfilament spacing is a major factor in the determination of length-dependent Ca^{2+} sensitivity the existence of other modulating influences cannot be ruled out. Almost all of our knowledge about modulation of Ca^{2+} sensitivity is based on results obtained with skinned muscle preparations and some caution is required in extrapolating the results of skinned fiber experiments to intact muscle. The skinning process leads to substantial swelling of the muscle fiber (Godt and Maughan, 1977). Therefore, the actual interfilament spacing in most published experiments is significantly different from the *in situ* physiological spacing. In principle one can restore the physiological lattice spacing by adding appropriate concentrations of osmotically active polymers and most investigators consider that in both cardiac and skeletal muscle immersion in 3-5% Dextran T-500 provides a close approximation to the in *situ* interfilament spacing (Allen and Moss, 1987; Kawai, et al, 1993; McDonald and Moss, 1995). However, inferences about interfilament spacing changes with osmotic compression are made on the basis of changes in fiber diameter and it is assumed that diameter and interfilament spacing are linearly related. X-ray diffraction measurements indicate that this assumption is valid for skinned rabbit psoas muscle fibers (Kawai, et al, 1993). It remains to be determined if this assumption applies to skinned cardiac muscle fibers over the entire physiological sarcomere length range.

A direct comparison, by means of X-ray diffraction, has recently been made between the sarcomere length-interfilament spacing relationships of intact and skinned rat cardiac trabeculae (Irving, et al, 2000). In the intact muscle constant volume behavior was confirmed, with the lattice spacing (D_{10} reflection) changing from 37.5nm at sarcomere length 1.8µm to 34nm at sarcomere length 2.2µm. In the skinned fibers the lattice volume deviated from constant volume behavior but nevertheless the spacing changed from 44.7nm to 41.5 nm over the same sarcomere length range. Thus with or without the lattice swelling there was a significant reduction in interfilament spacing with stretch over the physiological length range and the data from both intact and skinned muscle are consistent with the hypothesis that changes in interfilament spacing can modulate Ca^{2+} sensitivity. As the authors of this study point out, the lattice spacing in the relaxed state might not be an accurate reflection of the lattice spacing during Ca^{2+}-activated force generation. Indeed, it is known that during force generation there are small changes in interfilament spacing which may vary with sarcomere length, ionic strength, and muscle type (Millman, 1998).

Data from skeletal muscle raise the possibility that sarcomere length might have an independent effect on Ca^{2+} sensitivity. Martyn and Gordon (1988) measured the pCa_{50} values in skinned rabbit psoas muscle fibers at sarcomere lengths 2.3µm and 3.4µm under conditions in which interfilament spacing was free to change with sarcomere length and under conditions in which the spacing was was maintained relatively constant by osmotic compression. Under conditions of changing interfilament spacing, at pH 7.0, the ΔpCa_{50} was 0.25 whereas with constant spacing the ΔpCa_{50} was 0.06. Thus there appeared to be a small but significant independent effect of length. However, Martyn and Gordon also note that the interfilament spacing in the relaxed state may not be the same as it is during force generation. The picture is further complicated (see below) by recent experiments suggesting that titin extension may play a role in the modulation of Ca^{2+} sensitivity (Cazorla, et al, 1997, 1999). There is clearly a need for careful studies in which the Ca^{2+} sensitivity can be directly related to the actual actin-myosin separation as measured by x-ray diffraction.

6. STRONG-BINDING CROSS-BRIDGES AND Ca^{2+} SENSITIVITY

Further evidence that the the tight coupling between strong-binding cross-bridge interactions and Ca^{2+} sensitivity plays a key role in length-dependent Ca^{2+} activation comes from studies in which actin-myosin interactions are modified at different sarcomere lengths. As pointed out above, activation of the thin filament involves both the binding of Ca^{2+} to TnC and the cooperative effect of cross-bridges strongly-bound to actin. Bringing the actin and myosin closer together presumably increases the probability that force-generating interactions between myosin and actin will take place. Both biochemical and structural studies strongly indicate that Ca^{2+} binding to TnC, by itself, produces only partial activation of the thin filament (Swartz, et al, 1996; Vibert, et al, 1997).

To further examine the role of strong-binding cross-bridges Fitzsimons and Moss (1998) prepared single skinned ventricular myocytes from rat heart and obtained force-pCa curves in the presence and absence of N-ethylmaleimide-modified myosin subfragment 1 (NEM-S1). NEM-S1 binds strongly to actin even under relaxing conditions so it was possible to examine the effect of strongly-bound, non-force generating cross-bridges on the length-dependence of Ca^{2+} sensitivity. The addition of NEM-S1 caused a reversible increase in Ca^{2+} sensitivity at the short sarcomere length (~1.9µ) and almost completely eliminated the length-dependence of Ca^{2+} sensitivity. The authors' interpretation of these observations was that in the presence of NEM-S1 the myosin-mediated cooperative mechanism was fully "turned on", thereby eliminating the length-dependence of cooperative activation.

A similar conclusion emerged from a study of the effects of MgADP addition (0.1-5mM) to the solution bathing skinned rat trabeculae (Fukuda, et al, 2000a). With elevated [MgADP] the population of strong-binding A•M•ADP complexes is increased, thereby enhancing myosin-mediated thin filament activation (Hoar, et al 1987; Fukuda, et al, 1998). As with NEM-S1, MgADP increased Ca^{2+} sensitivity, the effect being more pronounced at short (1.9μm) than at long sarcomere length (2.3μm). Thus the ΔpCa_{50} was reduced in a concentration-dependent manner, reaching 20% of the control value at 5mM MgADP. The increase in Ca^{2+} sensitivity produced by 5% Dextran T-500 at short sarcomere length was attenuated to the same extent by 5mM MgADP. Thus this line of investigation strongly supports the hypothesis that myosin-mediated activation of the thin filament is markedly reduced at short sarcomere lengths.

An issue that has generated some controversy is whether the change in force that accompanies the change in length (or interfilament spacing) results simply from a change in the number of cross-bridges which are recruited or also involves changes in the kinetics of cross-bridge recruitment. The most prevalent method now used for evaluating cross-bridge kinetics is to measure the rate of force redevelopment (k_{tr}) in response to a quick release (to detach force-generating cross-bridges) followed by rapid re-stretch to the original length. Based on results obtained with this method, there seems to be agreement that in skinned skeletal and cardiac muscle elevation of Ca^{2+} concentration at fixed sarcomere length enhances the kinetics of force development (Brenner, 1988; Millar and Homsher, 1990; Wolff, et al, 1995; Vannier, et al, 1996; Palmer and Kentish, 1998). In skinned skeletal muscle, both fast and slow, increase in sarcomere length or osmotic compression also produced an increase in k_{tr} (McDonald, et al, 1997). On the other hand, in intact tetanized ferret papillary muscle no effect of either Ca^{2+} or muscle length on k_{tr} could be detected (Hancock, et al, 1993). Wannenberg et al (2000) used sinusoidal length perturbations with skinned cardiac muscle to determine the rates of cross-bridge attachment and detachment. They reported that cross-bridge kinetics was modulated by Ca^{2+} but not by sarcomere length. Whether these different results reflect the different experimental techniques used or are indicative of a fundamental distinction between cardiac and skeletal muscle is a subject which is still being debated (Hancock, et al, 1997; Gordon et al, 2000). A more detailed discussion of the kinetic aspects of thin filament activation will be found elsewhere in this volume.

The idea that cooperative activation by strong-binding cross-bridges is of lesser magnitude at short sarcomere length might account for length-dependent differences in the effects of osmotic compression. It has been observed in a number of studies in both skeletal and cardiac muscle that as the sarcomere length is increased the effect of a given concentration of osmotically active polymer on the Ca^{2+} sensitivity becomes smaller and ultimately disappears. In skinned skeletal muscle fibers a concentration of Dextran T-500 which causes a significant enhancement of Ca^{2+} sensitivity at sarcomere length 2.0-2.2μm has no effect at 2.7-2.8μm (Stienen, et al, 1985; deBeer, et al, 1988; Wang and Fuchs, 2000a). Similarly, in cardiac muscle a degree of osmotic compression which causes a significant increase in Ca^{2+} sensitivity at sarcomere length 1.8-1.9μm causes little or no change in Ca^{2+}

sensitivity at sarcomere length 2.2-2.3μm (Harrison, et al, 1988; Wang and Fuchs, 1995). Based on the work reviewed above, the lesser effectiveness of osmotic compression at the longer sarcomere length presumably reflects the higher probability for strong-binding cross-bridge attachment as compared to that at the shorter sarcomere length. In a recent study Wang and Fuchs (2000a) stretched skinned rabbit psoas fibers to sarcomere length 3.4μm and found that 5% Dextran T-500 caused a large decrease in Ca^{2+} sensitivity. The same concentration of Dextran caused an increase in Ca^{2+} sensitivity at 2.0μm and no significant change at 2.7μm. The increase in Ca^{2+} sensitivity at sarcomere length 2.0μm was reversed back to the control value with 10% Dextran (see also Godt and Maughan, 1981) and reduced below control value by 15% Dextan. The same results have been obtained with skinned cardiac muscle at sarcomere length 1.8-1.9μm (Wang and Fuchs, 2000b). These results bring out another point, namely, that there seems to be an optimal filament separation at which Ca^{2+} sensitivity is maximized. The reduced Ca^{2+} sensitivity of skeletal muscle seen with 5% Dextran at very long sarcomere length and with 15% Dextran at short sarcomere length is presumably a consequence of overcompression of the filament lattice with resultant mechanical interference with the cross-bridge power stroke (Zhao and Kawai, 1993; Adhikari and Fajer, 1996). Combined X-ray diffraction and mechanical studies will be needed to more precisely define optimal separation in relation to sarcomere length.

7. CROSS-BRIDGE INTERACTIONS AT SHORT SARCOMERE LENGTH

The data collected thus far are consistent with the hypothesis that as the sarcomere is shortened along the ascending limb of the force-length curve the access of cross-bridges for actin is progressively hindered. This weakening of the actin-myosin interaction is now attributed mainly to the increased interfilament spacing at short sarcomere length, with possibly a small contribution from the double overlap of actin filaments. It should be noted that even at a sarcomere length of 1.7-1.8μm roughly 75% of the cross-bridges will be in the zone of single overlap. It seems reasonable to assume that interfilament spacing will make a much more important contribution than filament overlap to the steep force-length relationship seen in the intact myocardium.

   Although the interfilament spacing as a function of sarcomere length can now be accurately measured (Irving, et al, 2000) there is as yet little definitive information on the nature of the cross-bridge interactions at the short sarcomere lengths characteristic of normal cardiac muscle. X-ray diffraction measurements on cardiac muscle indicate that in the diastolic state a considerable number of non-force generating cross-bridges are in close proximity to the thin filaments (Matsubara and Millman, 1974; Matsubara, et al, 1977; 1989). Cross-bridge interactions at low $[Ca^{2+}]_i$ might possibly play a role in the determination of diastolic cell length (Sollot, et al, 1996). It is reported that during maximum activation (at sarcomere length 2.1-2.2μm) of skinned papillary muscles 80% of the myosin heads are transferred to the thin filaments (Matsubara, et al, 1989). How diastolic or systolic cross-bridge interactions, as detected by X-ray diffraction, are related to sarcomere length is not yet known.

An attempt was made to determine how sarcomere length influences actin-myosin interactions in the rigor state (Fuchs and Wang, 1997), using the K^+-ATPase assay first employed by Cooke and Franks (1980). The basic premise behind this assay is that in the absence of divalent cations the binding of actin to myosin inhibits the myosin ATPase activity; thus the K^+-ATPase activity should increase as the population of rigor links with actin decreases. Confirming the observations of Cooke and Franks (1980) with rabbit psoas fibers, it was shown that reduction of the sarcomere length from 2.3µm to 1.6µm was associated with a roughly 2.5-fold increase in K^+-ATPase activity. This finding would appear to constitute at least qualitative evidence that rigor bridge attachment is impaired at short sarcomere length. This impairment would account for the reduced Ca^{2+}-cTnC affinity at short sarcomere length in the rigor state (Hofmann and Fuchs, 1987a; Fuchs and Wang, 1997). However, some caution is necessary in extrapolating this result to more physiological conditions where cross-bridge cycling and filament motion take place.

The Ca^{2+}-activated ATPase activity is considered to be a reflection of the rate of cross-bridge cycling and the way it might change with sarcomere length could provide information about cross-bridge interactions. In combined mechanical and enzymatic measurements with skinned rat cardiac muscle Kentish and Stienen (1994) found that with maximal Ca^{2+} activation reducing the sarcomere length from 2.4µm to 1.8µm caused the force to decline more steeply (~30%) than the ATPase activity (~10%). Thus, the authors attribute just part of the force decline to a reduction in cross-bridge cycling. There might also be a reduced force per cross-bridge because of distortion of the filament lattice at short lengths, as well as some contribution from internal restoring forces. An even more striking divergence between maximal force and ATPase activity was reported in a study with skinned skeletal muscle fibers (Stephenson, et al, 1989). The ATPase activity was constant over the sarcomere length range 1.5-2.5µm, although active force was reduced by 50% at the shorter length. Of greater physiological relevance to cardiac muscle would be data on the ATPase/force ratio with submaximal levels of Ca^{2+} activation but little can be found in the literature. It has been reported that in skinned pig heart muscle the normalized force-pCa and ATPase-pCa curves were quite similar and they underwent comparable shifts as the sarcomere length was increased from 2.1 to 2.4µm (Kuhn, et al, 1990).

Studies have also been carried out to determine the length sensitivity of myosin-mediated activation of the thin filament. As pointed out above, Ca^{2+} sensitivity is enhanced to the extent that cooperative activation of the thin filament by force-generating cross-bridges is greater. It is possible to quantitatively characterize myosin-mediated activation by bathing skinned fibers in a Ca^{2+}-free relaxing solution and gradually reducing the MgATP concentration (Brandt, et al, 1990; Metzger, 1995). With the lowering of the MgATP concentration the increasing population of strong-binding cross-bridges activates the thin filament, thereby allowing for the cooperative attachment of additional force-generating cross-bridges. If the actin-myosin interaction is weaker at the shorter sarcomere length one would expect that the MgATP concentration needed to maintain the relaxed state should be lower at short sarcomere length than at long sarcomere length. This should be expressed as a leftward shift of the force-pMgATP curve at the shorter length.

Two studies have been reported which addressed this question. In the experiments of Smith and Fuchs (2000) no significant difference could be found between the force-pMgATP curves at sarcomere lengths 1.9μm and 2.5μm. This result would appear to contradict the results of the rigor binding study cited above but a possible effect of sarcomere length inhomogeneities could not be ruled out. However, the expected difference could be elicited by the addition of low concentrations of phosphate analogs   (Vi, AlF$_4$.) which generate a population of weak-binding cross-bridges. The interpretation of the authors was that in their preparation the actin-myosin affinity was sufficiently strong that geometric factors had a negligible effect except in the presence of agents which reduced this affinity. The implications of this finding will be taken up in more detail in a later section.

Similar experiments were carried out by Kajiwara et al (2000), who found a small but significant difference (ΔpMgATP$_{50}$=0.12) when comparing the force-pMgATP curves of rat trabeculae at sarcomere lengths 1.9μm and 2.3μm. The reasons for this discrepancy are not evident. One possibility may reside in the different species used. The bovine ventricular muscle used by Smith and Fuchs (2000) expresses almost exclusively the β-type myosin heavy chain (MHC) whereas the rat ventricle used by Kajiwara, et al (2000) contains mostly the α-MHC. These two isoforms differ with respect to kinetic properties and average cross-bridge force (Van Buren, et al, 1995; Palmiter, et al, 1999). Skinned rat cardiac fibers in which either the α or β isoforms are expressed show significant differences in Ca^{2+} sensitivity (Metzger, et al, 1999). Also of potential importance are differences in cross-bridge flexibility (Weisberg and Winegrad, 1998). It is possible that under conditions in which actin and myosin bind strongly the more flexible β-MHC is less subject to the constraints imposed by lattice spacing. There is a need for careful studies of length-dependence of Ca^{2+} activation and myosin activation in skinned fibers of the same species where the two MHC isoforms are differentially expressed.

8. WEAK AND STRONG CROSS-BRIDGE INTERACTIONS

Although the tight coupling between strong-binding actin-myosin interactions and Ca^{2+} sensitivity seems to be well established it is important to emphasize that activation does not involve a simple one-step transition of the cross-bridge from a detached to a force-generating state. Studies in skeletal muscle have shown that there are intermediate weak-binding states which precede the force-generating state (Chalovich, et al, 1991), thus raising a question about the relative roles of weak-binding and strong-binding cross-bridges in length-dependent modulation of Ca^{2+} sensitivity. Reference was made in the previous section to the experiments of Smith and Fuchs (2000) on the length-dependence of myosin-mediated thin filament activation. Length dependence was observed only in the presence of phosphate analogs (Vi, AlF$_4$.) which form relatively stable weak-binding myosin intermediates (M•ADP•Vi, M•ADP•AlF$_4$.). These intermediates are structural analogs of the weak-binding M•ADP•Pi complex which represents the dominant cross-bridge state in the relaxed myofilament lattice (Phan, et al, 1997). This finding raises the possibility that the formation of a weak-binding intermediate is the actual length-sensitive step in the activation process.

A useful framework for analyzing the interrelationships between the Ca^{2+}-mediated and myosin-mediated activation pathways is provided by the three-state model of thin filament activation formulated by Geeves, Lehrer, and co-workers (McKillop and Geeves, 1993; Lehrer, 1994). According to this model, regulation can be described in terms of Ca^{2+}-dependent transitions between three thin filament states (blocked, closed, and open). These thin filament states are coupled to two cross-bridge states: a weak-binding actin-myosin complex (A-state) and a strong-binding actin-myosin complex (R-state). In the presence of physiological concentrations of MgATP but with regulatory binding sites on TnC unoccupied the tropomyosin molecules would be in their optimal position for blocking the access of myosin heads to actin. Hence most of the cross-bridges would be detached from the thin filament and the thin filament regulatory units would be in the "blocked" state. With the binding of Ca^{2+} to the regulatory sites on TnC there would be a partial movement of tropomyosin toward the center of the actin filament groove, thereby allowing myosin heads to form weak bonds with actin (the "closed" state). If Ca^{2+} is bound to TnC the weakly-bound cross-bridge (A-state) undergoes an isomerization to the strongly-bound, force-generating state (R-state), in which the thin filament regulatory units are now in the "open" state. In this state the cross-bridges push tropomyosin further into the groove of the actin double helix, thereby exposing more sites on actin at which force-generating cross-bridges can attach. This model is consistent with structural studies showing that the amount of tropomyosin movement produced by Ca^{2+} binding alone is less than that produced by the combined action of Ca^{2+} binding and cross-bridge attachment (Vibert, et al, 1997). As emphasized by Lehrer and Geeves (1998), the thin filament regulatory system can be characterized as an allosteric regulatory system in which myosin (M, M•ADP) is the true activating ligand and TnC (±Ca^{2+}) is the allosteric effector (positive when Ca^{2+} is bound, negative when Ca^{2+} is dissociated).

The results of Smith and Fuchs (2000) can be explained by assuming that at very low MgATP concentrations the actin-myosin affinity is so strong that most of the cross-bridges will be in the R-state and therefore most of the thin filament regulatory units will be in the open state. If it is the blocked ↔ closed equilibrium which is length-sensitive then the length modulation will be bypassed if there is a strong rightward shift favoring a transition of most of the thin filament regulatory units to the open state. The results with the phosphate analogs suggest that length modulation might require the presence of a critical number of A-state cross-bridges, that is, cross-bridges which are weakly bound to thin filament regulatory units in the closed state.

Support for this idea comes from experiments on the effects of ionic strength on the length-dependence of Ca^{2+} sensitivity. Head, et al (1995) have shown in solution studies that reducing the ionic strength to less than ~50mM (in the absence of Ca^{2+}) shifted the blocked ↔ closed equilibrium far to the right without causing a major change in the closed ↔ open equilibrium. Earlier X-ray diffraction and mechanical studies with skinned skeletal muscle fibers in the relaxed state had shown that at low ionic strength there is a labeling of the thin filaments with cross-bridges along with a small reduction in interfilament spacing and an increase in stiffness (Brenner, et al, 1984). A recent study analyzed the effects of variation in KCl concentration (5-100mM) on the length-dependence of Ca^{2+} sensitivity in skinned bovine ventricular muscle (Smith and Fuchs, 1999). Ca^{2+} regulation was essentially normal over the entire [KCl] range but length modulation of Ca^{2+} sensitivity disappeared at [KCl]<50mM (Fig. 5). Thus although the level of Ca^{2+} sensitivity is determined by the number of strong-binding cross-bridges it appears that length control is exercised at the step in which weak-binding cross-bridge attachments are formed. The length-dependence of Ca^{2+} binding to cTnC was also eliminated at low ionic strength, thus providing additional evidence that Ca^{2+} binding is linked to cross-bridge interactions rather than muscle length.

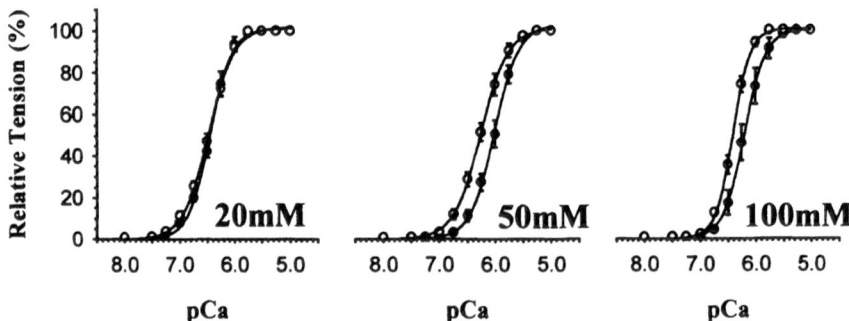

Figure 5. Normalized force-pCa curves of skinned bovine cardiac muscle obtained at sarcomere lengths 1.9µm (closed circles) and 2.5µm (open circles). Measurements were carried out at the different KCl concentrations, as indicated. Data redrawn from Smith and Fuchs (1999).

It is suggested in a recent review (Solaro and Rarick, 1998) that in the diastolic state the cardiac thin filament consists of a mixture of regulatory units in the blocked and closed configurations. It will be important to quantitate this distribution under physiological conditions and determine if the ratio of closed to blocked regulatory units varies with sarcomere length. The large population of non-force-generating cross-bridges in the vicinity of the thin filament in resting cardiac muscle (Matsubara et al, 1977, 1989) might very well be A-type cross-bridges weakly bound to thin filament regulatory units in the closed state.

9. PROTEIN-PROTEIN INTERACTIONS AND LENGTH-DEPENDENT ACTIVATION

It has been recognized since the work of Bremel and Weber (1972) that there must be complex protein-protein interactions within the thin filament which link actin-myosin interactions with changes in Ca^{2+} sensitivity and in the affinity of Ca^{2+} for the troponin C subunit. How these interactions are altered by changes in myofilament geometry is a subject still in need of investigation. Most studies until now have been concerned with determining which proteins play an essential role in length modulation. Initial studies involved the extraction of subunits from skinned muscle and replacement with different isoforms or with proteins which had genetically-engineered structural changes. There are numerous technical problems associated with extraction-substitution experiments (Moss, 1992) and more use is now being made of transgenic approaches.

The first experiments centered on the question of whether the cTnC subunit has a specific "length-sensing" function which could not be subserved by the fast skeletal isoform of troponin C (sTnC). Extraction-substitution type experiments produced conflicting results (Gulati, et al, 1991; Moss, et al, 1991). However, in myocytes prepared from transgenic mouse hearts in which the cTnC was replaced by sTnC the length-dependent shifts in Ca^{2+} sensitivity were identical in wild-type and transgenic mice (McDonald, et al, 1995). Thus it is unlikely that the length-dependent behavior of cardiac muscle, or any differences with respect to skeletal muscle, can be ascribed to a specific property of cTnC.

It would be of great interest to know whether the sTnC expressed in cardiac muscle exhibits the same change in Ca^{2+} binding with change in length (or force) as does cTnC. Measurements in skinned fast skeletal muscle fibers, by two different techniques, showed that length-dependent changes in force occurred without any apparent changes in Ca^{2+} binding to sTnC (Fuchs and Wang, 1991; Patel, et al, 1997). The study of Fuchs and Wang (1991) also verified an earlier finding (Fuchs, 1985) that force-generating cross-bridges at a fixed length had no measureable effect on the binding of Ca^{2+} to sTnC in fast skeletal muscle. However, the effect of cross-bridge interactions on the Ca^{2+}-binding properties of both cTnC and sTnC could very well be environmentally determined. Slow skeletal muscle contains the cTnC isoform. In skinned fibers prepared from the slow rabbit soleus muscle there was no effect of force generation or sarcomere length on the binding of Ca^{2+} to cTnC (Wang and Fuchs, 1994). Thus the effect of cross-bridges on the Ca^{2+}-binding properties of cTnC appears to depend on the protein environment in which it is expressed. Whether the same conclusion holds true for sTnC remains to be determined.

These data raise some fundamental questions about the physiological significance of the different Ca^{2+}-binding characteristics of cardiac and skeletal TnC, especially if one starts with the assumption that cardiac and skeletal muscle share a common mechanism for length modulation of Ca^{2+} sensitivity. Depending on muscle type, length-dependent changes in Ca^{2+} sensitivity can occur with or without measureable changes in Ca^{2+}-TnC affinity. As pointed out previously, rigor complexes do promote an increased Ca^{2+} binding to sTnC in skinned rabbit psoas muscle fibers (Fuchs, 1977; Cantino, et al, 1993). Why force-generating complexes do not do the same, as is the case in cardiac muscle, is still to be explained. One possibility is that force-generating complexes do promote an increase in Ca^{2+} binding to sTnC but the increase is difficult to detect because the fraction of cross-bridges engaged in force generation is much less than that attached in rigor (Gordon, et al, 2000). Another possibility, related to the phenomenon of shortening deactivation (Edman, 1975), is that in skeletal muscle changes in binding affinity are more closely related to filament motion than to force and would not be detected with the isotopic binding technique (Fuchs and Wang, 1991). The latter, because of its' poor time resolution, can only measure Ca^{2+} binding during steady-state force generation (Fuchs and Fox, 1982; Fuchs, 1985). Indeed, Edman (1975) had reported that the mechanical deactivation of skeletal muscle produced by rapid shortening was correlated with the amount of shortening rather than with the change in load. In intact skeletal muscle fibers loaded with a Ca^{2+} indicator changes in free Ca^{2+} in association with rapid shortening and force redevelopment have been observed and these have been interpreted as indicating that Ca^{2+}-sTnC affinity is a function of the developed force (Vandenboom, et al, 1998). An alternative interpretation is that Ca^{2+}-sTnC affinity is unchanged by force generation but filament motion facilitates Ca^{2+} dissociation (Caputo, et al, 1994). Recent evidence suggests that in both skeletal and cardiac muscle there is a cooperative inactivation of the thin filament linked to filament sliding and this inactivation might promote a more rapid dissociation of Ca^{2+} from the thin filament (McDonald, 2000). Experiments in which fluorescent probes are attached to sTnC subunits that have been substituted into skinned psoas fibers have yielded contradictory answers as to whether force-generating cross-bridges alter sTnC conformation in the way that Ca^{2+} binding or rigor bridge attachment does (Guth and Potter, 1987; Martyn, et al, 1999). The results of Martyn, et al (1999) were complementary to the Ca^{2+}-binding measurements (Hofmann and Fuchs, 1987a, 1988; Fuchs and Wang, 1991; Patel, et al, 1997) in the sense that conformational changes in sTnC in rabbit psoas fibers were detected in the rigor state but not during force generation. On the other hand, consistent with the results of Ca^{2+}-binding measurements in cardiac muscle, the conformation of the regulatory domain of cTnC was altered by the attachment to actin of either rigor bridges or force-generating cross-bridges (Martyn, et al, 2001). Thus the evidence points to a subtle difference between cardiac and skeletal muscle with respect to the link between the actin-myosin interaction and TnC conformation.

It is not implausible that length modulation of Ca^{2+} sensitivity, although outwardly similar in cardiac and skeletal muscle, may involve different mechanisms in the two muscle types, with different effects on Ca^{2+} binding. The force-pCa relation is highly cooperative. The sources of this cooperativity can include 1) cross-bridge mediated enhancement of the affinity of Ca^{2+} for TnC, 2) direct effects of strongly-bound cross-bridges on thin filament activation, and 3) nearest neighbor interactions between regulatory units along the length of the thin filament (Tobacman, 1996; Gordon, et al, 2000). The quantitative contributions of these mechanisms, and the way in which they are influenced by changes in length, may differ in different muscle types. Mathematical modeling of the activation process has shown how cross-bridge mediated changes in Ca^{2+}-TnC affinity might make different contributions to length-dependent activation in cardiac and skeletal muscle (Landesberg and Sideman, 1994; Hancock, et al, 1997). There is clearly a need for further research into the way in which protein-protein interactions and filament geometry modulate the Ca^{2+} binding properties of cTnC and sTnC.

The interaction between cTnC and the troponin I (TnI) subunit is crucial in Ca^{2+} regulation of cardiac contraction and there is strong evidence that TnI plays a role in length modulation of Ca^{2+} sensitivity. The TnI subunit expressed in adult cardiac muscle (cTnI) is similar to those expressed in slow (ssTnI) and fast (fsTnI) skeletal muscle except that it has a 32-amino acid N-terminal extension with two sites for phosphorylation by protein kinase A. Phosphorylation of these sites causes a reduction in myofilament Ca^{2+} sensitivity and in the affinity of Ca^{2+} for the regulatory binding site on cTnC (see Solaro and Van Eyk, 1996; Solaro and Rarick, 1998, for reviews). Studies on transgenic mice in which cTnI is replaced by ssTnI support the view that this phosphorylation contributes to the faster rate of cardiac relaxation caused by β-adrenergic stimulation (Fentzke, et al, 1999). Another result of replacement of cTnI with ssTnI was a significant increase in myofilament Ca^{2+} sensitivity. The length-dependence of Ca^{2+} sensitivity has been studied in the same transgenic mouse model (Arteaga, et al, 2000). Comparing sarcomere lengths 1.9μm and 2.3μm, there was an approximately 50% attenuation of the length sensitivity (expressed in terms of ΔpCa_{50}) in the presence of ssTnI. This change is in the same direction as that obtained either by adding NEM-S1 (Fitzsimons and Moss, 1998) or MgADP (Fukuda, et al, 2000). Thus it is very likely that isoform switching alters strong-binding actin-myosin interactions. Here again, it would be of great interest to find out if the Ca^{2+}-binding properties of cTnC in the transgenic mouse follow the pattern of normal cardiac muscle or of slow skeletal muscle. It is clear that cTnI plays a role in length modulation of Ca^{2+} sensitivity but additional studies will be required to pinpoint the signaling pathways that mediate this function. It will be recalled that cTnI is more effective than ssTnI in sensitizing the cardiac myofilaments to the effects of pH change on Ca^{2+} activation (Wattanapermpool, et al, 1995). Functional analysis of chimeric forms of TnI inserted into the cardiac myofilaments has shown that the pH-sensitive region of cTnI resides in the C-terminal domain (Westfall, et al, 2000). It will be of interest to determine if a common signaling pathway is involved in both length and pH modulation of force.

Tropomyosin plays a central role in Ca^{2+} activation and cardiac myocytes in which the cardiac isoform (α-Tm) is replaced (60-65% replacement) by the skeletal isoform (β-Tm) display several changes in contractile properties (Palmiter, et al, 1996; Wolska, et al, 1999). These include increases in rigor activation and myofilament Ca^{2+} sensitivity, reduced maximum tension, and a decrease in the rates of contraction and relaxation. However, the β-Tm had no effect on the length-dependence of Ca^{2+} sensitivity. It is of interest to compare these hearts with those containing the ssTnI isoform. In that case there was also an increase in Ca^{2+} sensitivity but it was associated with a significant loss of length sensitivity. This difference may reflect the different effects of these isoforms on thin filament activation. As suggested by Arteaga, et al (2000), the ssTnI may cause a rightward shift of the blocked ↔ closed equilibrium, thus bypassing what appears to be the length-sensitive step (Smith and Fuchs, 1999) in the activation process. On the other hand, one might speculate that if an α-Tm→β-Tm switch reduces the rate of the closed to open transition, as suggested by Wolska, et al (1999), maximum force would decrease, as was observed, but length control of Ca^{2+} sensitivity would be unchanged. Further studies are needed to relate the functional aspects of Tm isoform substitution to thin filament structure and cross-bridge kinetics. Indeed, recent structural analysis has shown that the skeletal and cardiac isoforms of Tm do preferentially occupy different positions in the grooves of the actin filament (Lehman, et al, 2000).

There are numerous isoforms of troponin T (TnT) expressed in striated muscle and there is evidence that isoform switching can alter the myofilament Ca^{2+} sensitivity in both skeletal and cardiac muscle (Schachat, et al, 1987; Tobacman and Lee, 1987, Nassar, et al, 1991). Little is known about TnT effects on length modulation of activation except for a study by Akella, et al (1995) on Ca^{2+} activation of cardiac muscle in diabetic rats. In the normal rat heart there are three isoforms expressed (TnT1, TnT2, TnT3), with TnT1 and TnT2 being dominant. In the diabetic rat (streptozotocin-treated) there was a large reduction in TnT1 content and a greater expression of TnT3. Associated with this isoform shift was a marked reduction in Ca^{2+} sensitivity at short sarcomere length (1.9μm), no change at long sarcomere length (2.4μm), and hence a significant increase in the ΔpCa_{50}. This result is consistent with the hypothesis that TnT3 had a suppressive effect on myosin-mediated thin filament activation. In the diabetic rat there is also a switch of α-MHC to β-MHC but control experments were done to indicate that this was not a contributing factor in the observed results. In another study with diabetic rat heart no change was seen in Ca^{2+} sensitivity at either long or short sarcomere length (Ishikawa, et al, 1999). Since the latter investigators provided no data on isoform expression in the control and diabetic rats a direct comparison of the two studies is not possible. In recent years there has been an enormous proliferation of studies of the functional consequences of the various TnT mutations associated with familial hypertrophic cardiomyopathy (FHC). Altered Ca^{2+} activation is associated with many of these mutations (Sweeney, et al, 1998; Tobacman, et al, 1999; Rust, et al, 1999) and there is a report of one TnT mutation which is associated with an enhanced length sensitivity of Ca^{2+} activation (Chandra, et al, 2001). Such studies of TnT and other FHC-associated mutant proteins may provide new insights into protein function as well as useful information on cardiac function in patients with this disorder.

10. EFFECTS OF SECOND MESSENGERS AND DRUGS

To what extent the length-dependent activation of the cardiac myofilaments can be modified by second messengers and pharmacological agents is a subject of both physiological and clinical interest in which research is just getting under way. Thus far the main focus has been on protein kinase A (PKA)-catalyzed phosphorylation.

Given the key role of cyclic-AMP (cAMP) in the modulation of both contraction and relaxation, an effect of β-agonists on length-dependent Ca^{2+} activation would not be surprising. Unfortunately, the results that have been obtained thus far do not permit any firm conclusions. Komukai and Kurihara (1997) analyzed length-dependent Ca^{2+} activation in aequorin-loaded intact ferret papillary muscles which were tetanized by repetitive stimulation after ryanodine treatment. The intracellular Ca^{2+} signal was calibrated and force-pCa curves were obtained at different muscle lengths. Following application of isoproteronol there was a substantial increase in the ΔpCa_{50} when comparing 100% and 90% L_{max}. More recently, the same group (Kajiwara, et al, 2000) has examined the effect of protein kinase A treatment on skinned rat trabeculae and found a moderate but significant decrease in the ΔpCa_{50}, comparing sarcomere lengths 1.9 and 2.3µm. In the discussion of their paper they report that skinned ferret muscle behaves in the same way as skinned rat muscle. No explanation is provided as to why the intact and skinned preparations yielded opposite results. In a study of skinned human cardiomyocytes PKA treatment was found to have no effect on the ΔpCa_{50} for the sarcomere length range 1.8-2.2µm (van der Velden, et al, 2000). Some questions might be raised about the state of this preparation since, contrary to what has been repeatedly observed (Solaro and Rarick, 1998), PKA treatment had no effect on the force-pCa relation at any sarcomere length (it did reduce Ca^{2+} sensitivity in myocytes from failing hearts). Finally, a recent preliminary communication reports on experiments in skinned mouse cardiac myocytes which demonstrate a marked enhancement of length-dependence of Ca^{2+} sensitivity following PKA treatment (Konhilas, et al, 2000). Given the species differences and the different experimental techniques used, it would be difficult to draw a conclusion about cAMP effects on the basis of the available data.

The only other known substrate for PKA in the myofibril is myosin binding protein C (Winegrad, 1999). Although this protein plays a role in the modulation of cross-bridge interactions there is no indication as yet that it is involved in length control of Ca^{2+} sensitivity. There are also sites on cTnI and TnT which can be phosphorylated by protein kinase C (PKC) but the physiological significance of this reaction is still uncertain (Solaro and Van Eyk, 1996) and there is no evidence that PKC is involved in any length-dependent effects.

Another major intracellular signaling pathway which merits further study is the phosphorylation of the myosin regulatory light chain (RLC) catalyzed by the Ca^{2+}-calmodulin-myosin light chain kinase (MLCK) system (see Sweeney, et al, 1993; Morano, 1999, for reviews). In both skeletal and cardiac muscle phosphorylation of RLC causes an increase in Ca^{2+} sensitivity. In skeletal muscle thick filaments RLC phosphorylation causes a disordering of the cross-bridge helical array that is thought to reflect an increased mobility of the cross-bridges and a greater potential for actin-myosin interaction (Levine, et al, 1996). In support of this hypothesis it is shown that the effects of RLC phosphorylation on Ca^{2+} sensitivity are virtually identical to the effects of osmotic compression or increase in sarcomere length (Yang, et al, 1998). In fibers already osmotically compressed RLC phosphorylation caused no further change in Ca^{2+} sensitivity and the effects of phosphorylation were greater at short length than at long length. It remains to be determined if these results are applicable to cardiac muscle. If so, one might predict that RLC phosphorylation would cause an attenuation of length-dependent Ca^{2+} activation in cardiac muscle. Cardiac muscle has less MLCK than skeletal muscle and phosphorylation proceeds much more slowly (Sweeney, et al, 1993) but the possibility exists that inotropic and chronotropic stimulation (Silver, et al, 1986), as would occur during exercise, could bring about a time-dependent shift in the force-length relation. In addition, RLC phosphorylation could be a confounding factor in experiments on length-dependent Ca^{2+} activation in intact cardiac muscle, where there may be variable levels of phosphorylation, especially under tetanizing conditions. Further studies of this signaling system could be very informative.

There are several pharmacological agents that interact with the myofilaments to sensitize them to Ca^{2+} and this group of compounds is currently under investigation as potential positive inotropic drugs (Lee and Allen, 1997). Some of these compounds interact directly with cTnC to enhance Ca^2 binding while others appear to act at the cross-bridge level to promote strong-binding interactions. Pimobendan is a Ca^{2+} sensitizer that increases Ca^{2+}-cTnC affinity (Fujino, et al, 1988). In skinned cardiac fibers it brings about equivalent increases in Ca^{2+} sensitivity at long and short sarcomere lengths with no change in the ΔpCa_{50} (Fukuda, et al, 2000a). On the other hand, the calmodulin antagonist calmidazolium (CDZ) also increases Ca^{2+}-cTnC affinity (El-Saleh and Solaro, 1987) yet it completely eliminated length-dependent Ca^{2+} activation (Arteaga, et al, 2000). Evidently it has a more complex mode of action. Two Ca^{2+} sensitizers which promote strong-binding actin-myosin interactions, EMD 57033 (Solaro, et al, 1993) and CGP-48506 (Wolska, et al, 1996), with no apparent effects on Ca^{2+}-cTnC affinity, markedly reduced the length-dependent change in Ca^{2+} sensitivity (Arteaga, et al, 2000).   The action of these drugs poses an interesting question: if they increase Ca^{2+} sensitivity by promoting strong actin-myosin interactions why is there no increase in Ca^{2+} binding to cTnC as a consequence of myosin-mediated activation of the thin filament? In fact, the same question could be raised about caffeine, another Ca^{2+} sensitizer which promotes actin-myosin interaction and suppresses length-dependent Ca^{2+} activation but has no effect on the binding of Ca^{2+} to cTnC (Wendt and Stephenson, 1983; de Beer, et al, 1988; Powers and Solaro, 1995). Conceivably, these compounds increase the force generated by the individual cross-bridge rather than increase the number of cross-bridges recruited (Wolska, et al, 1996; Kraft and Brenner, 1997). As more is learned about the mechanism of action of these drugs they may become useful tools in analyzing the interrelationships among myofilament geometry, thin filament activation, and Ca^{2+} binding.

11. TITIN - DOES IT PLAY A ROLE IN LENGTH-DEPENDENT ACTIVATION?

A new dimension in the analysis of length-dependent Ca^{2+} activation in cardiac muscle is provided by recent studies suggesting that titin plays an integral role in the link between sarcomere length and myofilament Ca^{2+} sensitivity. Titin is the giant elastic protein (MW cardiac titin ~2500kDa) which extends from the Z line to the M line. There is growing evidence that it is the primary determinant of diastolic compliance in the physiological sarcomere length range (Labeit, et al, 1997, Wu, et al, 2000). In the I band zone there are specific domains which can undergo a reversible unfolding with sarcomere stretch. Cazorla, et al (1997) reported that active force generation by guinea pig ventricular myocytes was more closely correlated with resting tension than with sarcomere length, the implication being that resting tension reflects the extent of conformational change in the extensible regions of the titin molecule. The same group (Cazorla, et al, 1999) then reported that trypsin treatment of skinned myocytes caused a selective degradation of titin which was accompanied by a decrease in myocyte stiffness and a reduction in the ΔpCa_{50} over the sarcomere length range 1.9-2.3µm. The fall in ΔpCa_{50} was correlated with the decrease in stiffness and reached about half the control value when the stiffness was about 20% of the control value. Thus Ca^{2+} sensitivity is considered by these investigators to be controlled by the combined effects of change in interfilament spacing and titin extension.

It is quite possible that there are multiple length-sensitive parameters that can modulate Ca^{2+} sensitivity and a role for titin is plausible given that it can interact with other thin filament proteins (Linke, et al, 1997). However, there are still some questions which must be resolved. An effect of trypsin on proteins other than titin, although not evident on electrophoretic gels, has not been rigorously excluded. In the study of Fuchs and Wang (1996) there was no effect of sarcomere length on pCa_{50} when muscle width was held constant. The preparation used in that study had been stored in glycerol and it is conceivable that it had already undergone some titin degradation, in which case the contribution of titin with stretch would not be detected. This criticism would not apply to the data of McDonald and Moss (1995), who used freshly prepared myocytes. When relating their results to the changes in fiber diameter produced by either stretch or osmotic compression they found only a slight "mismatch" between length-induced and compression-induced changes in Ca^{2+} sensitivity. Based on the data of Cazorla, et al (1999), an independent effect of titin should have been observable. Fukuda, et al (2000a) also observed only a slight mismatch between stretch and osmotic compression in terms of change in Ca^{2+} sensitivity. Their preparation was stored in glycerol for up to a week in the presence of a protease inhibitor. However, a recent preliminary communication from the same group states that a precise match between the effects of stretch and osmotic compression was seen only in fibers which had undergone proteolytic breakdown of titin (Fukuda, et al, 2000b). More studies will be needed in which titin degradation is monitored in conjunction with changes in Ca^{2+} sensitivity. As recognized by Cazorla, et al (1999), there is also a need to determine whether titin degradation alters the relationship between sarcomere length and lattice spacing. In fact, a more recent study presents strong evidence that the effect of titin on Ca^{2+} activation is related to its' role in determining the filament lattice spacing (Cazorla, et al, 2001).

12. LENGTH-DEPENDENCE OF CA^{2+} ACTIVATION IN INTACT MUSCLE FIBERS

The goal of cardiac research is to explain the function of the intact, living myocardium in terms of well-defined molecular and subcellular mechanisms. Concepts based on experiments with reconstituted contractile protein systems and skinned fibers should hopefully be testable in intact systems. The large body of evidence obtained with skinned fibers leads to two immediate questions: What is the actual *in situ* change in Ca^{2+} sensitivity that occurs with variation in the physiologically relevant range of sarcomere lengths (or end diastolic volumes)? How much of the active force change that occurs with a given length change is attributable to length-dependent modulation of Ca^{2+} sensitivity and how much is a function of other factors, including physical factors? So far only a few studies have been reported which address these questions and the answers they provide should be considered as tentative.

A main focus of all such studies is the characterization of the force-pCa relation in the intact cell. This requires the use of properly calibrated intracellular Ca^{2+} indicators. With the availability of techniques for tetanizing cardiac muscle it is possible to measure steady-state force as a function of steady-state Ca^{2+} concentration, thereby circumventing the problems caused by rapid changes of free Ca^2 and time-dependent changes in the height of the Ca^{2+} transient. Intracellular pCa can be varied by changing the extracellular free Ca^{2+} concentration, generally over the range of 1-10mM. From measurements of $[Ca^{2+}]_i$ and the associated force, normalized force-pCa curves can be constructed. A number of complicating factors, including possible uncertainties in Ca^{2+} indicator calibration, make it difficult to directly compare results obtained in skinned and intact preparations. As already pointed out, the lattice spacings and their sarcomere length dependence are not the same in the two preparations. Also, the chemical environments in which the myofilaments are bathed are not identical. For these reasons, and possibly others as well, the sensitivity and cooperativity of the force response may be altered by the skinning procedure. In a study in which force-pCa curves were generated in the same rat trabeculae before and after skinning it was found that in the intact muscle the Hill coefficients were significantly greater ($\sim$5-6) than in the skinned muscle ($\sim$2-4), as was the Ca^{2+} sensitivity (Gao, et al, 1994). In a later study the Hill coefficient in intact mouse cardiac muscle was found to be 9.9 (Gao, et al, 1998). Such steep force–pCa curves mean that force is regulated over a very narrow range of free Ca^{2+} concentrations and even very small changes in Ca^{2+} sensitivity could have disproportionately large effects on active force development. Hence it is probably not realistic to expect a close correspondence between functional effects and ΔpCa_{50} values in intact and skinned preparations.

Interestingly enough, the first study in a living preparation was carried out with an intact Langendorff-perfused ferret heart that was treated with ryanodine and tetanized (Stennett, et al, 1996). Aequorin was infused into a small region at the ventricular apex for measurement of $[Ca^{2+}]_i$. Systolic performance was expressed in terms of the isovolumic pressure developed at a given diastolic volume and $[Ca^{2+}]_i$.

Sarcomere length was determined from tissues fixed at different diastolic volumes. The data were then converted into volume-dependent stress-strain curves. At maximum strain (sarcomere length 2.18µm) the pCa_{50} was 6.38 and at 75% maximum strain (sarcomere length 1.86µm) the pCa_{50} was 6.25 (ΔpCa_{50}=0.13). Thus there was a reasonably good qualitative correspondence with skinned fiber data. A problem with this type of study, clearly recognized by the authors, is that the force-length relationship is characterized in terms of global properties of the intact ventricle (end diastolic volume and developed pressure) but $[Ca^{2+}]_i$ is sampled from a single small region of the ventricle. The assumption that the measured values of $[Ca^{2+}]_i$ are the same throughout the ventricle must be tested, especially since there are well-recognized inhomogeneities within the ventricle with respect to a number of physiological parameters. There are regional differences in electrophysiological properties (Fedida and Giles, 1991), as well as in cell compliance and force-length relationships (Cazorla, et al, 2000). It is not known that the myocytes in the localized region containing the Ca^{2+} indicator were stretched to the same extent as cells elsewhere in the ventricle. So far this appears to be the only study done with the intact heart. Future studies may remove some of these uncertainties, particularly with the development of optical mapping techniques for recording both electrical events and Ca^{2+} transients in various regions of the intact heart (Choi and Salama, 2000).

Komukai and Kurihara (1997) estimated the force-pCa relationships in aequorin-loaded ferret papillary muscles subjected to tetanic stimulation. Comparing $100\%L_{max}$ and $90\%L_{max,}$ they observed a greater Ca^{2+} sensitivity at the longer length, with a ΔpCa_{50} of 0.07. Sarcomere length was not measured. Given the prevailing uncertainties, the correspondence between results from the intact heart and the papillary muscle of the same species was reasonably good. Kentish and Wrzosek (1998) undertook a study to determine the quantitative contributions of Ca^{2+} sensitivity change and physical factors to length control of force in rat trabeculae loaded with fura-2. Physical factors could include the effect of double overlap of actin filaments as well as effects of changing interfilament separation on the force developed by individual cross-bridges (Bagni, et al, 1990). There might also be an influence of internal restoring forces at short sarcomere length (Kentish and Stienen, 1994). The relationship between force and length under conditions of maximum Ca^{2+} activation is considered to represent the contribution of physical factors. In this study calibration of the Ca^{2+} indicator was not done; instead, the 340nm/380nm fluorescence ratio was used as an index of $[Ca^{2+}]_i$. Comparing $100\%L_{max}$ and $90\%L_{max,}$ the force-(340/380) curve was clearly shifted to the left at the longer length. The authors estimated that about 40% of the length-dependent change in force could be attributed to physical factors and the remaining 60% reflected the change in Ca^{2+} sensitivity. Stennett et al (1996), in the maximally activated intact ventricle, reported a 30% drop in stress in going from a strain of 1.0 to 0.75. In skinned rat trabeculae Kentish and Stienen (1994) reported only a 15% fall in maximum Ca^{2+}-activated force in comparing $100\%L_{max}$ to $90\%L_{max}$. Thus the contributions of physical factors seem to differ quantitatively in the different preparations. The Hill coefficients determined by Kentish and Wrzosek were in the range of 7-8, confirming the other reports of high *in situ* values. These studies leave little doubt that changes in fiber length cause a physiologically meaningful shift in Ca^{2+} sensitivity in living cardiac muscle. This shift appears to account for at least 60% of the observed force change and might account for considerably more. The picture that has emerged will no doubt be refined by further studies with intact preparations.

So far two studies have been reported which took up the same question in intact skeletal muscle. In frog skeletal muscle fibers which had been loaded with fura-2 and calibrated for free Ca^{2+} the force-pCa curve at sarcomere length 2.9μm was shifted very markedly to the left of that at sarcomere length 2.2μm (Claflin, et al, 1998). Length-dependent changes in Ca^{2+} sensitivity were also observed by Balnave and Allen (1996) in mouse skeletal muscle loaded with indo-1, although in that study sarcomere length was not measured. An interesting observation made in the latter study was that the Ca^{2+} sensitivity reached a peak on the descending limb of the force-length curve (110-120%L_{max}) but then started to decline with high degrees of stretch (~140%L_{max}). Such a reversal of the pCa$_{50}$ value has not been in skinned skeletal muscle fibers subjected to corresponding increases in length. This phenomenon could conceivably be related to length-dependent differences in lattice spacing in skinned and intact preparations (Wang and Fuchs, 2000a).

13. SUMMARY

The past two decades have seen impressive progress made in our understanding of how myocardial force is regulated through changes in contractile state and filling pressure. While the distinction between contractility-dependent and length-dependent modulation of force may still have some didactic value in the teaching of cardiac function to students it has become clear that at the subcellular level this distinction disappears. In fact, this point was made very forcefully in the review by Jewell (1977), written at a time when relatively little was known about the molecular details of Ca^{2+} activation in cardiac muscle. If inotropic state is equated at the myofilament level with the level of Ca^{2+} activation then sarcomere length must be included among the determinants of inotropic state. Abundant evidence has been gathered which strongly indicates that the cross-bridges function as "length sensors" which, through myosin-mediated activation of the thin filament, adjust the Ca^{2+} sensitivity as a function of sarcomere length. The primary controlling factor however is not length itself but the length-dependent variation in the proximity between the cross-bridges and the actin filament. Based on newer concepts of thin filament activation it is now possible to discuss the contributions of Ca^{2+} and cross-bridges in terms of the equilibria between well-defined thin filament states. Cooperative activation of the thin filament by strong-binding cross-bridges is favored by a reduction in the lateral spacing between actin and myosin filaments. The contributions of other length-dependent influences, such as titin extension, is currently under investigation.

Relatively little is known about the signaling pathways involved in length modulation of Ca^{2+} sensitivity in either cardiac or skeletal muscle. This situation should be rectified with the continued application of molecular-genetic techniques which allow for the analysis of the functional effects of specific mutational changes in the proteins involved in both Ca^{2+}-activated and myosin-mediated activation of the thin filament. A role for cTnI seems to have been identified but further studies are needed on the possible roles of other myofibrillar proteins. Another broad area of investigation which has not yet been undertaken is analysis of the physiological consequences of altered myocardial length modulation with respect to the intact cardiovascular system. For example, what effect, if any, would a genetically engineered change in the relationship between sarcomere length and developed force have on cardiac output regulation and exercise capacity? What adaptations would take place in response to pressure and volume overloads? Studies along these lines will provide a common meeting ground for molecular and integrative physiologists interested in the function of the cardiovascular system.

Department of Cell Biology and Physiology
University of Pittsburgh School of Medicine
Pittsburgh, PA 15261, USA

14. REFERENCES

Adhikari, B.B.and Fajer, P.G. (1996) Myosin head orientation and mobility during isometric contraction: effects of osmotic compression, *Biophys. J.* **70**, 1872-1880.

Akella, A.B., Ding, X.-L., Cheng, R., and Gulati, J. (1995) Diminished Ca^{2+} sensitivity of skinned cardiac muscle contractility coincident with troponin T-band shifts in the diabetic rat, *Circ. Res.* **76**, 600-606.

Allen, D.G., Jewell, B.R., and Murray, J.W. (1974) The contribution of activation processes to the length-tension relation of cardiac muscle, *Nature* **248**, 509-513.

Allen, D.G. and Kentish, J.C. (1985) The cellular basis of the length-tension relation in cardiac muscle, *J. Mol. Cell. Cardiol.* **17**, 821-840.

Allen, D.G. and Kentish, J.C. (1988) Calcium concentration in the myoplasm of skinned ferret ventricular muscle following changes in muscle length, *J. Physiol.(Lond.)* **407**, 489-503.

Allen, D. G. and Kurihara, S. (1982) The effects of muscle length on intracellular calcium transients in mammalian cardiac muscle, *J. Physiol. (Lond.)* **327**, 79-94.

Allen, J.D. and Moss, R.L. (1987) Factors influencing the ascending limb of the sarcomere length-tension relationship in rabbit skinned muscle fibers, *J. Physiol. (Lond.)* **390**, 119-136.

Alvarez, B.V, Perez, N.G., Ennis, I.L., Camilion de Hurtado, M.C., and Cingolani, H.E. (1999) Mechanisms underlying the increase in force and Ca^{2+} transient that follow stretch of cardiac muscle: a possible explanation of the Anrep effect, *Circ. Res.* **85**, 716-722.

Arteaga, G.M., Palmiter, K.A., Leiden, J.M., and Solaro, R.J. (2000) Attenuation of length dependence of calcium activation in myofilaments of transgenic mouse hearts expressing slow skeletal troponin I, *J. Physiol (Lond.)* **526**, 541-549.

Bagni, M.A., Cecchi, G., and Colomo, F. (1990) Myofilament spacing and force generation in intact frog muscle fibres, *J. Physiol. (Lond.)* **430**, 61-75.

Balnave, C.D. and Allen, D.G. (1996) The effect of muscle length on intracellular calcium and force in single fibres from mouse skeletal muscle, *J. Physiol. (Lond.)* **492**, 705,-113.

Bluhm, W.F. and Lew, W.Y.W. (1995) Sarcoplasmic reticulum in cardiac length-dependent activation in rabbits, *Am. J. Physiol.* **269**, H965-H972.

Brandt, P.W., Roemer, D., and Schachat, F.H. (1990) Co-operative activation of skeletal muscle thin filaments by rigor cross-bridges. The effect of troponin C extraction. *J. Mol. Biol.* **212**, 473-480.

Bremel, R.D. and Weber, A. (1972) Cooperation within actin filament in vertebrate skeletal muscle, *Nature* **238**, 97-101.

Brenner, B. (1988) Effect of Ca^{2+} on cross-bridge turnover kinetics in skinned single rabbit psoas fibers: implications for regulation of muscle contraction, *Proc. Nat. Acad. Sci. USA* **85**, 3265-3269.

Brenner, B., Yu, L.C., and Podolsky, R.J. (1984) X-ray diffraction evidence for cross-bridge formation in relaxed muscle fibers at various ionic strengths, *Biophys. J.* **46**, 299-306.

Calaghan, S.C and White, E. (1999) The role of calcium in the response of cardiac muscle to stretch, *Prog. Biophys. Mol. Biol.* **71**, 59-90.

Cantino, M.E., Allen, T.S., and Gordon, A.M. (1993) Subsarcomeric distribution of calcium in demembranated fibers of rabbit psoas muscle. *Biophys. J.* **64**, 211-222.

Caputo, C., Edman, K.A.P., Lou, F. and Sun, Y.-B. (1994) Variation in myoplasmic Ca^{2+} concentration during contraction and relaxation studied by the indicator fluo-3 in frog muscle fibres, *J. Physiol. (Lond.)* **478**, 137-148.

Cazorla, O., Pascarel, C., Garnier, D., and LeGuennec, J.-Y. (1997) Resting tension participates in the modulation of active tension in isolated guinea pig ventricular myocytes, *J. Mol. Cell. Cardiol.* **29**, 1629-637.

Cazorla, O., Vassort, G., Garnier, D., and LeGuennec, J.-Y. (1999) Length modulation of active force in rat cardiac myocytes, *J. Mol. Cell. Cardiol.* **31**, 1215-1227.

Cazorla, O., Le Guennec, J.-Y., and White, E. (2000) Length-tension relationships of sub-epicardial and sub-endocardial single ventricular myocytes from rat and ferret hearts, *J. Mol. Cell. Cardiol.* **32**, 735-744.

Cazorla, O., Wu, Y., Irving, T.C., and Granzier, H. (2001) Titin-based modulation of calcium sensitivity of active tension in mouse skinned cardiac myocytes, *Circ. Res.* **88**, 1028-1035.

Chalovich, J.M., Yu, L.C., and Brenner, B. (1991) Involvement of weak binding cross-bridges in force production in muscle, *J. Muscle Res. Cell Motil.* **12**, 503-506.

Chandra, M., Rundell, V.L.M., Tardiff, J.C., Leinwand, L.A., DeTombe, P.P., and Solaro, R.J. (2001) Ca^{2+} activation of myofilaments from transgenic mouse hearts expressing R92Q mutant cardiac troponin T, *Am. J. Physiol.* **280**, H705-H713.

Choi, B.-R. and Salama, G. (2000) Simultaneous maps of optical action potentials and Ca^{2+} transients in guinea pig hearts: mechanisms underlying concordant alternans, *J. Physiol. (Lond.)* **529**, 171-188.

Claflin, D.R., Morgan, D.L., and Julian, F.J. (1998) The effect of length on the relationship between tension and intracellular $[Ca^{2+}]$ in intact frog skeletal muscle fibres, *J. Physiol. (Lond.)* **508**, 179-186.

Close, R.I. (1972) The relations between sarcomere length and characteristics of isometric twitch contractions of frog sartorius muscle, *J.Physiol. (Lond.)* **220**, 745-762.

Cooke, R. and Franks, K. (1980) All myosin heads form bonds with actin in rigor rabbit muscle, *Biochemistry* **19**, 2265-2269.

Crozatier, B. (1996) Stretch-induced modifications of myocardial performance: from ventricular function to cellular and molecular mechanisms, *Cardiovasc. Res.* **32**, 25-37.

de Beer, E.L., Grundeman, R.L.F., Wilhelm, A.J., Caljouw, C.J., Klepper,D., and Schiereck, P. (1988) Caffeine suppresses length dependency of Ca^{2+} sensitivity of skinned striated muscle, *Am. J. Physiol.* **254**, C491-C497.

de Beer, E.L., Grundeman, R.L.F., Wilhelm, A.J., van den Berg, C., Caljouw, C.J., Klepper, D., and Schiereck, P. (1988) Effect of sarcomere length and filament lattice spacing on force development in skinned cardiac and skeletal muscle preparations from the rabbit, *Bas. Res. Cardiol.* **83**, 410-423.

Edman, K.A.P. (1975) Mechanical deactivation induced by active shortening in isolated muscle fibers of the frog, *J. Physiol. (Lond.)* **246**, 255-275.

Edman, K.A.P. and Kiessling, A. (1971) The time course of the active state in relation to sarcomere length and movement studied in single skeletal muscle fibers of the frog, *Acta Physiol. Scand.* **81**, 182-196.

El-Saleh, S.C. and Solaro, R.J. (1987) Calmidazolium, a calmodulin antagonist, stimulates calcium-troponin C and calcium-calmodulin-dependent activation of striated muscle myofilaments, *J. Biol. Chem.* **262**, 17240-17246.

Endo, M. (1972) Stretch-induced increase in activation of skinned muscle fibers by calcium, *Nature* **237**, 211-213.

Fabiato, A. and Fabiato, F. (1975) Dependence of the contractile activation of skinned cardiac cells on the sarcomere length, *Nature* **256**, 54-56.

Fabiato, A. and Fabiato, F. (1978) Myofilament-generated tension oscillations during partial calcium activation and activation dependence of the sarcomere length-tension relation of skinned cardiac cells, *J. Gen. Physiol.* **72**, 667-699.

Fedida, D. and Giles, W.R. (1991) Regional variations in action potentials and transient outward current in myocytes isolated from rabbit left ventricle, *J. Physiol. (Lond)* **442**, 191-209.

Fentzke, R.C., Buck, S.H., Patel, J.R., Lin, H., Wolska, B.M., Stojanovic, M.O., Martin, A.F., Solaro, R.J., Moss, R.L., and Leiden, J.M. (1999) Impaired cardiomyocyte relaxation and diastolic function in transgenic mice expressing slow skeletal troponin I in the heart, *J. Physiol. (Lond.)* **517**, 143-157.

Fitzsimons, D.P. and Moss, R.L. (1998) Strong binding of myosin modulates length-dependent Ca^{2+} activation of rat ventricular myocytes, *Circ. Res.* **83**, 602-607.

Frank, O. (1895) Zur Dynamik des Herzmuskels, *Z. Biol.* **32**, 370-447.

Fuchs, F. (1977) The binding of calcium to glycerinated muscle fibers in rigor: the effect of filament overlap, *Biochim. Biophys. Acta* **491**, 523-531.

Fuchs, F. (1985) The binding of calcium to detergent-extracted rabbit psoas muscle fibres during relaxation and force generation, *J. Muscle Res. Cell Motil.* **6**, 477-486.

Fuchs, F. (1995) Mechanical modulation of the Ca^{2+} regulatory protein complex in cardiac muscle, *NewsPhysiol. Sci.* **10**, 6-12.

Fuchs, F. and Fox, C. (1982) Parallel measurements of bound calcium and force in glycerinated rabbit psoas muscle fibers, *Biochim. Biophys. Acta* **679**, 110-115.

Fuchs, F. and Wang, Y.-P. (1991) Force, length, and Ca^{2+}-troponin C affinity in skeletal muscle, *Am. J. Physiol.* **261**, C787-C792.

Fuchs, F. and Wang, Y.-P. (1996) Sarcomere length vs. interfilament spacing as determinants of cardiac myofilament Ca^{2+} sensitivity and Ca^{2+} binding, *J. Mol. Cell. Cardiol.* **28**, 1375-1383.

Fuchs, F. and Wang, Y.-P. (1997) Length-dependence of actin-myosin interaction in skinned cardiac muscle fibers in rigor, *J. Mol. Cell. Cardiol.* **29**, 3267-3274.

Fujino, K., Sperelakis, N., and Solaro, R.J. (1988) Differential effects of d-and l-pimobendan on cardiac myofilament Ca^{2+} sensitivity, *J. Pharmacol. Exp. Therap.* **247**, 519-523.

Fukuda, N., Fujita, H., Fujita, T., and Ishiwata, S. (1998) Regulatory roles of MgADP and calcium in tension development of skinned cardiac muscle, *J. Muscle Res. Cell Motil.* **19**, 909-921.

Fukuda, N., Kajiwara, H., Ishiwata, S., and Kurihara, S. (2000a) Effects of MgADP on length dependence of tension generation in skinned rat cardiac muscle, *Circ.Res.* **86**, e1-e6.

Fukuda, N., Sasaki, D., Ishiwata, S., and Kurihara, S. (2000b) Understanding of sarcomere length (SL)-dependent tension generation in skinned cardiac muscle, *Biophys. J.* **78**, 141A.

Gao, W.D., Backx, P.H., Azan-Backx, M., and Marban, E. (1994) Myofilament Ca^{2+} sensitivity in intact versus skinned rat ventricular muscle, *Circ. Res.* **74**, 408-415.

Gao, W.D., Perez, N.G., and Marban, E. (1998) Calcium cycling and contractile activation in intact mouse cardiac muscle, *J. Physiol. (Lond.)* **507**, 175-184.

Godt, R.E. and Maughan, D.W. (1977) Swelling of skinned muscle fibers of the frog, *Biophys. J.* **19**, 103-116.

Godt, R.E. and Maughan, D.W. (1981) Influence of osmotic compression on calcium activation and tension in skinned muscle fibers of the rabbit, *Pflugers Arch.* **391**, 334-337.

Gordon, A.M., Homsher, E., and Regnier M. (2000) Regulation of contraction in striated muscle, *Physiol. Rev.* **80**, 853-924.

Gordon, A.M., Huxley, A.F., and Julian, F.J. (1966) The variation in isometric tension with sarcomere length in vertebrate muscle fibres, *J. Physiol. (Lond.)* **184**, 170-192.

Gulati, J., Sonnenblick, E., and Babu, A. (1991) The role of troponin C in the length dependence of Ca^{2+}-sensitive force of mammalian skeletal and cardiac muscles, *J. Physiol. (Lond.)* **441**, 305-324.

Guth, K. and Potter, J.D. (1987) Effect of rigor and cycling cross-bridges on the structure of troponin C and on the Ca^{2+} affinity of the regulatory sites in skinned rabbit psoas fibers, *J. Biol. Chem.* **262**, 13627-13635.

Hancock, W.O., Martyn, D.A., and Huntsman, L.L. (1993) Ca^{2+} and segment length dependence of isometric force kinetics in intact ferret cardiac muscle, *Circ. Res.* **73**, 603-611.

Hancock, W.O., Huntsman, L.L., and Gordon, A.M. (1997) Models of calcium activation account for differences between skeletal and cardiac force redevelopment kinetics, *J. Muscle Res. Cell Motil.* **18**, 671-681.

Harrison, S.M., Lamont, C., and Miller, D.J. (1988) Hysteresis and the length dependence of calcium sensitivity in chemically-skinned rat cardiac muscle, *J. Physiol. (Lond.)* **401**, 115-143.

Head, J.G., Ritchie,M.D., and Geeves, M.A. (1995) Characterization of the equilibrium between blocked and closed states of muscle thin filaments, *Eur. J. Biochem.* **227**, 694-699.

Hibberd, M.G. and Jewell, B.R. (1982) Calcium and length-dependent force production in rat ventricular muscle, *J. Physiol. (Lond.)* **329**, 527-540.

Hoar, P.E., Mahoney, C.W., Kerrick, W.G.L., and Montague, D. (1987) MgADP increases maximum tension and Ca^{2+} sensitivity in skinned rabbit soleus fibers, *Pflugers Arch.* **410**, 30-36.

Hofmann, P.A. and Fuchs, F. (1987a) Effect of length and cross-bridge attachment on Ca^{2+} binding to cardiac troponin C, *Am. J. Physiol.* **253**, C90-C96.

Hofmann, P.A. and Fuchs, F. (1987b) Evidence for a force-dependent component of calcium binding to cardiac troponin C, *Am. J. Physiol.* **253**, C541-C546.

Hofmann, P.A. and Fuchs, F. (1988) Bound calcium and force development in skinned cardiac muscle bundles: effect of sarcomere length, *J. Mol. Cell. Cardiol.* **20**, 667-677.

Holroyde, M.J., Robertson, S.P., Johnson, J.D., Solaro, R. J., and Potter, J.D. (1980) The calcium and magnesium binding sites on cardiac troponin and their role in the regulation of myofibrillar adenosine triphosphatase, *J. Biol. Chem.* **255**, 11688-11693.

Hongo, K., White, E., LeGuennec, J.-Y., and Orchard, C.H. (1996) Changes in $[Ca^{2+}]_i$, $[Na^+]_i$, and Ca^{2+} current in isolated rat ventricular myocytes following an increase in cell length, *J. Physiol. (Lond.)* **491**, 599-609.

Huntsman, L.L. and Stewart, D.K. (1977) Length-dependent calcium inotropism in cat papillary muscle, *Circ. Res.* **40**, 366-371.

Huxley, H.E. (1963) Electron microscope studies on the structure of natural and synthetic protein filaments from striated muscle, *J. Mol. Biol.* **7**, 281-308.

Huxley, H.E. and Hanson, J. (1954) Changes in the cross-striations of muscle during contraction and stretch and their structural interpretation, *Nature* **173**, 973-976.

Irving, T.C., Konhilas, J., Perry, D., Fischetti, R., and DeTombe, P.P. (2000) Myofilament lattice spacing as function of sarcomere length in isolated rat myocardium, *Am. J. Physiol.* **279**, H2568-H2573.

Ishikawa, T., Kajiwara, H., and Kurihara, S. (1999) Alterations in contractile properties and Ca^{2+} handling in streptozotocin-induced diabetic rat myocardium, *Am. J. Physiol.* **277**, H2185-H2194.

Jewell, B.R. (1977) A reexamination of the influence of muscle length on myocardial performance, *Circ. Res.* **40**, 221-230.

Julian, F.J., Sollins, M.R., and Moss, R.L. (1976) Absence of a plateau in length-tension relationship of rabbit papillary muscle when internal shortening is prevented, *Nature* **260**, 340-342.

Kajiwara, H., Morimoto, S., Fukuda, N., Ohtsuki, I., and Kurihara, S. (2000) Effect of troponin I phosphorylation by protein kinase A on length-dependence of tension activation in skinned cardiac muscle fibers, *Biochem. Biophys. Res. Comm.* **272**, 104-110.

Kaufmann, R.L., Bayer, R.M., and Harnasch, C. (1972) Autoregulation of contractility in the myocardial cell: displacement as a controlling parameter, *Pflugers Arch.* **332**, 96-116.

Kawai, M., Wray, J.S., and Zhao, Y. (1993) The effect of lattice spacing change on cross-bridge kinetics in chemically skinned rabbit psoas muscle fibers. 1. proportionality between the lattice spacing and the fiber width, *Biophys. J.* **64**, 187-196.

Kentish, J.C. and Stienen, G.J.M. (1994) Differential effects of length on maximum force production and myofibrillar ATPase activity in rat skinned cardiac muscle, *J. Physiol. (Lond.)* **475**, 175-184.

Kentish, J.C. and Wrzosek, A. (1998) Changes in force and cytosolic Ca^{2+} concentration after length changes in isolated rat ventricular trabeculae, *J. Physiol.(Lond.)* **506**, 431-444.

Kentish, J.C., ter Keurs, H.E.D.J., Ricciardi, L., Bucx, J.J.J., and Noble, M.I.M. (1986) Comparison between the sarcomere length-force relations of intact and skinned trabeculae from rat right ventricle: Influence of calcium concentrations on these relations, *Circ. Res.* **58**, 755-768.

Komukai K. and Kurihara, S. (1997) Length dependence of Ca^{2+}-tension relationship in aequorin-injected ferret papillary muscles, *Am.J. Physiol.* **273**, H1068-H1074.

Konhilas, J.P., Wolska, B.M., Martin, A.F., Solaro, R.J., and deTombe, P.P. (2000) PKA modulates length-dependent activation in murine myocardium, *Biophys. J.* **78**, 108A.

Kraft, T. and Brenner, B. (1997) Force enhancement without changes in cross-bridge turnover kinetics: the effect of EMD 57033, *Biophys. J.* **72**, 272-281.

Kuhn, H.J., Bletz, C., and Ruegg, J.C. (1990) Stretch-induced increase in the Ca^{2+} sensitivity of myofibrillar ATPase activity in skinned muscle fibres from pig ventricles, *Pflugers Arch.* **415**, 741-746.

Kurihara, S. and Komukai, K. (1995) Tension-dependent changes of the intracellular Ca^{2+} transients in ferret ventricular muscles, *J. Physiol. (Lond.)* **489**, 617-625.

Labeit, S., Kolmerer, B, and Linke, W.A. (1997) The giant protein titin. Emerging roles in physiology and pathophysiology, *Circ. Res.* **80**, 290-294.

Lakatta, E.G. (1991) Length modulation of muscle performance: Frank-Starling law of the heart, in Fozzard, H.A., Jennings, R.B., Haber, E., Katz, A.M., and Morgan H.E. (eds.) *The Heart and Cardiovascular System, Vol. II,* Raven Press, New York, pp. 1325-1351.

Lakatta, E.G. and Jewell, B.R. (1977) Length-dependent activation. Its effect on the length-tension relation in cat ventricular muscle, *Circ. Res.* **40**, 251-257.

Landesberg, A. and Sideman, S. (1994) Coupling calcium binding to troponin C and cross-bridge cycling in skinned cardiac cells, *Am. J. Physiol.* **266**, H1260-1271.

Lee, J.A. and Allen, D.G. (1997) Calcium sensitizers: mechanisms of action and potential usefulness as inotropes, *Cardiovasc. Res.* **36**, 10-20.

Lehman, W., Hatch, V., Korman, K., Rosol, M., Thomas, L., Maytum, R., Geeves, M.A.,Van Eyk, J.E., Tobacman, L.S., and Craig, R. (2000) Tropomyosin and actin isoforms modulate the localization of tropomyosin strands on actin filaments, *J.Mol. Biol.* **302**, 593-606.

Lehrer, S.S. (1994) The regulatory switch of the muscle thin filament: Ca^{2+} or myosin heads? *J. Muscle Res. Cell. Motil.* **15**, 232-236.

Lehrer, S.S. and Geeves, M.A. (1998) The muscle thin filament as a classical cooperative/allosteric regulatory system, *J. Mol. Biol.* **277**, 1081-1089.

Levine, R.J.C., Kensler, R.W., Yang, Z., Stull, J.T., and Sweeney, H.L. (1996) Myosin light chain phosphorylation affects the structure of rabit skeletal muscle thick filaments, *Biophys. J.* **71**, 898-907.

Linke, W.A., Ivemeyer, M., Labeit, S., Hinssen, H., Ruegg, J.C., and Gautel, M. (1997) Actin-titin interaction in cardiac myofibils: probing a physiological role, *Biophys. J.* **73**, 905-919.

Martyn, D.A. and Gordon, A.M. (1988) Length and myofilament spacing-dependent changes in calcium sensitivity of skeletal fibres: effects of pH and ionic strength, *J. Muscle Res. Cell Motil.* **9**, 428-445.

Martyn, D.A., Freitag, C.J., Chase, P.B., and Gordon, A.M. (1999) Ca^{2+} and cross-bridge induced changes in troponin C in skinned skeletal muscle fibers: effects of force inhibition, *Biophys. J.* **76**, 1480-1493.

Martyn, D.A., Regnier, M., Xu, D., and Gordon, A.M. (2001) Ca^{2+} and crossbridge dependent changes in N- and C- terminal structure of troponin C in cardiac muscle, *Biophys. J.* **80**, 360-370.

Matsubara, I. and Millman, B.M. (1974) X-ray diffraction patterns from mammalian heart muscle, *J.Mol. Biol.* **82**, 527-536.

Matsubara, I., Suga, H., and Yagi, N. (1977) An X-ray diffraction study of the cross-circulated canine heart, *J.Physiol. (Lond.)* **270**, 311-320.

Matsubara, I., Maughan, D.W., Saeki, Y., and Yagi, N. (1989) Cross-bridge movement in rat cardiac muscle as a function of calcium concentration, *J. Physiol. (Lond.)* **417**, 555-565.

McDonald, K.S. (2000) Ca^{2+} dependence of loaded shortening in rat skinned cardiac myocytes and skeletal muscle fibres, *J. Physiol. (Lond.)* **525**, 169-181.

McDonald, K.S. and Moss, R.L. (1995) Osmotic compression of single cardiac myocytes eliminates the reduction in Ca^{2+} sensitivity of tension at short sarcomere length, *Circ. Res.* **77**, 199-205.

McDonald, K.S., Wolff, M.R., and Moss, R.L. (1997) Sarcomere length dependence of the rate of tension redevelopment and submaximal tension in rat and rabbit skinned skeletal muscle fibres, *J. Physiol. (Lond.)* **501**, 607-621.

McDonald, K.S., Field, L.J., Parmacek, M.S., Soonpaa, M., Leiden, J.M., and Moss, R.L. (1995) Length dependence of Ca^{2+} sensitivity of tension in mouse cardiac myocytes expressing skeletal troponin C, *J. Physiol. (Lond.)* **483**, 131-139.

McKillop, D.F.A. and Geeves, M.A. (1993) Regulation of the interaction between actin and myosin subfragment 1:evidence for three states of the thin filament, *Biophys. J.* **65**, 693-701.

Metzger, J.M. (1995) Myosin binding-induced cooperative activation of the thin filament in cardiac myocytes and skeletal muscle fibers, *Biophys. J.* **68**, 1430-1442.

Metzger, J.M., Wahr, P.A., Michele, D.E., Albayya, F., and Westfall, M.V. (1999) Effects of myosin heavy chain isoform switching on Ca^{2+}-activated tension development in single adult cardiac myocytes, *Circ. Res.* **84**, 1310-1317.

Millar , N.C. and Homsher, E. (1990) The effect of phosphate and calcium on force generation in glycerinated rabbit skeletal muscle fibers: a steady state and transient kinetic study, *J. Biol. Chem.* **265**, 20234-20240.

Millman, B.M. (1998) The filament lattice of striated muscle. Physiol. Rev. 78, 359-391.

Morano, I. (1999) Tuning the human heart molecular motors by myosin light chains, *J. Mol. Med.* **77**, 544-555.

Moss, R.L. (1992) Ca^{2+} regulation of mechanical properties of striated muscle. Mechanistic studies using extraction and replacement of regulatory proteins, *Circ. Res.* **70**, 865-884.

Moss, R.L., Nwoye, L.O., and Greaser, M.L. (1991) Substitution of cardiac troponin C into rabbit muscle does not alter the length dependence of Ca^{2+} sensitivity of tension. *J. Physiol. (Lond.)* **440**, 273-289.

Nassar, R., Malouf N.N., Kelly, M.B., Oakeley, A.E., and Anderson P.A. (1991) Force-pCa relation and troponin T isoforms of rabbit myocardium, *Circ. Res.* **69**, 1470-1475.

Nichols, C.G., Hanck, D.A., and Jewell, B.R. (1988) The Anrep effect: an intrinsic myocardial mechanism, *Can. J. Physiol. Pharmacol.* **66**, 924-929.

Page, S.G. and Huxley, H.E. (1963) Filament lengths in striated muscle, *J. Cell Biol.* **19**, 369-390.

Palmer, S. and Kentish, J.C. (1998) Roles of Ca^{2+} and cross-bridge kinetics in determining the maximum rates of Ca^{2+} activation and relaxation in rat and guinea pig skinned trabeculae, *Circ. Res.* **83**, 179-186.

Palmiter, K.A., Kitada, Y., Muthuchamy, M., Wieczorek, D.F., and Solaro, R.J. (1996) Exchange of β- for α-tropomyosin in hearts of transgenic mice induces change in thin filament response to Ca^{2+}, strong cross-bridge binding, and protein phosphorylation, *J. Biol. Chem.* **271**, 11611-11614.

Palmiter, K.A., Tyska, M.J, Dupuis, D.E., Alpert, N.R., and Warshaw, D.M. (1999) Kinetic differences at the single molecule level account for the functional diversity of rabbit cardiac myosin isoforms, *J. Physiol. (Lond.)* **519**, 669-678.

Parmley, W.W. and Chuck, L. (1973) Length-dependent changes in myocardial contractile state, *Am. J. Physiol.* **224**, 1195-1199.

Patel, J.R., McDonald, K.S., Wolff, M.R., and Moss, R.L. (1997) Ca^{2+} binding to troponin C in skinned skeletal muscle fibers assessed with caged Ca^{2+} and a Ca^{2+} fluorophore: invariance of Ca^{2+} binding as a function of sarcomere length, *J. Biol. Chem.* **272**, 6018-6027.

Patterson, S.W. and Starling, E.H. (1914) On the mechanical factors which determine the output of the ventricles, *J. Physiol. (Lond.)* **48**, 357-379.

Phan, B.C., Peyser, Y.M., Reisler, E., Muhlrad, A. (1997) Effect of complexes of ADP and phosphate analogs on the conformation of the Cys707-Cys697 region of myosin subfragment 1, *Eur. J. Biochem.* **243**, 636-642.

Powers, F.M. and Solaro, R.J. (1995) Caffeine alters myofilament activity and regulation independently of Ca^{2+} binding to troponin C, *Am. J. Physiol.* **268**, C1348-C1353.

Rassier, D.E., MacIntosh, B.R., and Herzog, W. (1999) Length dependence of active force production in skeletal muscle, *J. Appl. Physiol.* **86**, 1445-1457.

Roszek, B., Baan, G.J.,and Huijing, P.A. (1994) Decreasing stimulation frequency-dependent length-force characteristics of rat muscle, *J.Appl. Physiol.* **77**, 2115-2124.

Rust, E.M., Albayya, F.P., and Metzger, J.M. (1999) Identification of a contractile deficit in adult cardiac myocytes expressing hypertrophic cardiomyopathy-associated mutant troponin T proteins. *J. Clin. Invest.* **103**, 1459-1467.

Saeki, Y., Kurihara, S., Hongo, K., and Tanaka, E. (1993) Alterations in intracellular calcium and tension of activated ferret papillary muscle in response to step length changes, *J. Physiol. (Lond.)* **463**, 291-306.

Sarnoff, S.J., Mitchell, J.P., Gilmore, J.P., and Remensnyder, J.P. (1960) Homeometric autoregulation in the heart, *Circ. Res.* **8**, 1077-1091.

Schachat, F.H., Diamond, M.S., and Brandt, P.W. (1987) Effect of different troponin t-tropomyosin combinations on thin filament activation, *J. Mol. Biol.* **198**, 551-554.

Silver, P.J., Buja, L.M., and Stull, J.T. (1986) Frequency-dependent myosin light chain phosphorylation in isolated myocardium, *J. Mol. Cell. Cardiol.* **18**, 31-37.

Smith, S.H. and Fuchs, F. (1999) Effect of ionic strength on length-dependent Ca^{2+} activation in skinned cardiac muscle, *J. Mol. Cell. Cardiol.* **31**, 2115-2125.

Smith, S.H. and Fuchs, F. (2000) Length-dependence of cross-bridge mediated activation of the cardiac thin filament, *J. Mol. Cell. Cardiol.* **32**, 831-838.

Solaro, R.J. and Rarick, H.M. (1998) Troponin and tropomyosin: proteins that switch on and tune in the activity of cardiac myofilaments, *Circ. Res.* **83**, 471-480.

Solaro, R.J. and Van Eyk, J. (1996) Altered interactions among thin filament proteins modulate cardiac function, *J. Mol. Cell. Biol.* **28**, 217-230.

Solaro, R.J., Gambassi, G., Warshaw, D.M., Keller, M.R., Spurgeon, H.A., Beier, N., and Lakatta, E.G. (1993) Stereoselective actions of thiadiazinones on canine cardiac myocytes and myofilaments, *Circ.Res.* **73**,981-990.

Sollot, S.J., Ziman, B.D., Warshaw, D.M., Spurgeon, H.A., and Lakatta, E.G. (1996) Actomyosin interaction modulates resting length of unstimulated cardiac venticular cells, *Am. J. Physiol.* **271**, H896-H905.

Stienen, G.J.M., Blange, T., and Treijtel, B.W. (1985) Tension development and calcium sensitivity in skinned muscle fibers of the frog, *Pflugers Arch.* **405**, 19-23.

Stennett, R., Ogino, K., Morgan, J.P., and Burkhoff, D. (1996) Length-dependent activation in intact ferret hearts: study of steady-state Ca^{2+}-stress-strain interrelations, *Am. J. Physiol.* **270**, H1940-H1950.

Stephenson, D.G. and Wendt, I.R. (1984) Length dependence of changes in sarcoplasmic calcium concentration and myofibrillar calcium sensitivity in striated muscle fibres, *J. Muscle Res. Cell Motil.* **5**, 243-272.

Stephenson, D.G., Stewart, A.W., and Wilson, G.J. (1989) Dissociation of force from myofibrillar MgATPase and stiffness at short sarcomere lengths in rat and toad skeletal muscle, *J. Physiol. (Lond.)* **410**, 351-366.

Swartz, D.R., Moss, R.L., and Greaser, M.L. (1996) Calcium alone does not fully activate the thin filament for S1 binding to rigor myofibrils, *Biophys. J.* **71**, 1891-1904.

Sweeney, H.L., Bowman, B., and Stull, J.T. (1993) Myosin light chain phosphorylation in vertebrate striated muscle: regulation and function, *Am. J. Physiol.* **264**, 1085-1095.

Sweeney, H.L., Feng, H.S., Yang, Z., and Watkins, H. (1998) Functional analyses of troponin T mutations that cause hypertrophic cardiomyopathy: insights into disease pathogenesis and troponin function, *Proc.Nat. Acad. Sci. USA* **95**, 14406-14410.

ter Keurs, H.E.D.J. (1996) Heart failure and Starling's law of the heart, *Can. J. Cardiol.* **12**, 1047-1057.

ter Keurs, H.E.D.J. and Noble, M.I.M., eds. (1988) *Starling's Law of the Heart Revisited,* Kluwer Academic Publishers, Dordrecht.

ter Keurs, H.E.D.J., Rijnsburger, W.H., van Heuningen, R., and Nagelsmit, M.J. (1980) Tension development and sarcomere length in rat cardiac trabeculae: evidence of length-dependent activation, *Circ. Res.* **46**, 703-714.

Tobacman, L.S. (1996) Thin filament-mediated regulation of cardiac contraction, *Ann. Rev. Physiol.* **58**, 447-481.

Tobacman, L.S. and Lee, R. (1987) Isolation and functional comparison of bovine cardiac troponin T isoforms, *J. Biol. Chem.* **262**, 4059-4064.

Tobacman, L.S., Lin D., Butters, C., Landis, C., Back, N., Pavlov, D., and Homsher, E. (1999) Functional consequences of troponin T mutations found in hypertrophic cardiomyopathy, *J. Biol. Chem.* **274**, 28363-28370.

Tucci, P.J.F., Bregagnollo, E.A., Spadaro, J., Cicogna, A.C., and Ribeiro, M.C.L. (1984) Length dependence of activation studied in the isovolumic blood perfused dog heart, *Circ. Res.* **55**, 59-66.

Van Buren, P., Harris, D.E., Alpert, N.R., and Warshaw, D.M. (1995) Cardiac V1 and V3 myosins differ in their hydrolytic and mechanical activities *in vitro, Circ. Res.* **77**, 439-444.

Vandenboom, R., Claflin, D.R., and Julian, F.J. (1998) Effects of rapid shortening on rate of force regeneration and myoplasmic [Ca^{2+}] in intact frog skeletal muscle fibres, *J. Physiol. (Lond.)* **511**, 171-180.

van der Velden, J., de Jong, J.W., Owen, V.J., Burton, P.B.J., and Stienen, G.J.M. (2000) Effect of protein kinase A on calcium sensitivity of force and its sarcomere length dependence in human cardiomyocytes, *Cardiovasc. Res.* **46**, 487-495.

Vannier, C., Chevassus, H., and Vassort, G. (1996) Ca-dependence of isometric force kinetics in single skinned ventricular cardiomyocytes from rats, *Cardiovasc. Res.* **32**, 580-586.

Vibert, P., Craig, R., and Lehman, W. (1997) Steric-model for activation of muscle thin filaments, *J. Mol. Biol.* **266**, 8-14.

Wang, Y.-P. and Fuchs, F. (1994) Length, force, and Ca^{2+}-troponin C affinity in cardiac and slow skeletal muscle, *Am. J. Physiol.* **266**, C1077-C1082.

Wang, Y.-P. and Fuchs, F. (1995) Osmotic compression of skinned cardiac and skeletal muscle bundles: effects on force generation, Ca^{2+} sensitivity, and Ca^{2+} binding, *J. Mol. Cell. Cardiol.* **27**, 1235-1244.

Wang, Y.-P. and Fuchs, F. (2000a) Length-dependent effects of osmotic compression on skinned rabbit psoas muscle fibers, *J. Muscle Res. Cell Motil.* **21**, 313-319.

Wang, Y.-P. and Fuchs, F. (2000b) Correlations among lattice spacing, Ca sensitivity, and Ca-troponin C affinity in skinned cardiac muscle, *Biophys. J.* **78**, 230A.

Wannenburg, T., Heijne, G.H., Geerdink, J.H., Van Den Pool, H.W., Janssen, P.M.L., and De Tombe, P.P. (2000) Cross-bridge kinetics in rat myocardium: effect of sarcomere length and calcium activation, *Am. J. Physiol.* **279**, H779-H790.

Wattanapermpool, J., Reiser, P.J., and Solaro, R.J. (1995) Troponin I isoforms and differential effects of acidic pH on soleus and cardiac myofilaments, *Am. J. Physiol.* **268**, C323-C330.

Weisberg, A. and Winegrad, S. (1998) Relation between crossbridge structure and actomyosin ATPase activity in rat heart, *Circ. Res.* **83**, 60-72.

Wendt, I.R. and Stephenson, D.G. (1983) Effects of caffeine on Ca-activated force production in skinned cardiac and skeletal muscle fibres of the rat, *Pflugers Arch.* **398**, 210-216.

Westfall, M.V., Albayya, F.P., Turner, I.I., and Metzger, J.M. (2000) Chimera analysis of troponin domains that influence Ca^{2+}-activated myofilament tension in adult cardiac myocytes, *Circ. Res.* **86**, 470-477.

Winegrad, S. (1999) Cardiac myosin binding protein C, *Circ. Res.* **83**, 1117-1126.

Wolff, M.R., McDonald, K.S., and Moss, R.L. (1995) Rate of tension development in cardiac muscle varies with level of activator calcium, *Circ. Res.* **76**, 154-160.

Wolska, B.M., Kitada, Y., Palmiter, K.A., Westfall, M.V., Johnson, M.D., and Solaro, R.J. (1996) CGP-48506 increases contractility of ventricular myocytes and myofilaments by effects on actin-myosin reaction, *Am. J. Physiol.* **270**, H24-H32.

Wolska, B., Keller, R.S., Evans, C.E., Palmiter, K.A., Phillips, R.M., Muthuchamy, M., Oehlenschlager, J., Wieczorek, D.F., de Tombe, P.P., and Solaro, R.J. (1999) Correlation between myofilament response to Ca^{2+} and altered dynamics of contraction and relaxation in transgenic cardiac cells that express β-tropomyosin, *Circ. Res.* **84**, 745-751.

Wu, Y., Cazorla, O., Labeit, D., Labeit, S.,and Granzier, H. (2000) Changes in titin and collgen underlie diastolic stiffness diversity of cardiac muscle, *J. Mol. Cell. Cardiol.* **32**, 2151-2161.

Yang, Z., Stull, J.T., Levine, R.J.C., and Sweeney, H.L. (1998) Changes in interfilament spacing mimic the effects of myosin regulatory light chain phosphorylation in rabbit psoas fibers, *J. Struct. Biol.* **122**, 139-148.

Zhao, Y. and Kawai, M. (1993) The effect of lattice spacing change on cross-bridge kinetics in chemically skinned rabbit psoas muscle fibers. II. elementary steps affected by the spacing change, *Biophys. J.* **64**, 197-210.

EARL HOMSHER

DETERMINANTS OF UNLOADED SHORTENING VELOCITY IN STRIATED MUSCLE

Unloaded Shortening Velocity

1. INTRODUCTION

When a muscle is activated and shortens, the rate of shortening, the shortening velocity (v), varies inversely with the load (P) against which it shortens. This relationship is hyperbolic and is given by Hill's classic equation (43):

$$(P + a)v = b(P_o - P) \qquad (1)$$

where P_o is the isometric force, and *a* and *b* are constants in units of P_o and muscle lengths/sec respectively. Typical values for frog skeletal muscle *a* and *b* at 0°C are 0.25 P_o and 0.33 l_o/s respectively. Total muscle fiber length, l_o, is usually reported at optimal thick and thin filament overlap (2.2-2.5 µm per sarcomere). This is important because shortening occurs by the relative sliding of the thick and thin filaments past each other in all the sarcomeres in the muscle; i.e., each sarcomere in series shortens about the same amount. Thus, the longer a muscle fiber or cell is (i.e., the greater the number of sarcomeres arranged in series), the faster the total length of the fiber or cell decreases for a given afterload. Maximal overlap for mammalian muscle cells occurs at about 2.4 µm/ sarcomere (thin filaments each are 1.1 µm long and the pseudo H-zone is 0.2µm long while the thick filament in all vertebrate striated muscle is about 1.65-1.7 µm). Thus if one sarcomere shortens against an afterload, P, at 2.4 µm/s, then 10 sarcomeres arranged in series shortening against the same load P, will shorten at a rate of 24 µm/s. Consequently the measured shortening velocity is referenced to the muscle fiber length at optimal overlap to normalize the observed values. To obtain the speed with which individual thin filaments slide past the thick filaments, the sarcomere shortening speed is divided by 2 and the result expressed as µm/half-sarcomere/s (µm/hs/s).

When the force against which a muscle shortens is zero (unloaded), the equation reduces to

$$v = bP_o/a \qquad (2)$$

R.J. Solaro and R.L. Moss (eds), Molecular Control Mechanisms in Striated Muscle Contraction, 417-442 and therefore for frog muscle at 0-2°C the "unloaded shortening velocity", V_u, is

$$V_u = 1.32 \, l_o/s \qquad (3)$$

The designation V_u is used here instead of V_{max} to avoid confusion with enzymatic rates. Originally V_u was measured estimates of V_u from by fitting the afterload and shortening velocity data to eq. 1 (43). In 1979 Edman (26) introduced a simpler method, the "slack test", for estimating V_u. In a tetanically stimulated muscle fiber after isometric force reached a steady state, he rapidly released the muscle fiber by stepping one end toward the other a distance sufficient to reduce to force to zero (allowing the fiber to become "slack"). During this release force falls to zero and, when the "slack" is taken up by sarcomere shortening, force begins to rise at the new shorter isometric length. The time taken to begin force redevelopment following the release is called the "slack time." The greater the distance released, the greater the "slack time". Plots of the distance released against the "slack time" in fully activated fibers are linear, and the slope of this line directly estimates the unloaded shortening velocity (see Fig. 1). We will call this estimate of unloaded

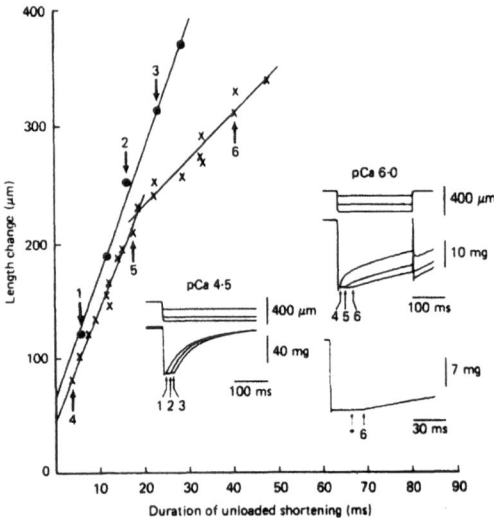

Figure 1- *Protocol for measurement of unloaded shortening velocity using the slack test (26) at pCa 4.5 and 6.0. The **insert** shows the protocol with fiber length and force Vs time records at each pCa. In each record at the given pCa, the fiber was quickly released a distance Δx at zero time and the time required for force to begin redeveloping (slack taken up in the fiber) was measured (indicated by the numbered arrows). For the pCa 6 data, records are shown at higher force sensitivity. In the **graph** the length change (Δx) is plotted against the duration of time required for the unloaded shortening to take up the slack, force to begin to redevelop. The numbered data points 1, 2, 3 are for pCa 4.5 and 4, 5, 6 for pCa 6.0. The slope of this line is the unloaded shortening velocity. Note that at shorter distances of shortening, the velocity (slope) is independent of Ca^{2+}, but that at shortenings greater than 10%, the unloaded shortening velocity for pCa 6.0 is substantially slower than at pCa 4.5.*

shortening velocity V_o to distinguish it from V_u, measured by fits to eq. 1. Measurements of V_u and V_o give similar values with differences between them attributed to inaccuracies of the measurement techniques (26,60) (extrapolation in the case of V_u and errors in estimating the time of departure of the force trace from zero for V_o). To understand the mechanisms underlying variation in unloaded shortening velocity, we first review the mechanism of ATP hydrolysis by actomyosin and how it produces the sliding of the thin filaments past the thick

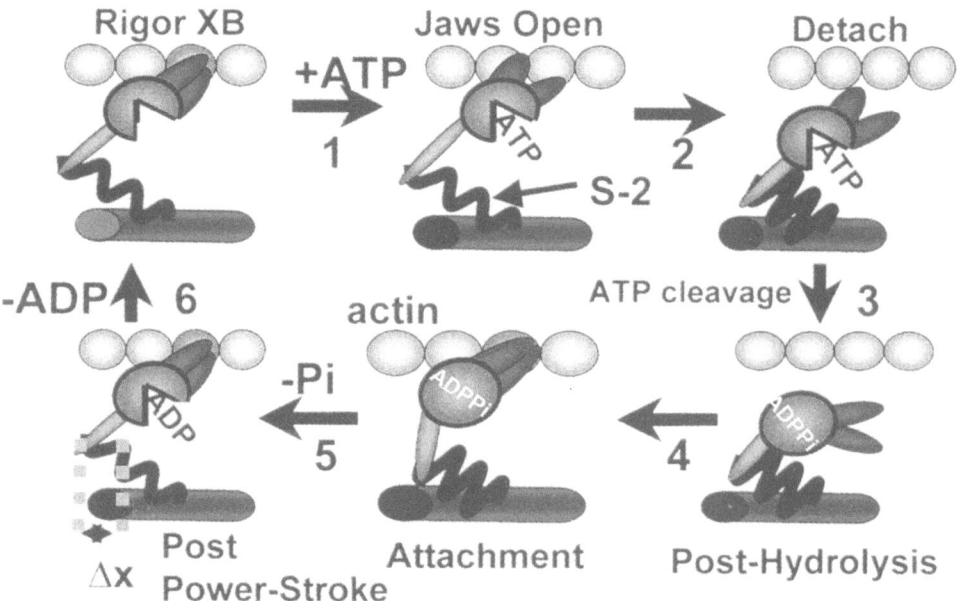

Figure 2-*Cartoon representation of the crossbridge cycle in striated muscle. The light gray spheres represent individual g-actin monomers of the thin filament which is attached to a z-line to the left of the thin filament. The magenta rod is the backbone of the thick filament and the S-2 segment is labeled. The red-green-cyan object is the globular portion of the S-1 head. The red represents the upper and lower portions (actin binding) of the 50 kD domain, the green, the catalytic (25 kD domain), and the cyan the neck (20 kD domain) or "lever arm" of the S-1 molecule. The crossbridge cycle occurs in a clockwise direction with the addition of ATP (+ATP) and the release of products (-Pi, -ADP) given at the steps at which they occur. When a g-actin monomer is colored yellow the crossbridge strongly bound to it.*
Color version on page 466.

filaments during shortening. Figure 2 is a cartoon representation of the crossbridge (the globular head of myosin, the S-1 fragment) cycle as the crossbridge attaches and detaches from thin (actin) filament. The myosin crossbridge head contains: a 25 kD catalytic domain (shown in green, containing the ATP binding pocket, shown as a wedge-shaped opening); a 50 kD domain which forms an upper and lower jaw (in red in Fig. 2), containing the actin binding sites; and a 20 kD α-helical "neck" region (shown in blue) containing a converter region which amplifies movement near the ATP binding site and produces movement of 10 nm at the end of the neck. The end of the neck region is attached to the backbone of the thick filament (in magenta) by a springy helical region called the S-2 segment. In the absence of ATP, the upper and lower jaws of the crossbridge bind strongly to actin, forming the so-called rigor linkage (Fig. 2, Rigor XB). This is a convenient place to begin the crossbridge cycle. At physiological [ATP] (3-5 mM), ATP binds to the catalytic site of the rigor crossbridge (step 1) in very rapid ($>>1000$ s^{-1}) and essentially irreversible step. After ATP binds, the upper and lower jaws of the 50 kD domain separate (Jaws Open in Fig. 2). This detaches the myosin from actin and is the detachment of actin from the AM•ATP complex (step 2) forming M•ATP and actin. This step is also very rapid. Next the MgATP in the binding site is hydrolyzed which produces a "flexing" or bending of the myosin neck region with respect to the 50 kD domain (Hydrolysis, Fig. 2). Step 3 occurs with a rate of ~150 s^{-1} at 20°C and is very temperature sensitive and falling to about 20 s^{-1} at 5 °C. Following ATP cleavage, myosin again first binds weakly to actin at a high rate and then strongly (labeled Attachment in Fig. 2). In the absence of [Ca^{2+}], the troponin/tropomyosin (Tn/Tm) regulatory complex sterically blocks access of the myosin head to strong binding sites on actin. When Ca^{2+} binds to troponin, Tn detaches from actin, allowing the Tm/Tn complex to roll or slide back and forth over the thin filament surface, transiently exposing weak and strong binding sites on actin to the complementary regions in myosin's 50 kD domain. The greater the [Ca^{2+}], the greater the fraction of time the Tm/Tn complex allows myosin access to strong binding sites on actin. An actin monomer strongly attached to the crossbridge is represented in yellow in Fig. 2. The **rate** of strong cross-bridge attachment, the flux through step 5, is thus dependent on [Ca^{2+}] and Tm position. Strong binding of myosin to actin is thought to be associated with movement of the upper and lower 50 kD sub-domains toward each other (or closing the jaws). This movement may allow the neck region of myosin to extend, opening a pathway for inorganic phosphate (Pi) release from the ATP binding pocket in myosin. Alternatively, closing the jaws might promote Pi release from the binding pocket, which then allows the extension of myosin's neck region. In any event, myosin neck extension, shown in step 5, is the power stroke which, in isometric muscle, stretches an elastic element (represented here as the S2 segment) by ~10 nm (represented by the Δx in the Post Power Stroke state) and produces a force of ~2 pN/cross-bridge (69). When the load on the muscle (and hence the crossbridge) is less than that exerted by the stretched S-2 segment), the S-2 segment shortens, dragging the thin filament past the thick filament. Step 6, the release of ADP, is strain sensitive; i.e., when the force on the cross-bridge is large (as in isometric contractions) k_{+6} is slow, 3-10 s^{-1}, and is the rate-limiting step for the cross-bridge cycle. However when the strain on the cross-bridge is low, as during rapid shortening, k_{+6} rises to 100 s^{-1}. Step 6 produces the Rigor XB state. During isometric contractions the slowness of k_{+6} allows the steady state population of the A•M •ADP state to rise, and with it, force. Cross-bridges attach and exert force constantly during steps 5,6, and 1 during isometric contraction and force drops to zero only when the cross-bridge detaches in step 2. During shortening, as the filaments slide past each other, the strain on the cross-bridge is reduced and step 6 occurs more rapidly. This causes the rate of the ATP hydrolysis to rise above the isometric rate and accounts for the Fenn effect and is an important feature in optimizing the efficiency of contraction (42). As the muscle shortens, the crossbridge is dragged to the right by the thin filament during shortening, decreasing the stretch of the S-2 segment and the force exerted on the thin filament. Once the crossbridge has moved 10 nm the S-2 segment no longer exerts force on the thin filament. This model implies that the rate of movement of actin past the crossbridge site on the thick filament will be directly dependent on how far the crossbridge moves (about 10 nm) during the power stroke which may require about 2 ms at 15°C (104). Thus the maximum speed with which the rabbit MHC-IIB crossbridge can pull the thin filament past the thick filament will be 10 nm/2 ms=5 μm/s assuming that by the end of each power stroke, another crossbridge on the thick filament has attached to the thin filament and can pull the thin filament another 10 nm. On the other hand the rate at which of crossbridge releases ADP, binds ATP, and detaches from the thin filament will also limit the shortening velocity. This is because when a crossbridge remains attached to f-actin for a time **longer** than it takes to slide 10 nm, the S-2 segment will be compressed or stretched in a direction which opposes further movement of the thick and thin filaments and puts a load on the other crossbridges trying to propel the thin filament.

Gordon, Huxley and Julian and Edman (26,227, 36) provided compelling support for the sliding filament mechanism by showing that as the muscle was stretched beyond a sarcomere spacing of 2.2 μm, the isometric force declined in direct proportion to the reduction in thick and thin filament overlap (27,36). However, unlike the isometric force, V_u does **not** change with extent of thin filament overlap, so that the unloaded shortening velocity is *independent of thick and thin filament overlap* (26,36). V_o, measured in intact skeletal muscle fibers within the range of sarcomere lengths from 1.7 μm to 2.7 μm (where significant resting force begins to develop in single skeletal muscle fibers), is independent of sarcomere length (26). At sarcomere lengths less than 1.7 μm V_o declines (26). These results compare favorably with measurements of V_u during controlled releases (37). Furthermore, V_u is independent of the extent of thick and thin filament overlap from 2.2 to 3.1 μm/sarcomere (26,36). In this range the thick and thin filament overlap ranges from 0.75 μm/half-sarcomere (100% or optimal overlap) to 0.30. This demonstrates that the unloaded shortening velocity in the intact, maximally activated muscle fiber is independent of the number of cross-bridges pulling on the thin filament (at least between 40-100% of those available). This is important because it implies that in maximally activated muscle cells there is no retarding muscle forces other than those involved with the crossbridge behavior. Unfortunately, measurements of V_u

and V_o are not available at sarcomere lengths greater than ~3.1 μm where filament overlap is < 40% of optimal. This is because passive tension (passive recoil of elastic elements such as titin (which attaches the z-disk to the thick filaments and connective tissue) beyond 3.1 μm is large and contributes to the apparent shortening velocity, complicating the analysis. At sarcomere lengths < 1.7 μm, V_o, like V_u, also decreases markedly. This occurs presumably because of a large resistive force generated by the compression of the thick filaments by the z-band.

V_o can be measured in electrically stimulated living muscle fibers and in mechanically or chemically "skinned" muscle fibers. Skinned muscle fibers are individual muscle cells, whose cell membrane has been mechanically or chemically removed. They can be attached to force or displacement transducers. Removal of the cell membrane allows one to attach the fiber or cell to transducers while the contractile proteins are bathed in a solution comparable to the physiological solution (ionic strength of about 200 mM, MgATP at 1-5 mM, pH near 7.0, and a calcium ion concentration between 1×10^{-9} - 1×10^{-5} (pCa 9 and 5 respectively). In this way the effects of changes in the bathing solution, alterations of proteins in the fibers, and physical changes on contractile function (force, shortening velocity, ATPase, force-pCa curve, etc.) can be characterized. V_o can also be measured in a motility assay in which purified contractile proteins (myosin, actin, and regulatory proteins) are used to make mechanical measurements without the structural constraints imposed by the muscle fiber lattice. In this case, nitrocellulose-coated cover slips are attached via 75-150 μm spacers to a microscope slide forming a flow chamber (41,104). A myosin solution is then introduced into the flow cell and allowed to incubate for several minutes to coat the nitrocellulose surface with myosin molecules many of whose S-1 heads project into the solution of the flow cell. After washing away excess myosin and coating the rest of the surface with BSA to block any non-specific binding sites on the slide surface, f-actin labeled with rhodamine-phalloidin is introduced into the chamber in the absence of ATP. The labeled thin filaments then bind to the myosin heads. When viewed under a fluorescent microscope, the labeled thin filaments are bound to the surface but not moving. If physiological concentration of MgATP (1-5 mM) is added to the flow cell, the labeled thin filaments begin to move over the surface in a random paths with a speed, V_f (the subscript **f** for *thin filament*), comparable to that seen in shortening unloaded muscle fibers. As these filaments are not attached to any real load, they are considered "unloaded" thin filaments. By attaching the filaments to microneedles or to small latex beads held in an optical trap, one can measure the force crossbridges exert on the labeled thin filaments. Table 1 compares the unloaded shortening velocity of intact muscle fibers, skinned muscle fibers and isolated thin filaments (in vitro motility studies) in muscles having different myosin isoforms and at different temperatures. These results illustrate basic characteristics of the unloaded shortening velocity.

Table 1 – Comparison of Unloaded Shortening Velocity

Animal	Muscle (type)	Temp. (°C)	Living Muscle (μm/hs/s)	Skinned Fiber (μm/hs/s)	Motility Assay (μm/s)
Mouse	EDL(II)	35	23.8[64], 30.3[17]		
		30	13.3*[109]		
		25	6.7[6], 7.9[5] 9.9[109], 11.8[12]		
		20	6.8*[109], 7.6[23], 9.8[12]		
		15	4.1[f109]	7.7[11]	
		10	2.5*[109]		
		5		2.9[87]	
	Soleus (I)	35	11.6[64], 15.8[17]		
		25	3.8[6], 4.2[5], 5[12]		
		20	2.3[23], 1.9[19]		
		5		1.6[94]	
Rat	EDL (II)	35	15.6[83], 21.3[16]		10.2[51]
		30	11.7[83]		7.9[51]
		25	8.8[f83]		6.1[51]
		20	6.0[83]		4[51]
		15	4.3[83]		3.1[51]
		10	2.6[83]	2.2[8], 2.4[66]	2[51]
		5		1.7[94]	
	Soleus (I)	35	8.4[83], 9.1[16]		2.5[51]
		30	6.1[83]		1.8[51]
		25	4.3[83]		1.3[52], 1.4[52]
		20	2.9[83]		0.8[51]
		15	1.5[83]	2.1[103]	0.4[51]
		10	0.8[83]	0.8[8], 1.1[66]	0.2[51]
		5		1.1[94]	
Rabbit	Psoas (II)	30		7.5[f81]	11.2[41], 7.6[49], 7.3[37]
		25		5.6[f81]	4.3[49], 7.4[47], 9.4[56]
		20		4.1[81], 13[102]	3[102], 2.4[49]
		15		3.2[73], 3.3[81], 4[68], 4.2[68]	1[f49]
		10		1.3[73], 1.6[z98], 2.2[81] 2.5[99], 2.6[48]	
		5		1[101], 1.1[94], 1.3[f81]	
	Soleus (II)	15		0.7[92], 0.9[87,88] 1.0[73]	
		10		0.5[99], 0.7[14]	
		5		0.8[94]	

*Velocities given as μm/hs/s assuming that in sarcomere length was 2.5 υm unless specified otherwise. *-velocities adjusted using Q10 values given in paper. Superscript numbers are reference number in the bibliography from which the value was taken.*

2. UNLOADED SHORTENING VELOCITY IS DIRECTLY DEPENDENT ON THE MYOSIN HEAVY CHAIN COMPOSITION

In adult striated muscle two broad isoforms of myosin heavy chain are expressed: slow (type I) and fast (type II) MHC isoforms. These forms were first characterized by differences in the positive and negative dP/dt (rate of change of force), twitch times, shortening velocity, and actin-activated myosin ATPase rate of muscles expressing them (4,16-18,39,40). These distinctions followed from the seminal observation of Barany that both myosin and actomyosin ATPase rate are proportional to the unloaded shortening velocity over 3 orders of magnitude (4).

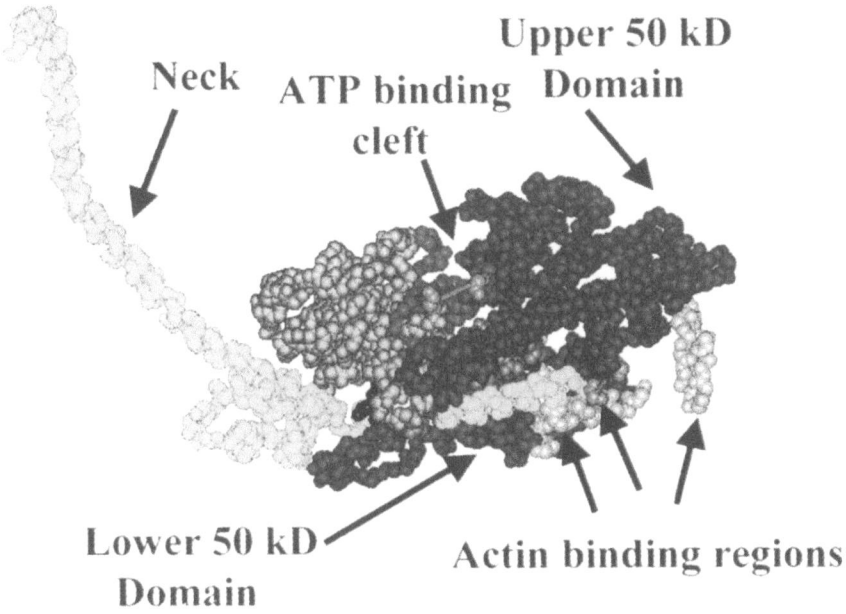

Figure 3- A space-filling model of chicken skeletal myosin S-1 to show the various portions of the molecule in greater detail. The amino acid side chains have been deleted to better show the spatial relationships between different portions of the molecule and it has been color coded to represent the following features: the 25 kD or catalytic domain is given in green (residues 4-204), the 50 kD domain is red (residues 217-626), and the 20 kD domain in cyan (residues 646-843). There are two major loops whose structures were not determined: loop1 (residues 205-216) and loop 2 (residues 627-646) and these are represented as solid gray connectors between the 25-50 kD domains and the 50-20 kD domains. The residues binding the phosphate and a nucleotide (residues 179-186, 187-199, 218-233, and 667-687) are colored magenta and represent the ATP binding pocket. The amino acid sequences which interact with actin (residues 143-147, 403-416, 567-578, and 626-646 are colored gray), the sequence bordering the 50 kD cleft (that between the upper and lower domains, residues 464-495) is colored orange, and the so-called converter domain (residues 697-707) is colored black. WebLabViewerPro (Molecular Simulations, San Diego, CA 92121) was used to construct this image from the PDB file 2MYS. Color version on page 467.

Myosin is a hetero-hexameric protein consisting of 2 heavy chains (MHC), two regulatory light chains (LC2), and two alkali light chains. One LC2 and one alkali light chain are associated with each head. The globular head of myosin is composed of the first 843 amino acid residues from the amino terminus of the MHC and is represented as a space filling model (without side chains) in Fig. 3. Beginning at the NH$_2$ terminus, the first of amino acids is the 25 kD catalytic domain (containing the majority of the nucleotide binding residues, colored dark green), then a 50 kD domain containing the actin binding sites (colored red), and an α-helical sequence (colored cyan) containing a converter domain (colored black) at its NH2 end and a neck region to which the LC's bind at its COOH end. The neck protrudes from the globular head portion of the S-1 molecule by 8.5 nm (84,93). The neck region attaches to a sequence of ~1100 amino acids forming an α-helical coil, which forms the backbone of the thick filament (the light-meromyosin (LMM) region and a putative elastic connection between the globular head and LMM region. While both type I and II can be subdivided into different sub-classes of myosin isoforms, whole skeletal muscle rarely express any one of the isoforms exclusively. For example soleus muscles have fibers expression about 98% type I fibers and 2% type II, while EDL contain 79% of type IIB (or IIX) those whose unloaded shortening velocity is the fastest, 3% of type I fibers, and 19% of type IIA whose unloaded shortening velocity is intermediate between types I and IIB (39). The data in Table I shows the rates of unloaded shortening velocity in intact stimulated muscle fiber, the unloaded shortening velocity of skinned muscle fibers and the unloaded thin filament sliding velocity in *in vitro* motility assays. The data is expressed as μm/half-sarcomere/s for the fibers and μm/s for in vitro motility assays to directly compare the results. The data confirm the original observations that the unloaded shortening velocity is directly related to the isoform expressed. The data given is for fibers expressing either type I or type IIB MHC. The fibers expressing MHC-IIB shorten at a speed 2-4 times greater than MCH-I, and this is true across species at different temperatures. Although the basis for these differences is likely to be found in the globular head structure of the MHC (the portion containing the actin binding and nucleotide binding sites), the mechanism is not yet clear. Comparisons of the amino acid sequence of type I and IIB show that the MHC-I and IIB are highly conserved (80% amino acid identity in the globular head region shown in Fig. 3) and that the sequences forming the nucleotide phosphate binding loop (G179-Thr186, colored magenta) and the helices forming the nucleotide binding pocket (Lys185-Ile199, Leu218-Gly233, and Thr667-Glu687, colored magenta) are 100% conserved. There is also 100% conservation of several sequences involved in actin binding (see Fig. 4): the hypertrophic cardiomyopathy (HCM) loop (Arg403-Lys415, light gray), a small loop (Arg143-Arg147, light gray), and a hydrophobic rigor binding sequence (Pro529-Pro543, light gray). Further a sequence (Asp464-Met495, colored orange) forming a cleft in the 50 kD domain and thought to communicate acting binding to the ATP binding regions and another (Cys697-Cys707, colored black) believed to convert activity in the ATP binding pocket to movement of the neck region are also ~100% conserved. There are, however, three flexible loops present in all myosin which show significant diversity. These loops may act to modulate actomyosin

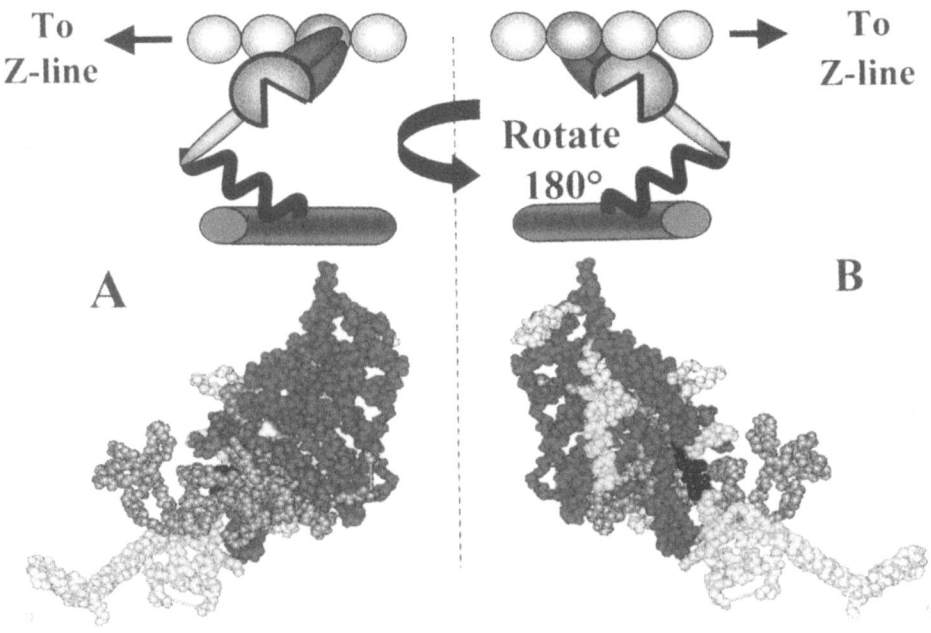

Figure 4-The approximate positioning of the myosin S-1 space-filling structure with respect to the thin filament. In the upper panel are two views of the cartoon arrangement of the crossbridge as viewed from a position (upper left) with the S-1 interposed between the viewer and the actin filament and a second view (upper right) with the thin filament interposed between the viewer and the myosin S-1. This constitutes a movement of the viewer of 180° between the two views. The lower images are the space filling images of the myosin S-1 molecule as viewed from the position of the viewer corresponding to the upper view, but with the actin molecule removed. Thus the left corresponds to a view from outside the S-1 molecule, and the right is the S-1 surface presented to the actin molecule. The space filling S-1 molecule in this figure is the extended form corresponding to the "Post Power Stroke" form in Figure 2. When ATP binds to the S-1 molecule shown here and the ATP is split, the neck (cyan region) in the left side (exterior view) rotates clockwise by about 70° producing the "bent" S-1 structure given in Fig. 2 as the "Attachment" or "Hydrolysis" state. The color code for the amino acid residues is as in Fig. 3. Color version on page 468.

behavior and thus shortening velocity. Loop 1 joins the 25 kD to the 50kD domain (1D204-G217) and loop 2 (S/G627-F651) (both colored dark gray) join the 50 kD to the 20 kD domain. Sweeney et al (100) have suggested that loop 1 may limit access of ATP to the nucleotide binding pocket and thus may limit the rate of actomyosin dissociation. Recent experiments, however, call this idea into question (91). Uyeda et al (105) have shown that modification of loop 2 can alter the myosin ATPase rate. A third loop that may play a role is the secondary actin-binding loop (Lys567-His578, also colored dark gray) contains positive charges whose alteration may affect binding to actin and thus shortening velocity (108). Thus while the basis of the different shortening velocity can not be directly linked to specific sequences that control unloaded shortening velocity between types I and IIB fibers, there are some intriguing possibilities.

3. UNLOADED SHORTENING VELOCITY OF A SPECIFIC FIBER TYPE IS INVERSELY PROPORTIONAL TO THE BODY MASS

Over 50 years ago A.V. Hill (44) noted that muscles from smaller animals seemed to shorten more rapidly than those of larger animals do. This trend is seen in the data in Table I in that the average body weights of adult mice, rats, and rabbits are 20 g, 350g, and 5 kg respectively. Indeed in studies of specific fiber types in mammals whose mass ranges from the 20 g (mouse) to 450 kg (horse), the shortening velocity (V_o) is given for type I and type IIB at 15° C by the following equations (89):

$$V_o=5.01 \ M_b^{-0.73} \quad \text{(Type II-B)} \qquad (4)$$
$$V_o=0.99 \ M_b^{-0.179} \quad \text{(Type I)}$$

Where Mb is the average body weight (in kg). In both cases, the regression coefficient (R^2) was 0.99. The mechanism producing this behavior remains to be determined, but is likely to stem from differences in the myosin heavy chain structure.

4. UNLOADED SHORTENING VELOCITY INCREASES AS TEMPERATURE INCREASES WITH A Q_{10} RANGING FROM 2-2.5.

Measurement of V_u in living muscle fibers over the temperature range of 10-25°C has shown that its activation energy is 40-60 kJ ($Q_{10} \approx$ 2-2.5) (13,26,82). Several studies have also shown that in the temperature range of 20-35°C the activation energy is less than 40 kJ (Q_{10} <2) while in the range of 5-20°C it can be >100 kJ (Q_{10} >3) (1,49,82). Studies of the temperature dependence of thin filament sliding speed propelled by rabbit MCH-IIB *in vitro* motility assays (1,49) have shown that the Q_{10} is near 2 for temperatures above 20°C and is much greater at temperatures <20°C. This behavior suggests that above 20°C, V_u, V_o, and V_f are limited by a specific crossbridge transition (for example the rate of ADP release from the

AM.ADP state, step 6 in Fig. 2). However, at temperatures <20°C a different crossbridge transition (perhaps the power stroke, step 5 in Fig. 2) limits the shortening velocity. Moreover, V_o and/or V_u have been measured in a muscles whose shortening velocity varies by more than a factor of 10^4 and it has been seen that V_u is roughly proportional to the actin-activated S-1/HMM/myosin ATPase rate within that range (4). Thus the rate of the power stroke (that part of the cross-bridge cycle during which force and/or displacement of the thin filament occurs) is proportional to the maximal ATPase rate in the unloaded condition. This observation does not, however, mean the power stroke is the rate-limiting step of the cross-bridge ATPase cycle. These results suggest that maximal shortening velocity is dependent on the cross-bridge cycling rate and is independent of the number of cross-bridges pulling on the thin filament.

5. V_u AND V_o ARE THE SAME FOR A GIVEN MUSCLE

Several studies have directly compared V_u to V_o in both intact and skinned muscles. They are in general agreement that the two estimates of the maximal shortening velocity are the same, but there is potential for obtaining values that do not completely agree. For example in whole muscles, the extrapolation to V_u by curve fitting will represent the contribution of most cells in the muscle and is thus a weighted average. Estimates of V_o, however, tend to give a value that represents that of the fastest fibers in the muscle. This is because the time at which the slack is taken up by the fastest fibers in the muscle and will thus bias the data to a higher velocity. There is also striking agreement between the speeds of the V_u or V_o of the electrically stimulated muscle and the skinned muscle fibers for the rat and mouse fast and slow muscle. The data included here was selected with the notion that there should be backup mechanism to prevent ADP levels in the skinned muscle fiber from rising and to activate the muscle within a 100 ms by jumping the temperature from a low value to a high value (by transferring the single cell from a low temperature bath to a higher temperature bath). This prevents the build up of ADP, H+, or Pi that might alter the thin filament sliding velocity. The implication of this result is that neither soluble proteins nor filament lattice spacing markedly influence V_o or V_u near the plateau of the length-tension curve. Finally in the cases in which data is available for V_f and V_u/V_o in the skinned muscle fiber, there is reasonable agreement for rabbit and rat muscle for temperatures >20°C, however, there is a significantly lower V_f than V_u or V_o for soleus muscle and shortening at low temperatures. This suggests several possibilities: a.) the motor proteins used in the these assays have under gone some proteolysis in during the isolation procedure; b.) at low temperatures the proteins undergo a structural change that does not occur in the fiber lattice; c.) the thin filaments may need to be stabilized by the addition of regulatory proteins (Tn/Tm); (It has been shown that regulatory proteins increase the thin filament sliding speed of f-actin filaments by 20-50% (50); d.) the ionic strength conditions of the assay alter the interaction of the motor proteins and the thin filaments; e.) actin·from the same source as the motor protein may be necessary (rabbit actin is used in practically all of these assays).

6. FACTORS DETERMINING UNLOADED SHORTENING VELOCITY

The current view is then that at V_u is primarily determined by the rate at which the crossbridge throw occurs and by drag imposed by strongly attached crossbridges. However, additional factors may alter V_u. These include: (1) intrinsic viscous drag on the filaments; (2) cytoskeletal elements that resist sliding; (3) drag from strongly attached cross-bridges; (4) drag from weakly bound crossbridges; or (5) the number of active cross-bridges.

6.1. Viscous drag

During sarcomere shortening at V_o, viscous drag is produced as the thin filaments slide through the cytosol past the thick filament. Both theoretical and experimental evaluations have shown that the drag offered by such sliding is very small (15,52,54). Using equations described by Hunt et al (53) the axial viscous drag/half-thick filament/2 thin filaments sliding at 6µm/, is 1.4×10^{-14} N/half-thick filament/2 thin filaments. If 300 S-1 heads are available to attach /half-thick filament (110) and 40% (20,62) attach and exert 2 pN of force/cross-bridge during an isometric contraction (31,69), the active positive force exerted by a thick filament is 2.4×10^{-10} N, or 20,000 times the viscous drag. Thus the force needed to over come viscous drag at V_o is <0.01% of the force the cross-bridges can exert. Thus viscous drag of sliding thin filaments is unlikely to limit shortening velocity.

6.2. Structural or cytoskeletal factors may limit shortening

At sarcomere lengths < 1.7 µm, energy is dissipated in deforming both the sarcomere and thick filament structures as well as changing filament separation. This is evidenced by the tendency for muscles, which had shortened, to these lengths to re-extend toward their initial length. At sarcomere lengths 1.7-3.1 µm other factors limit V_o, and there are two major sources of energy dissipation. As a fiber shortens, the filament lattice expands (29) and this expansion may be resisted by cytoskeletal links between myofibrils and attachments to the muscle basement membrane. This resistance is must be very small because skeletal muscles relaxing following shortening to sarcomere lengths from 1.7 to 2.5 µm do not re-extend. Thus there is negligible restoring force exerted during relaxation to return the muscle to its initial sarcomere length.

6.3. Resistive drag produced by strongly attached cross-bridges

A.F. Huxley (53) suggested that the drag from attached but negatively strained cross-bridges opposes thin filament sliding and limit V_o. He imagined that cross-bridges, driven by thermal energy, oscillate axially on the thick filament about an equilibrium position at $x = 0$. As they oscillate toward the Z line (at $x > 0$), they could attach in a conformation from which they would exert positive force which pulls the thin filament toward the M-line. Assuming a cross-bridge stiffness (κ) of 5×10^{-4} N/m (77,80, 94), an initial cross-bridge attachment to actin on the Z-line side of the cross-bridge equilibrium position would generate a force $\kappa \bullet x$. During muscle shortening, attached thin filaments slide toward the center of the sarcomere ($x \rightarrow 0$), the attached cross-bridge heads move toward their equilibrium position and attached cross-bridge force decreases. At its equilibrium position, cross-bridge force = 0, and the work done by the cross-bridge moving from its attachment point, x nm from the equilibrium position, will be ($\kappa \bullet x^2$)/2. As thick and thin filament sliding (shortening) continues, the attached cross-bridge is dragged **past** its equilibrium position into a negative force-bearing region where force is $-\kappa \bullet x$. A.F. Huxley (53) hypothesized that the accumulation of attached cross-bridges in the negative force-bearing region produces a force opposing continued sliding. When the force exerted by "negatively" strained cross-bridges is equal and opposite to the force exerted by cross-bridges generating positive force, the sliding velocity reaches its maximum steady-state velocity, V_u. Thus the detachment rate of the cross-bridges in the negatively strained region (x < 0 is what limits unloaded shortening velocity. This view is supported by the independence of V_o on thick and thin filament overlap at sarcomere lengths greater than 2.2 μm. Further interventions, which slow cross-bridge detachment from the thin filament (e.g., increasing [ADP] or decreasing [ATP], decrease V_o (14,21,22,30,86). The effects of [ATP] and [ADP] are predicted by several strain-dependent models (28,78).

Experiments supporting the idea that the drag imposed by negatively strained strongly attached cross-bridges limits V_o were presented by Josephson and Edman (58). They measured V_o early in a tetanus (in the first 30 ms after the beginning of stimulation before tension had risen to 30% of maximal), during the plateau of the tetanus, and during relaxation from the tetanus. At first glance A.F. Huxley's (53) view of V_o would seem to imply that V_o would be the same at each point in the tetanus. However, Josephson and Edman (58) found that at the start of the tetanic stimulation V_o was about 30% greater than that seen during the plateau of the tetanus. Further, during the initial stages of relaxation from the tetanus, V_o fell to values significantly less than those measured during the tetanus plateau. Josephson and Edman (58) argued that such behavior was predicted by Huxley's model (53). They noted that at the beginning a tetanus, before tension has reached the plateau, cross-bridges attach and exert force. If shortening begins before a steady state of attached positively strained cross-bridges is attained, then as cross-bridges are dragged into the negatively strained region, even more cross-bridges will be attaching in the positively strained region. Thus, at early times during tetanic contraction the rate of attachment of positively strained cross-bridges (from x = 0 to x) will exceed the rate of formation of negatively strained cross-bridges and unloaded shortening velocity will be high. If, however, shortening begins during the plateau of the tetanus the negatively strained and positively strained cross-bridges will be balanced and, consequently, V_o will be constant and independent of activation. During relaxation, as calcium concentration falls, the rate of formation of positively strained cross-bridges declines and the rate of entry of cross-bridges into the positively strained state lags behind the rate of entry of cross-bridges into the negatively strained state. This produces a slower V_0 than that seen at the plateau of shortening. Josephson and Edman (58) showed that a modification of A.F. Huxley's equations (53) accurately accounted for their experimental observations

6.4. Drag imposed by weakly attached crossbridges

During rapid shortening the number of weakly attached crossbridges increases (because the number of strongly attached crossbridges is significantly reduced (33,61,62,96). Basically during rapid shortening, M•ADP•Pi cross-bridges rapidly attach and detach from the thin filament (bind weakly) and so produce a resistive force (10). In the first studies of the passive force exerted by such cross-bridges during rapid stretches of relaxed muscle fibers, it was concluded that such forces were too small to constitute a significant drag (45). A recent re-examination of this question by Stehle and Brenner (96) produced a different view. They allowed muscle fibers to shorten against various loads (from 2.5-20% P_0) and then stretched the muscle fibers at speeds ranging from 0.5-30 µm/hs/s. The stiffness measured during the stretch was then plotted against the relative load. The zero load intercept of the stiffness as a function of relative load represented the stiffness of the muscle shortening against zero afterload. A plot of the zero-load stiffness against stretch velocity fell 2 orders of magnitude below a comparable plot for the isometrically contracting muscle. However after normalizing this plots the zero-load muscles showed an elevated stiffness at higher rates of stretch (as opposed to the isometrically contracting muscle) which suggested that during rapid shortening there may be a fraction of weakly bound crossbridges that contributed significant stiffness. The most simple interpretation of this data is that at zero load, there is a population of strongly bound crossbridges and a second population of weakly bound crossbridges that can contribute very significant stiffness (96). Measurement of relaxed muscle stiffness in 5 mM MgATP or 10 mM MgATP-γ-S (a nucleotide hydrolyzed so slowly that almost all crossbridges exist as MgATP-γ-S), there is practically no stiffness seen on rapid stretch. This result indicates that M.ATP crossbridges attach and detach so rapidly (10,000/s) that little or no stiffness is observed. However if the pCa is raised to 4.5, the stiffness of the MgATP-γ-S fibers rose to levels equivalent to that seen in rapidly shortening muscle fibers even though no force was developed. This result suggests that Ca^{2+} increases the crossbridge attachment rate (or decreases the detachment rate) in the muscle fiber. From this these data Stehle and Brenner conclude that during unloaded shortening, as much as 80% of the stiffness comes from weakly attached crossbridges while the number of strongly attached crossbridges is between 1-5% of that attached during isometric contraction. These results imply that weakly attached crossbridges may provide a significant drag on thin filament sliding during rapid shortening.

6.5. The effect of cross-bridge numbers on thin filament shortening velocity

Assuming that the cross-bridge throw (*d*) is 10 nm, that a cross-bridge generates positive force during the power stroke at rapid speeds for 5 ms (t_s), and that the maximum ATPase rate (k_{cat}) at 10°C is 8 s^{-1} (86), the equation relating V_o to the number of attached cross-bridges (N) (104) is

$$V_o = d/t_s \times (1 - (1 - t_s \times k_{cat})^N)$$ (5)

This equation shows that to produce shortening at 0.95 (1.9 μm/s) of the V_o, 73 actin monomers/per thin filament (27% of those available in a thin filament at optimal sarcomere overlap) must be attached to pulling cross-bridges. Similarly 17 cross-bridges must be bound per thin filament (6% of those in a thin filament at optimal sarcomere length) to generate a speed of about 0.5 V_o, and 7 cross-bridges must be bound per thin filament (3% of available actin monomers) to generate a speed of 0.25 V_o. Given the constraints of in vitro motility systems (the number of crossbridge heads pulling on a given thin filament) these constraints may well be important in limiting the observed V_f. However, for intact single muscle fibers or skinned muscle fibers the situation is very different. Given the number of crossbridges present in the thick filament, the number of actin monomers in reach of those crossbridges, and conservative assumptions of the fraction of crossbridges attached during isometric and unloaded contractions, it has been estimated that during shortening at V_o at maximum activation, 1 cross-bridge is strongly attached and propelling the thin filament per 11-26 actin monomers per thin filament (38). This implies that at V_o only 3-8 cross-bridges are pulling each thin filament toward the center of the thin filament. If there are 1000 thin filaments rigidly attached to the Z line in each myofibril, 3000-8000 cross-bridges produce the Z line movement at full activation, and the shortening velocity will be 2 μm/half-sarcomere/s. Even at 10% activation, the number of attached cross-bridges per Z line in each myofibril, be at least 300 cross-bridges, more than enough to reach V_o. This analysis is valid only if the attachment of thin filaments to the Z line is **rigid** and moves as a single unit. In summary, the evidence in intact skeletal muscle fibers supports the conclusion that the unloaded shortening velocity is independent of the number of attached cross-bridges, but is probably limited by the intrinsic drag from attached cross-bridges.

7. THE ROLE OF CALCIUM IN CONTROLLING V_u, V_o, AND V_f

If calcium regulates contraction by controlling only the rate of entry into the force-generating portion of the cross-bridge cycle, then *steady state* V_o should be independent of the extent of muscle activation. This conclusion assumes that even if the cross-bridge throw in the power-stroke probably is only 5-10 nm, there are enough cross-bridges pulling on the thin filament to assure that it is constantly moving (see above). Edman (26) tested the idea that V_o is independent of activation in single living muscle fibers by grading the intracellular $[Ca^{2+}]$ by Na-dantrolene inhibition of Ca^{2+} release from the sarcoplasmic reticulum. Using the slack test and releasing the muscle by <9% of the initial length, he found that, when the isometric force is reduced to as low as 10% of maximal by exposure of the fiber to Na-dantrolene, V_o was within 10% of its value at full activation. Although the result seems compelling, dantrolene treatment may not uniformly inhibit the release of calcium over the fiber cross-section; i.e., it may inhibit calcium release more in the outer fiber annulus than the inner core. If so, force may be reduced to zero in the outer annulus, but be normal in a core region having maximal Ca^{2+} activation. In such a case, the fiber force would be reduced but V_o would be unchanged. Thus experiments should be directed at examining the $[Ca^{2+}]$ profile throughout the fiber in the presence of dantrolene (106). Data exists suggesting that rapid shortening itself may deactivate the fiber (90), possibly through the detachment of strongly attached cross-bridges influencing Ca^{2+} binding to the thin filament or activation of the thin filament. The data of Vandenboom *et al.* (105) suggests that at maximal calcium activation, although the shortening may influence the rate of force development, it does not affect V_o, which in living fibers is not greatly sensitive to the level of activation.

7.1. The effect of $[Ca^{2+}]$ on unloaded shortening velocity in skinned muscle fibers

The use of skinned muscle fibers permits variation of parameters not easily altered in the intact fiber, such as $[Ca^{2+}]$ and the TnC content. The force exerted by skinned muscle fibers and the unloaded shortening velocity are similar to those measured in the intact fiber (59). If $[Ca^{2+}]$ simply regulated the number of attached crossbridges, reduction of $[Ca^{2+}]$ in skinned fibers would have an effect on V_o similar to the reduction of thick and thin filament overlap in intact fibers at saturating $[Ca^{2+}]$; i.e., little or no effect. Slack test measurements of V_o at sub-maximal $[Ca^{2+}]$, reveal that plots of the distance released against the slack time interval no longer form a single straight line [see Fig. 1, represented by crosses from data at pCa 6] and are better characterized by the sum of two straight lines. For releases < 6-8% of l_o (60-80 nm/hs), plots of the distance released vs "slack time" have a single slope, the unloaded shortening velocity of phase 1, V_{o1} which is independent of $[Ca^{2+}]$. At sub-maximal Ca^{2+} and releases of > 8% of l_o (phase 2), plots of distance released vs slack time exhibit a reduced slope, V_{o2}, indicative of a reduced unloaded shortening velocity. Plots V_{o1} and V_{o2} as a function of $[Ca^{2+}]$ (Fig. 5), show that V_{o1} decreases only slightly as Ca^{2+} is decreased until pCa is >6.5 (P/P_o <0.2), but then falls precipitously as Ca^{2+} is further decreased. V_{o2} on the other hand decreases roughly linearly with P/P_o (see Fig. 6). Thus at a pCa (6.2) producing a P/P_o of 0.5, V_{o1} is ~0.9 of the maximal V_o, while V_{o2} is 0.5 of maximal V_o. The data in Fig. 5 were obtained from skinned rabbit psoas fibers containing either skeletal TnC or cardiac TnC. As seen in Fig. 5 in neither case was there a significant difference in either maximal V_o or those at higher pCa. The behavior of V_{o2} is different from that seen in the intact fiber where V_o is independent of P/P_o values down to 0.4, when force is decreased by decreasing filament overlap. However, V_{o1} (at sub-maximal pCa)

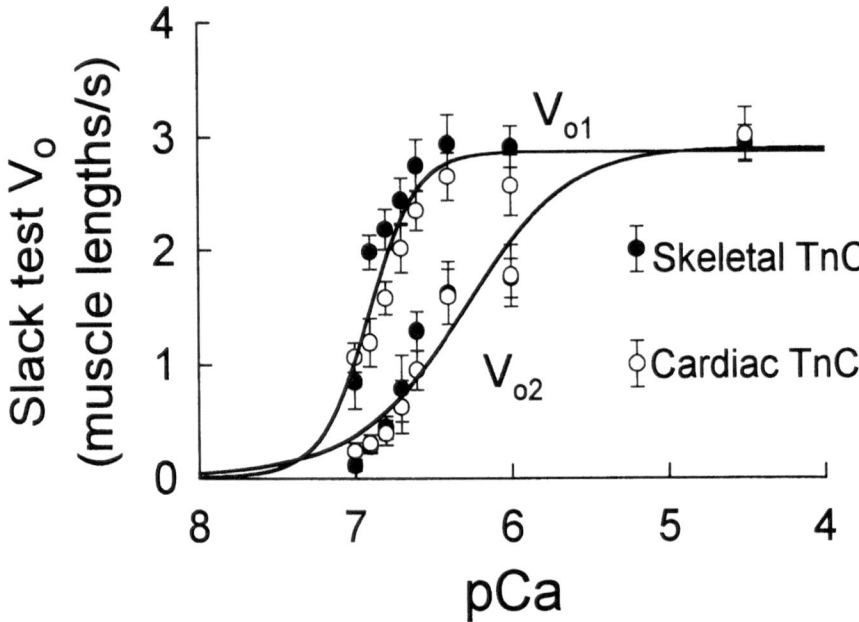

Figure 5-*A plot of V_{o1} and V_{o2} as function of pCa. Experiments were performed using skinned rabbit psoas muscle fibers contracting at $\Gamma/2$ of 200 mM, pH 7.1, and 15°C. V_{o1} and V_{o2} were measured using the "slack test" as shown in Fig. 1. Data given as mean$\pm$sem. Filled circles are data from fibers containing sTnC; open circles from fibers in which bovine cardiac TnC replaced the endogenous sTnC. The solid lines are fits to the ($V=V_o/(1+10^{\wedge}$ $^{(n*(x-pCa50))}$), the Hill equation where n is the Hill coefficient, pCa_{50} is the pCa at which V/V_o is 0.5, and x is the pCa at which the measurement was made. The solid lines are the least squares fit the data. For V_{o1}, $V_o=2.87\pm0.12$ muscle lengths/s, n= 2.80$\pm$0.59, and $pCa_{50}=6.91\pm0.03$ (mean $\pm$ sem) and $r^2=0.989$. For V_{o2}, $V_o=2.91\pm0.18$ muscle lengths/s, n=1.24$\pm$0.20, and $pCa_{50}=6.32\pm0.08$ (mean$\pm$sem) and $r^2=0.99$. The data is from the Ph.D. thesis of Carl Morris ("The Effects of Varying the Level of Thin Filament Activation in the Regulation of Muscle Contraction"), Physiology Department, UCLA School of Medicine, 1998.*

behaves in a manner similar to V_o in the fiber whose overlap is reduced. How closely they correspond can not be stated since there is no unambiguous data at thick and thin filament overlap less than 40% of optimal in the fully active fiber. The similarity between V_o in intact fibers and V_{o1} in skinned fibers is supported also by the data discussed above of Edman (26) suggesting that V_o in intact fibers may depend little on the level of Ca^{2+} activation.

The behavior of V_{o2} has no analogy in the fully active living fiber contracting at reduced overlap. In those cases in which shortening velocity was measured as a function of filament overlap using the slack test (26,105), releases of <9% were made. Thus the releases were not large enough to produce the V_{o2} result seen in skinned muscle fibers. Thus there may be no discrepancy between intact and skinned muscle fibers *vis a vis* the calcium dependence of V_o. Measurement of *in vitro* motility sliding speed, V_f, reveals little change with reductions in $[Ca^{2+}]$ until pCa is raised to 6.5 (37,47,50). At still greater pCa, V_f declines toward zero. Thus the behavior of regulated thin filaments in the *in vitro* motility assay using fast muscle myosin is more like V_{o1} than V_{o2}. Explanations for the effects of $[Ca^{2+}]$ on V_{o1} and V_{o2} include the trivial (the skinned fiber is a poor model for the behavior of the living fiber) and the significant: (a) a reduction in $[Ca^{2+}]$ reduces the rate of the cross-bridge power stroke; (b) a reduction in $[Ca^{2+}]$ reduces the rate of cross-bridge detachment from the thin filament during shortening; or, (c) a reduction in $[Ca^{2+}]$, along with shortening *per se*, reduces the rate of cross-bridge formation or the rate of cross-bridge detachment.

Gradation of activation of the thin filament, independent of $[Ca^{2+}]$, can be achieved by removal of endogenous sTnC (74) or its replacement with a mutant cardiac TnC, CBMII (50,70,71). CMBII replaces TnC in troponin but can not bind Ca^{2+} at its N-terminal Ca^{2+} binding trigger site and thus is not activated by Ca^{2+} Variation of relative isometric force, P/P_o, at saturating $[Ca^{2+}]$ indicates (Fig. 6) that

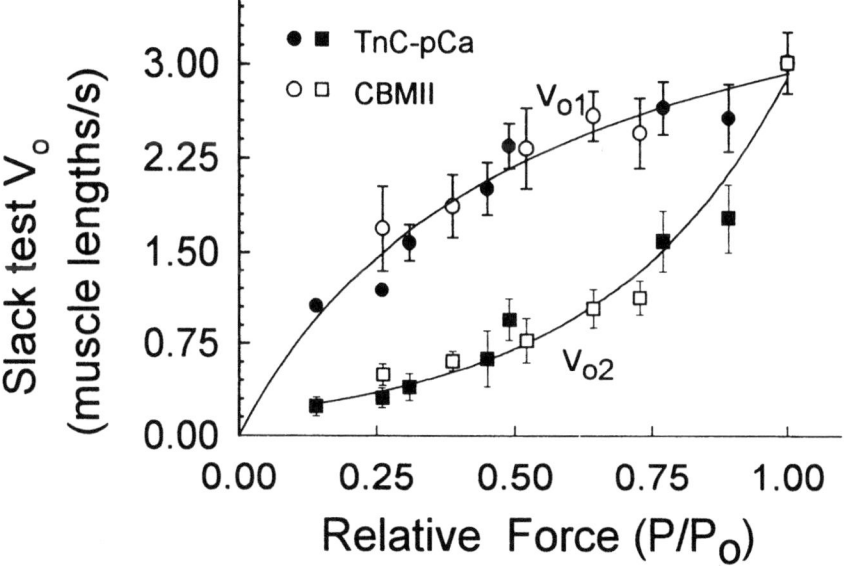

Figure 6 is a plot of Vo1 and Vo2 as functions of the relative isometric force varied either by varying pCa or by changing the fraction of the muscles TnC by replacing cTnC with varying amounts of CBMII TnC while holding pCa at 4.5. Experiments were performed as described in Fig. 6. The data is from the Ph.D. thesis of Carl Morris ("The Effects of Varying the Level of Thin Filament Activation in the Regulation of Muscle Contraction"), Physiology Department, UCLA School of Medicine, 1998.

P/P_o is inversely proportional to the amount of TnC removed or replaced by the CBMII (71, 72). The behavior of V_o following TnC removal, or replacement by CBMII is indistinguishable from that seen when $[Ca^{2+}]$ is varied (71,72,74); i.e., as isometric force declines, V_{o1} is little affected until force is reduced to less than 20% of maximal. However, V_{o2} declines in direct proportion to the reduction in relative isometric force (Fig. 6). These latter results suggest that the calcium dependent changes in V_{o2} are a consequence of regulatory protein interaction with the structural arrangement of the contractile proteins in the muscle fiber. The constancy of V_{o1}, even when force decreases significantly suggest that the rate of cross-bridge attachment and detachment are largely unaffected by the extent of thin filament activation. For displacements of < 80 nm/hs, the rate of cross-bridge attachment, detachment, and the size of the power stroke are unaffected by thin filament activation. Thus the first phase of unloaded shortening velocity (V_{o1}) is largely independent of the effects of $[Ca^{2+}]$, thin filament activation, or the number of thick and thin filament interactions. The behavior of V_{o1} is similar to the behavior of the regulated thin filament sliding speed as pCa is varied and/or CBMII content of the thin filament is varied. The behavior of the in vitro unloaded, regulated thin filament sliding speed in the in vitro motility assay, V_f, under reduced $[Ca^{2+}]$ or increased thin filament CBMII can be accounted for by changes in the number of cross-bridges pulling on the thin filament (37,47).

The behavior of V_{o2} is, however, different from what is seen in the in vitro motility assay and must involve a different explanation. In the above experiments pCa was constant and saturating. Since V_{o2} decreases as the proportion of CBMII bound to the thin filament increases (i.e., as the extent of thin filament activation was reduced) at saturating pCa, changes in V_{o2} can not be the result of a Ca^{2+} mediated change in binding to troponin or a regulatory light chain. Other possible explanations have been suggested.

Moss et al. (74) suggested that cross-bridges attached to thin filaments remain attached and be dragged long distances (60-80 nm) past their equilibrium position before they are forcibly detached, and thus produce a drag which slows the muscle shortening velocity. This hypothesized effect would be exacerbated by the reduction of the extent of thin filament activation. However, if V_o is 5 µm/hs/s, this hypothesis requires that the cross-bridge remains attached for up to 16 ms (80 nm/5µm/hs/s) as opposed to the 1-3 ms usually supposed in the Huxley model (54). Further this hypothesis requires that, over short ranges of cross-bridge travel (–2 to + 5 nm), the cross-bridge stiffness is 5 x 10^{-4} N/m but between -2 and -80 nm is > 100 times less to account for V_{o1}. and then increases markedly at distances > 80 nm. These requirements seem unrealistic.

Another possibility is that during rapid shortening the thin filament is deactivated by the rapid detachment of cross-bridges. Iwamoto (57) found that at sub-maximal activation, during a repetitive series of rapid shortenings at velocities near V_o followed by quick re-stretches to the initial muscle length, there is a progressive slowing of shortening velocity; i.e., the V_{o1} phase of shortening is lost and the V_{o2} phase predominates. Iwamoto found that a 300 ms period of isometric contraction must be interposed between releases to reverse this inhibition. During rapid shortening at saturating pCa, the stiffness of the muscle fiber decreases by 60-95% indicating that the number of attached cross-bridges is markedly reduced (61,62,96). At saturating Ca^{2+} enough cross-bridges are attached to forestall significant deactivation of the thin filament. However during shortening near V_o and sub-maximal $[Ca^{2+}]$, the number of strongly attached cross-bridges (which promote thin filament activation) may fall below a level required for full activation of the thin filament (25,67,68,96). A thin filament deactivation could manifest itself in two ways: (1) an increase in the rate of cross-bridge detachment, or (2) a decrease in the rate of cross-bridge attachment. Either effect would result in a reduced number of cross-bridges attached and interacting with the thin filament. If the rate of cross-bridge detachment increased, the unloaded shortening velocity would increase (because the number of negative force generating cross-bridges would decrease). On the other hand if the rate of cross-bridge attachment decreased, the unloaded shortening velocity would decrease in a fashion similar to that described by Josephson and Edman (59) during relaxation from a tetanus. Further Stehle and Brenner (96) have suggested that weakly attached crossbridges (AM.ADP.Pi) can produce a significant amount of stiffness at high shortening velocities which contributes to the decline in unloaded shortening velocity.

A paradoxical aspect of the reduction in V_{o2} at low $[Ca^{2+}]$ is the effect of added inorganic Pi. Metzger (68) found that at sub-maximal levels of $[Ca^{2+}]$ there is an inhibition of V_{o2}, but the addition of 10-30 mM Pi, returns V_{o2} toward control values of V_o. This behavior seems inconsistent with the changes in V_{o2} arising from structural constraints on the shortening velocity (an imposed load other than that arising from the cross-bridges themselves) because it implies that such a restraint is Pi dependent. A possible explanation for this behavior is that an elevated [Pi] reduces the fraction of negatively strained AM·ADP cross-bridges by increasing the fraction of cross-bridges in the AM·ADP·Pi, a weakly bound cross-bridge. If Pi binding to a negatively strained crossbridge were stronger than positively strained cross-bridges, then the drag force will be reduced and velocity of shortening would increase. If true, at saturating calcium, addition of Pi would increase V_o beyond that of control. Data on this point are equivocal (68, 79).

In summary, the unloaded shortening velocity of muscle is altered by many variables; the size of the cross-bridge throw, the duration of the cross-bridge duty cycle (the time the cross-bridges stays attached to the thin filament per ATP hydrolysis cycle), the number of cross-bridges interacting with the thin filament, the temperature, and both the rate of cross-bridge attachment (at least transiently) and detachment. The effects of Ca^{2+} on the V_o seem explicable by changes in the rate of cross-bridge attachment, which in turn are set by the extent of the access of the cross-bridge to the thin filament.

8. ACKNOWLEDGEMENTS

Supported by NIH NIAMSD grant AR30988 to EH

Physiology Department
UCLA School of Medicine
Center for Health Sciences
UCLA
Los Angeles, CA 90095-1751

9. REFERENCES

1. Anson, M.. Temperature dependence and Arrhenius activation energy of f-actin velocity generated *in vitro* by skeletal myosin *J. Mol. Biol.* 224: 1029-1038, 1992.
2. Anson, M, M. Geeves, E. Kurzawa, and D. Manstein. Myosin motors with artificial lever arms. *EMBO J.* 15: 6069-6074, 1996.
3. Araujo, A., and J. Walker. Phosphate release and force generation in cardiac myocytes investigated with caged phosphate and caged calcium. *Biophys. J.* 70: 2316-2326, 1996.
4. Barany, M. ATPase activity of myosin correlated with speed of muscle shortening. *J. Gen. Physiol.* 50: 197-218, 1967.
5. Barclay, C. A weakly coupled version of the Huxley crossbridge model can simulate energetics of amphibian and mammalian skeletal muscle. *J. Muscle Res. Cell Motil.* 20:163-176, 1999.
6. Barclay, C. Mechanical efficiency and fatigue of fast and slow muscles of mouse. *J. Physiol.* 497: 781-794, 1996.
7. Bottinelli, R., and C. Reggiani. Human Skeletal muscle fibers: Molecular and functional diversity. *Prog.Biophys. & Mol. Biol.* 73: 195-262, 2000.
8. Bottinelli, R., S. Schiaffino, and C. Reggiani. Force-velocity relations and myosin heavy chain isoform composition of skinned fibres from rat skeletal muscle. *J. Physiol.* 437: 665-672, 1991.
9. Bottinelli, R., R. Betto, S. Schiaffino, and C. Reggiani. Unloaded shortening velocity and myosin heavy chain and alkali light chain isoform composition in rate skeletal muscle fibres. *J. Physiol.* 478: 341-349, 1994.
10. Brenner, B. Rapid dissociation and reassociation of actomyosin cross-bridges during force generation: a newly observed facet of cross-bridge action in muscle. *Proc. Natl. Acad. Sci. U. S. A.* 88: 10490-4, 1991.
11. Brooks, S., and J. Faulkner. Isometric, shortening, and lengthening contractions of muscle fiber segments from adult and old mice. *Am. J. Physiol.* 267: C507-C513, 1994.
12. Brooks, S. and J. Faulkner, Contractile properties of skeletal muscle from young, adult, and aged mice. *J. Physiol.* 404: 71-82, 1988.
13. Cecchi, G., F. Colomo, and V. Lombardi. Force-velocity relation in normal and nitrate-treated frog single muscle fibres during rise of tension in an isometric tetanus. *J. Physiol.* 285: 257-73, 1978.
14. Chase, P.B. and M. Kushmerick Effects of physiological ADP concentrations on contraction of single skinned fibers from rabbit fast and slow muscles. *Am. J. Physiol. Cell Physiol.* 268: C480-C489, 1995.
15. Chase, P.B., T. Denkinger, and M. Kushmerick. Effect of viscosity on mechanics of single, skinned fibers from rabbit psoas muscle. *Biophys. J.* 74: 1428-1438, 1998.
16. Close, R. Dynamic properties of fast and slow skeletal muscle of the rat during development. *J. Physiol.* 173: 74-95, 1964.
17. Close, R. Force:velocity properties of mouse muscles. *Nature.* 206: 718-719, 1965.
18. Close, R. Dynamic properties of mammalian skeletal muscles. Physiol. Rev. 52: 129-197, 1972.
19. Coulton, G., N. Curtin, J. Morgan, and T. Partridge. The mdx mouse skeletal muscle myopathy: II. Contractile properties. *Neuropathol. Appl. Neurobiol.* 14: 299-314, 1988.
20. Cooke, R. Actomyosin interaction in striated muscle. *Physiol. Rev.* 77: 671-97, 1997.

21. Cooke, R. and W. Bialek. Contraction of glycerinated muscle fibers as a function of the ATP concentration. *Biophys. J.* 28: 241-258, 1979.

22. Cooke, R., and E. Pate. The effects of ADP and phosphate on the contraction of muscle fibers. *Biophys. J.* 48: 789-798, 1985.

23. Crow, M. and M. Kushmerick. Correlated reduction of velocity of shortening and the rate of energy utilization in mouse fast-twitch muscles during continuous tetanus. *J. Gen. Physiol.* 82: 703-720, 1983.

24. Dantzig, J., Y. Goldman, N. Millar, J. Lacktis, and E. Homsher. Reversal of the crossbridge force-generating transition by photogeneration of phosphate in rabbit psoas muscle fibres. *J. Physiol.* 451: 247-278, 1992.

25. Edman, K.A.P Mechanical deactivation induced by active shortening in isolated muscle fibers of the frog. *J. Physiol.* 246: 255-275, 1975.

26. Edman, K. A. P. The velocity of unloaded shortening and its relation to sarcomere length and isometric force in vertebrate muscle fibres. *J. Physiol.* 291: 143-159, 1979.

27. Edman, K. A. P., and K.-E. Andersson. The variation in active tension with sarcomere length in vertebrate skeletal muscle and its relation to fibre width. *Experientia* 24: 134-136, 1968.

28. Eisenberg, E., T. L. Hill, and Y.-D. Chen. Cross-bridge model of muscle contraction. Quantitative analysis. *Biophys. J.* 29: 195-227, 1980.

29. Elliott, G. F., J. Lowy, and C. R. Worthington. An X-ray and light diffraction study of the filament lattice of striated muscle in the living state and in rigor. *J. Mol. Biol.* 6: 295-305, 1963.

30. Ferenczi, M, Y. Goldman, and R. Simmons. The dependence of force and shortening velocity on substrate concentration in skinned muscle fibres from *Rana temporaria. J. Physiol.* 350: 519-543, 1984.

31. Finer, J. T., R. M. Simmons, and J. A. Spudich. Single myosin molecule mechanics: piconewton forces and nanometre steps. *Nature* 368: 113-119, 1994.

32. Ford, L. E., A. F. Huxley, and R. M. Simmons. The relation between stiffness and filament overlap in stimulated frog muscle fibres. *J. Physiol.* 311: 219-49, 1981.

33. Ford, L. E., A. F. Huxley, and R. M. Simmons. Tension transients during steady shortening of frog muscle fibres. *J. Physiol.* 361: 131-50, 1985.

34. Goldman, Y. E., and R. M. Simmons. The stiffness of frog skinned muscle fibres at altered lateral filament spacing. *J. Physiol.* 378: 175-194, 1986.

35. Gordon, A. M., R. E. Godt, S. K. Donaldson, and C. E. Harris. Tension in skinned frog muscle fibers in solutions of varying ionic strength and neutral salt composition. *J. Gen. Physiol.* 62: 550-74, 1973.

36. Gordon, A. M., A. F. Huxley, and F. J. Julian. The variation in isometric tension with sarcomere length in vertebrate muscle fibres. *J. Physiol.* 184: 170-92, 1966.

37. Gordon, A. M., M. LaMadrid, Y. Chen, Z. Luo, and P.B. Chase. Calcium regulation of skeletal muscle thin filament motility in in vitro. *Biophys. J.* 72: 1295-1307, 1997.

38. Gordon, A. M., E. Homsher, and M. Regnier. Regulation of Contraction in Striated Muscle. *Physiol. Rev.* 80: 853-924, 2000.

39. Gundersen, K., E. Leberer, T. Lomo, D. Pette, and R. Staron. Fibre types, calcium-sequestering proteins, and metabolic enzymes in denervated and chronically stimulated muscle of the rat. *J. Physiol.* 398: 177-189,1988.

40. Guth, L., and F. Samaha. Qualitative differences of actomyosin ATPase of slow and fast mammalian muscles. *Exp. Neurol.* 25: 138-152, 1969.

41. Harada, Y., K. Sakurada, T. Aoki, D. D. Thomas, and T. Yanagida. Mechanochemical coupling in actomyosin energy transduction studied by in vitro movement assay. *J. Mol. Biol.* 216: 49-68, 1990.

42. He, Z., R. Chillingworth, M. Brune, J. Corrie, M. Webb, and M. Ferenczi. The efficiency of contraction in rabbit skeletal muscle fibres, determined from the rate of release of inorganic phosphate. J. Physiol. 517: 839-854. 1999.

43. Hill, A. V. The heat of shortening and the dynamic constants of muscle. *Proc. Roy. Soc. Lond. B. Biol. Sci.* 126: 136-95, 1938.

44. Hill, A.V. The dimensions of animals and their muscular dynamics. *Sci. Progr. Twent. Cent.* 38: 209-230, 1950.

45. Hill, D. K. Tension due to interaction between the sliding filaments in resting striated muscle. The effect of stimulation. *J. Physiol.* 199: 637-684, 1968.

46. Holmes, K. and M. Geeves. The structural basis of muscle contraction. *Philos. Trans. R. Soc. Lond. B Biol. Sci.* 355: 419-431, 2000.

47. Homsher, E., B. Kim, A. Bobkova, and L. S. Tobacman. Calcium regulation of thin filament movement in an in vitro motility assay. *Biophys. J.* 70: 1881-92, 1996.

48. Homsher, E., J. Lacktis, and M. Regnier. Strain-dependent modulation of phosphate transients in rabbit skeletal muscle fibers. *Biophys. J.* 72: 1780-91, 1997.

49. Homsher, E., F. Wang, and J. R. Sellers. Factors affecting movement of F-actin filaments propelled by skeletal muscle heavy meromyosin. *Am. J. Physiol.* 262: C714-23, 1992.

50. Homsher, E., D. Lee, Morris, and L. S. Tobacman. Regulation of force and unloaded sliding speed in single thin filaments: effects of regulatory proteins and calcium. *J Physiol.* 524: 233-243, 2000.

51. Hook, P. and L. Larsson. Actomyosin interaction in a novel single muscle fiber in vitro motility assay. *J. Muscle Res. Cell Motil.* 21: 357-365, 2000.

52. Hook, P., X. Li, J. Sleep, S. Hughes, and L. Larsson. In vitro motility speed of slow myosin extracted from single soleus fibres from young and old rats. *J. Physiol.* 520: 463-471, 1999

53. Hunt, A. J., F. Gittes, and J. Howard. The force exerted by a single kinesin molecule against a viscous load. *Biophys. J.* 67: 766-81, 1994.

54. Huxley, A. F. Muscle structure and theories of contraction. *Prog. Biophys. Biophys. Chem.* 7: 255-318, 1957.

55. Huxley, A. F. *Reflections on muscle.* Princeton, NJ: Princeton University Press, 1980.

56. Ishijima, A., H. Kojima, H. Higuchi, Y. Harada, T. Funatsu, and T. Yanagida. Multiple- and single-molecule analysis of the actomyosin motor by nanometer-piconewton manipulation with a microneedle: unitary steps and forces. *Biophys. J.* 70: 383-400, 1996.

57. Iwamoto, H. Thin filament cooperativity as a major determinant of shortening velocity in skeletal muscle fibers. *Biophys. J.* 74: 1452-64, 1998.

58. Johnson, J., S. Charlton, and J. Potter. A fluorescence stopped flow analysis of Ca^{2+} exchange with troponin C. *J. Biol. Chem.* 254: 3497-3502, 1979.

59. Josephson, R. K., and K. A. Edman. Changes in the maximum speed of shortening of frog muscle fibres early in a tetanic contraction and during relaxation. *J. Physiol.* 507: 511-25, 1998.

60. Julian, F. J. The effect of calcium on the force-velocity relation of briefly glycerinated frog muscle fibres. *J. Physiol.* 218: 117-145, 1971.

61. Julian, F. J., and D. L. Morgan. Variation of muscle stiffness with tension during tension transients and constant velocity shortening in the frog. *J. Physiol.* 319: 193-203, 1981.

62. Julian, F. J., and M. R. Sollins. Variation of muscle stiffness with force at increasing speeds of shortening. *J. Gen. Physiol.* 66: 287-302, 1975.

63. Linari, M., I. Dobbie, M. Reconditi, N. Koubassova, M. Irving, G. Piazzesi, and V. Lombardi. The stiffness of skeletal muscle in isometric contraction and rigor: the fraction of myosin heads bound to actin. *Biophys. J.* 74: 2459-73, 1998.

64. Luff, A. Dynamic properties of the inferior rectus, extensor digitorum longus, diaphragm, and soleus muscles of the mouse. *J. Physiol.* 313: 161-171, 1981.

65. Marechal, G. and G. Beckers-Bleukx. Effect of nitric oxide on the maximal velocity of shortening of mouse skeletal muscle. *Pflugers Arch.* 436:906-913, 1998.

66. McDonald, K. Ca^{2+} dependence of loaded shortening velocity in rat skinned cardiac myocytes and skeletal muscle fibres. *J. Physiol.* 525: 169-181, 2000.

67. McKillop, D. F., and M. A. Geeves. Regulation of the interaction between actin and myosin subfragment 1: evidence for three states of the thin filament. *Biophys. J.* 65: 693-701, 1993.

68. Metzger, J. M. Effects of phosphate and ADP on shortening velocity during maximal and submaximal calcium activation of the thin filament in skeletal muscle fibers. *Biophys. J.* 70: 409-17, 1996.

69. Millar, N. and E. Homsher. The effects of phosphate and calcium on force generation in glycerinated rabbit skeletal muscle fibers. *J. Biol. Chem.* 265: 20234-20240, 1991.

70. Molloy, J. E., J. E. Burns, J.-J. Kendrick, R. T. Tregear, and D. C. White. Movement and force produced by a single myosin head. *Nature* 378: 209-12, 1995.

71. Morris, C. A., L. S. Tobacman, and E. Homsher. Modulation of thin filament activation using an inactivated cardiac troponin C in skinned skeletal muscle fibers. *Biophys. J.* 74: A347, 1998.

72. Morris, C., L.S. Tobacman, and E. Homsher. Modulation of contractile activation in skeletal muscle by a calcium-insensitive troponin C mutant. *J. Biol. Chem.* 276: 20245-20251, 2001.

73. Moss, R. L. The effect of calcium on the maximum velocity of shortening in skinned skeletal muscle fibres of the rabbit. *J. Muscle Res. Cell Motil.* 3: 295-311, 1982.

74. Moss, R. L. Effects on shortening velocity of rabbit skeletal muscle due to variations in the level of thin-filament activation. *J. Physiol.* 377: 487-505, 1986.

75. Moss, R. L., G. G. Giulian, and M. L. Greaser. The effects of partial extraction of TnC upon the tension-pCa relationship in rabbit skinned skeletal muscle fibers. *J. Gen. Physiol.* 86: 585-600, 1985.

76. Moss, R., G. Diffee, M. Greaser. Contractile properties of skeletal muscle fibers in relation to myofibrillar protein isoforms. *Rev. Physiol. Biochem. Pharmacol.* 126:1-63, 1995.

77. Pate, E., and R. Cooke. A model for the interaction of muscle cross-bridges with ligands which compete with ATP. *J. Theor. Biol.* 118: 215-230, 1986.

78. Pate, E., and R. Cooke. A model of crossbridge action: the effects of ATP, ADP and P_i. *J. Muscle Res. Cell Motil.* 10: 181-196, 1989.

79. Pate, E., K. L. Nakamaye, K. Franks-Skiba, R. G. Yount, and R. Cooke. Mechanics of glycerinated muscle fibers using nonnucleoside triphosphate substrates. *Biophys. J.* 59: 598-605, 1991.

80. Pate, E. F., and C. J. Brokaw. Cross-bridge behavior in rigor muscle. *Biophys. Struct. Mech.* 7: 51-63, 1980.

81. Pate, E., G. Wilson, M. Bhimani, R. Cooke. Temperature dependence of the inhibitory effects of orthovanadate on shortening velocity in fast skeletal muscle fibers. *Biophy. J.* 66: 1554-1562, 1994.

82. Ranatunga, K. W. Temperature-dependence of shortening velocity and rate of isometric tension development in rat skeletal muscle. *J. Physiol.* 329: 465-83, 1982.

83. Ranatunga, K. Temperature dependence of mechanical power output in mammalian (rat) skeletal muscle. *Exp. Physiol.* 83:371-376, 1998.

84. Rayment, I., W. Rypniewski, K. Schmidt-Base, R. Smith, D. Tomchick, M. Benning, D. Winkelmann, G. Wesenberg, and H. Holden. Three-dimensional structure of myosin subfragment-1: a molecular motor. *Science*, 261:50-58, 1993

85. Rayment, I., H. Holden, M. Whittaker, C. Yohn, M. Morenz, K. Holmes, and R. Milligan. Structure of the actin myosin complex and its implications for muscle contraction. *Science*, 261: 58-65, 1993.

86. Regnier, M., D. M. Lee, and E. Homsher. ATP analogs and muscle contraction: mechanics and kinetics of nucleoside triphosphate binding and hydrolysis. *Biophys. J.* 74: 3044-58, 1998.

87. Reiser, P., R. Moss, G. Giuliani, M. Greaser. Shortening velocity in single fibers from adult rabbit soleus muscles is correlated with myosin heavy chain composition. *J. Biol. Chem.* 260: 9077-9080, 1985.

88. Reiser, P., C. Kasper, M. Greaser, and R. Moss. Functional significance of myosin transitions in single fibers of developing soleus muscles. *Am. J. Physiol.* 23: C605-613, 1988.

89. Rome, L., A. Sosnick, and D. Goble. Maximum velocity of shortening of 3 fiber types from horse soleus muscle: implications for scaling with body size. *J. Physiol.* 431: 173-185, 1990.

90. Ruppel, K., and J. Spudich. Structure-function studies of the myosin motor domain: importance of the 50 kD cleft. *Mol. Biol. Cell*, 7: 1123-1136, 1996.

91. Sant'Ana Pereira, J., D. Pavlov, M. Nili, M. Greaser, E. Homsher, and R. Moss. Kinetic differences in cardiac myosins with identical loop 1 sequences. *J. Biol. Chem.* 276:4409-4415, 2001.

92. Sciote, J., and J. Kentish. Unloaded shortening velocities of rabbit masseter fibres expressing skeletal or α-cardiac myosin heavy chains. *J. Physiol.* 492:659-667, 1996.

93. Sellers, J. Myosin. Oxford University Press, Oxford, New York. 237 ppg., 1999.

94. Seow, C., and L. Ford. Shortening velocity and power output of skinned muscle fibers from mammals having a 25,000-fold range of body mass. *J. Gen. Physiol.* 97: 541-560, 1991.

95. Shih, W., Z. Gryczynski, J. Lakowicz, J. Spudich. A FRET-based sensor reveals large ATP hydrolysis-induced conformational changes and three distinct state of the molecular motor myosin. *Cell.* 102: 683-694, 2000.

96. Stehle, R., and B. Brenner. Cross-bridge attachment during high-speed active shortening of skinned fibers of the rabbit psoas muscle: implications for cross-bridge action during maximum velocity of filament sliding. *Biophys. J.* 78: 1458-1473, 2000.

97. Smith, C. and I. Rayment. Active site comparisons highlight structural similarities between myosin and other P-loop proteins. *Biophys. J.* 79:1590-1602, 1996.

98. Sun, Y-B., K. Hilber, and M. Irving. Effect of active shortening on the rate of ATP utilization by rabbit psoas muscle fibers. *J. Physiol.* 531:781-791, 2001.

99. Sweeney, H., M. Kushmerick, M. Marbuchi, F. Sreter, and J. Gergely. Myosin alkali light chain and heavy chain variation correlate with altered shortening velocity of isolated skeletal muscle fibers. *J. Biol. Chem.* 263: 9034-9039, 1988.

100. Sweeney, H., S. Rosenfeldt, F. Brown, L. Faust, J. Smith, J. Xing,, L. Stein, and J. Sellers. Kinetic tuning of myosin via flexible loop adjacent to the nucleotide binding pocket. *J. Biol. Chem.* 273:6262-6270. 1998.

101. Tesi, C., F. Colomo, S. Nencini, N. Piroddi, and C. Poggesi. Modulation by substrate concentration of maximal shortening velocity and isometric force in single myofibrils from frog and rabbit fast skeletal muscle fibers. *J. Physiol.* 566:847-853, 1999.

102. Thedinga, E., N. Karim, T. Kraft, and B. Brenner. A single-fiber in vitro motility assay. In vitro sliding velocity of F-actin vs unloaded shortening velocity in skinned muscle fibers. *J. Muscle Res. Cell Motil.* 20:785-796, 1999.

103. Thompson, L. and M. Brown. Age-related changes in contractile properties of single skeletal fibers from the soleus muscle. *J. Appl. Physiol.* 86: 881-886, 1999.

104. Uyeda, T., S. Kron, and J. Spudich. Myosin step size. Estimation from slow sliding movement of actin over low densities of heavy meromyosin. *J. Mol. Biol.* 214: 699-710, 1990.

105. Uyeda, T., K. Ruppel, and J. Spudich. Enzymatic activities correlate with chimeric substitutions at the actin-binding face of myosin. *Nature*, 368:567-569, 1994.

106. Vandenboom, R., D. R. Claflin, and F. J. Julian. Effects of rapid shortening on rate of force regeneration and myoplasmic [Ca^{2+}] in intact frog skeletal muscle fibres. *J. Physiol.* 511: 171-80, 1998.

107. Vergara, J., and M. DiFranco. Imaging of calcium transients during excitation-contraction coupling in skeletal muscle fibers. *Adv. Exp. Med. Biol.* 311: 227-36, 1992.

108. Weiss, A., S. Schiaffino, and L. Leinwand. Comparative sequence analysis of the complete human sarcomeric myosin heavy chain family: implications for functional diversity. J. Mol. Biol. 290: 61-75, 1999.

109. Westerblad, H., J. Bruton, and J. Lännergren. The effect of intracellular pH on contractile function of intact single fibres of the mouse muscle declines with increasing temperature. *J. Physiol.* 500: 193-204, 1997.

110. Woledge, R. C., N. A. Curtin, and E. Homsher. Energetic aspects of muscle contraction. *Monogr Physiol Soc* 41: 1-357, 1987.

AUTHOR INDEX

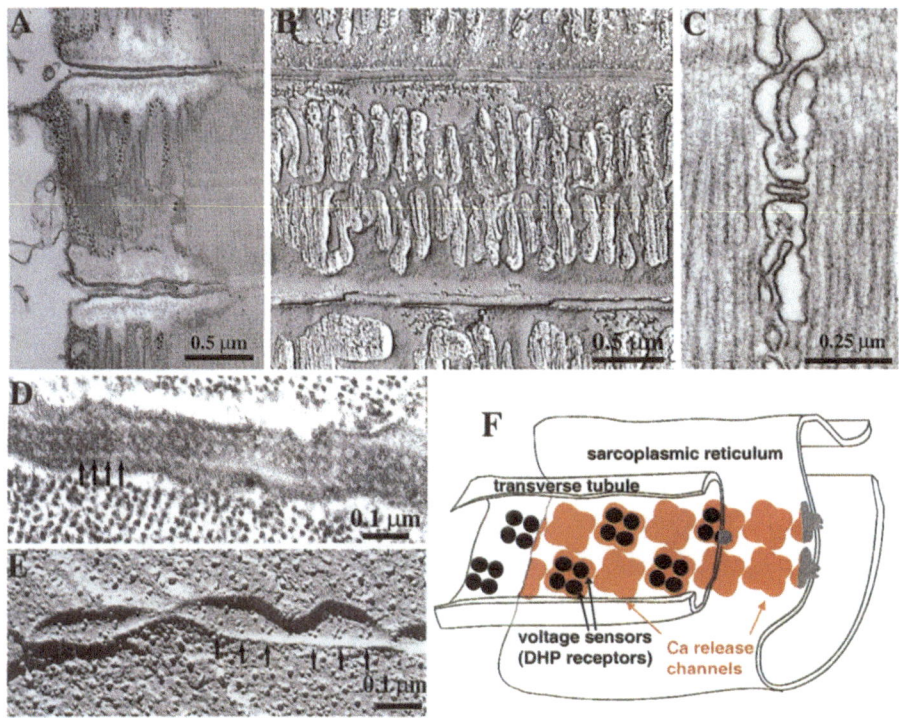

Figure 4 – Page 11

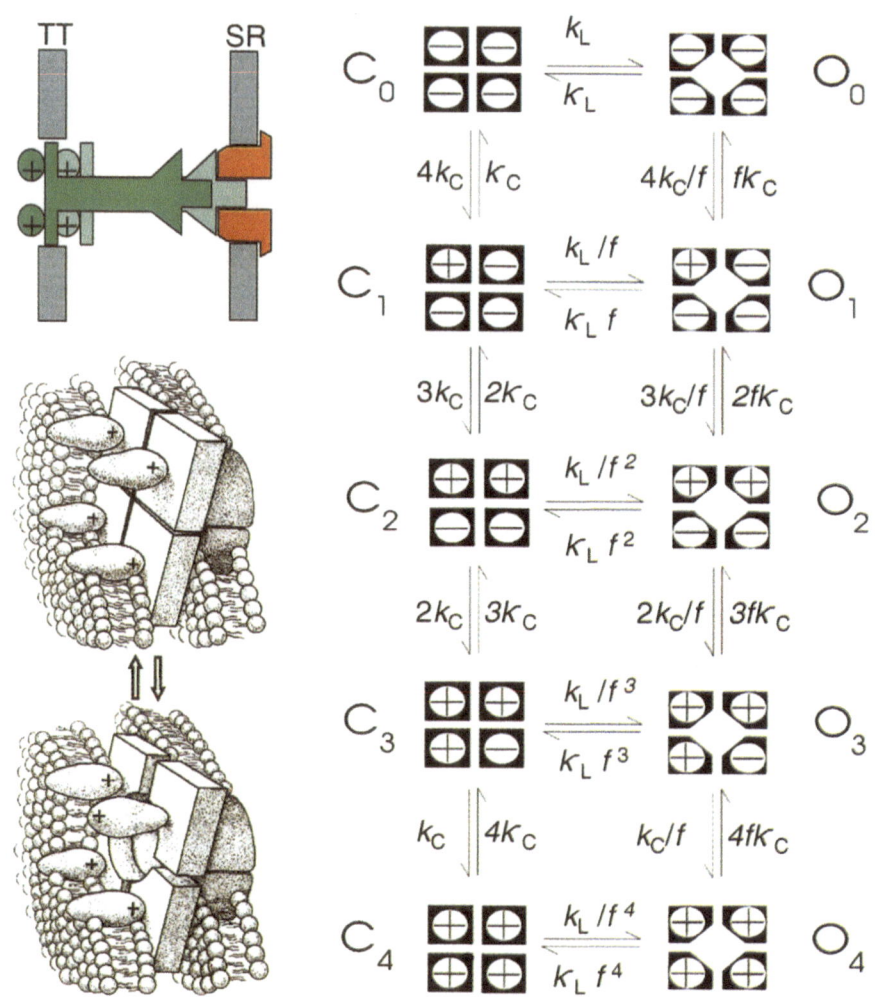

Figure 8 – Page 20

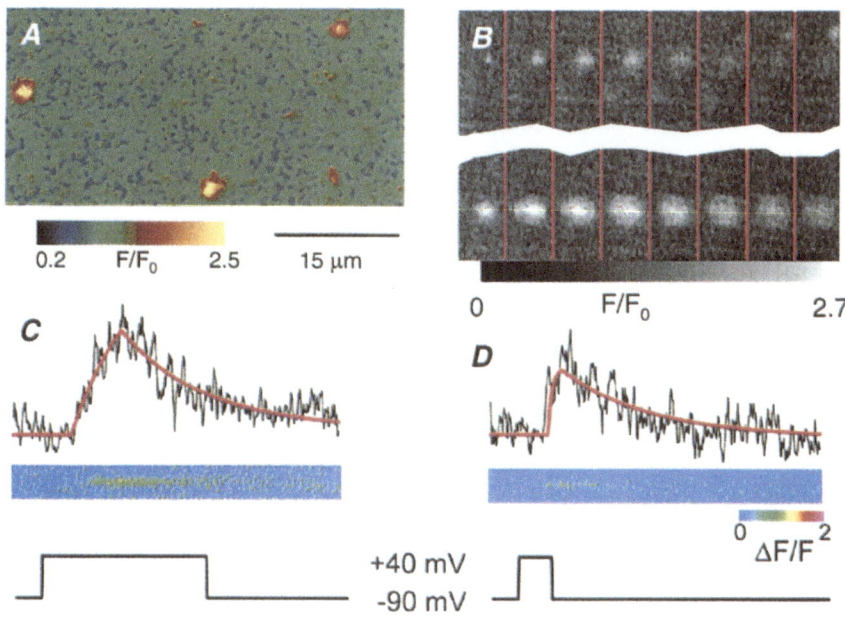

Figure 9 – Page 23

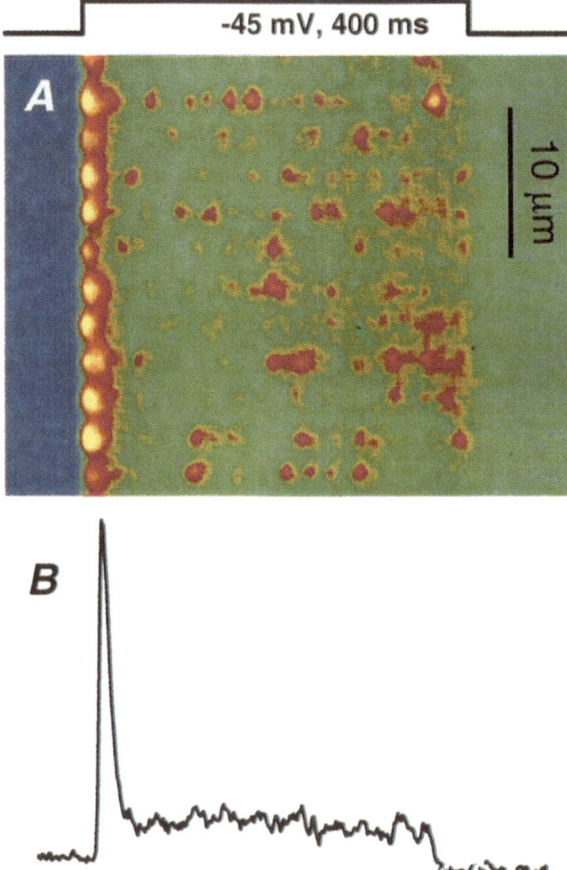

Figure 10 – Page 25

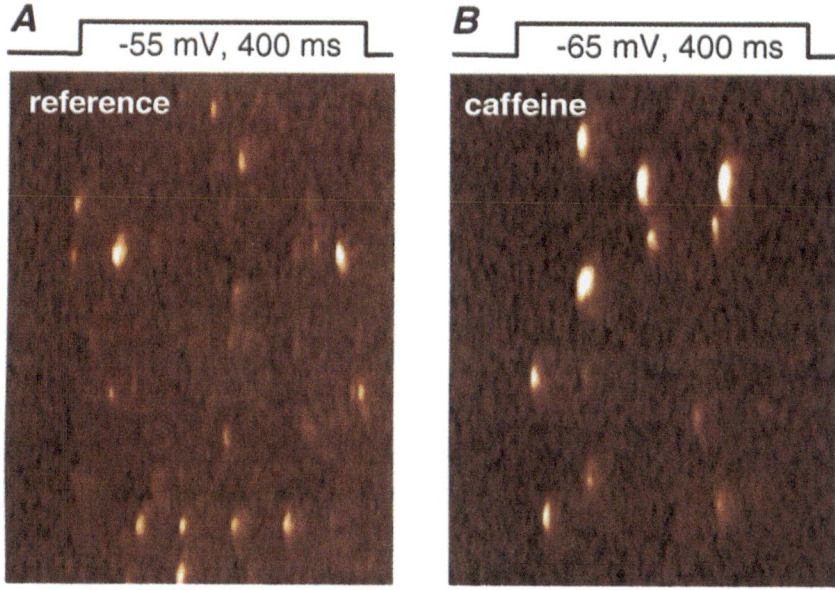

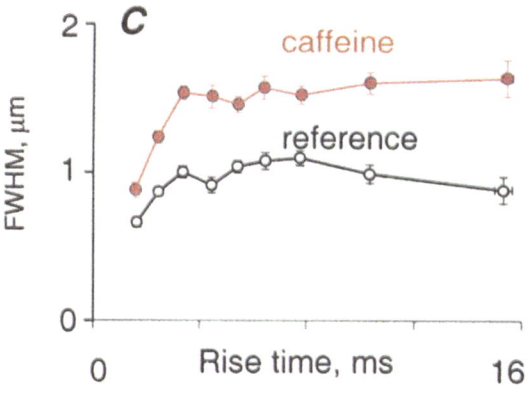

Figure 11 – Page 28

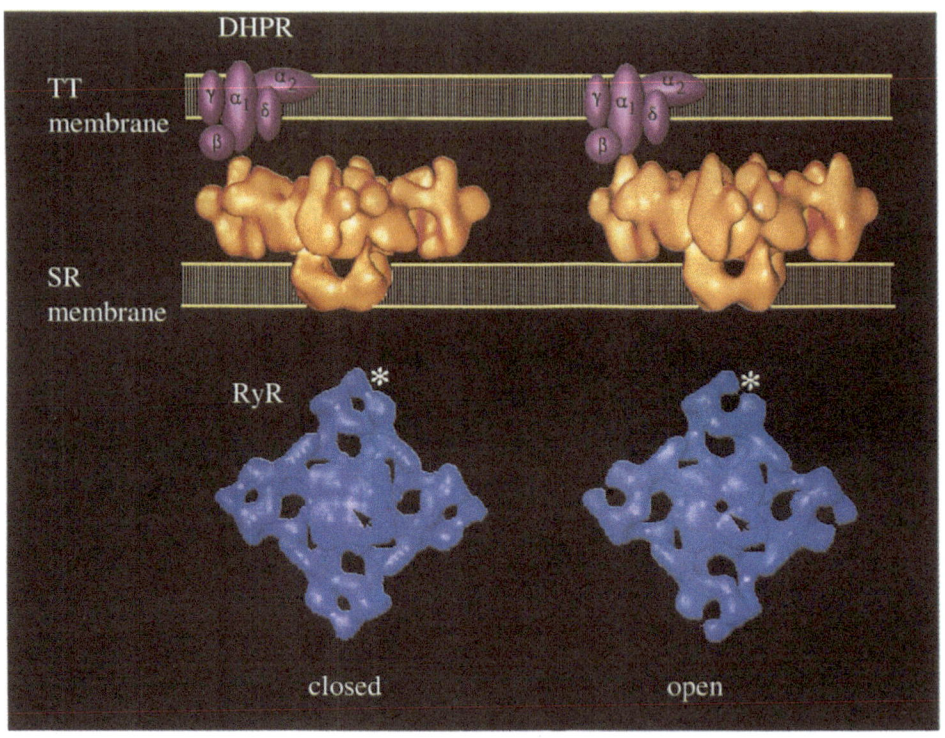

Figure 12 – Page 30

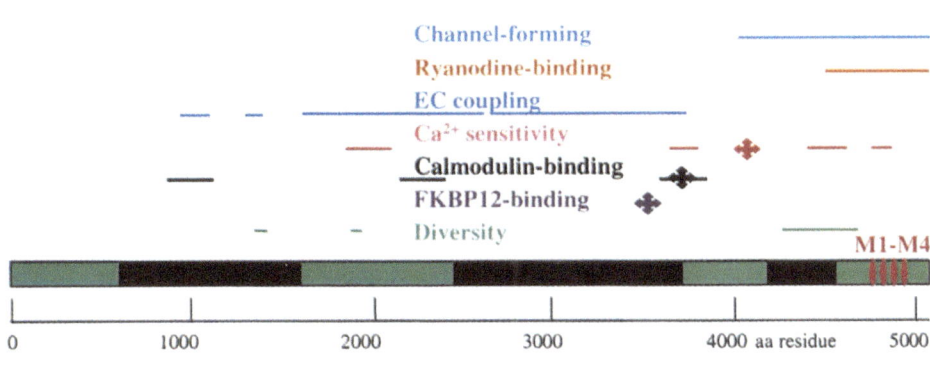

P34.TIF

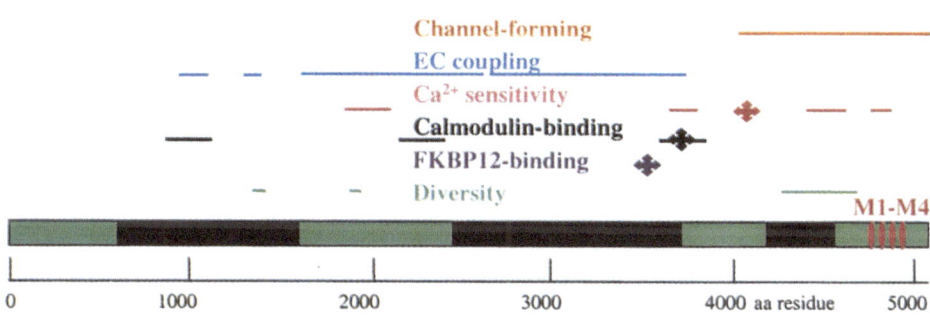

Figure 13 – Page 34

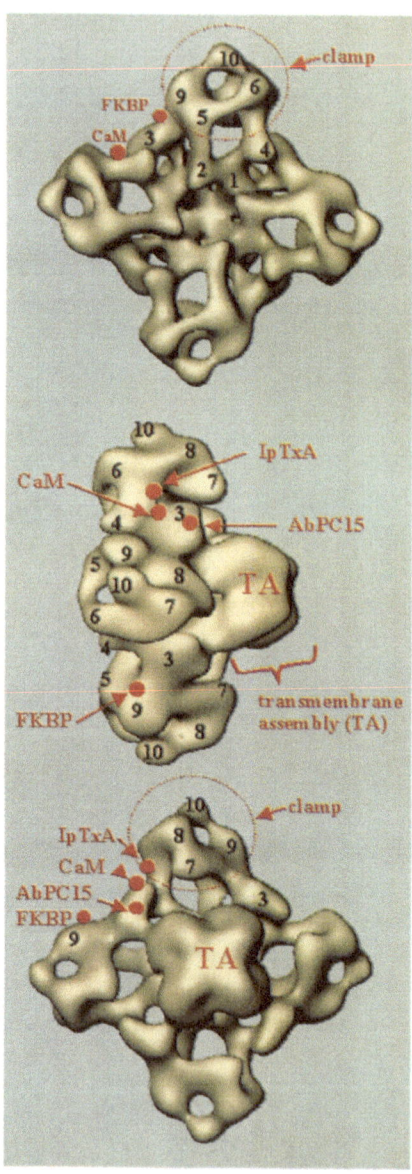

Figure 14 – Page 35

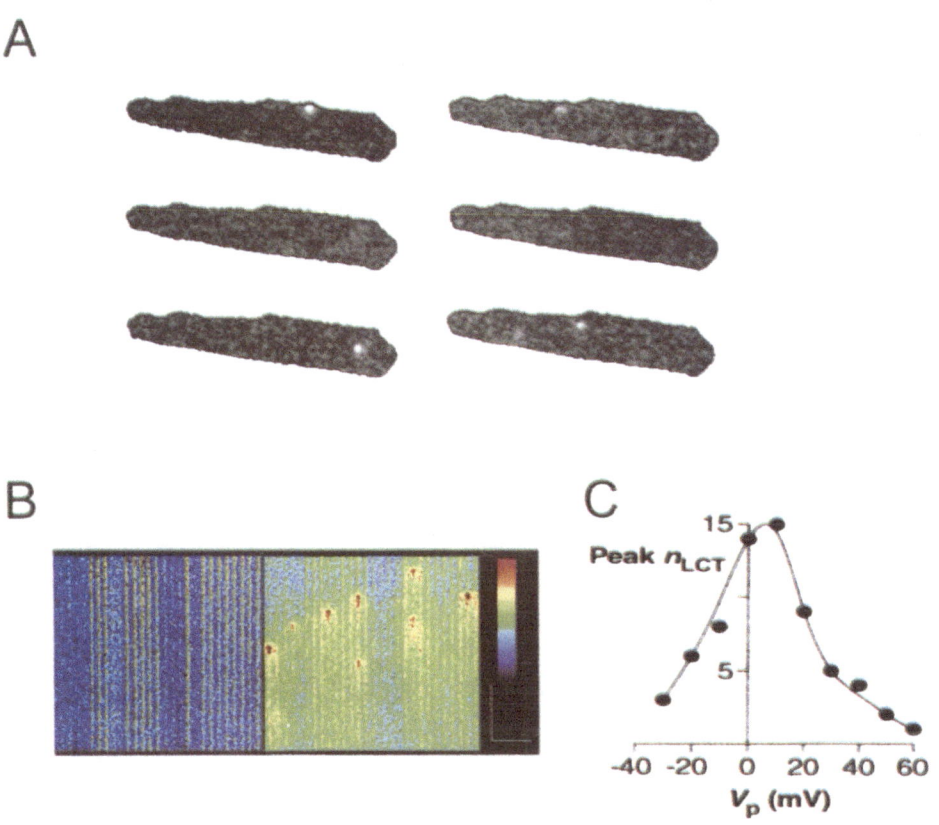

Figure 8 – Page 65

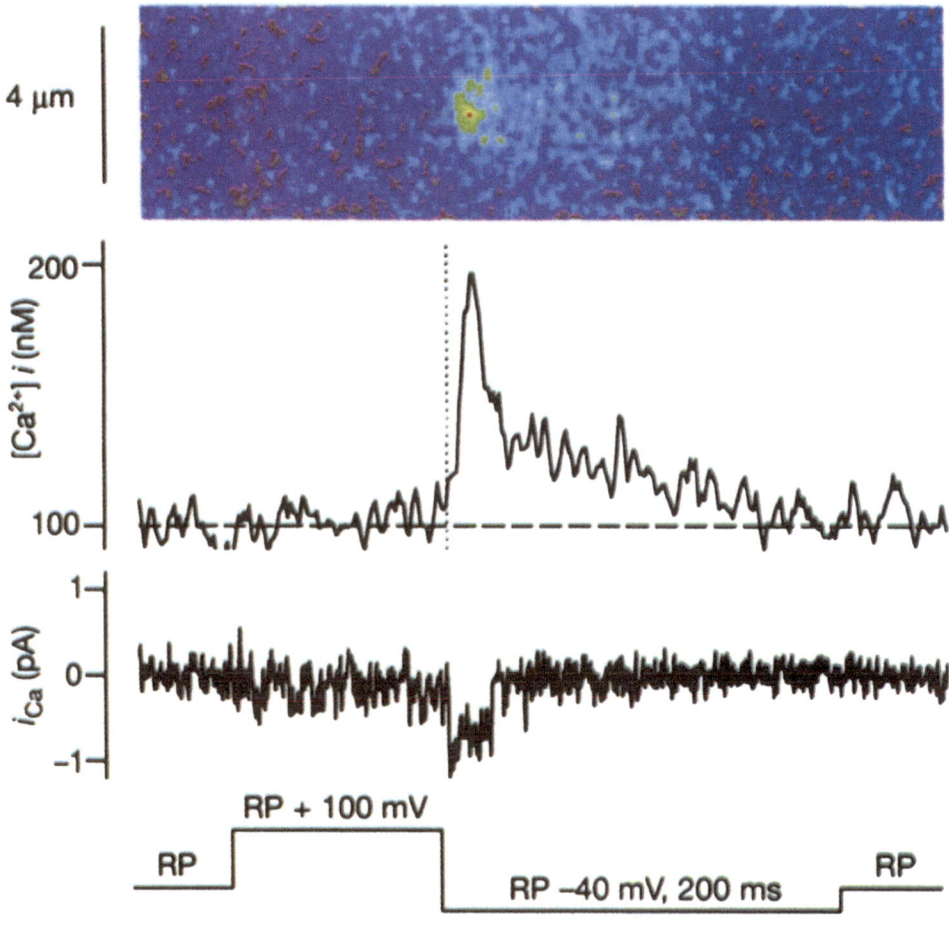

Figure 9 – Page 66

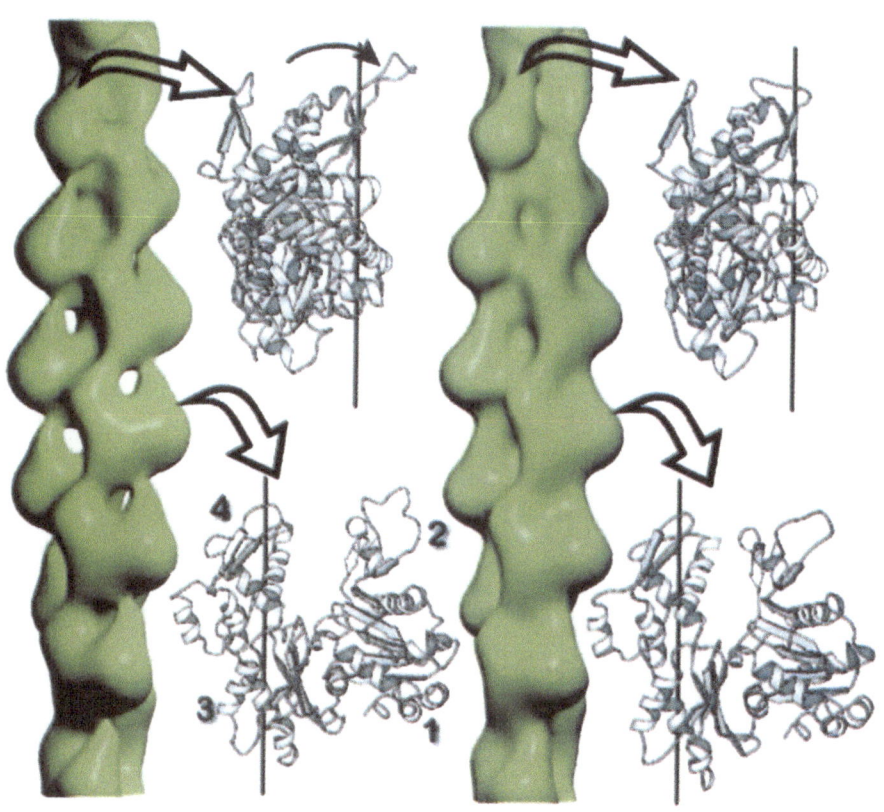

Figure 1 – Page 144

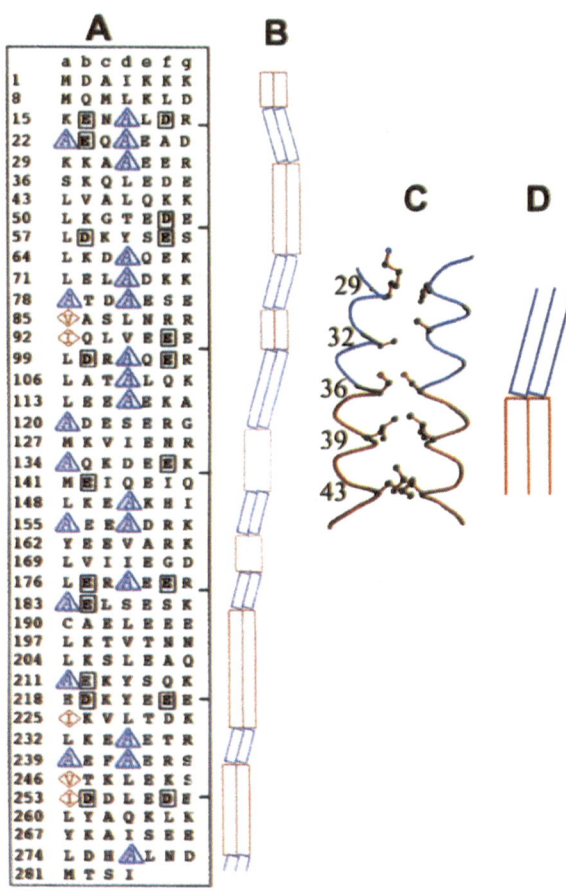

Figure 2 – Page 146

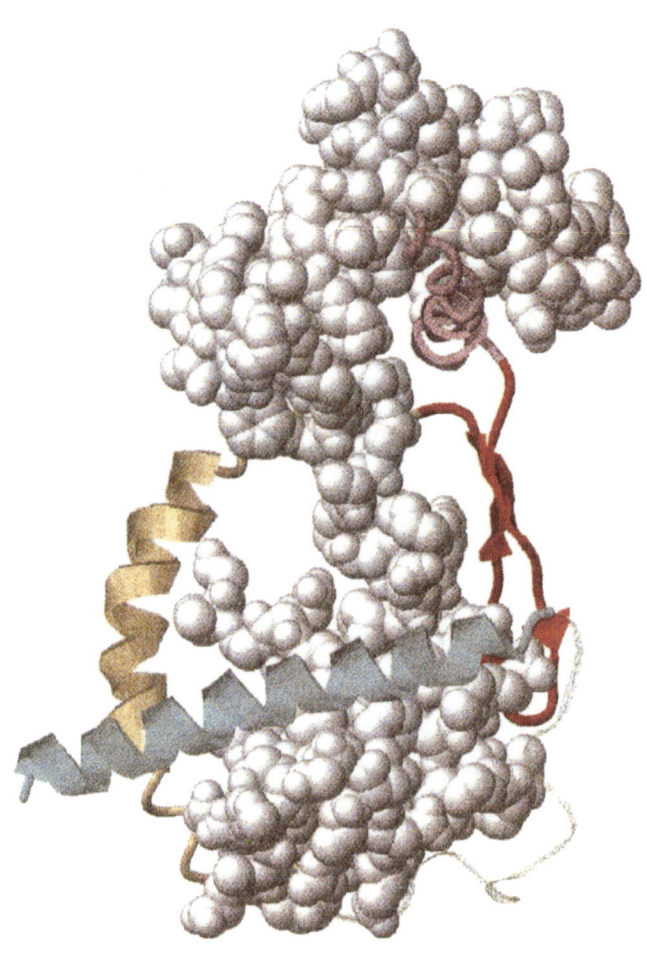

Figure 3 – Page 148

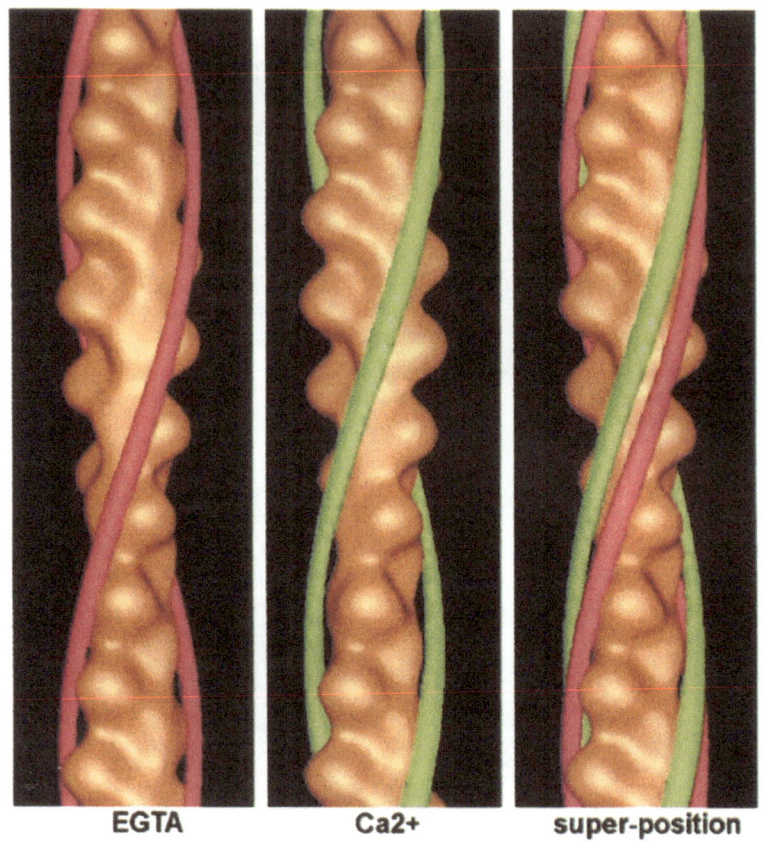

EGTA   **Ca2+**   **super-position**

Figure 4 – Page 151

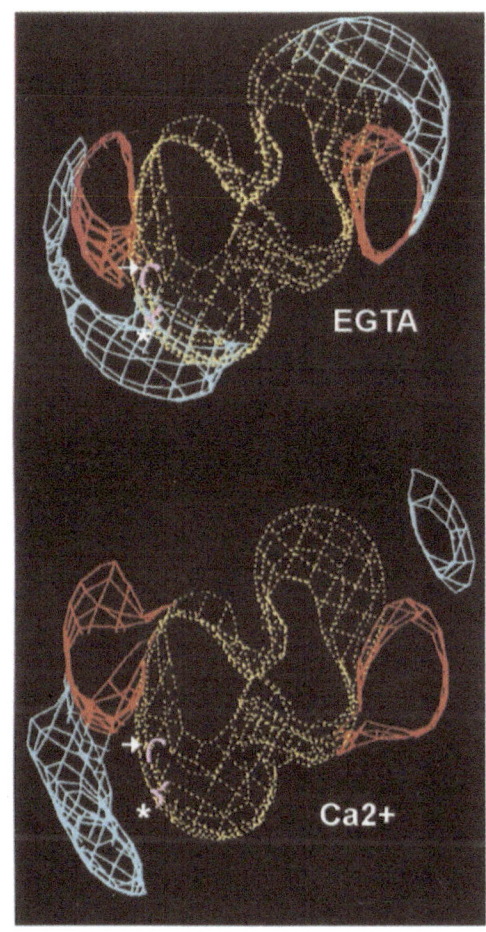

Figure 5 – Page 153

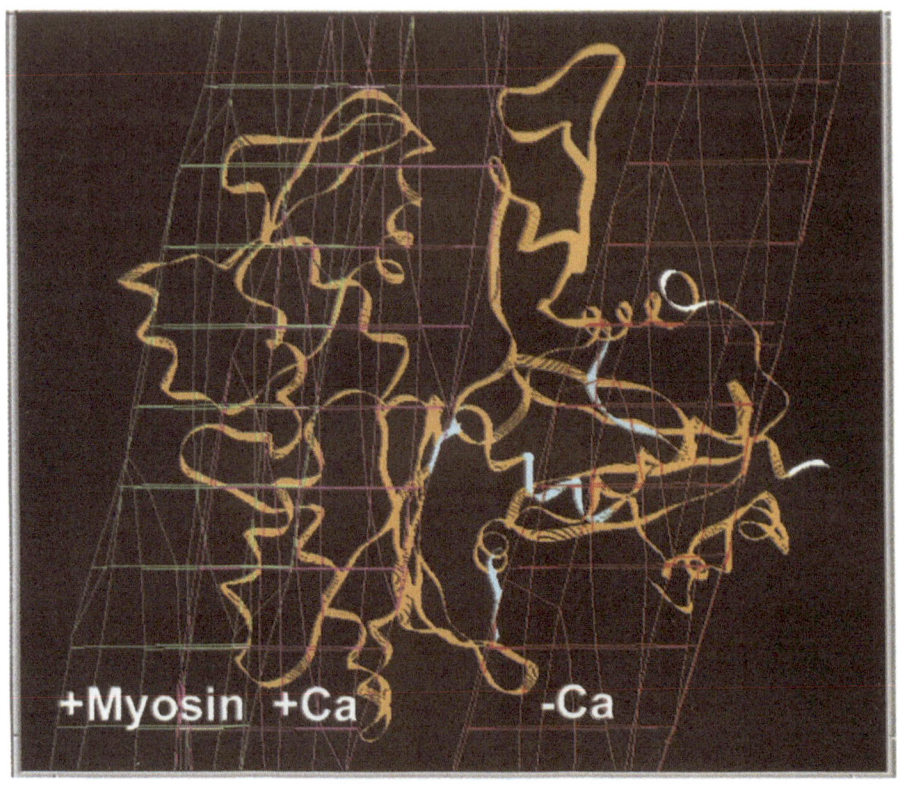

Figure 6 – Page 155

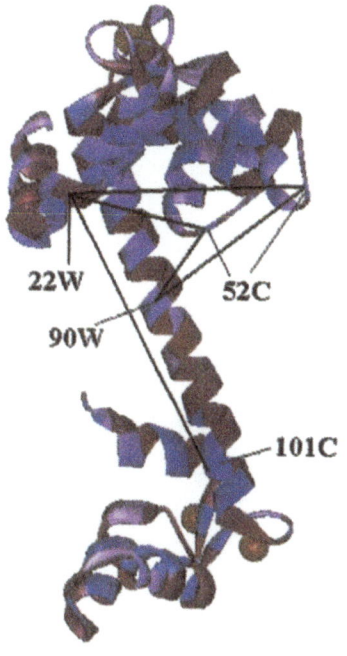

Figure 3 – Page 207

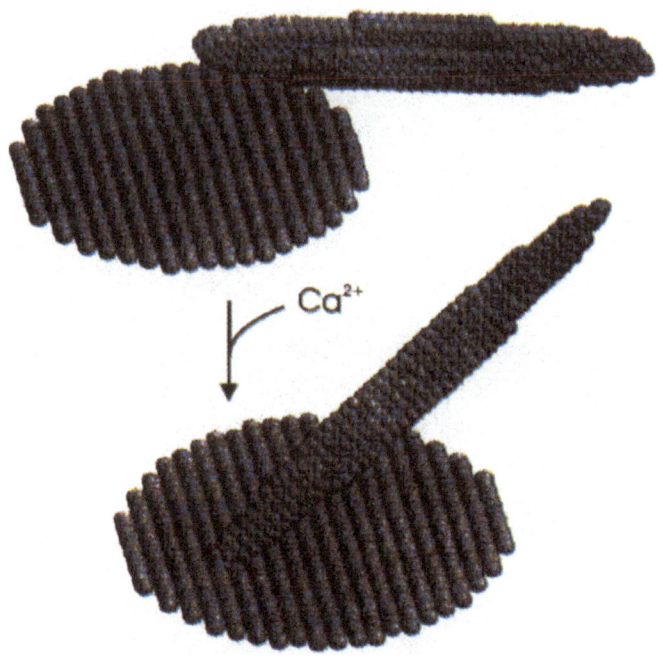

Figure 7 – Page 219

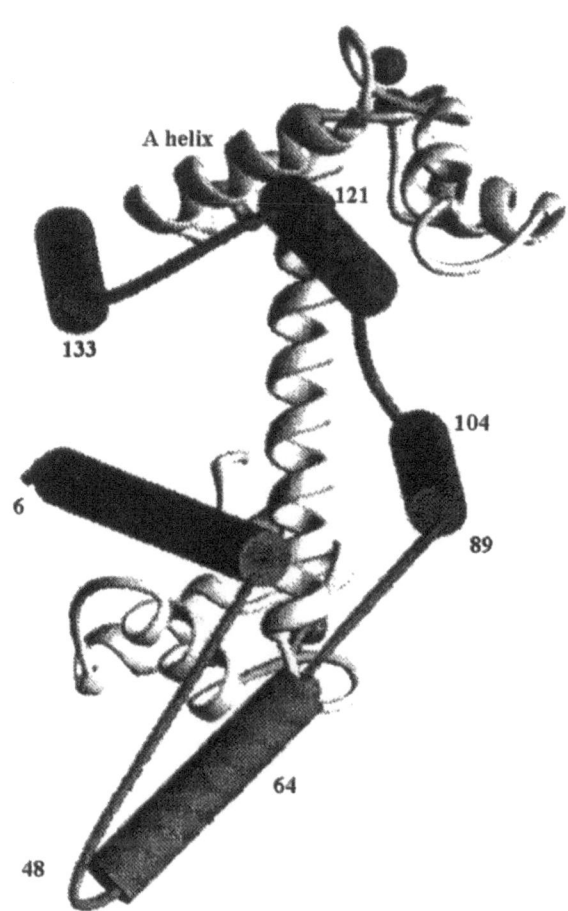

A helix

133

6

48

121

104

89

64

Figure 8 – Page 220

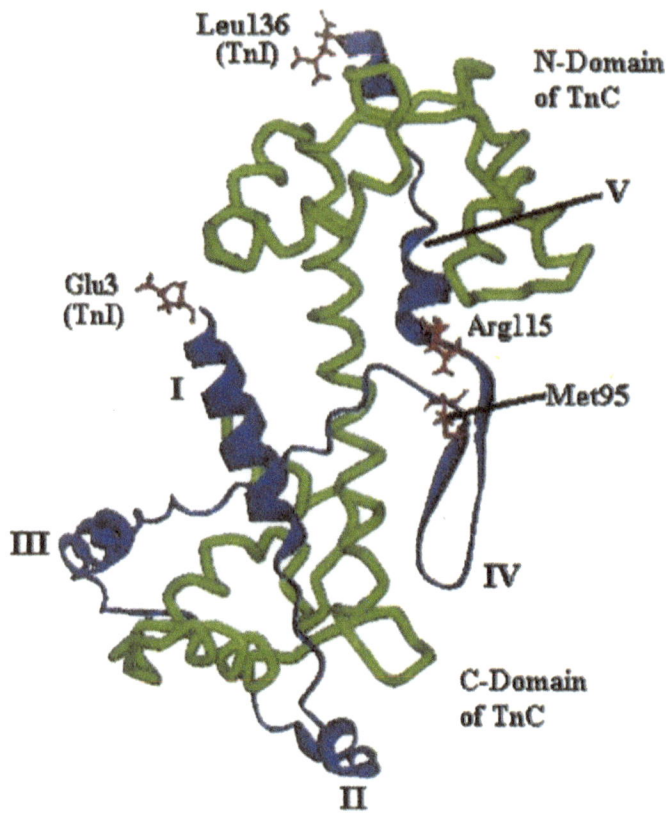

Figure 9 – Page 222

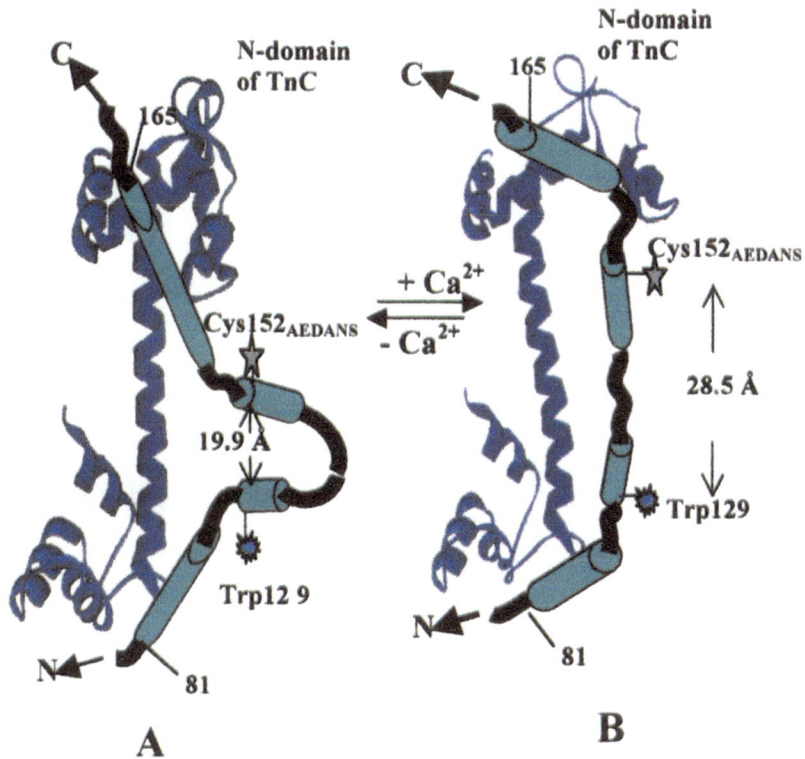

Figure 12 – Page 239

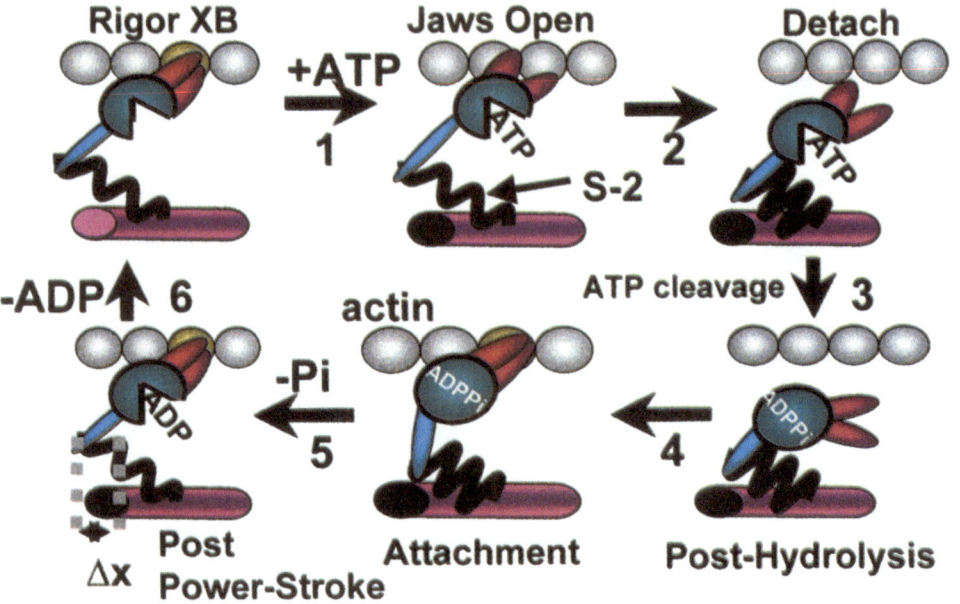

Figure 2 – Page 419

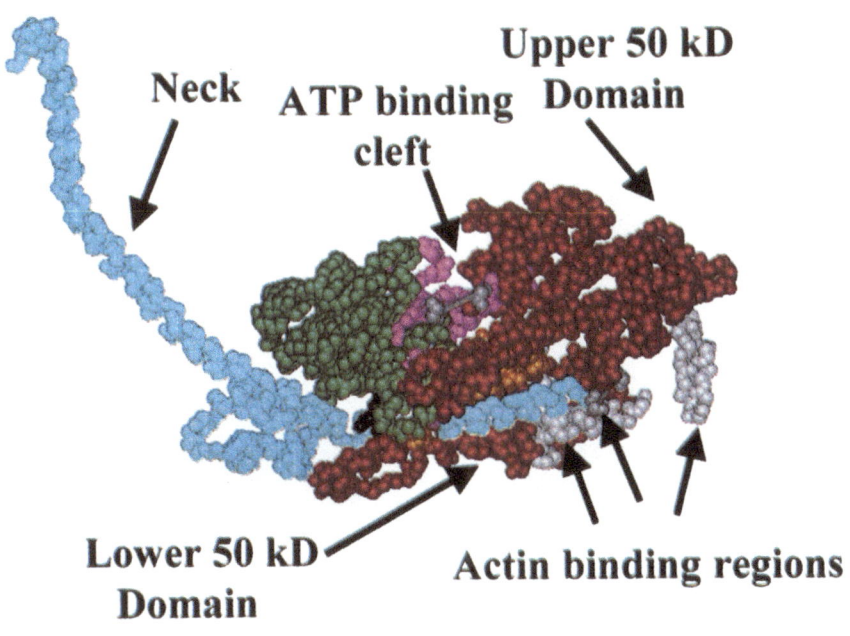

Figure 3 – Page 424

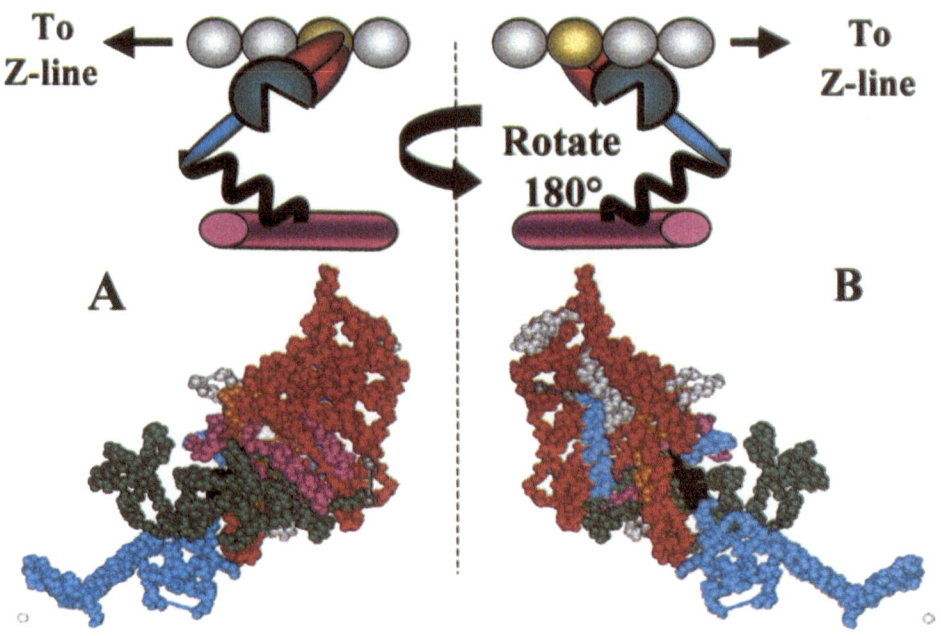

Figure 4 – Page 426